Organoselenium Chemistry

Organoselenium Chemistry

Edited by

DENNIS LIOTTA

Emory University
Atlanta, Georgia

A Wiley-Interscience Publication

JOHN WILEY & SONS

New York · Chichester · Brisbane · Toronto · Singapore

Library of Congress Cataloging in Publication Data:

Organoselenium chemistry.

"A Wiley-Interscience publication."

Includes index.
1. Organoselenium compounds. I. Liotta, Dennis.

QD412.S5073   1987   547'.05724   86-5582
ISBN 0-471-88867-2

Printed in the United States of America

10 9 8 7 6 5 4 3 2 1

Contributors

THOMAS G. BACK, Department of Chemistry, University of Calgary, Calgary, Alberta, Canada

DAVID BROWN, Department of Chemistry, Emory University, Atlanta, Georgia

D. A. CLAREMON, Merck Sharp and Dohme Research Laboratories, West Point, Pennsylvania

FRANK S. GUZIEC, JR., Department of Chemistry, New Mexico State University, Las Cruces, New Mexico

STEVEN V. LEY, Department of Chemistry, Imperial College, South Kensington, London, England

DENNIS LIOTTA, Department of Chemistry, Emory University, Atlanta, Georgia

ROBERT MONAHAN, Department of Chemistry, Emory University, Atlanta, Georgia

K. C. NICOLAOU, Department of Chemistry, University of Pennsylvania, Philadelphia, Pennsylvania

N. A. PETASIS, Department of Chemistry, University of Pennsylvania, Philadelphia, Pennsylvania

HANS J. REICH, Department of Chemistry, University of Wisconsin, Madison, Wisconsin

LILADHAR WAYKOLE, Department of Chemistry, Emory University, Atlanta, Georgia

FRED WUDL, Institute for Polymers and Organic Solids, Department of Physics, University of California at Santa Barbara, Santa Barbara, California

Preface

The development of new methods that permit efficient manipulation of functional groups in molecules represents one of the most important aspects of modern organic chemistry. As synthetic strategies become more daring, chemists increasingly require reactions which are both efficient and highly selective (chemoselective, regioselective, and/or stereoselective). In addition, it is desirable that these processes be operationally simple and able to be carried out under mild conditions. This search for new and efficient methods has increasingly led synthetic chemists outside the traditional realm of organic chemistry. Such has been the case with the development of modern organoselenium chemistry.

Although the chemistry of selenium and its derivatives has been well studied since its discovery in the nineteenth century, prior to the 1970s many organic chemists considered selenium-derived reagents to be of little synthetic utility. However, during the last 10 years, many investigators have described important chemical transformations that were efficiently achieved using organoselenium reagents. In fact, by the mid 1970s, many organoselenium reagents were regarded as commonplace, rather than esoteric. In large part, this is because they provided chemists with the ability to consistently perform important transformations simply and in high overall yields.

This book chronicles the important developments in organoselenium chemistry since 1970. The authors of the various chapters have attempted to strike a reasonable balance between synthetic and mechanistic results. As a result this book should prove to be valuable to both the novice and the seasoned organic chemist.

DENNIS LIOTTA

Atlanta, Georgia
November 1986

Contents

1

Electrophilic Selenium Reactions

THOMAS G. BACK

Department of Chemistry, University of Calgary,
Calgary, Alberta, Canada

CONTENTS

1. INTRODUCTION

Although organoselenium chemistry has been studied for more than a century, its utility in synthesis remained largely unexploited until the early 1970s. In the past decade, recognition of the selenoxide *syn* elimination as a powerful olefin forming method prompted a renewed interest in this area that led to the discovery of other valuable selenium-based reagents and reactions. It also became evident that the success of such methodology was contingent upon the availability of adequate procedures for the efficient introduction of selenium into a wide range of organic substrates. The use of both nucleophilic and electrophilic selenium reagents for this purpose has been intensely scrutinized in recent years and it is the intent of the present chapter to review the nature of electrophilic selenium reactions and to indicate their synthetic applications and limitations.

Detailed coverage of the literature from 1973 to 1982 is provided. Relevant articles that appeared during the preparation of this work in the first half of 1983 have been included wherever possible. A thorough review of earlier work in organoselenium chemistry was presented in 1973 by Klayman and Günther.[1] Literature references contained therein have been included in the present work only when required for the purpose of background and illustration. Several excellent review articles on organoselenium chemistry have also appeared in the 1970s.[2–6]

Electrophilic selenium species exist in many forms. The free element is itself mildly electrophilic and reacts with strong carbon nucleophiles such as Grignard reagents and organolithium compounds. The products of these reactions are in turn selenium nucleophiles, which are discussed in Chapter 4.

A number of selenium halides are known, but these are of limited synthetic value. Selenium monochloride, $\underline{1}$, and tetrachloride $\underline{2}$, are available from the direct combination of the elements. Their tendency to disproportionate according to Eqs. (1) and (2)[7–9] makes careful maintenance of the desired stoichiometry essential during their preparation. These properties, as well as the hydrolytic instability of such reagents, create additional difficulties in their prolonged storage. Similar considerations apply to the bromide derivatives.

$$Se_2Cl_2 \; \rightleftharpoons \; SeCl_2 \; + \; Se \tag{1}$$
$$\underline{1}$$

$$2\,SeCl_2 \; \rightleftharpoons \; SeCl_4 \; + \; Se \tag{2}$$
$$\underline{2}$$

The majority of electrophilic selenium reactions involve selenenic, Se (II), compounds. The selenenyl halides and certain pseudohalides of the general structure RSeX have been thoroughly studied and have earned a secure place on the reagent shelves of many organic chemistry laboratories. The selenenyl chlorides (RSeCl) and bromides (RSeBr) are particularly useful, as are diselenides (RSeSeR). It is common and expedient to employ aryl instead of alkyl derivatives as the former possess greater ease of handling, increased stability, lower volatility, and less offensive odors. Diphenyl diselenide (PhSeSePh), benzeneselenenyl chloride (PhSeCl), and benzeneselenenyl bromide (PhSeBr) are all commercially available, crystalline solids amenable to long-term storage. An Organic Syntheses preparation of PhSeSePh on a 1-mole scale has been reported.[10] Its conversion to PhSeCl or PhSeBr is easily accomplished by treatment with an equimolar amount of chlorine[10–12] or bromine [Eq. (3)].[11,13] The selenenyl bromide is often prepared *in situ* just prior to use. Sulfuryl chloride can be used in place of chlorine in the preparation of PhSeCl, although removal of the by-product sulfur dioxide is sometimes desirable as complications from its presence have been noted.[12]

$$
\text{PhSeSePh} \quad
\begin{cases}
\xrightarrow{\substack{Cl_2 \ or \\ SO_2Cl_2}} & 2 \ PhSeCl \\
\xrightarrow{Br_2} & 2 \ PhSeBr
\end{cases}
\tag{3}
$$

Selenenic acids (RSeOH) and certain related compounds also function as electrophiles. The parent acids are unstable and the identification of several originally claimed to be isolable have recently been proved in error.[14,15] Some selenenic acids readily form the corresponding anhydrides (RSeOSeR) or,[14,15] in one case, the isomeric selenolseleninate [RSeSe(=O)R].[16] Benzeneselenenic acid (PhSeOH) disproportionates according to the overall process in Eq. (4), producing the diselenide and benzeneseleninic acid, 3.[17–19] The anhydrides, 4 and 5, are plausible intermediates in the disproportionation and they are themselves electrophiles. Their existence may, however, be fleeting and their concentrations minute at any given time.[20] Caution must therefore be exercised in any attempt to precisely identify the active electrophile in the reactions of these equilibrating systems.

$$
3 \ PhSeOH \ \rightleftharpoons \ PhSeSePh \ + \ \underset{\underline{3}}{Ph\overset{O}{\overset{\|}{Se}}OH} \ + \ H_2O
$$

$$
\underset{\underline{4}}{PhSeOSePh} \qquad\qquad \underset{\underline{5}}{Ph\overset{O}{\overset{\|}{Se}}OSePh}
\tag{4}
$$

Despite such ambiguities, electrophilic reactions of selenenic acids and their congeners have a number of useful applications and several methods for their *in situ* generation have been developed. Selenenic acids are also formed as by-products in selenoxide eliminations [Eq. (5)] and other processes where they sometimes cause unwelcome side reactions.[18,19] Methods for their continuous removal from such systems are therefore of equal importance to methods for their deliberate production in other situations.

$$
\tag{5}
$$

Selenenyl pseudohalides RSeX, where X is a nonhalide leaving group, also perform electrophilic reactions. The nature and departing ability of X⁻ modulate the reactivity of the compound. Examples of special interest and methods for their preparation are provided by structures 6–12 in Scheme 1 and Eqs. (6)–(10). Most such compounds can be obtained by the displacement of halide ion from PhSeCl or PhSeBr, as well as from other selenenyl halides, with an appropriate nucleophile (Scheme 1).

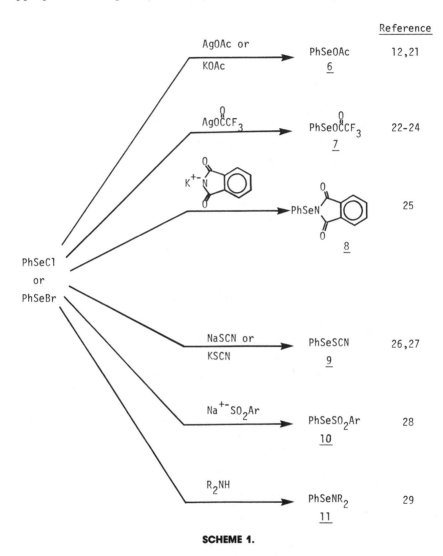

SCHEME 1.

A more satisfactory procedure for the preparation of aryl selenocyanates employs the reaction of a diazonium salt, obtained from an aniline precursor, with potassium selenocyanate (Eq. (6)).[30,31] The metathesis of PhSeCl with trimethylsilyl cyanide [Eq. (7)][32] or trimethylsilyl isothiocyanate [Eq. (8)][33] provides convenient access to phenyl selenocyanate, 12, or benzeneselenenyl thiocyanate, 9, in virtually quantitative yield. Trimethylsilyl chloride is the only by-product formed and its volatility permits its easy removal. Improved procedures for obtaining selenosulfonates 10, from the oxidations of sulfonylhydrazides[34] or sulfinic acids[35,36] with seleninic acid, 3, have recently been reported [Eqs. (9) and (10)].

$$\text{ArNH}_2 \xrightarrow[\text{2. KSeCN}]{\text{1. HNO}_2} \text{ArSeCN} \tag{6}$$

$$\text{PhSeCl} + \text{Me}_3\text{SiCN} \longrightarrow \underset{12}{\text{PhSeCN}} + \text{Me}_3\text{SiCl} \tag{7}$$

$$\text{PhSeCl} + \text{Me}_3\text{SiNCS} \longrightarrow \underset{9}{\text{PhSeSCN}} + \text{Me}_3\text{SiCl} \tag{8}$$

$$\text{ArSO}_2\text{NHNH}_2 + \text{PhSeO}_2\text{H} \longrightarrow \underset{10}{\text{ArSO}_2\text{SePh}} + \text{N}_2 + 2\text{H}_2\text{O} \tag{9}$$

$$2 \text{ArSO}_2\text{H} + \text{PhSeO}_2\text{H} \longrightarrow \underset{10}{\underline{10}} + \text{ArSO}_3\text{H} + \text{H}_2\text{O} \tag{10}$$

The selenenyl acetate, 6, and trifluoroacetate, 7, are typically prepared *in situ*. Selenocyanate, 12, and selenenyl thiocyanate, 9, are isolable but are best used promptly. *N*-(Phenylseleno)phthalimide, 8, and selenosulfonates, 10, are stable, crystalline, and amenable to prolonged storage.

The electrophilic reactions of Se (IV) compounds have been less studied than those of their Se (II) counterparts. Selenium (IV) halides such as 2 or the tetrabromide 13 act as halogenating agents instead of, or as well as, selenium electrophiles. A few electrophilic reactions of these compounds, as well as those of selenium oxychloride, 14, and selenium trichlorides, 15, are described at the end of this chapter. Selenium dioxide, benzeneseleninic acid, 3, and its corresponding anhydride, 16, act principally as oxidants in their reactions with organic compounds. The selenium imide, 17, is chiefly employed as an aminating agent. Although such reagents may be considered to function as electrophiles in certain situations, their chemistry is beyond the scope of the present chapter. The reactions of seleninic acid, 3, and anhydride, 16, provide the subject for Chapter 3; those of selenium dioxide have been reviewed elsewhere.[37]

$$\underset{13}{\text{SeBr}_4} \qquad \underset{14}{\text{SeOCl}_2} \qquad \underset{15}{\text{RSeCl}_3} \qquad \underset{16}{\overset{\text{O O}}{\text{PhSeOSePh}}} \qquad \underset{17}{\text{TsN=Se=NTs}}$$

2. α-SELENENYLATION OF CARBONYL COMPOUNDS

Several early studies of selenoxide fragmentations indicated that olefins could be prepared under remarkably mild conditions via Eq. (5).[38-42] The extension of this work to the synthetically important conversion of carbonyl compounds to their α,β-unsaturated derivatives [Eq. (11)] was independently reported by Clive,[22] Reich, et al.[43] and Sharpless et al.[13] in 1973. This firmly established the utility of the selenoxide elimination and illustrated the accompanying need for convenient procedures for introducing a selenium-containing group α to a carbonyl moiety. To date, examples of such reactions number in the hundreds, reflecting their impact on diverse areas of organic synthesis.

$$\text{(11)}$$

2.1. Ketones and Aldehydes

2.1.1. DIRECT SELENENYLATION

The direct selenenylation of a ketone was first reported by Rheinboldt and Perrier,[44] who demonstrated that areneselenenyl thiocyanates with electron-withdrawing groups react with acetone to afford the α-arylseleno derivatives quantitatively [Eq. (12)].

$$\text{(12)}$$

Subsequently Sharpless et al.[13] prepared the α-phenylseleno derivatives of a series of ketones and aldehydes by treating them with PhSeCl in ethyl acetate at room temperature. The products were converted directly to the corresponding enones by oxidation with hydrogen peroxide, peracetic acid, or sodium periodate [Eq. (13)]. Overall yields ranged from 34–84%. Only the product with the (E)-configuration was produced in those enones capable of forming geometrical isomers. Other oxidants that later proved effective in selenoxide eliminations include m-chloroperbenzoic acid (MCPBA), ozone, and t-butyl hydroperoxide, although hydrogen peroxide has been the most frequently employed.[a]

[a] Although it is not the author's intent to review selenoxide eliminations per se, such processes are often used in conjunction with electrophilic selenium reactions as part of an overall transformation, as above. Some general considerations of the oxidation–elimination step will therefore be included where appropriate.

$$\text{(13)}$$

Benzeneselenenyl chloride (PhSeCl) continues to be the reagent of choice for the direct selenenylation of ketones. The selenenyl bromide is less suitable and can result in competing bromination of the substrate.[13,44] Diphenyl diselenide is not sufficiently reactive for this purpose. Methyl acetate,[45] carbon tetrachloride,[46] and N-N-dimethylformamide (DMF)[47] have been used as the solvent in place of ethyl acetate but do not appear to offer any general advantage. The selenenylations are acid catalyzed and presumably proceed via the enol. Although HCl is a by-product of the process, the addition of concentrated HCl at the start of the reaction is sometimes expedient.[13,48-51] The use of a catalytic amount of Dowex 50-X8 (H$^+$) resin[b] has also been reported.[52] The method is compatible with a variety of functional groups. Ketones have been selenenylated in the presence of esters[2,45,53-56] (ethyl acetate is the usual solvent), lactones,[50,51,57-61] carbamates,[48,49] ethers,[2,48,49,62] silyl ethers,[51] ketals,[63,64] epoxides,[52] alcohols,[2,56,65] carboxylic acids,[66,67] and carbon–carbon double bonds.[2,55,65,68,69] The selenenylation of aldehydes can be performed under similar conditions.[13,70]

Ketone selenenylations often display high chemo-, regio-, and stereoselectivity. For example, one carbonyl group of the trione, <u>18</u>, was selectively functionalized [Eq. (14)].[71] The norperistylane dione, <u>19</u>, afforded a unique bis-selenenylated stereoisomer [Eq. (15)].[54] 3-Ketosteroids normally,[13,47] but not without exception,[53] form 2α-phenylseleno products, which yield Δ^1-enones upon oxidation. 17-Ketosteroids react at the 16α-position.[2,65] Enones generally undergo α'-selenenylation and form cross conjugated dienones when oxidized to their selenoxides.[56,58,60,61,63,72] In some instances the monoselenenylation of ketones is accompanied by the formation of bisselenenylated products and/or the persistence of unreacted started material.[13,45,50] Baeyer–Villiger reactions have been observed during the oxidation–elimination step, providing α,β-unsaturated lactones instead of enones.[2] Several illustrative examples of direct ketone selenenylations and accompanying transformations are shown in Eqs. (14)–(22). In addition to this, such reactions have been employed in synthetic approaches to a variety of compounds that include: (+)-codeine,[48]

$$\text{(14) (Ref. 71)}$$

[b] Dowex is a registered trademark of the Dow Chemical Co.

lycoramine,[62] prodigiosins,[45] (−)-axisonitrile-3,[64] iridoids,[55] crassin,[51] various pseudoguaianolides,[50,57,68,69] and other products.[46,59,73]

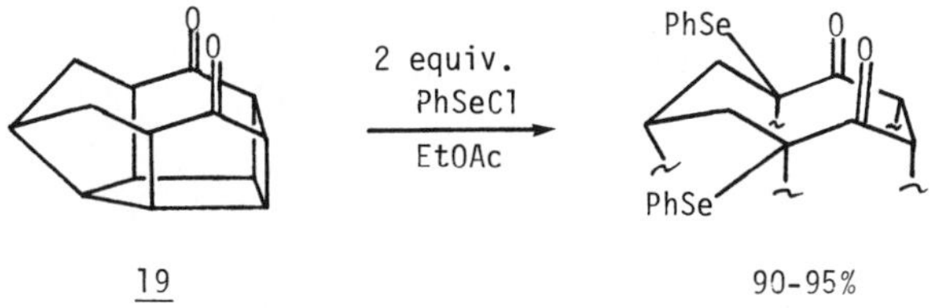

(15) (Ref. 54)

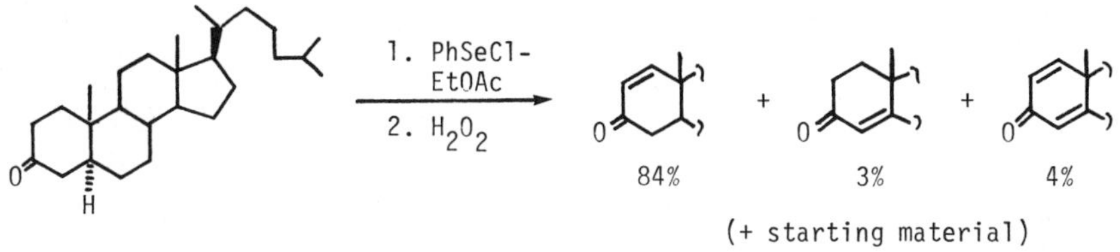

19   90-95%

84%   3%   4%

(+ starting material)

(16) (Ref. 13)

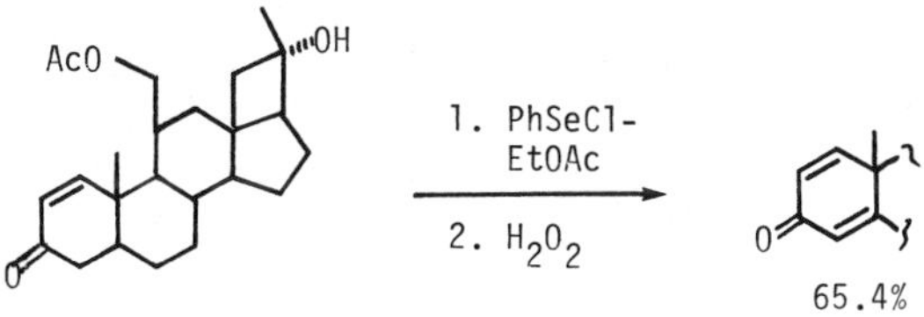

65.4%

(17) (Ref. 56)

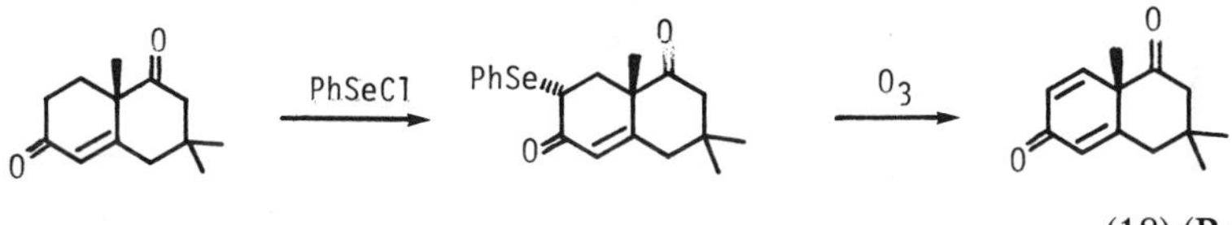

(18) (Ref. 72)

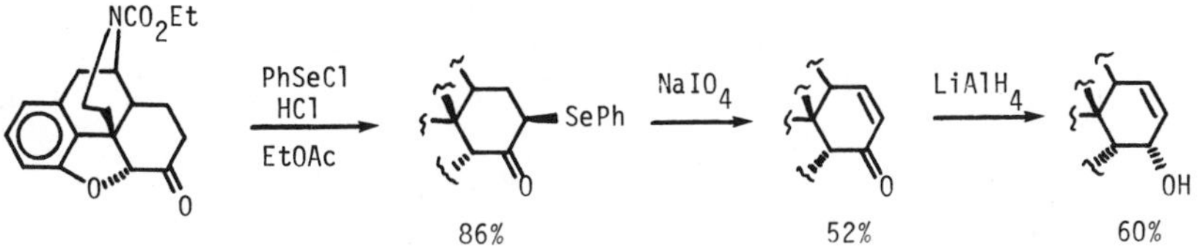

86%   52%   60%

(19) (Ref. 49)

78-87%

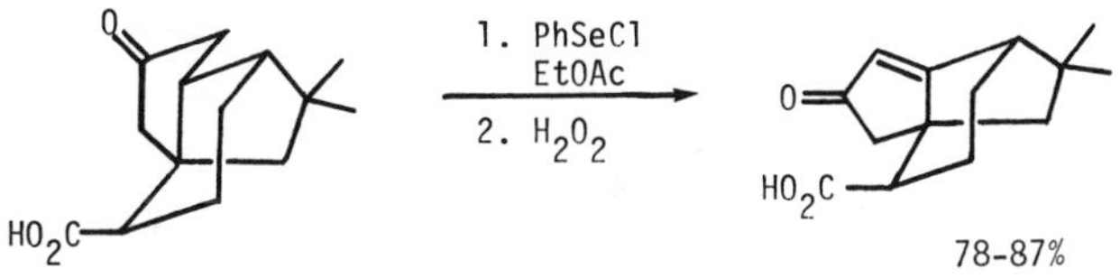

(20) (Refs. 66, 67)

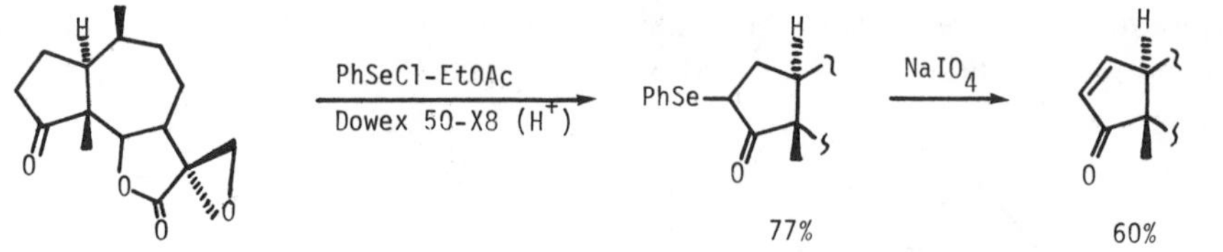

PhSeCl-EtOAc
————————————→   PhSe
Dowex 50-X8 (H⁺)

NaIO₄
————————→

77%   60%

(21) (Ref. 52)

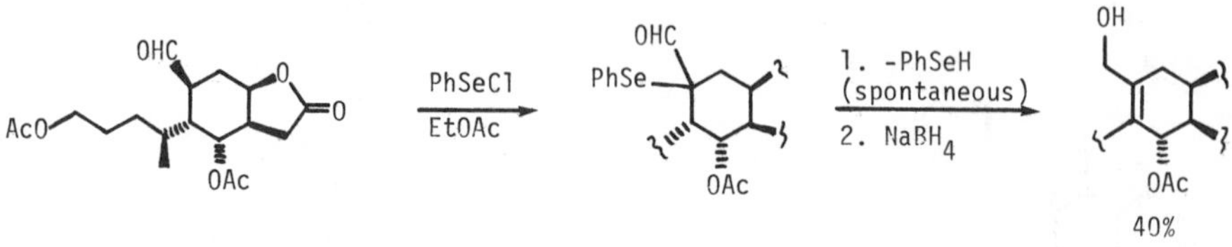

PhSeCl
————————→   PhSe
EtOAc

1. -PhSeH
(spontaneous)

2. NaBH₄

40%

(22) (Ref. 70)

Diphenyl diselenide may be converted *in situ* into more reactive selenium electrophiles by chemical or electrochemical methods. Cyclic and acyclic ketones as well as aldehydes are selenenylated when treated with a mixture of PhSeSePh, selenium dioxide, and a catalytic amount of sulfuric acid [Eq. (23)].[74] Under these conditions redox processes generate electrophilic selenenic species that act as the equivalent of the cation PhSe⁺.

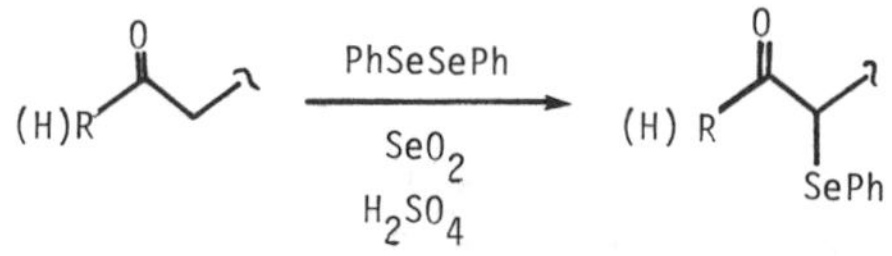

$$(H)R \xrightarrow[\substack{SeO_2 \\ H_2SO_4}]{PhSeSePh} (H) R \diagdown SePh$$

(23)

α-Phenylseleno ketones have also been prepared by the electrolysis of mixtures of the ketone, PhSeSePh, and magnesium bromide in polar solvents such as methanol or acetic acid containing tetraethylammonium bromide [Eq. (24)].[75] The two-electron anodic oxidation of Br⁻ generates Br⁺, which attacks the diselenide to produce PhSeBr. The latter electrophile then reacts with the ketone via its enol in the usual manner. The magnesium salt not only provides a source of bromide ion, but also catalyzes the enolization. The presence of sulfuric acid is sometimes beneficial for the same reason. In one case the selenenylation was successfully performed in the absence of magnesium bromide, indicating that other mechanisms can also operate.[75]

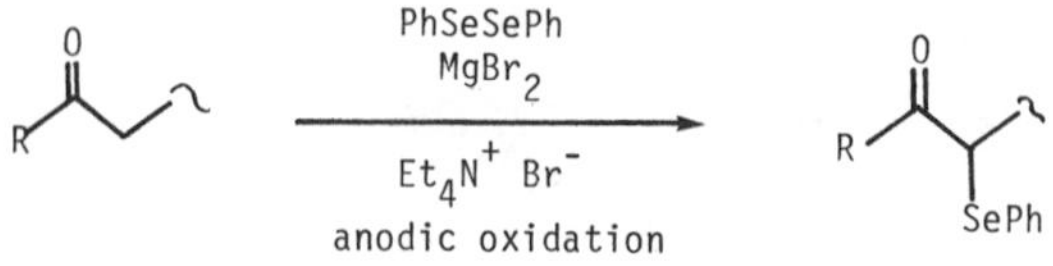

$$R \xrightarrow[\substack{Et_4N^+ Br^- \\ anodic\ oxidation}]{\substack{PhSeSePh \\ MgBr_2}} R \diagdown SePh$$

(24)

Several variations involving more specialized selenenylating reagents have been reported. A polymer supported selenenyl chloride, 20, has been used to convert 4-methylcyclohexanone to its enone via Eq. (25).[76]

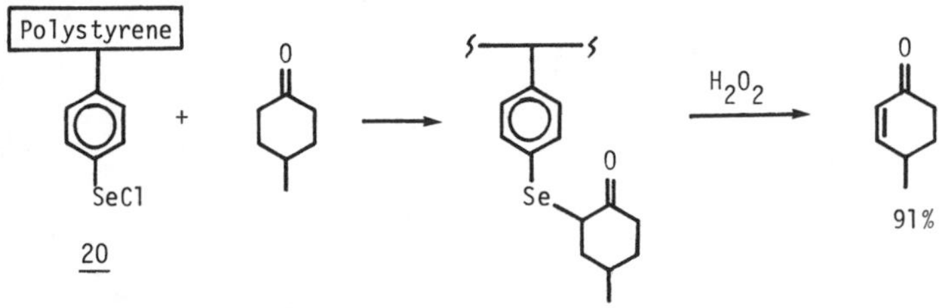

$$(25)$$

2-Pyridylselenenyl bromide, 21, can be prepared *in situ* from bromination of the corresponding diselenide. It acts as an effective selenenylating agent for both ketones and aldehydes under acidic conditions [Eq. (26)].[77] The *syn* elimination of the corresponding selenoxides is particularly efficient and works smoothly even for cycloheptenone and cyclooctenone. In contrast, cyclic enones with these ring sizes are particularly difficult to prepare by oxidation–elimination of the α-phenylseleno ketones.[11,51,78,79] (Very recently, several cycloheptenones were obtained in high yield by treatment of the α-phenylseleno ketone derivatives wth ozone at −78°C, followed by pyrolysis of the resulting selenoxides in refluxing carbon tetrachloride containing diisopropylamine.)[80]

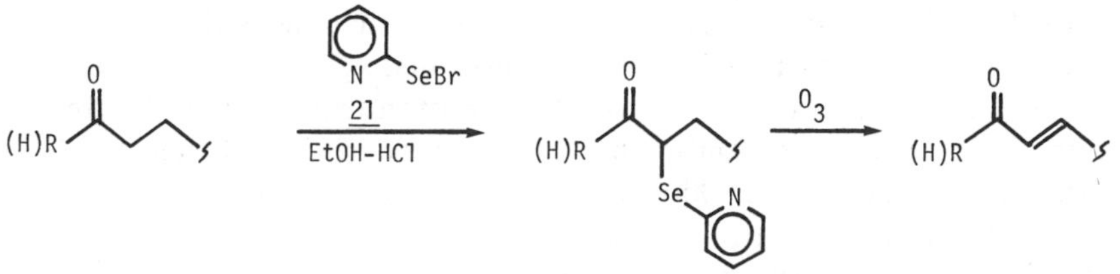

$$(26)$$

2.1.2. SELENENYLATION OF ENOLATES

The direct selenenylation technique of the previous section is particularly well suited for ketones and aldehydes that are not labile towards acid. An alternative approach that employs basic conditions was developed by Reich and co-workers.[11,43,78] The ketone is first converted to its lithium enolate, usually with lithium diisopropylamide (LDA) in tetrahydrofuran (THF) solution, and is then selenenylated with PhSeCl or PhSeBr (Scheme 2). The reaction is very rapid even at low temperatures and is generally carried out at −78°C. Diphenyl diselenide has been used as the electrophile,[81] but generally gives unsatisfactory results because the equilibrium in Eq. (27) does not favor formation of the

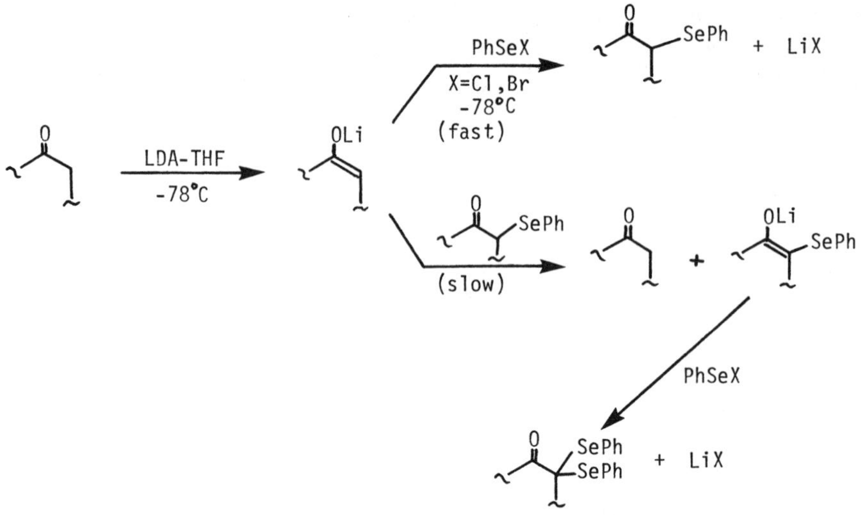

SCHEME 2.

desired α-phenylseleno ketone.[11] The presence of hexamethylphosphoric triamide (HMPA) is sometimes beneficial.[11,43,82–84] Complications are few in
those substrates that form enolates smoothly. Small amounts of unreacted ketones (usually $<10\%$)[11] are sometimes recovered, probably because of proton
transfer from the product α-phenylseleno ketone to the original enolate. This
is possible because the α-hydrogen in the product is rendered more acidic by
the presence of the phenylseleno moiety. Fortunately, however, the very rapid
rate of selenenylation compared to that of proton transfer prevents this from
being a serious complication in the vast majority of cases (Scheme 2).[11] Bis-
selenenylated products are seldom significant, but in a few exceptional cases[50,85]
they are produced by the further selenenylation of the α-phenylseleno enolate
formed by proton transfer as described previously.

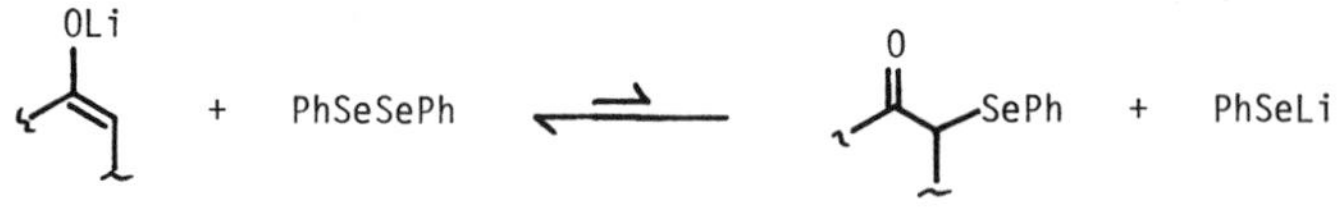

$$\tag{27}$$

Unmasked hydroxyl[56,86–91] and carboxyl[92] groups are tolerated, providing
that at least two equivalents of base are employed to ensure complete generation of the enolate. Double[65,81,89,92–98] and triple[99] carbon–carbon bonds are
unaffected.

Since kinetic enolates are produced from unsymmetrical ketones under these
conditions, the phenylseleno group is introduced into the less substituted position, usually with high regioselectivity. Enolates derived from enones normally
react at the α'-position,[11,84,90,100–106] although γ-selenenylation has been ob-

served in cases where the α'-site is unavailable [e.g., Eqs. (28)–(30)].[99,107] α'-Phenylseleno enones undergo oxidation–elimination to dienones, which aromatize to phenols when γ-hydrogens are available for tautomerization [Eqs. (28) and (29)].[105,106] An attempt to selectively selenenylate the γ-position of a β-keto lactone dianion was only partly successful as competing α-selenenylation was significant [Eq. (31)].[82,83] The increased acidity of α-phenylseleno ketones compared to their parent ketones permits their regiospecific alkylation prior to oxidation. An enone product containing an additional α-substituent is thus obtained [e.g., Eq. (32)].[108–110] Intramolecular selenenylations of methyl ketones have been performed in the presence of milder bases such as pyridine or triethylamine [Eq. (33)].[111] A number of other recent examples of selenenylations of ketone enolates have been reported,[80,112–132] and several are depicted in Eqs. (34)–(38).

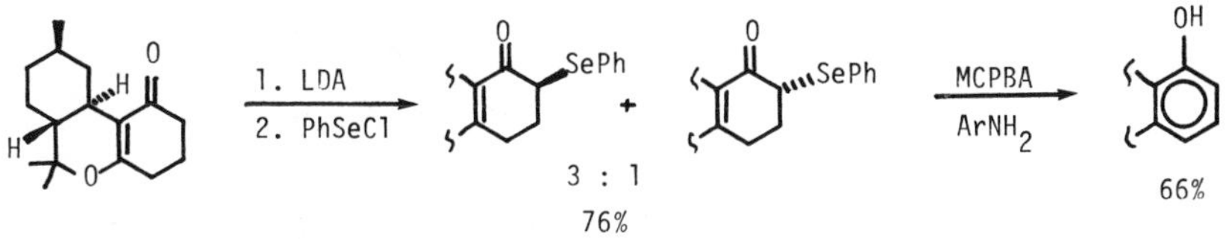

(28) (Ref. 106)

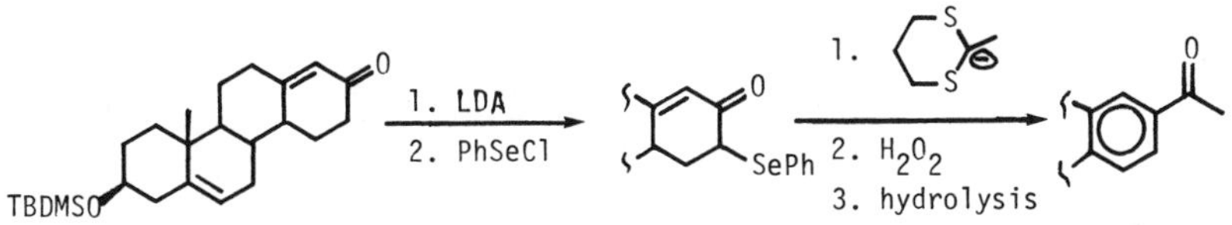

(29) (Ref. 105)

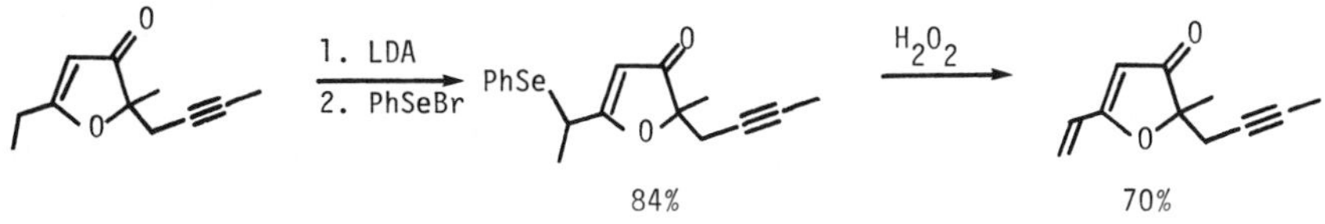

(30) (Ref. 99)

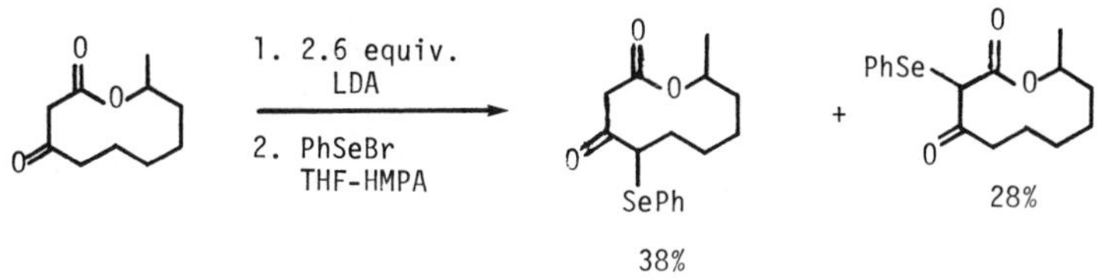

(31) (Refs. 82, 83)

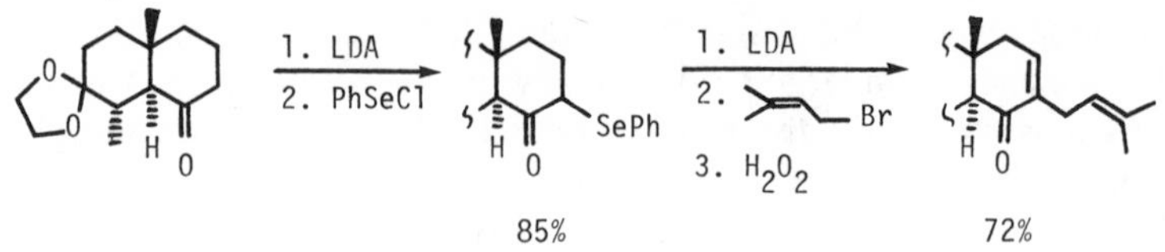

(32) (Ref. 110)

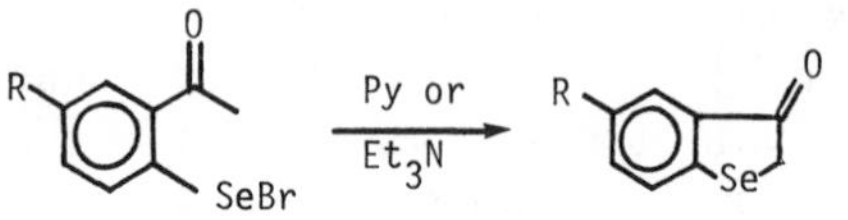

(33) (Ref. 111)

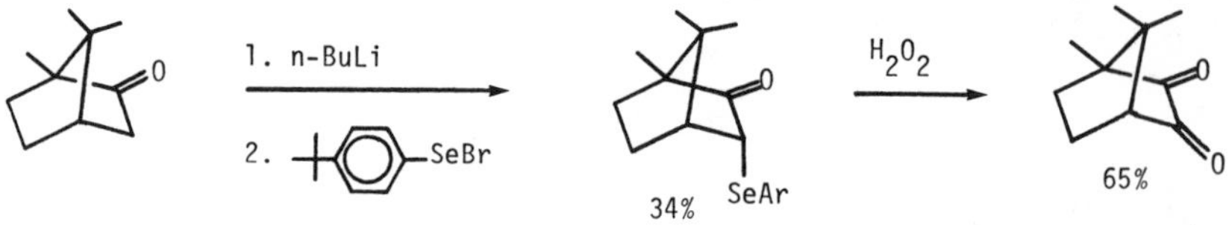

(34) (Ref. 129)

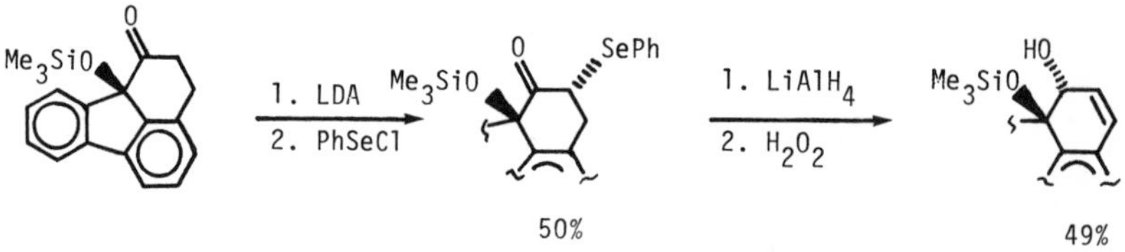

(35) (Ref. 122)

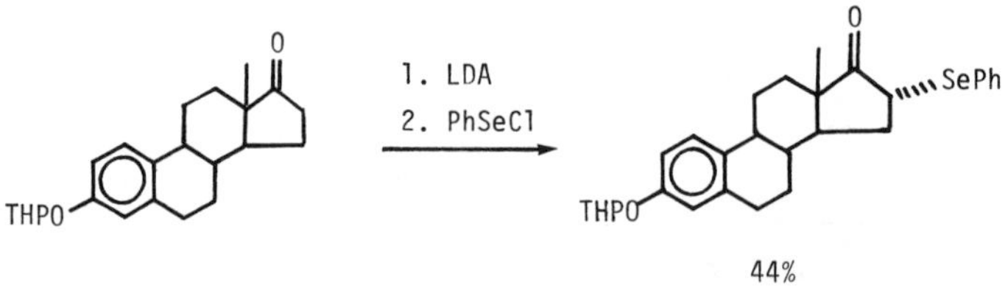

(36) (Ref. 121)

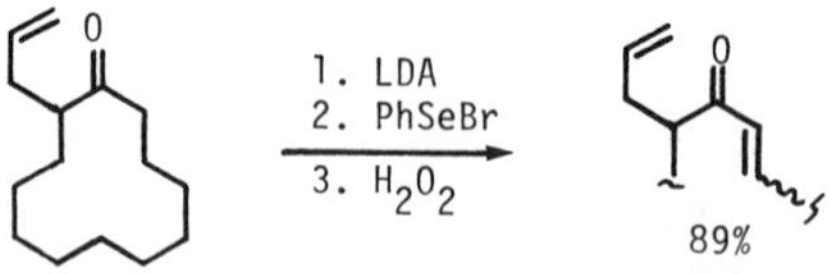

(37) (Ref. 126)

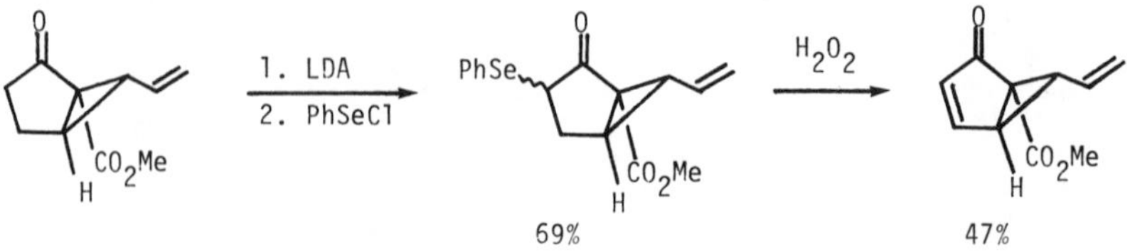

(38) (Ref. 131)

In a variation reported by Liotta et al.,[133] kinetic enolates were reacted sequentially with elemental selenium and methyl iodide to afford α-methylseleno ketones in yields of 80–90% [Eq. (39)]. The products could be converted to enones in the usual manner. The lower cost of elemental selenium compared to other selenenic electrophiles is an advantage of this method.

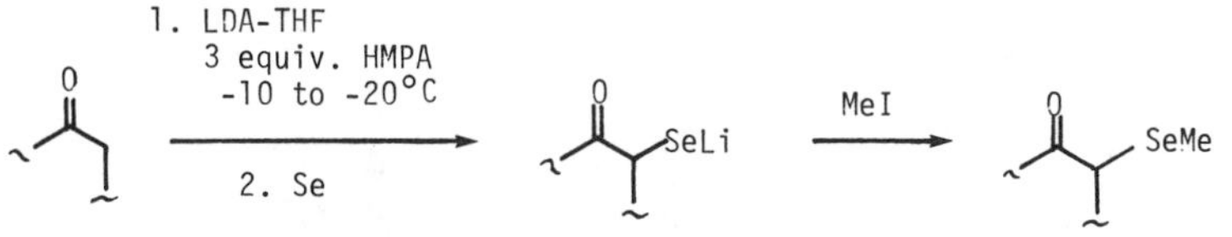

$$
\begin{array}{c}
\text{1. LDA-THF} \\
\text{3 equiv. HMPA} \\
\text{-10 to -20°C} \\
\hline
\text{2. Se}
\end{array}
\qquad
\xrightarrow{\text{MeI}}
\qquad (39)
$$

The enolates required for selenenylation can be prepared by methods other than deprotonation of the parent ketones. Enolate, 22, was formed from cyclo-octatetraene epoxide by treatment with LDA. Addition of PhSeCl produced the acetophenone derivatives, 25 and 26, presumably via valence tautomer, 24 (Scheme 3).[134] Alternatively, inverse addition of the enolate to PhSeBr afforded the α-phenylseleno trienone, 23, as the chief product.

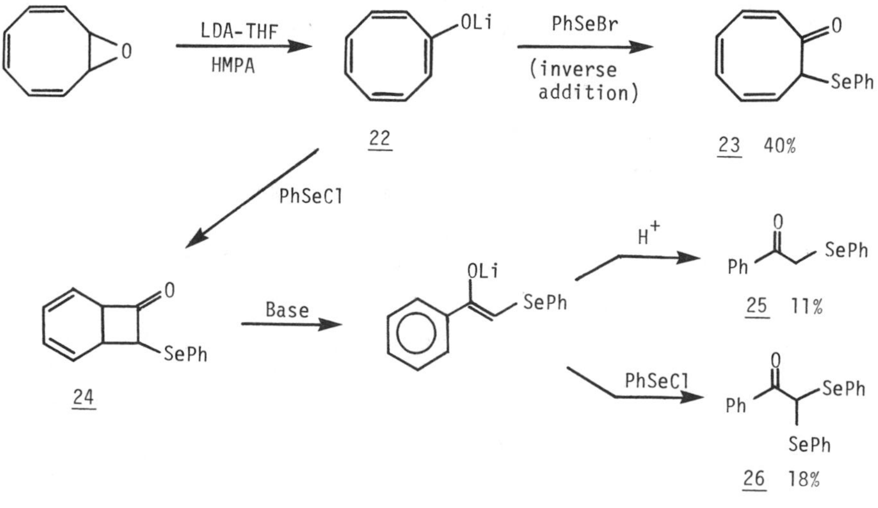

SCHEME 3.

The Michael addition of organocuprates to enones produces enolates that can be treated *in situ* with PhSeBr.[11,78,135] The inclusion of PhSeSePh aids in suppressing competing halogenation of the enolate. Oxidation–elimination of the resulting β-alkyl (or aryl) α-phenylseleno ketones affords the products of overall β-alkylation (or arylation) of the starting enone. Organozirconium[136] and organoaluminum[136] nucleophiles as well as Grignard reagents in the presence of cuprous salts have been similarly employed.[137–139] Several examples are given in Eqs. (40)–(43).

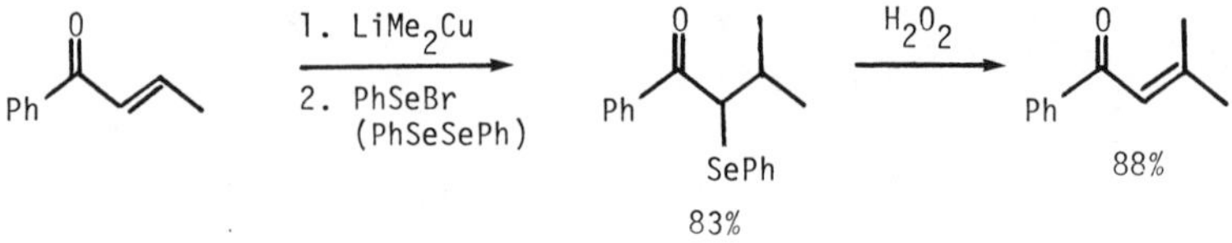

(40) (Refs. 11, 78)

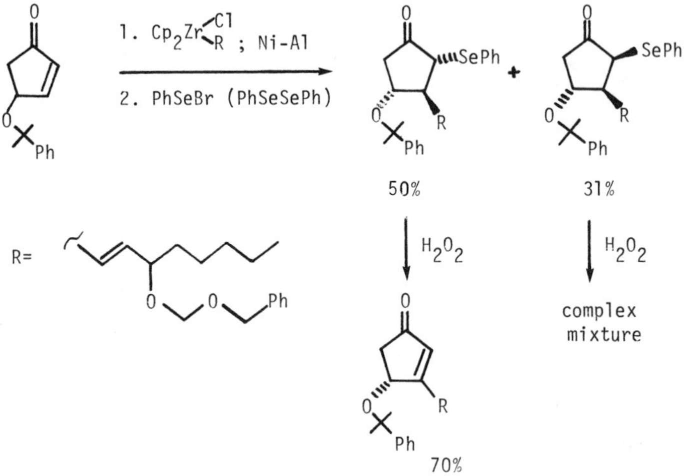

(41) (Ref. 136)

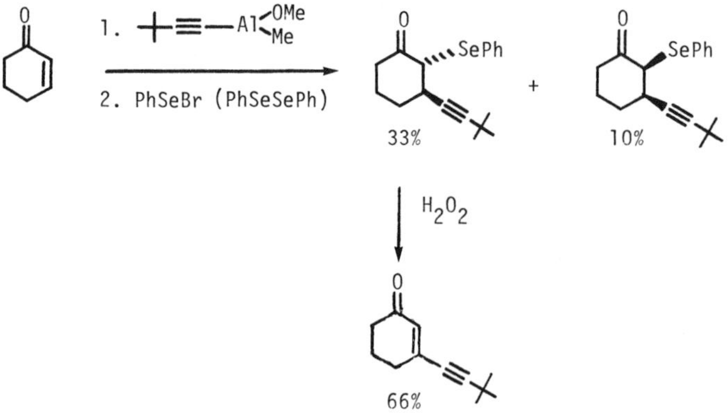

(42) (Ref. 136)

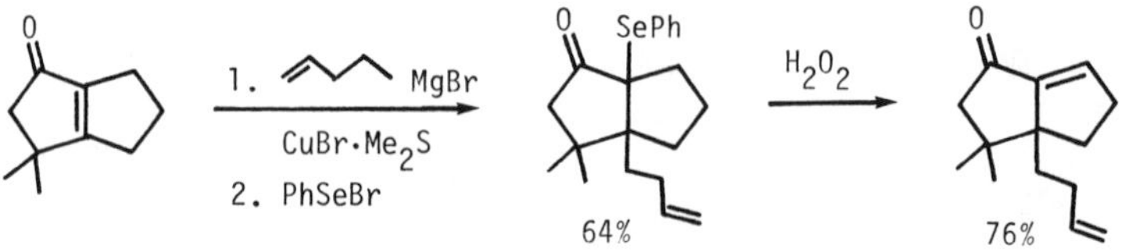

(43) (Ref. 138)

The nucleophilic character of the nitrogen atom of selenenamides, 11 results in the formation of enolates from enones by conjugate addition. Reich and Renga[29] have suggested that product formation ensues from the intramolecular selenenylation of intermediate 27 (Scheme 4). Further transformations of the adducts, 28, include amine and selenoxide eliminations.

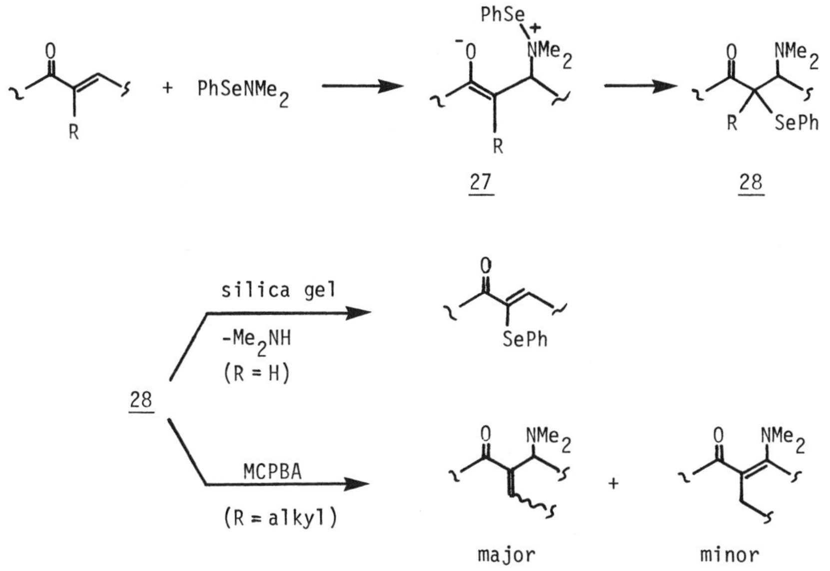

SCHEME 4.

A related process permits the one-step conversion of enones to their α-phenylseleno derivatives. Zima and Liotta[140] treated enones with PhSeCl and pyridine to form the desired products, probably via an enolate intermediate, 29 [Eq. (44)]. The method is applicable to aldehydes as well as ketones. The synthetic utility of α-phenylseleno enones has been demonstrated in a new approach to dihydrojasmone[141] and in other endeavors.[142] The use of excess selenenylating agent (PhSeCl or PhSeBr) in this process affords α-halo enones in generally high yield according to Eq. (45).[143]

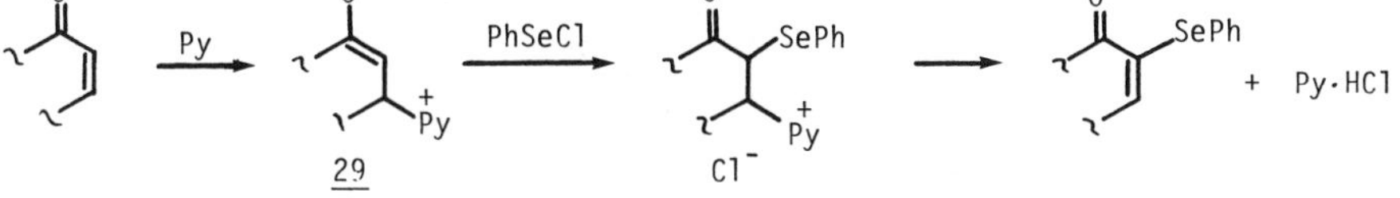

$$(44)$$

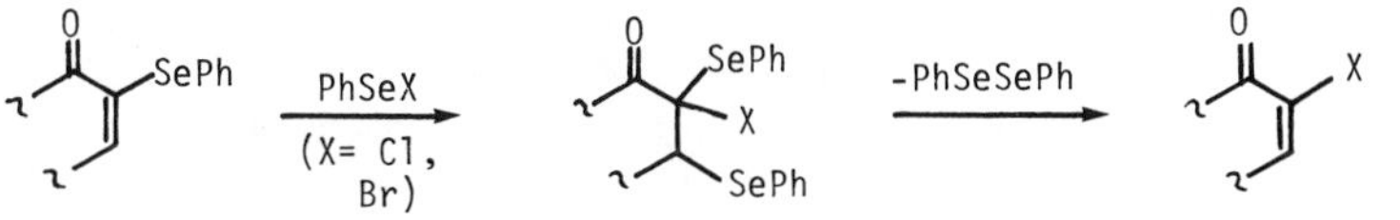

$$(45)$$

The cleavage of the α,β C—C bond was observed in one case where an enone was treated with LDA and PhSeBr [Eq. (46)].[104] A Michael addition of the amide base to the enone, followed by a retro-Mannich reaction was proposed to rationalize this result. Michael additions of LDA to α,β-unsaturated esters have also been studied (*vide infra*).

$$\text{Ph} \overset{O}{\diagdown} \quad \xrightarrow[\text{2. PhSeBr}]{\text{1. LDA}} \quad \text{Ph} \overset{O}{\diagdown} \text{SePh} \quad \quad (46)$$
$$59\%$$

Selenenamides are sufficiently basic to generate enolates from aldehydes, which then undergo selenenylation as in Eq. (47).[144] The greater kinetic acidity of aldehydes compared to ketones permits their selective selenenylation (e.g., product 30). Product 31 was formed by the rearrangement of the initially formed α-phenylseleno diester isomer.

$$\text{R} \overset{}{\diagdown} \text{CHO} + \text{PhSeNEt}_2 \longrightarrow \text{R} \overset{O^-}{\underset{H}{\diagdown}} + \text{PhSeNHEt}_2^+ \longrightarrow \text{R} \overset{\text{SePh}}{\underset{}{\diagdown}} \text{CHO} + \text{Et}_2\text{NH} \quad (47)$$

30   72%   31   60%

N-(Phenylseleno)morpholine, 32, has been investigated in a similar context and also produces α-phenylseleno aldehydes, although in some cases by way of isolable enamine intermediates [e.g., Eqs. (48) and (49)].[145]

$$\text{R} \overset{}{\diagdown} \text{CHO} + \text{PhSeN} \overset{}{\diagup} \text{O} \longrightarrow \text{R} \overset{\text{SePh}}{\diagdown} \text{CHO} \xrightarrow{32} \text{R} \overset{\text{PhSe}\quad\text{SePh}}{\diagdown} \text{CHO} \quad (48)$$

R= Et 75%   R= Et 65%
R= Ph 85%   R= Ph 50%

$$\text{PhSe} \overset{}{\diagdown} \text{CHO} \xrightarrow{32} \overset{\text{PhSe}}{\underset{\text{PhSe}}{\diagdown}} \overset{H}{\underset{N}{\diagdown}} \overset{}{\diagup} O \xrightarrow[\text{silica-gel}]{1\% \text{ HCl or}} \text{PhSe} \overset{\text{SePh}}{\diagdown} \text{CHO} \xrightarrow{32} (\text{PhSe})_3\text{CCHO}$$

85%   75%   60%

$$(49)$$

α-Phenylseleno ketones are also produced when their parent ketones react with α-phenylseleno phosphoranes, 33.[146] Proton transfer occurs from the ketone to the phosphorane, followed by selenenylation of the resulting enolate with the simultaneously formed selenophosphonium salt, 34 (Scheme 5). As expected, the same products are formed when the enolate is prepared from the ketone and LDA, and is allowed to react directly with the phosphonium salt.

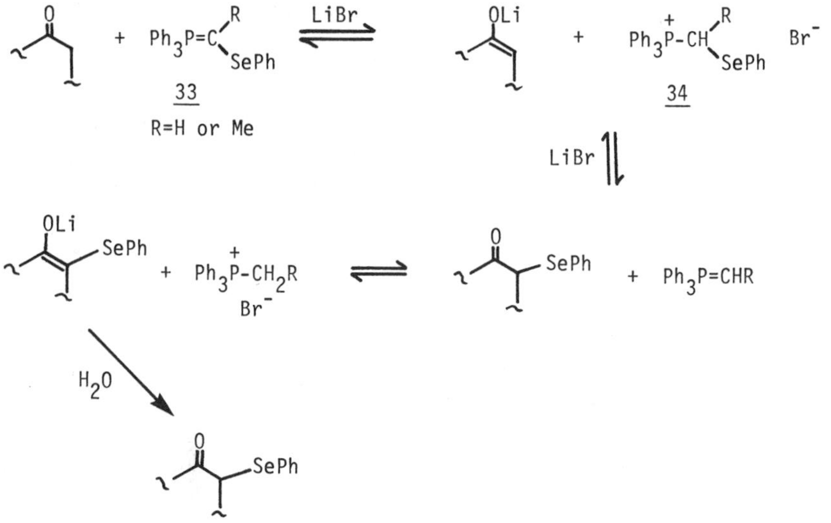

SCHEME 5.

Since ketone enolates are reasonably effective leaving groups, their displacement from α-phenylseleno ketones by nucleophilic attack at selenium is not surprising. The reverse of the process in Eq. (27) provides one such example where the nucleophile is PhSe⁻. If the nucleophile is the enolate of the α-phenylseleno ketone itself, then the overall effect is one of transselenenylation between this ketone and its enolate. Liotta and co-workers[141,147] found that α-alkyl-α-phenylseleno ketones undergo virtually quantitative transselenenylation to the less substituted α'-position when treated with 0.5 equivalents of LDA in THF–HMPA. The reaction is presumably driven in the observed direction by a cascade of intermolecular phenylseleno group and proton migrations to form increasingly stable enolate species.[147] A possible sequence of this type is shown in Scheme 6. Transselenenylation fails in the presence of a full equivalent of LDA as all of the α-phenylseleno ketone is then converted to its enolate, thus preventing the possibility of nucleophilic displacement at selenium. It is interesting that the presence of a β- or γ-substituent is also required for the reaction to proceed favorably. The application of this method to the synthesis of cis-jasmone, 36,[141] is shown in Eq. (50). The oxidation–elimination of 35 afforded the exocyclic double bond exclusively and so necessitated transselenenylation prior to oxidation.

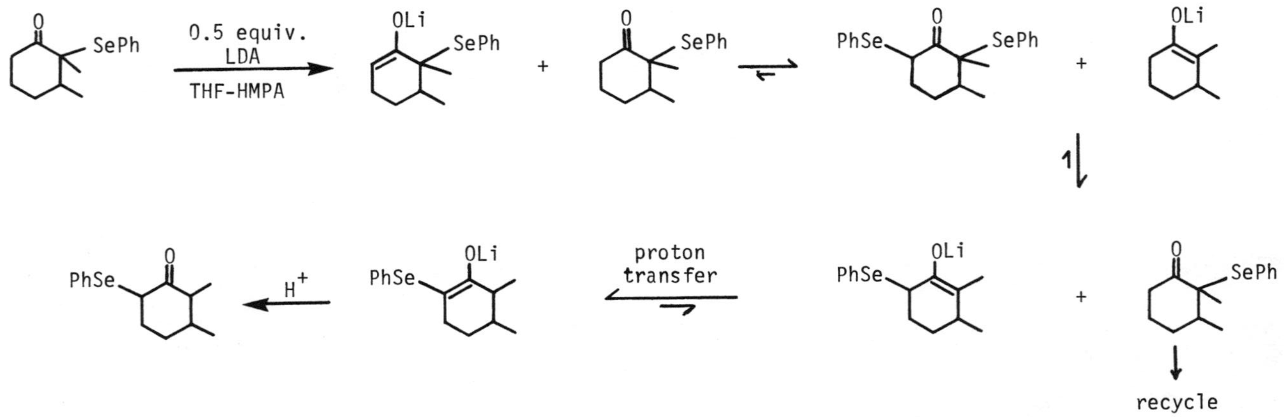
SePh
0.5 equiv.
LDA
THF-HMPA
OLi
SePh
+
SePh
PhSe
SePh
+
OLi
PhSe
H+
PhSe
OLi
proton
transfer
PhSe
OLi
+
SePh
recycle

SCHEME 6.

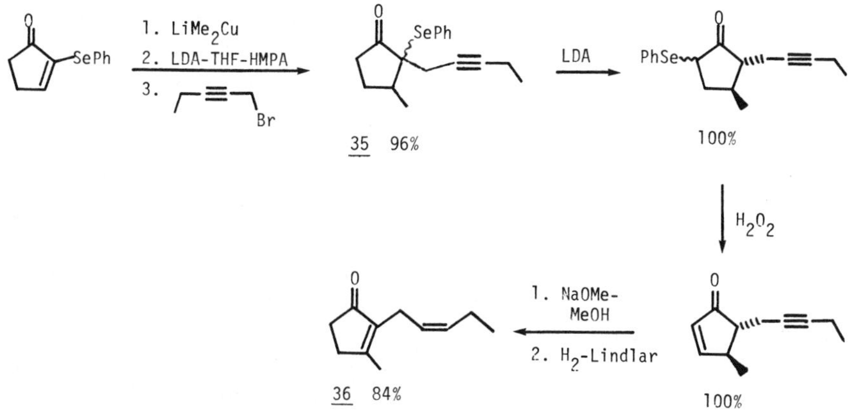

$$(50)$$

2.2. Esters and Lactones

General methods for the direct selenenylation of esters and lactones are un-available and prior formation of the corresponding enolate is required. A pro-cedure developed independently by Reich and co-workers[11,43,78] and Sharpless et al.[13] resembles that employed with ketones. The enolate is generated at low temperature (usually $-78°C$) by means of LDA in THF and is then quenched with an appropriate selenenic electrophile [Eq. (51)]. The latter include PhSeSePh as well as PhSeCl and PhSeBr. The use of the diselenide is possible with ester and lactone enolates because the equilibrium in Eq. (27) is shifted further towards the right than in the case of ketones.[11] The passage of oxygen through the reaction mixture during work-up removes PhSe$^-$ oxidatively, and has been reported to enhance the yield by driving the reaction even further to completion.[148] Elemental selenium reacts with ester and lactone enolates in a manner analogous to Eq. (39).[133] Other hindered lithium amide bases,[148–155] as well as potassium hydride in THF containing DMF,[156] have been used in place of LDA. Hexamethylphosphoric triamide is frequently included.[28,148, 157–164] Excess base permits enolate selenenylations to take place in the presence of sulfonamide,[153] carboxylic acid,[165] or hydroxyl[156,166] functionalities.

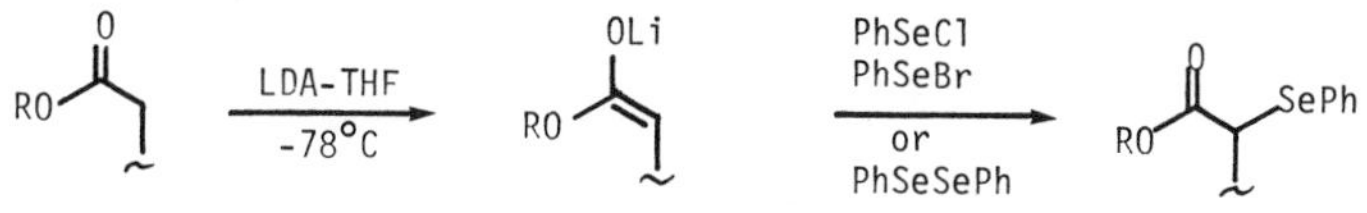

$$(51)$$

As with ketones, the selenenylation is usually employed in conjunction with selenoxide elimination to furnish α,β-unsaturated derivatives. Alkylations of α-phenylseleno esters with aldehydes [e.g., Eq. (52)],[167] or of α-phenylseleno lactones with alkyl halides[157,168] have been performed prior to oxidation [e.g.,

Eq. (53)]. Grieco and Miyashita[157] demonstrated that the order of selenenylation and alkylation can determine the regiochemical outcome of the subsequent
selenoxide elimination [Eqs. (54) and (55)]. Since the substrate was a relatively
rigid fused-ring lactone, both alkylation and selenenylation occurred stereoselectively, with the last substituent being introduced into the less hindered *exo*
position. The *exo*-phenylseleno group in 37 is cis to the ring fusion proton and
syn elimination afforded principally the *endo* olefin, 38. On the other hand,
endocyclic *syn* elimination in 39 is conformationally difficult and so the *exo*
olefin, 40, was formed instead. Similar strategies involving selenenylation and
oxidation have been exploited in introducing the exocyclic double bond of a
number of other α-methylene lactones [e.g., Eq. (56)].[128,159,161,162,166,169−171]
In other cases [e.g., Eqs. (57) and (58)] predominantly endocyclic elimination
was observed.[148,160,169,172−174] δ-Lactones can be converted to furans by
selenenylation, oxidation, and reduction with diisobutylaluminum hydride
(DIBAL) [Eq. (59)].[158] In addition to the examples cited, the selenenylation of
various other esters[175−189] and lactones[190−198] has been reported. Several other
illustrative examples are provided in Eqs. (60)–(62).

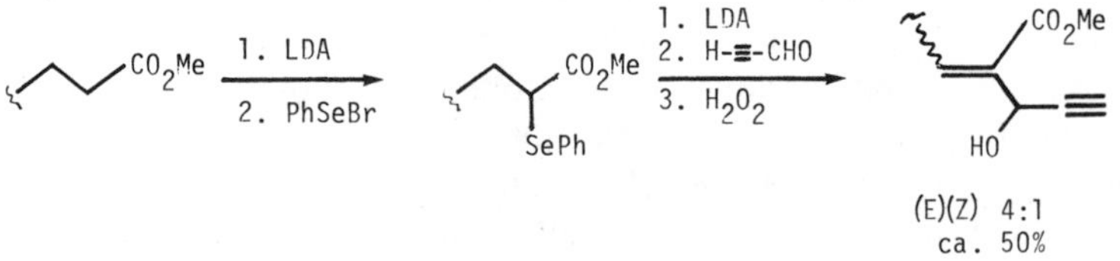

(E)(Z) 4:1
ca. 50%

(52) (Ref. 167)

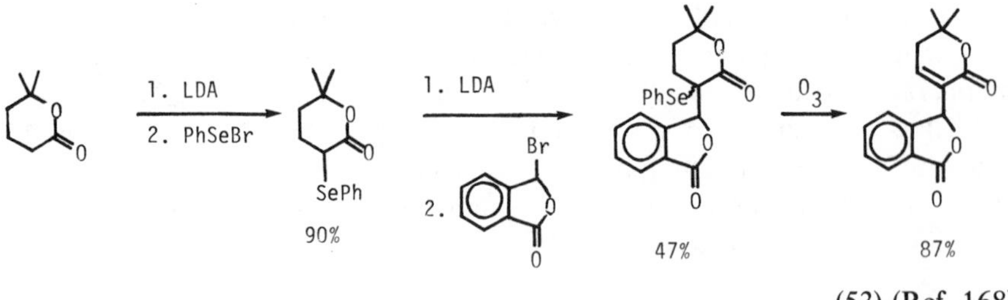

90%   47%   87%

(53) (Ref. 168)

37 88%   38   40

9 : 1

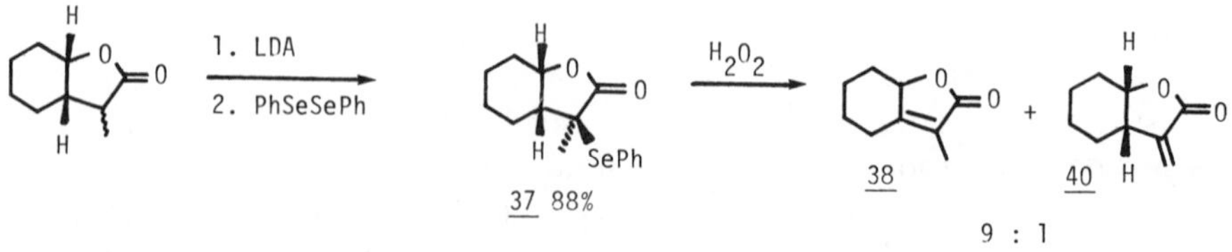

(54) (Ref. 157)

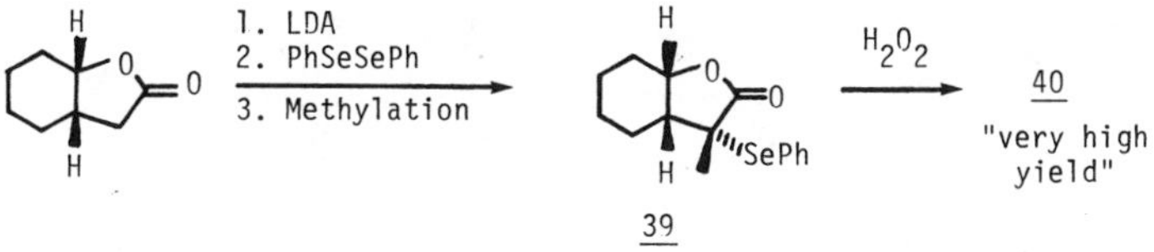

$$\text{(55) (Ref. 157)}$$

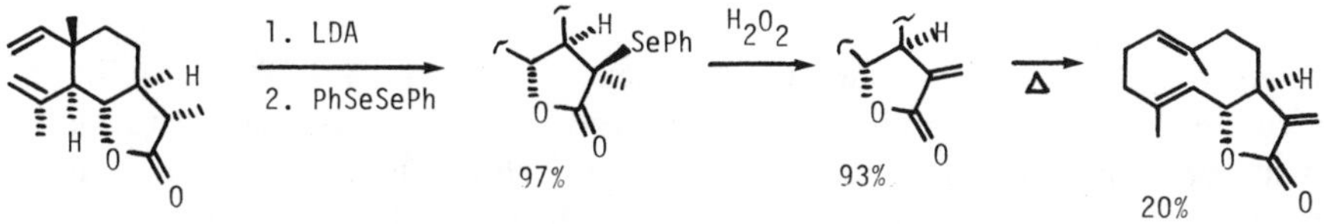

$$\text{(56) (Ref. 159)}$$

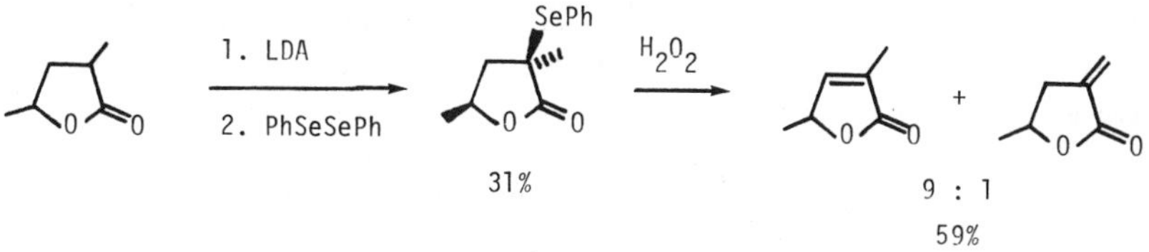

$$\text{(57) (Ref. 172)}$$

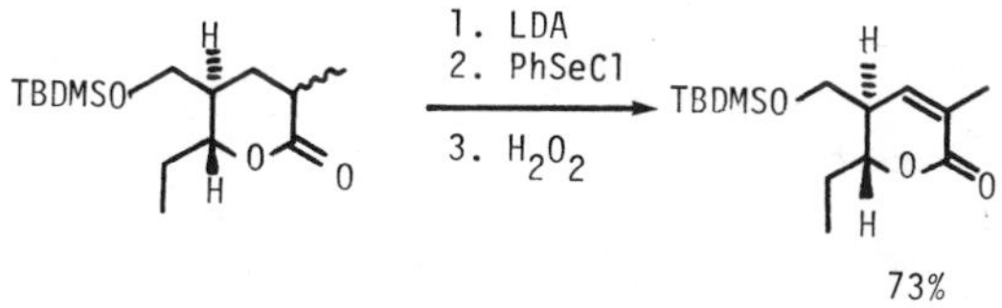

$$\text{(58) (Ref. 174)}$$

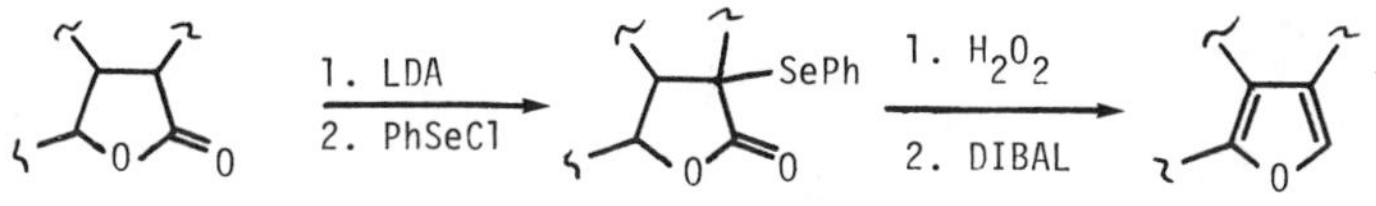

$$\text{(59) (Ref. 158)}$$

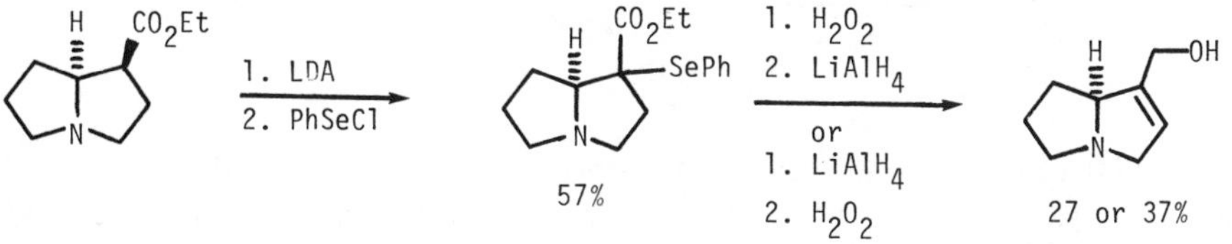

$$\text{(60) (Refs. 186, 187)}$$

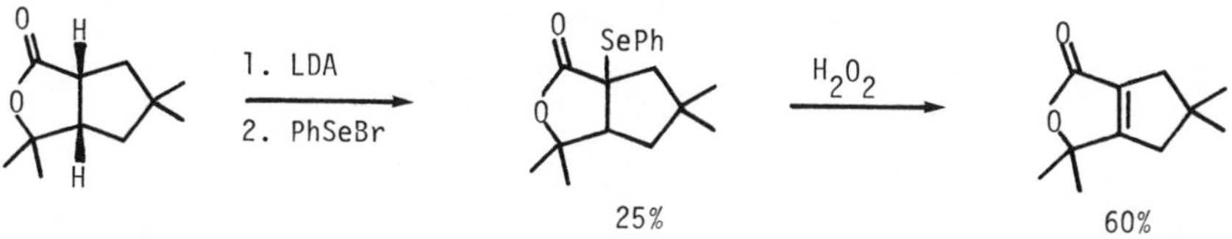

$$\text{(61) (Ref. 194)}$$

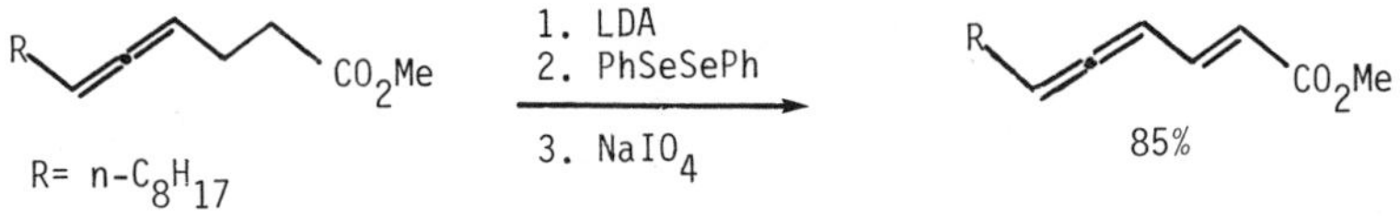

$$(62) \text{ (Ref. 178)}$$

α,β-Unsaturated esters undergo α-selenenylation with LDA and PhSeBr, providing that they are relatively unhindered and of the (E) configuration.[104] A mechanism involving Michael addition, selenenylation, and elimination has been proposed [Eq. (63)] for this reaction. More hindered substrates, as well as those with (Z) configurations, are deprotonated by LDA and selenenylated at the γ-position to give (E) products [Eq. (64)]. The latter products were similarly produced from allylic esters.[104] The α,β-unsaturated esters in Eq. (65) formed γ-selenenylated products that were converted to the corresponding pyrroles by selenoxide elimination[199]

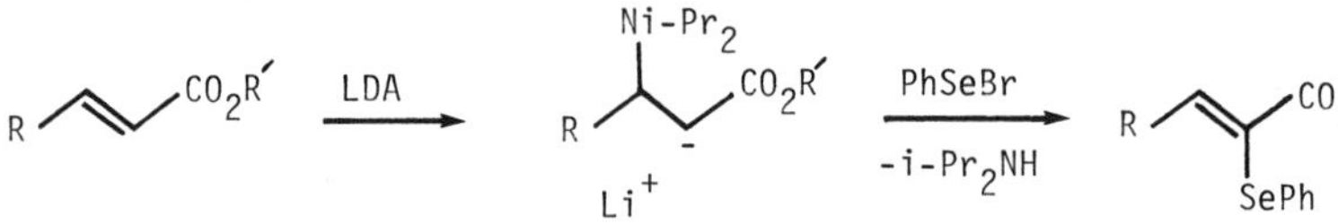

$$(63)$$

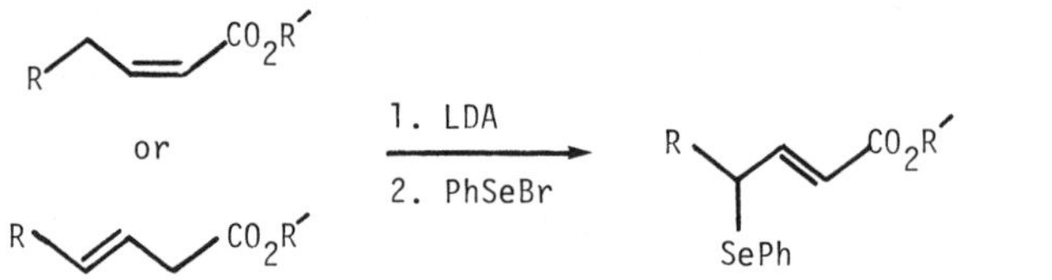

$$(64)$$

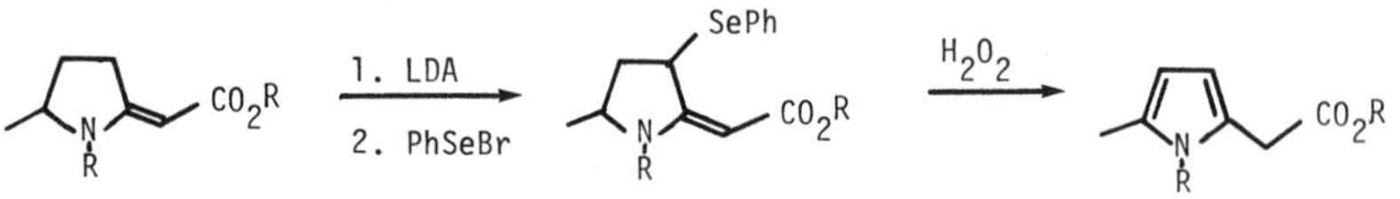

$$(65)$$

Oxidative coupling occurred when ethyl acrylate was treated with PhSeBr in the presence of catalytic amounts of palladium (II) chloride and triphenylphosphine[200] [Eq. (66)].

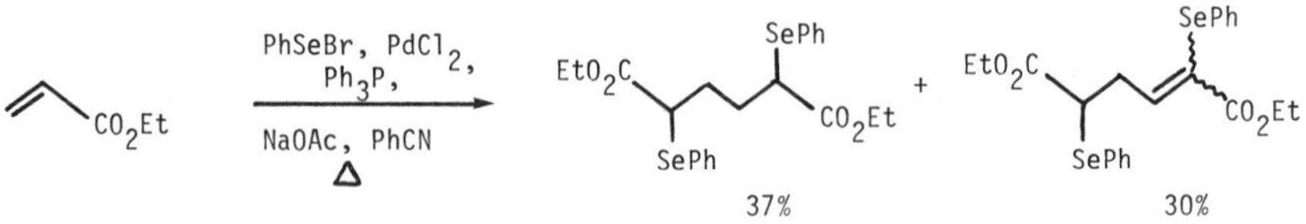

$$(66)$$

2.3. Amides and Lactams

Compared to the large number of examples of α-selenenylations of ketone and ester enolates, similar reactions of metalated amides and lactams are relatively scarce. The anion of 1-methyl-2-pyrrolidinone undergoes either mono- or bis-selenenylation depending on the amount of LDA employed in its formation.[201] With one equivalent of base the latter dominates to give **41**. Unlike the case with ketone or ester enolates (where bisselenenylation is rare), proton transfer from the monosubstituted lactam, **42**, to the anion, **43**, must be faster than the selenenylation of **43** (Scheme 7). The selenium-stabilized anion, **44**, thus generated reacts further with PhSeCl to give the observed product, **41**. With two equivalents of LDA, both the unsubstituted and monosubstituted lactams are deprotonated and the respective anions, **43** and **44**, compete for the remaining PhSeCl. Since **43** is the more reactive species, monoselenenylation dominates to produce **42**. Similar considerations apply to the corresponding sulfenylations.

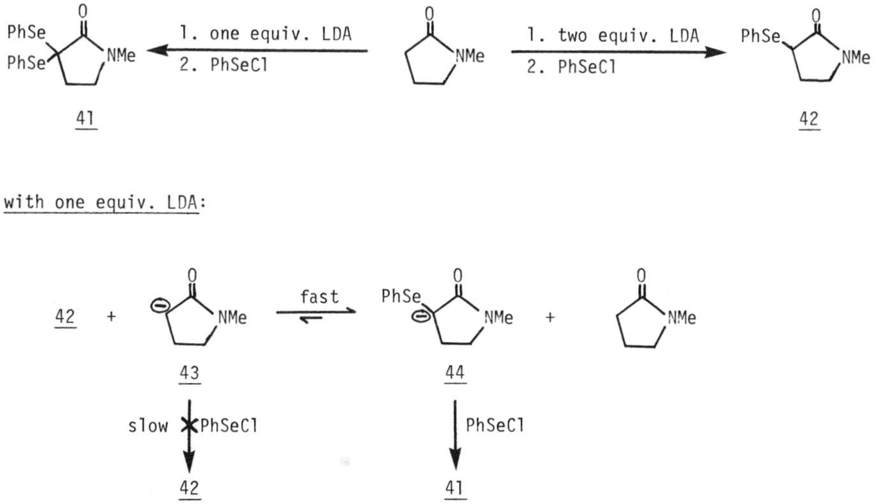

with one equiv. LDA:

SCHEME 7.

Other examples of lactam monoselenenylations[202–204] are provided in Eqs. (67) and (68). The bisselenenylation in Scheme 8 was performed in the presence

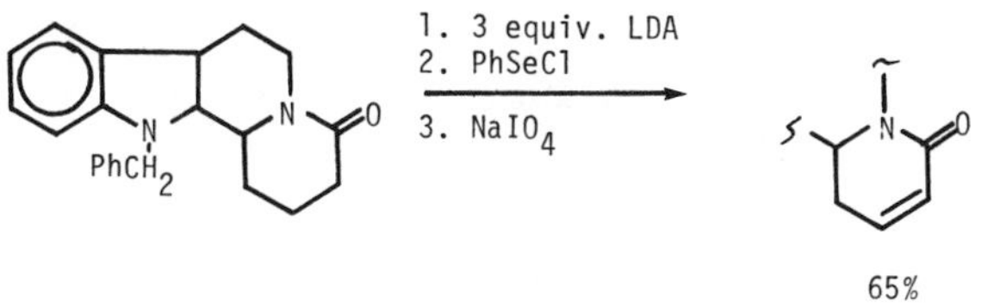

(67) (Ref. 202)

65%

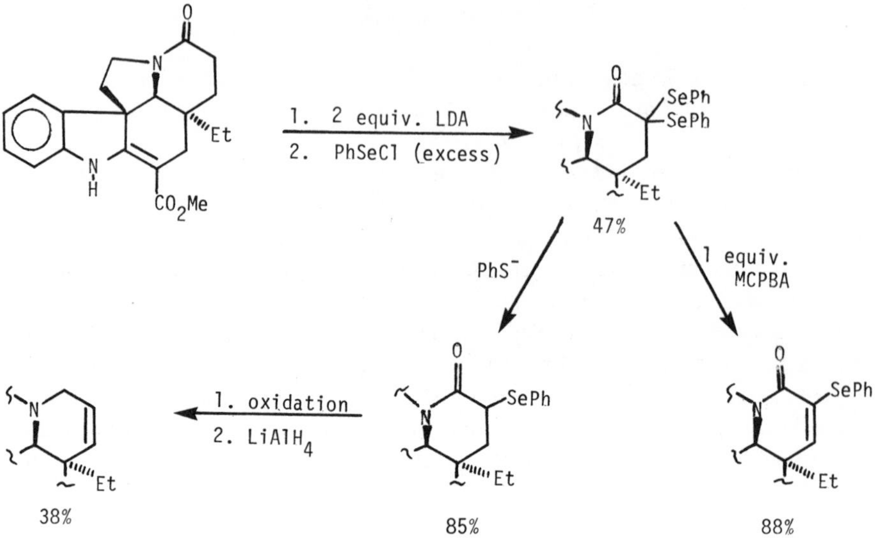

SCHEME 8.

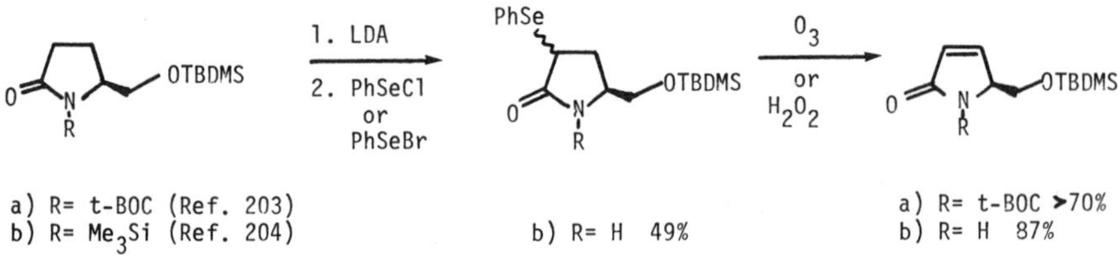

a) R= t-BOC (Ref. 203)
b) R= Me₃Si (Ref. 204)

b) R= H   49%

a) R= t-BOC >70%
b) R= H   87%

$$(68)$$

of one effective equivalent of base (another equivalent was consumed by the NH proton).[205] The monoseleno product was obtained by cleavage of one PhSe group with benzenethiolate anion. A dimetalated succinamide derivative was sequentially alkylated and selenenylated in Eq. (69). Significant stereoselectivity in favor of the threo isomer was observed.[206] An example of the α-selenenylation of an α,β-unsaturated amide has been reported [Eq. (70)].[207–208]

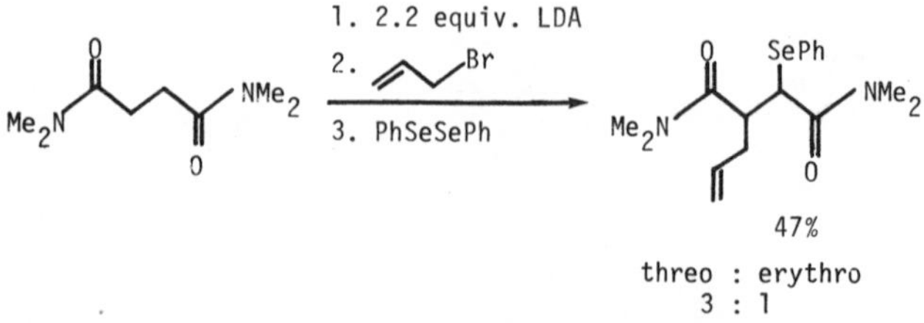

threo : erythro
3 : 1

$$(69) \text{ (Ref. 206)}$$

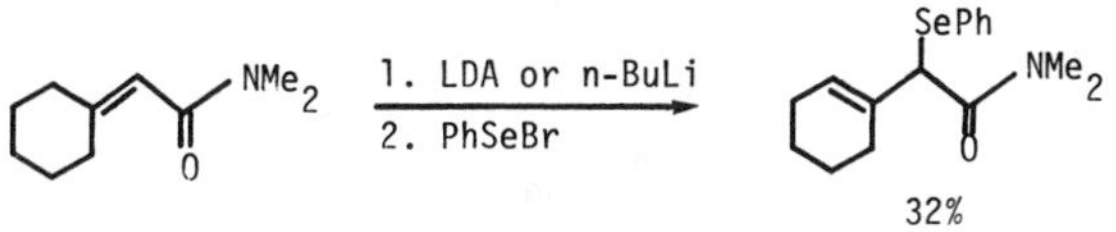

(70) (Refs. 207, 208)

32%

2.4. β-Dicarbonyl Compounds

A general method for the α-selenenylation of β-dicarbonyl compounds was developed by Reich and co-workers[11,78,209] and later utilized by other groups.[124,210–216] The substrate is converted to its enolate with sodium hydride, usually in THF, and then treated with PhSeCl or PhSeBr. Alternative bases include n-butyllithium,[54,217] LDA,[218] sodium methoxide,[219] potassium fluoride on Celite,[c,220] pyridine,[221] and triethylamine.[222] Diketones,[11,78,209, 216,218,220,221,223] ketoaldehydes,[221,222] ketoesters (or lactones),[11,78,124,209–213,215,221] diesters,[54,217,219] or lactam esters[214] can all be selenenylated by such methods. The direct selenenylation of a ketoaldehyde with selenenamide, **11** (R = Et) has been reported.[29] There are no other reports of direct selenenylation techniques, possibly because such processes are reversible and the equilibria unfavorable in the absence of a hydrogen halide scavenger. Illustrative examples are provided in Eqs. (71)–(76). It should be pointed out that the procedures for the selenenylation of ketones and aldehydes shown in Eqs. (23), (24), and (39) are also applicable to β-dicarbonyl compounds.[74,75,224]

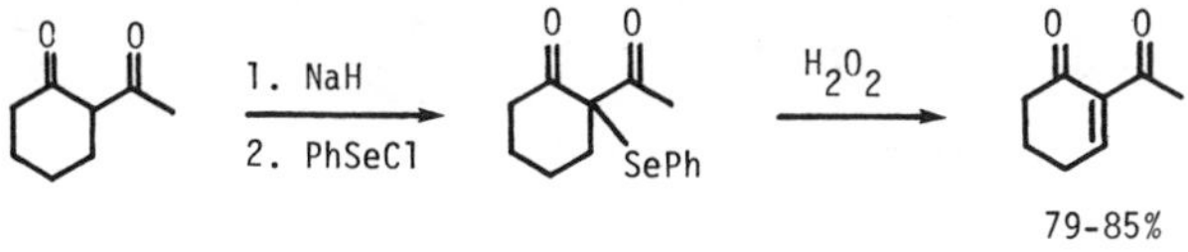

(71) (Ref. 209)

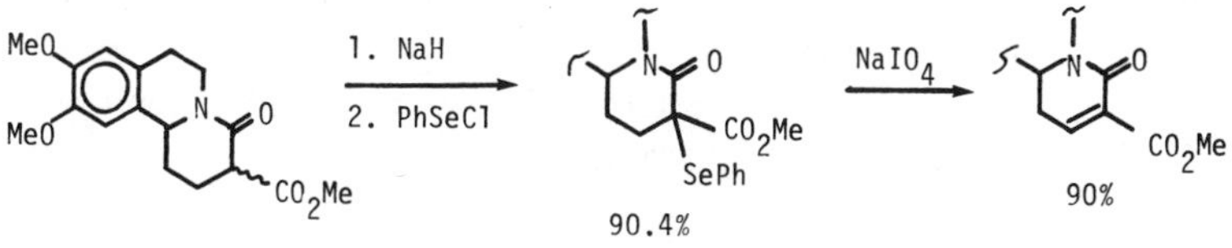

79-85%

(72) (Ref. 214)

90.4%   90%

21%   50%

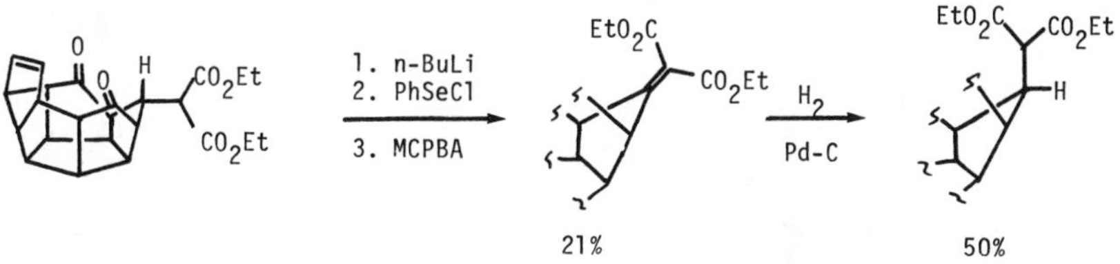

(73) (Ref. 217; cf. Ref. 54)

---

[c] Celite is a registered trademark of the Johns Manville Co.

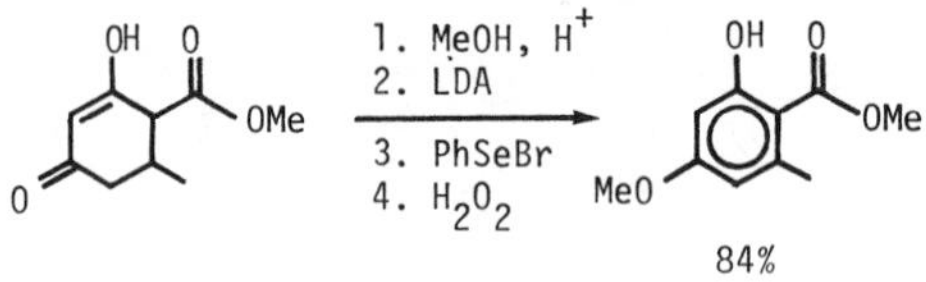

(74) (Ref. 218)

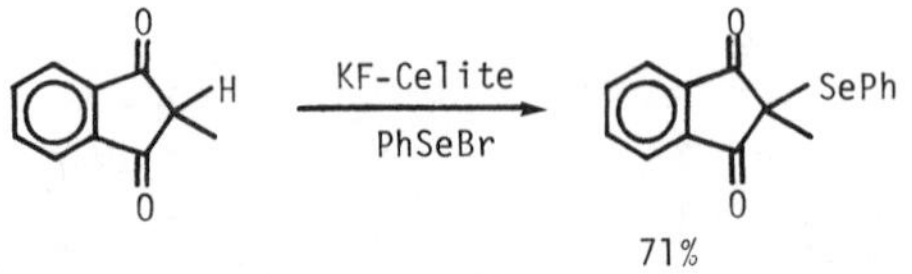

(75) (Ref. 220)

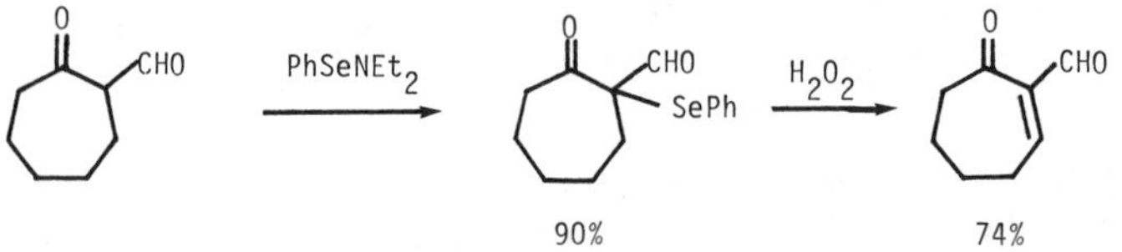

(76) (Ref. 29)

The product selenides can usually be smoothly converted to α,β-unsaturated derivatives by the usual oxidation techniques. Moreover, a nonoxidative elimination takes place when certain α-phenylseleno-β-dicarbonyl compounds are treated with excess PhSeCl in pyridine [Eq. (77)].[221]

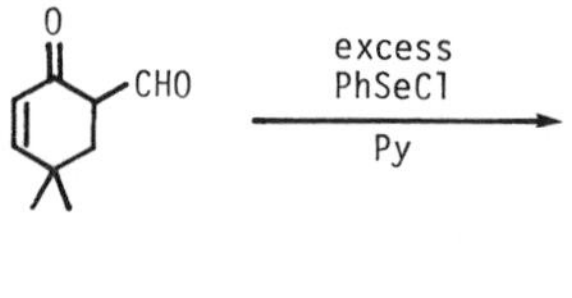

(77)

Equations (78) and (79) provide examples where decarboxylation accompanied selenenylation. The rearrangement of the phenylseleno group from the two to the six position in the ketoester in Eq. (80) occurred in the presence of excess LDA, sodium hydride, or pyrrolidine.[227]

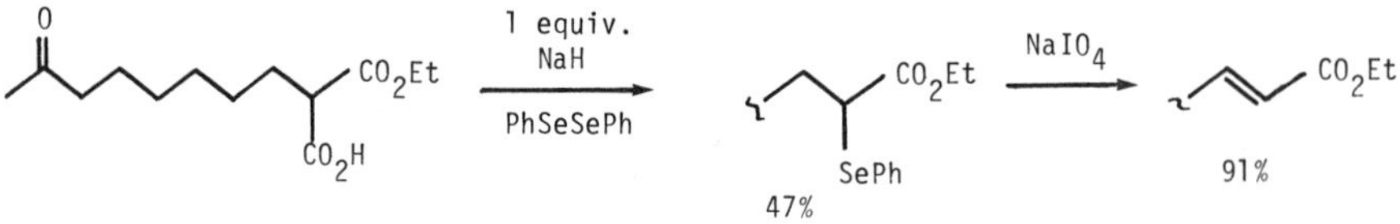

(78) (Ref. 225)

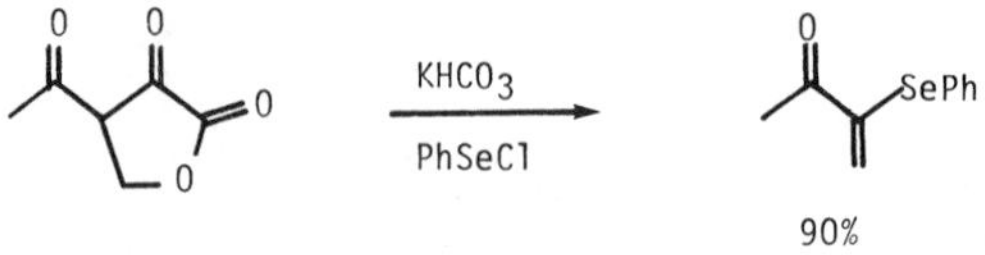

(79) (Ref. 226)

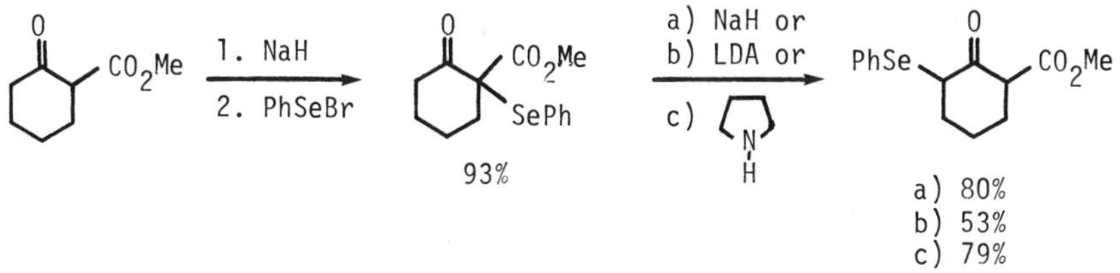

(80)

3. SELENENYLATION OF OTHER FUNCTIONAL GROUPS

3.1. Enol Acetates, Enol Silyl Ethers, and Enol Boranes

In 1973, Clive[22] reported that enol acetates react with PhSeBr and silver tri-fluoroacetate to afford the corresponding α-phenylseleno ketones [Eq. (81)]. Other examples of this process were provided by Reich and co-workers,[11,24,78] who also demonstrated that the selenenyl trifluoroacetate, 7, is the active electrophile under these conditions.[24] Other reagents that have been employed in the conversion of enol acetates to α-seleno ketones are selenenyl acetates[228] [from RSeBr and potassium acetate; Eq. (82), PhSeCl in aqueous acetonitrile [Eq. (83)],[229] and PhSeCl followed by work-up with aqueous sodium bicarbonate.[230]

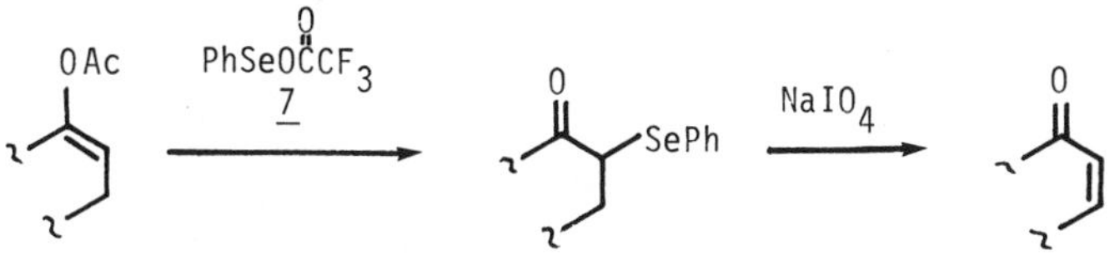

(81)

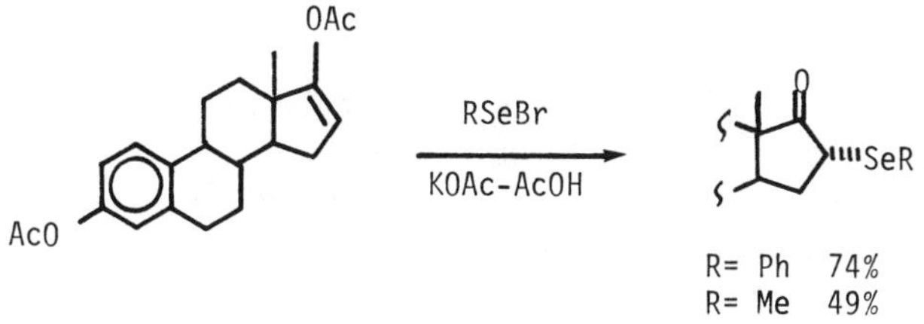

(82)

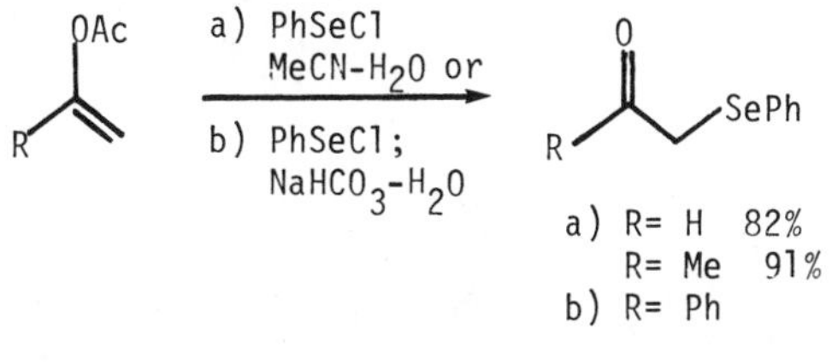

(83)

An alternative approach employs the cleavage of the enol acetate with methyllithium to generate the corresponding enolate, which is then selenenylated as described earlier [Eq. (84)].[11,65,78]

$$
\begin{array}{ccccc}
\text{OAc} & \xrightarrow{\text{MeLi}} & \text{OLi} & \xrightarrow[\text{PhSeBr}]{\text{PhSeCl or}} & \overset{\text{O}}{\underset{}{\|}}\text{SePh}
\end{array} \qquad (84)
$$

An electrochemical method developed by Torii et al.[231] produced 2-phenylselenocyclohexanone from the corresponding enol acetate by electrolysis in the presence of PhSeSePh and tetraethylammonium bromide [Eq. (85)]. The mechanism probably involves the two-electron oxidation of Br^- to Br^+, followed by formation of PhSeBr, as in Eq. (24). A refinement of this method permits the direct conversion of an enol acetate to an enone and employs a catalytic amount of the diselenide in the presence of magnesium sulfate instead of bromide ion [Eq. (86)].[232] Benzeneselenenic acid (PhSeOH) or a related electrophile functions as the selenenylating agent and the initially formed α-phenylseleno ketone undergoes further anodic oxidation to the enone.

$$
\begin{array}{ccc}
\text{OAc} & \xrightarrow[\text{anodic oxidation}]{\substack{\text{PhSeSePh} \\ \text{Et}_4\text{N}^+ \text{ Br}^-}} & \overset{\text{O}}{\underset{}{\|}}\text{SePh}
\end{array} \qquad (85)
$$

$$
95\%
$$

$$
\begin{array}{ccc}
\text{OAc} & \xrightarrow[\substack{\text{MeCN-H}_2\text{O} \\ \text{anodic oxidation}}]{\substack{\text{PhSeSePh (cat.)} \\ \text{MgSO}_4}} &
\end{array} \qquad (86)
$$

$$
81\%
$$

An important feature in the use of enol acetates (or enol silyl ethers) is that they generally produce α-phenylseleno ketones that are regioisomers of those obtained from the selenenylation of kinetic enolates [e.g., Eq. (87)].[11,78] The two techniques are thus complementary.

$$
\underset{60\%}{\overset{\text{O}}{\|}}\text{Ph} \xleftarrow[\text{3. H}_2\text{O}_2]{\substack{\text{1. LDA} \\ \text{2. PhSeCl}}} \overset{\text{O}}{\|}\text{Ph} \longrightarrow \overset{\text{OAc}}{}\text{Ph} \xrightarrow[\text{3. H}_2\text{O}_2]{\substack{\text{1. MeLi} \\ \text{2. PhSeBr}}} \underset{81\%}{\overset{\text{O}}{\|}}\text{Ph}
$$

$$
(87)
$$

Enol silyl ethers react rapidly with PhSeCl or PhSeBr to produce α-phenylseleno ketones[233-236] or aldehydes [Eq. (88)].[233,237] The volatile trimethylsilyl

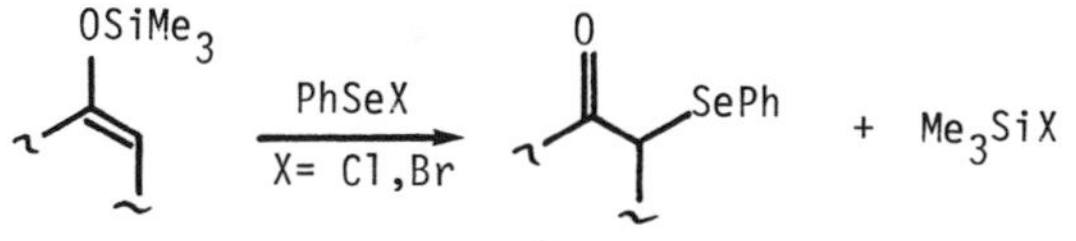

$$
\text{(88)}
$$

halide by-product can be easily removed with the solvent. Danishefsky's diene reacts selectively at the four position and affords a product that can be resilylated [Eq. (89)] and which is of interest in Diels–Alder cycloadditions.[238,239] An example of a reaction between an enol silyl ether and selenium dichloride has been reported [Eq. (90)].[16] Several enol boranes have been converted to α-phenylseleno ketones with PhSeCl [Eq. (91)].[240] Enol alkyl ethers may also be used as precursors of α-phenylseleno carbonyl compounds. These and other reactions and applications of such systems will be discussed in Section 7.

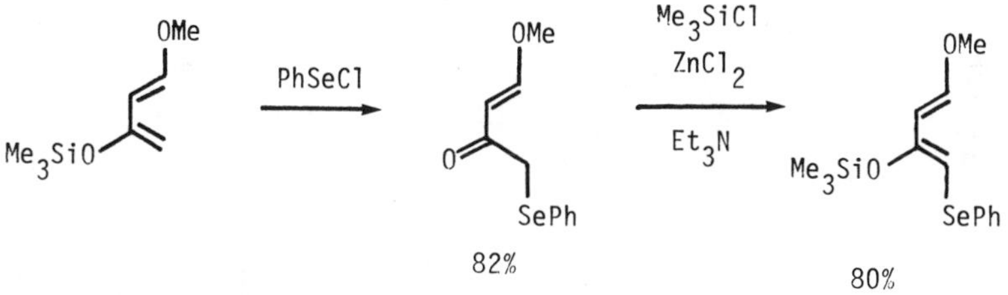

$$
\text{(89)}
$$

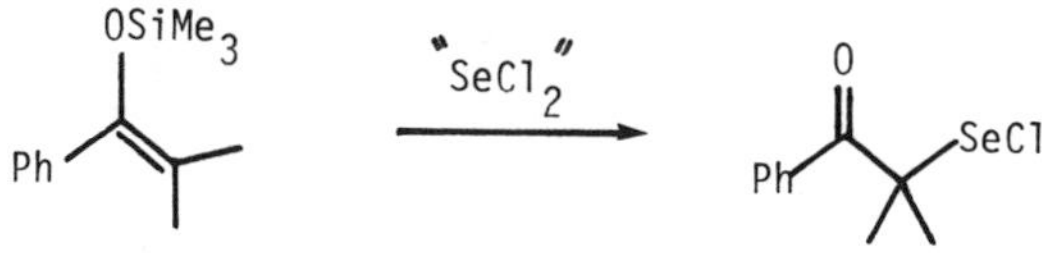

82%   80%

$$
\text{(90)}
$$

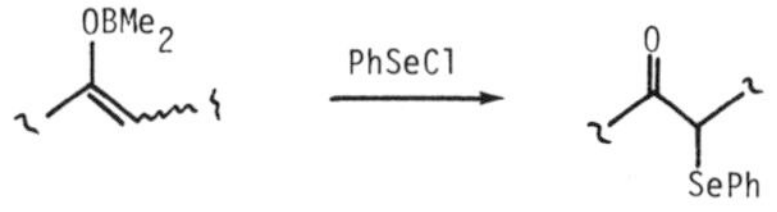

$$
\text{(91)}
$$

3.2.   Enamines and Enamides

Enamines derived from aldehydes react with PhSeCl at $-110°C$ to furnish α-phenylseleno aldehydes after hydrolysis [Eq. (92)].[241,242] 2-n-Butyl-1,2-dihydropyridine and its N-lithio derivative underwent aromatization with PhSeCl [Eq. (93)].[243] The transselenenylation of an α'-phenylseleno enamine has been reported [Eq. (94); cf. Eq. (80)].[227]

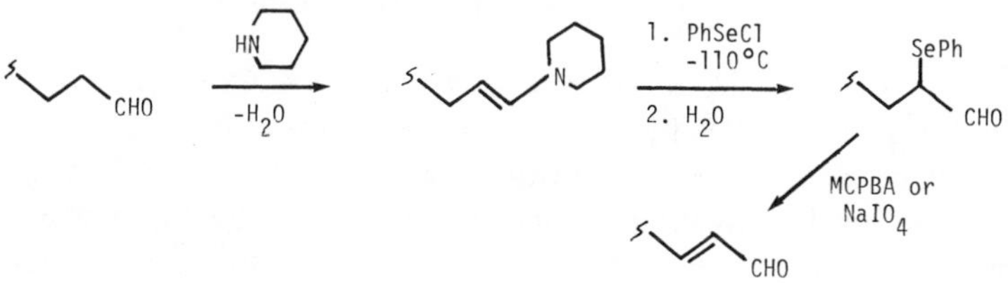

$$
\text{(92)}
$$

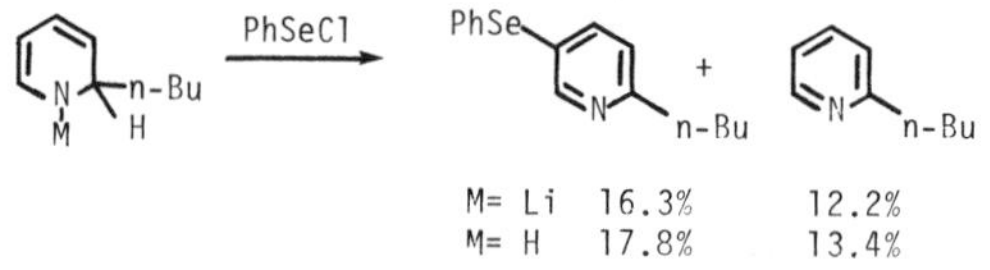

$$(93)$$

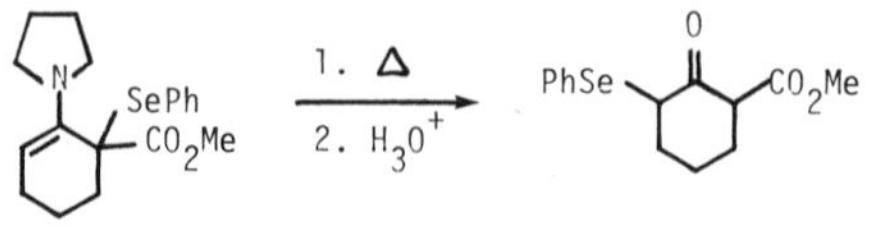

M= Li   16.3%   12.2%
M= H   17.8%   13.4%

$$(94)$$

The less nucleophilic 4-azasteroidal enamides in Eq. (95) react rapidly with PhSeCl at room temperature to furnish the corresponding 6-phenylseleno derivatives (Eq. (95)).[244,245] The products may be further converted to carbinol-amides by oxidation with excess meta-chloroperbenzoic acid (MCPBA), or to selenoxides with one equivalent of the oxidant. One such compound was successfully separated from its diastereomer to afford the first stable, optically pure selenoxide.[245] Similarly, uracil gave the 5-phenylseleno derivative and its corresponding selenoxide as shown in Eq. (96).[246]

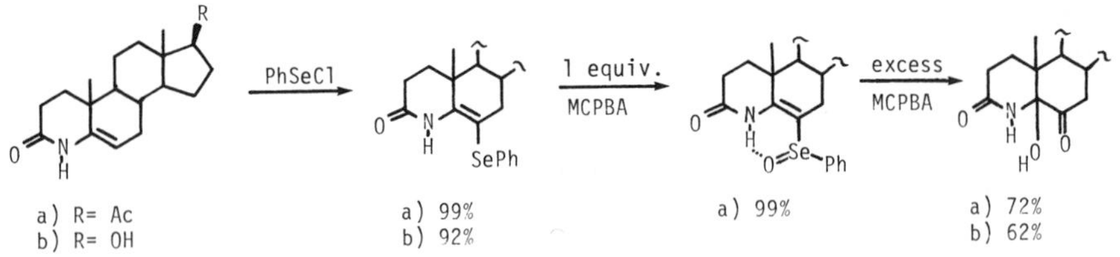

a) R= Ac   a) 99%   a) 99%   a) 72%
b) R= OH   b) 92%   b) 62%

$$(95)$$

76%   66%

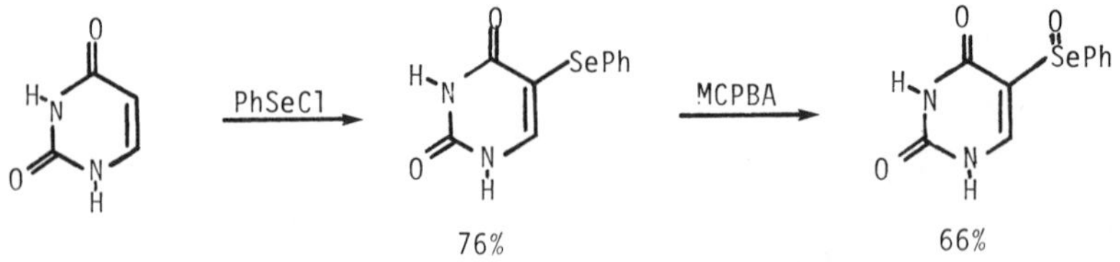

$$(96)$$

3.3. Nitriles, Nitro, and N-Nitroso Compounds

Brattesani and Heathcock[247] studied the selenenylation of α-lithiated octano-nitrile with PhSeBr, PhSeSePh, EtSeBr, and PhSeOAc. As in the case of amides and lactams (cf. Scheme 7), rapid proton transfer from the product to the un-reacted anion resulted in the persistence of unreacted starting material when only one equivalent of base was employed. Similar equilibration was also observed in the analogous sulfenylation. The use of excess base again proved expedient. Two synthetic applications are shown in Eqs. (97) and (98).

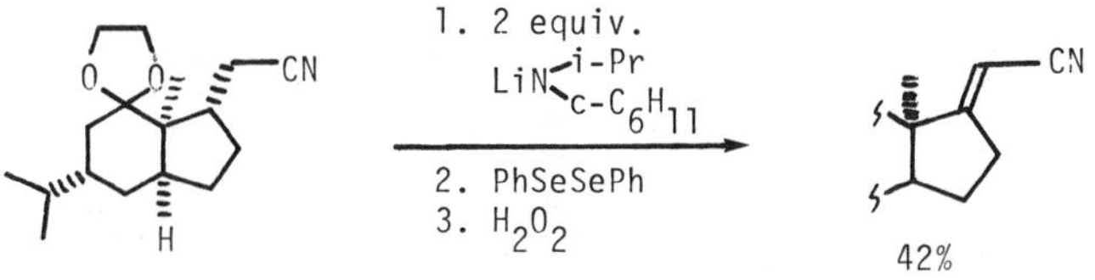

(97) (Ref. 248)

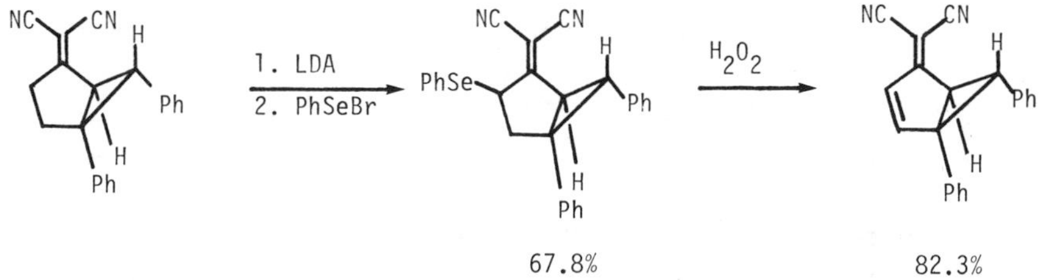

42%

67.8%   82.3%

(98) (Ref. 249)

The α,β-dehydrogenation of nitro[250-252] and N-nitroso[253,254] compounds can be carried out by metalation, selenenylation, and oxidation–elimination [Eqs. (99) and (100)]. α-Phenylseleno nitro compounds can be hydroxymethylated with formaldehyde prior to oxidation [Eq. (100)].[251] They are relatively labile compounds, decomposing when chromatographed on silica gel[250] and regenerating the saturated nitro compound when exposed to nucleophiles such as water, methanol, or amines, or in the presence of strong acids.[252]

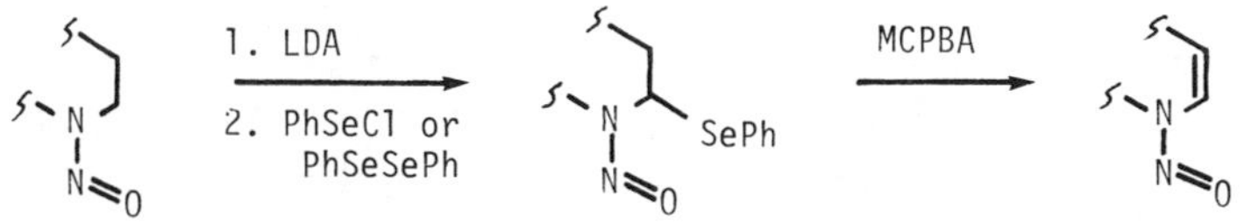

(99)

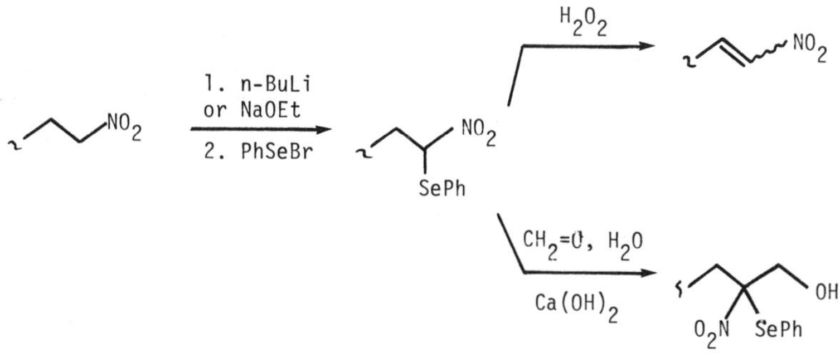

(100)

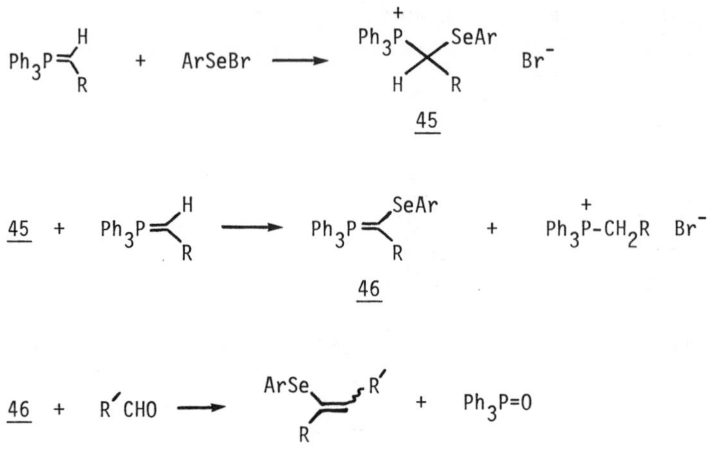

$$Ph_3P=\overset{H}{\underset{R}{<}} \quad + \quad ArSeBr \quad \longrightarrow \quad \overset{+}{\underset{H}{Ph_3P}}\overset{SeAr}{\underset{R}{\times}} \quad Br^-$$

$$\underline{45}$$

$$\underline{45} \quad + \quad Ph_3P=\overset{H}{\underset{R}{<}} \quad \longrightarrow \quad Ph_3P=\overset{SeAr}{\underset{R}{<}} \quad + \quad \overset{+}{Ph_3P}-CH_2R \quad Br^-$$

$$\underline{46}$$

$$\underline{46} \quad + \quad R'CHO \quad \longrightarrow \quad \overset{ArSe}{\underset{R}{\diagdown}}\diagup^{R'} \quad + \quad Ph_3P=O$$

$R=H$, Me or CO_2Et; Ar=Ph or 2,4-dinitrophenyl

SCHEME 9.

3.4. Phosphorus-Containing Functional Groups

Petragnani and co-workers[146,255,256] found that areneselenenyl bromides react with phosphoranes to form α-arylseleno phosphonium salts, $\underline{45}$, initially, followed by proton transfer to a second molecule of the starting phosphorane to give α-arylseleno phosphoranes, $\underline{46}$ (Scheme 9). The latter compounds undergo Wittig reactions with aldehydes to produce vinyl selenides[256] but act as selenenylating agents toward ketones,[146] as described previously (cf. Scheme 5). When the α-seleno phosphonium salt has no α-hydrogen, proton transfer obviously cannot occur, and the perchlorate salts of such species, $\underline{47}$, can be converted to the corresponding vinyl phosphonium salts, $\underline{48}$, by oxidation with MCPBA [Eq. (101)].[257] The bisselenenylation of a phosphonium salt has also been reported [Eq. (102)].[257]

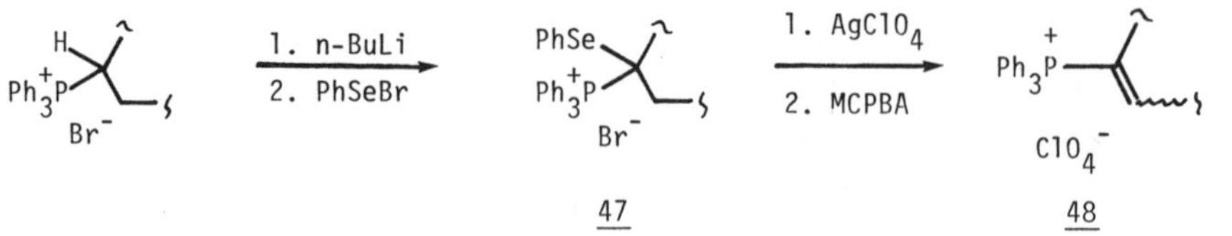

$$\underset{Br^-}{\overset{H}{\underset{Ph_3^+P}{\times}}} \quad \xrightarrow[\text{2. PhSeBr}]{\text{1. n-BuLi}} \quad \underset{Br^-}{\overset{PhSe}{\underset{Ph_3^+P}{\times}}} \quad \xrightarrow[\text{2. MCPBA}]{\text{1. AgClO}_4} \quad \underset{ClO_4^-}{\overset{+}{Ph_3P}}\diagup$$

$$\underline{47} \qquad\qquad \underline{48}$$

$$(101)$$

$$\underset{Br^-}{\overset{+}{Ph_3P}}\diagdown \quad \xrightarrow[\text{2. 2 PhSeBr}]{\text{1. 2 n-BuLi}} \quad \underset{Br^-}{\overset{+}{Ph_3P}}\overset{SePh}{\underset{SePh}{\times}}$$

$$93\%$$

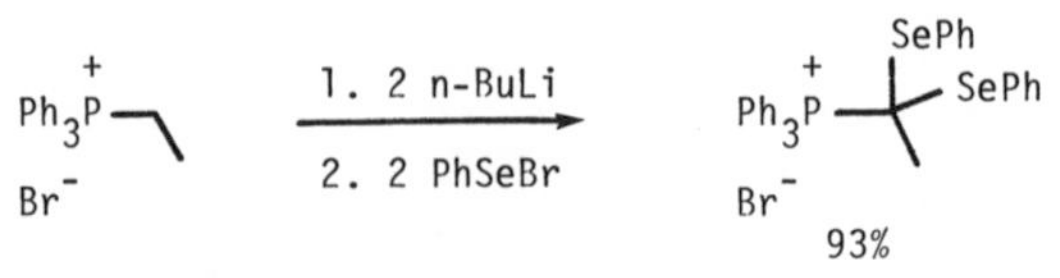

$$(102)$$

Metalated phosphine oxides[257,258] and phosphonates[259,260] can be α-selenenylated and oxidized to give unsaturated derivatives. Examples are shown in Eqs. (103)–(105). The synthetic utility of ethyl 2-diethylphosphonoacrylate, 49, stems from its ability to undergo Michael additions with enolates, followed by Wadsworth–Emmons reactions.[259]

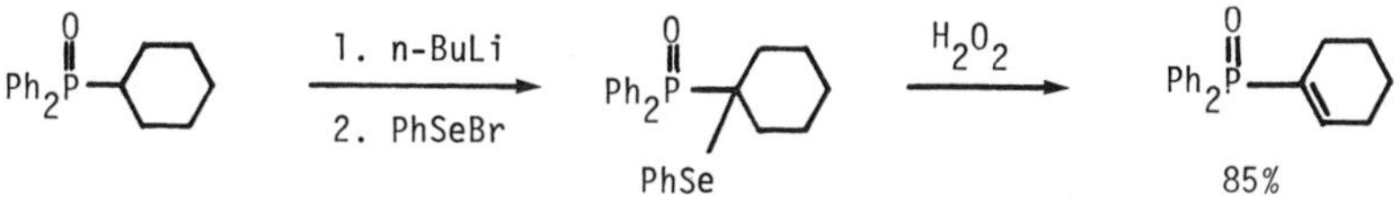

(103) (Ref. 257)

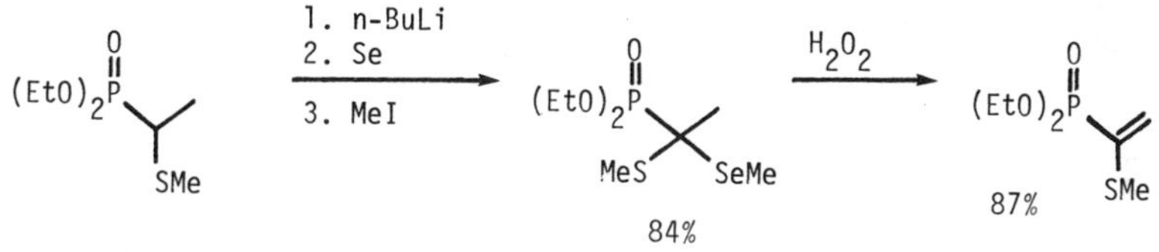

(104) (Ref. 260)

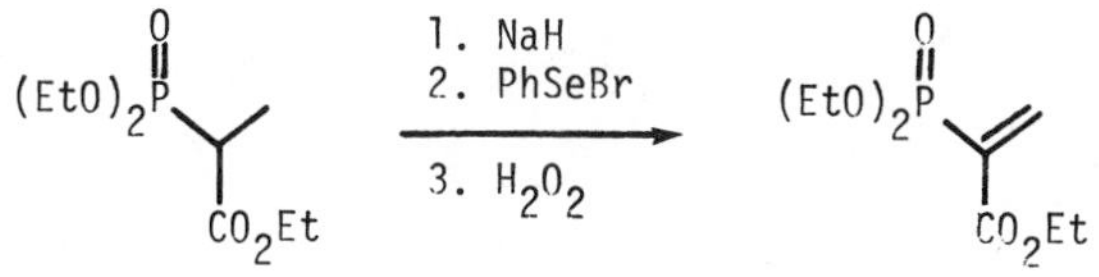

49 82%

(105) (Ref. 259)

3.5. Sulfur, Selenium, and Tin-Containing Functional Groups

Metalated sulfides,[261,262] selenides,[263,264] β-ketosulfoxides[11] and selenoxides,[5,78,265] sulfones,[266–268] and stannanes[269] can be α-selenenylated. In the case of selenoxides, it is necessary to perform the metalation–selenenylation procedure at low temperatures to prevent premature selenoxide elimination. Examples are given in Eqs. (106)–(112).

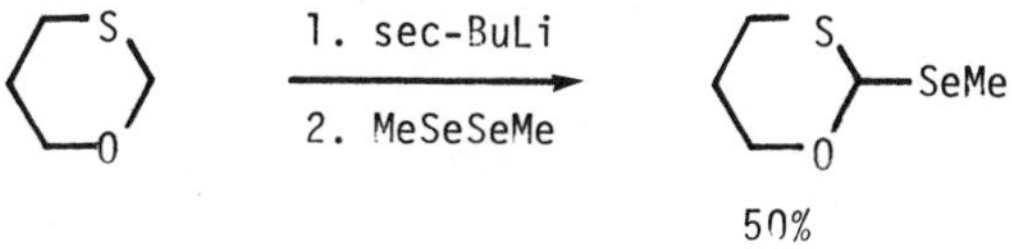

(106) (Ref. 262)

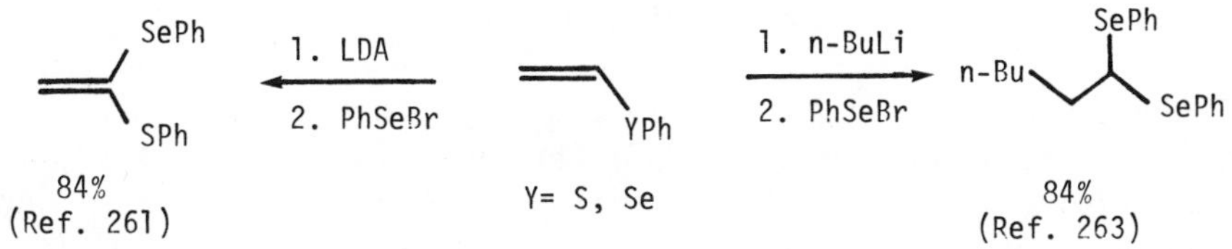

(107)

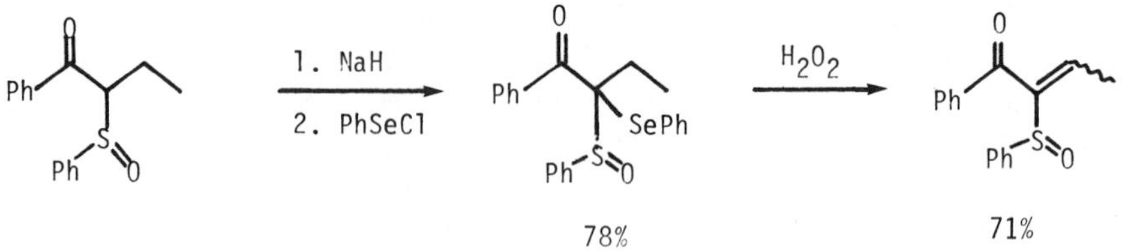

(108) (Ref. 11)

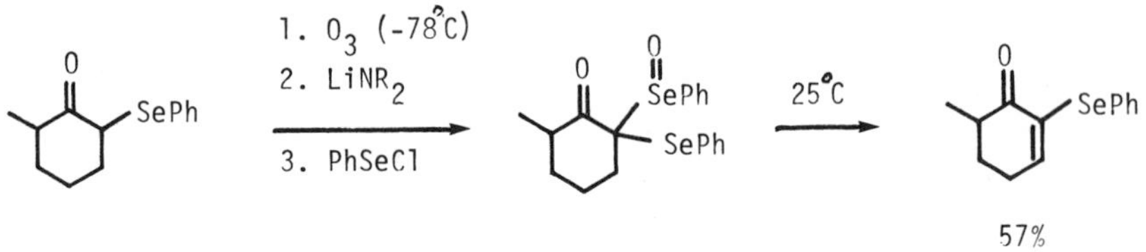

(109) (Ref. 78)

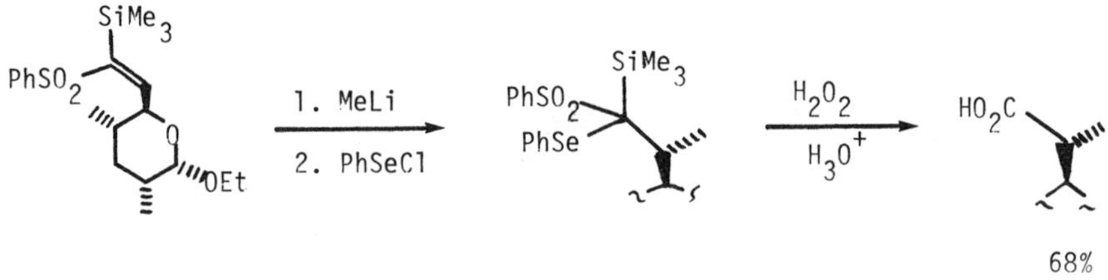

(110) (Ref. 268)

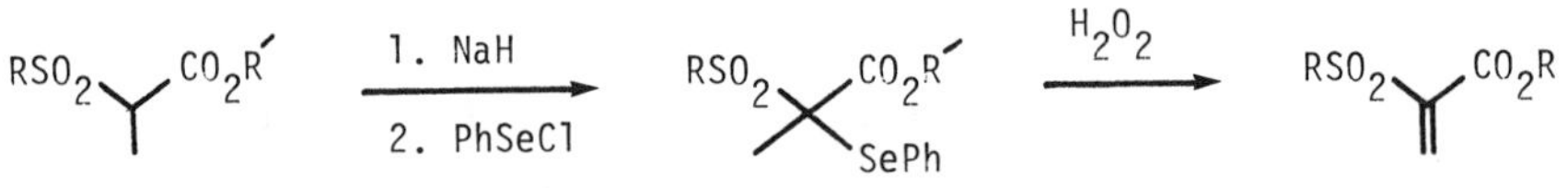

(111) (Ref. 266)

$$Ph_3SnCH_2I \xrightarrow{\text{n-BuLi}} Ph_3SnCH_2Li \xrightarrow{\text{PhSeBr}} \underset{40\%}{Ph_3SnCH_2SePh}$$

(112) (Ref. 269)

3.6. Diazo Compounds

Certain electrophilic selenium species react with diazo compounds to give insertion products according to Eq. (113). Thus, Petragnani et al.[256] found that PhSeBr reacts with diazomethane or diazoethane to afford the corresponding α-bromo selenides [Eq. (114)]. α-Diazoketones form 2-chloro-2-phenylseleno derivatives quantitatively when treated with PhSeCl [Eq. (115)].[270] The products serve as convenient precursors of 2-chloro- or 2-phenylselenoenones via selenoxide elimination or dehydrohalogenation, respectively. Similar insertion reactions have been performed with 6-diazopenicillinates, 53 [Eq. (116)].[271,272]

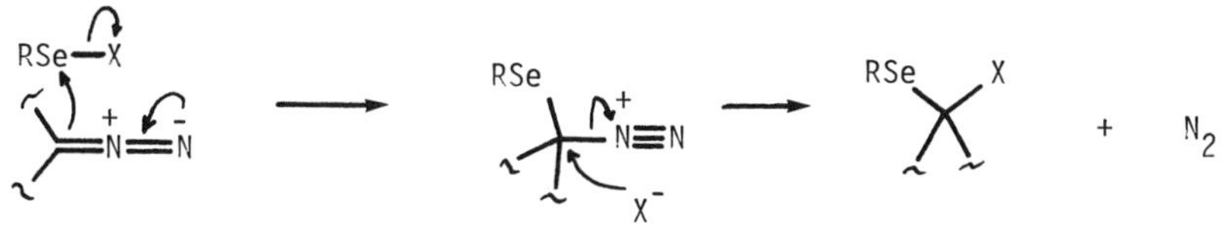

$$(113)$$

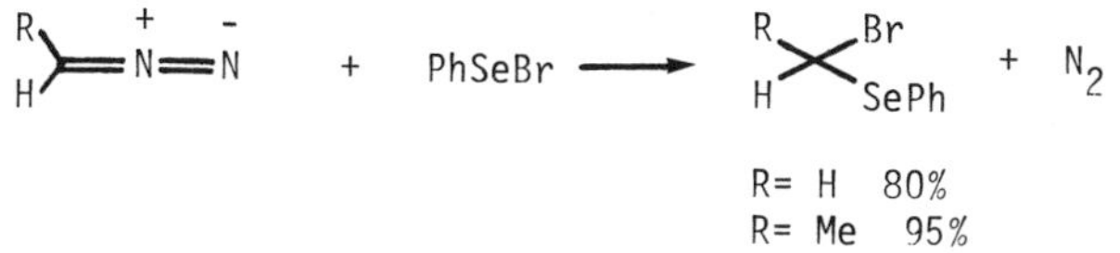

$$R = H \quad 80\%$$
$$R = Me \quad 95\%$$

$$(114)$$

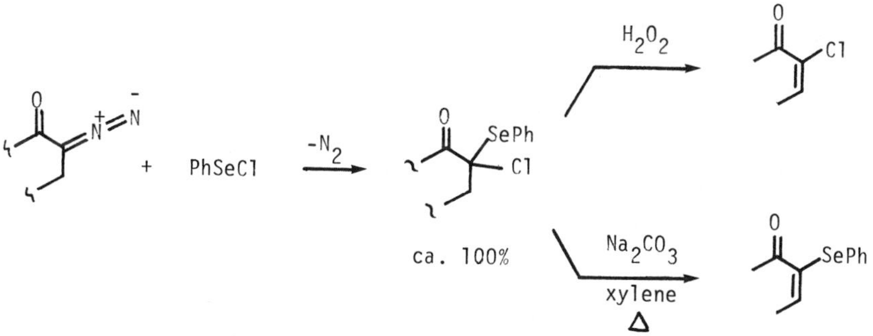

ca. 100%

$$(115)$$

53

$$R = CH_2CCl_3 \quad 64\%$$
$$R = CH_2Ph \quad 26\%$$

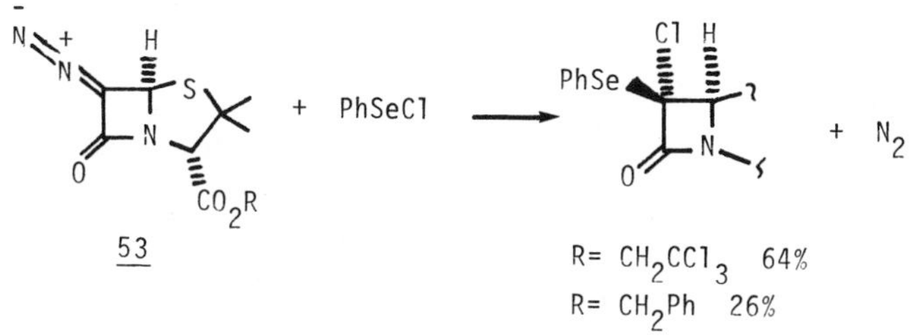

$$(116)$$

The reactions of diazo compounds with various selenenyl pseudohalides have been studied by Back and Kerr.[33,267] In the absence of light, diazomethane attacks Se-phenyl p-tolueneselenosulfonate (10; Ar = p-tolyl) at selenium, displacing the sulfinate anion and generating the diazonium ion, 50. Since the sulfinate anion is an ambident nucleophile, both the sulfinate ester, 51, and the sulfone, 52, are produced by the further displacement of nitrogen from 50 [Eq.

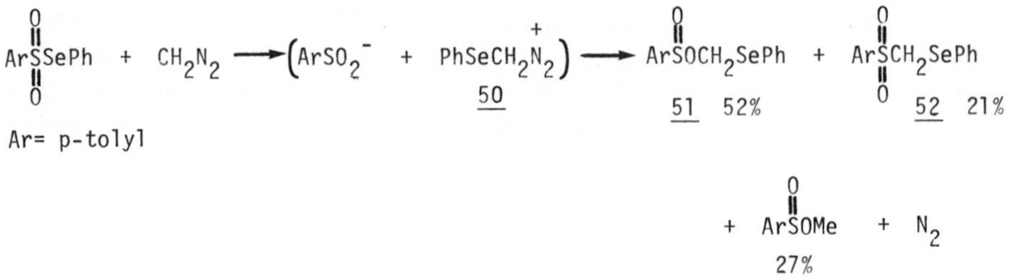

$$\text{ArSSePh} + CH_2N_2 \longrightarrow \left(ArSO_2^- + PhSeCH_2\overset{+}{N_2} \right) \longrightarrow ArSOCH_2SePh + ArSCH_2SePh$$

$$\underline{50} \qquad \underline{51} \; 52\% \qquad \underline{52} \; 21\%$$

Ar= p-tolyl

$$+ \; ArSOMe \; + \; N_2$$

27%

(117)

(117)].[267] In the presence of light, a free-radical chain reaction occurs instead; see Chapter 7. Selenenyl thio- and selenocyanates display similar, though more vigorous ionic behavior, with the incorporation of SCN⁻ or SeCN⁻ occurring

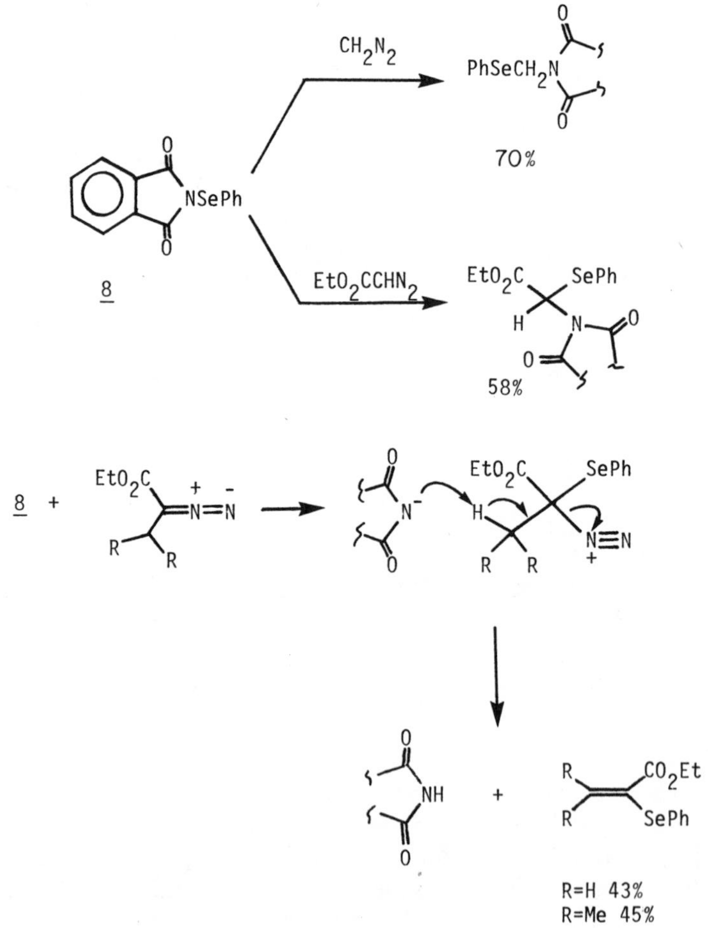

SCHEME 10.

principally through nitrogen [Eqs. (118) and (119)].[33] Phenyl selenocyanate, 12, is unreactive towards diazomethane at room temperature.[33] N-(Phenyl-seleno)phthalimide, 8, affords insertion products with diazomethane or ethyl diazoacetate, but produces vinyl selenides by elimination when substituted diazo esters are employed (Scheme 10).[33]

$$PhXYCN \ + \ CH_2N_2 \ \longrightarrow \ PhXCH_2NCY \ + \ N_2$$

$$
\begin{array}{l}
X=Y=S \quad 72\% \\
X=Y=Se \quad 68\% \\
X=Se; \ Y=S \quad 85\%
\end{array}
$$

(118)

$$
\begin{array}{l}
R= H \quad 70\% \qquad 18\% \\
R=i-Pr \quad 65\% \qquad 7\% \\
R= Me \quad 60\% \qquad 4\%
\end{array}
$$

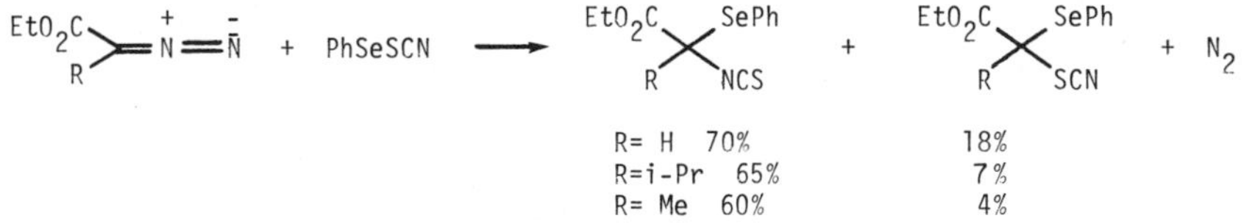

(119)

It is appropriate to mention that a number of other reactions exist that formally resemble the insertion process in Eq. (113), although they apparently do not proceed by electrophilic mechanisms. Diphenyl diselenide reacts with diazomethane,[273] α-diazo esters,[274] and 6-diazopenicillinates[271,272] to provide diselenoacetal or ketal products (Scheme 11). These reactions require irradiation[273] or catalysis with copper bronze[274] or boron trifluoride etherate.[271,272,274] In some cases, carbenes or carbenoids are possible intermediates.[273,274] Selenoesters and selenocarbonates form (phenylseleno)- or (methylseleno)methyl ketones or esters when treated with diazomethane and copper powder or cuprous iodide [Eq. (120)],[275] possibly via rearrangement of a tetrahedral intermediate. Thioesters fail to react under similar conditions. An interesting reaction has been observed between 6-diazopenicillinates and allyl phenyl selenide (or sulfide) in the presence of boron trifluoride etherate [Eq. (121)].[272,276] The product is formed by the [2,3]sigmatropic rearrangement of the initially formed ylide, 54.

(120)

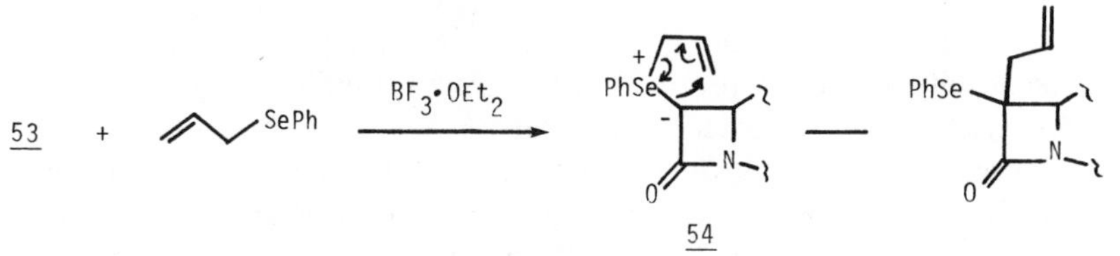

(121)

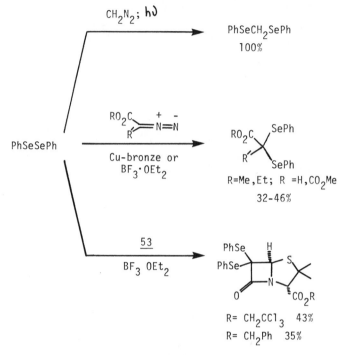

SCHEME 11

4. REACTIONS OF Se(II) ELECTROPHILES WITH ACETYLENIC, VINYLIC, ALLENIC, AND BENZYLIC CARBANIONS

The reactions of simple carbon nucleophiles such as Grignard reagents with elemental selenium[277] or selenenyl halides[17] have been known for many years. Variations of such processes have provided a traditional route to selenols and selenides, which in turn are important precursors of many other classes of organoselenium compounds.[1] More recently, these early studies have been extended to the title species.

The reaction of sodium acetylides with elemental selenium produces alkynylselenolates that may be further alkylated [Eq. (122)].[278–281] Lithium acetylides generated from terminal acetylenes with n-butyllithium provide acetylenic selenides when reacted with PhSeBr[256,282,283] or diselenides.[121] The products are convenient precursors of vinyl selenides since they can be stereoselectively reduced to either E or Z products with lithium aluminum hydride[256,283] or dicyclohexylborane,[282] respectively (Scheme 12). Tomoda et al. developed two alternative methods for the preparation of acetylenic selenides that circumvent the need for a strong base such as n-butyllithium. In one procedure[284] terminal acetylenes react with PhSeCN in the presence of triethylamine and catalytic amounts of cuprous bromide or cyanide to afford the products via

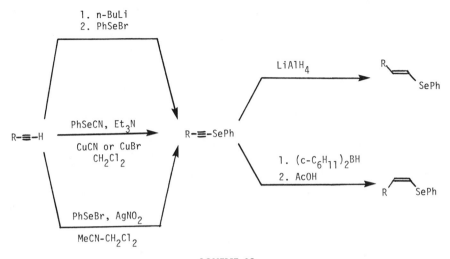

SCHEME 12

postulated copper acetylide intermediates. The presence of free hydroxyl or ester groups is tolerated. The second method[285] employs PhSeBr and silver nitrite as the selenenylating agent; the mechanism is unclear (Scheme 12).

$$R-\!\!\equiv\!\!-Na \xrightarrow{\text{Se}} R-\!\!\equiv\!\!-SeNa \xrightarrow{E^+} R-\!\!\equiv\!\!-SeE \qquad (122)$$

The selenenylations of various metalated vinylic,[261,263,264,286] allenic,[287,288] and benzylic[289] carbanions are shown in Eqs. (107) and (123)–(128). The allenic selenides produced in Eqs. (126) and (127) undergo [2,3]sigmatropic rearrangement when oxidized with MCPBA.[287,288]

$$
\begin{array}{c}
R\diagdown\!\!\diagup\!\!\diagdown_{MgBr} \xrightarrow[\text{PhSeSePh}]{\text{ArSeBr or}} R\diagdown\!\!\diagup\!\!\diagdown\!\!\diagdown_{SeAr} \\
60\text{-}92\% \\
R=\text{H, Me} \\
Ar=\text{Ph or }m\text{-CF}_3\text{Ph}
\end{array} \qquad (123)\text{ (Ref. 264)}
$$

$$
\begin{array}{c}
\overset{SeAr}{\underset{Ar=\;m\text{-}CF_3Ph}{\diagup\!\!\diagdown_{Li}}} \xrightarrow{\text{PhSeSePh}} \overset{SeAr}{\underset{89\%}{\diagup\!\!\diagdown_{SePh}}}
\end{array} \qquad (124)\text{ (Ref. 264)}
$$

$$
\begin{array}{c}
\text{(benzothiophene-Br)} \xrightarrow[\text{2. MeSeMeSe}]{\text{1. }n\text{-BuLi}} \text{(benzothiophene-SeMe)} \\
79.5\%
\end{array} \qquad (125)\text{ (Ref. 286)}
$$

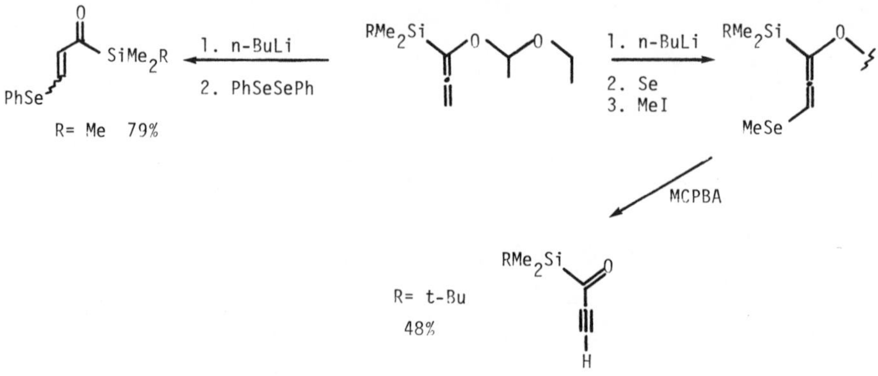

(126) (Ref. 287)

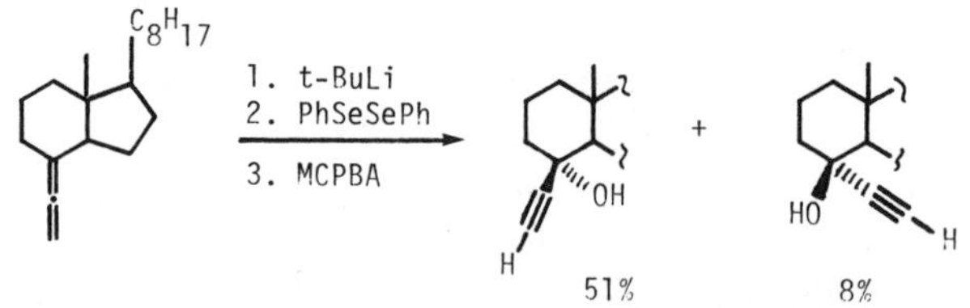

(127) (Ref. 288)

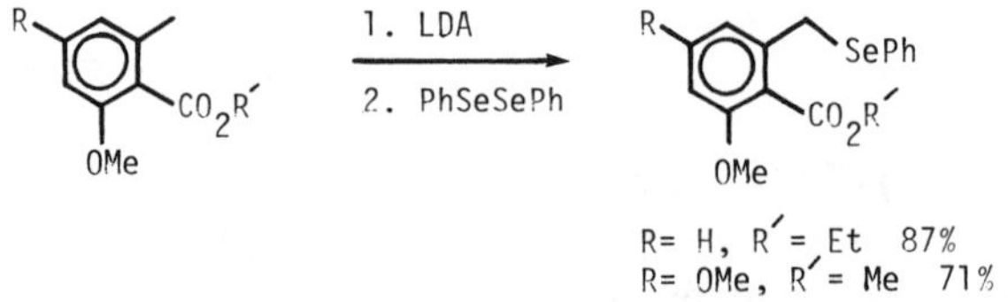

R= H, R′ = Et 87%
R= OMe, R′ = Me 71%

(128) (Ref. 289)

5. REACTIONS OF Se(II) ELECTROPHILES WITH HETEROATOM NUCLEOPHILES

5.1. With Oxygen Nucleophiles

Selenenyl halides are easily hydrolyzed,[17] generally providing mixtures of the corresponding diselenide and seleninic acid via disproportionation of the initally formed selenenic acid [Eq. (4)]. Alcoholyses are relatively slow[290] and PhSeCl is reported to be stable in methanol solution for several days at room temperature.[27] Selenenic esters (RSeOMe) can be prepared from the methanolysis of selenenyl chlorides in the presence of silver acetate[291,292] or silver oxide.[293] The selenenyl acetate (ArSeOAc) is an intermediate in the former case. Selenenic esters are hydrolytically unstable and are also capable of undergoing transesterification in the presence of a different alcohol.[16] They have few applications in synthesis. Reich et al.[294,295] prepared selenenic esters such as 55 *in situ* from the corresponding allylic alcohol and selenenyl halides in the presence of triethylamine [Eq. (129)]. The products were required in an attempt to prepare the corresponding cyclohexadiene by [2,3]sigmatropic rearrangement of 55 to their selenoxides, followed by *syn* elimination. In connection with a study of the intermediates produced in the oxidation of olefins by selenium dioxide, Sharpless and Lauer[296] formed the selenenic ester, 56, from the corresponding allylic alcohol using PhSeBr and silver acetate [Eq. (130)].

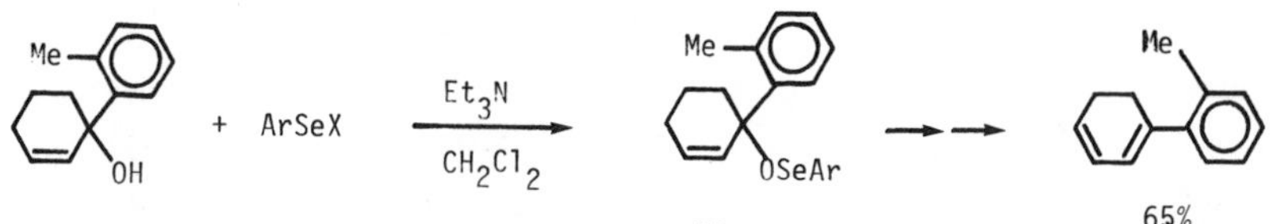

Ar= o-nitro-p-tolyl

55

65%

(129)

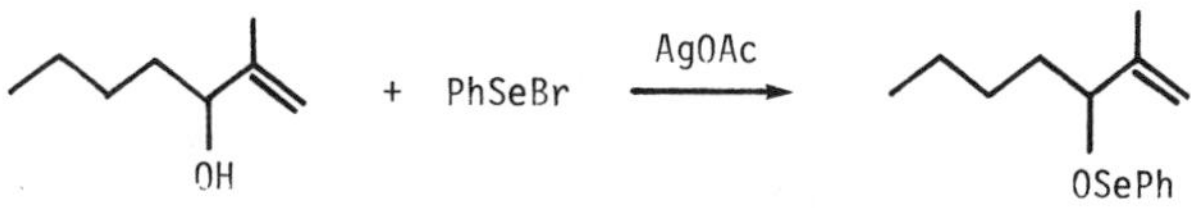

56

(130)

Benzhydrol reacts with PhSeCl and triethylamine in refluxing benzene to afford benzophenone and PhSeSePh,[298] presumably via the selenenic ester, 57 [Eq. (131)]. Various alcohols have also been oxidized to the corresponding aldehydes or ketones with diaryl diselenides and *t*-butyl hydroperoxide.[297,298] However, the electrophile that attacks the alcohol could also be the seleninic anhydride, 16, rather than a selenenic species under these circumstances.[20,298]

$$Ph_2CHOH \xrightarrow{PhSeCl} Ph_2CHOSePh \xrightarrow{Et_3N} Ph_2C=O + PhSeSePh$$

57

(131)

The reactions of PhSeCl and PhSeBr with silver acetate in acetic acid[12,21] or with silver trifluoroacetate in ether, benzene, or methylene chloride[22−24] provide convenient methods for the *in situ* generation of the selenenyl acetate, 6, and trifluoroacetate, 7, as shown in Scheme 1.

Phenols react with selenenyl halides[299,300] or with other selenenic electrophiles[301,302] to furnish products of aromatic selenenylation[299,301,302] or aromatic chlorination[300] instead of selenenic esters [Eq. (132)].

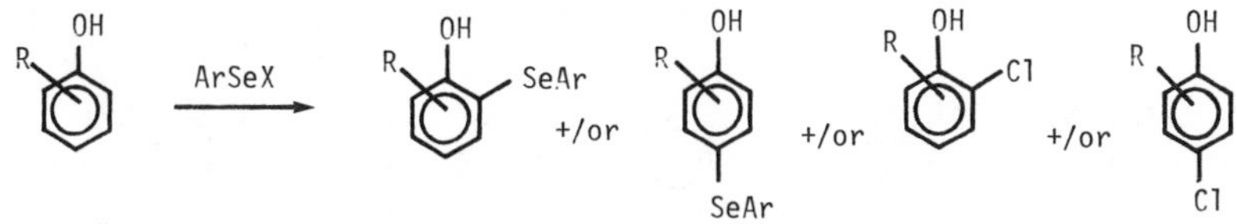

(132)

5.2. With Nitrogen Nucleophiles

Reich and Renga[29] prepared selenenamides, $\underline{11}$ (Scheme 1), from PhSeCl and two equivalents of the corresponding amine (R = Me, 62%; R = Et, 59% R = i-Pr, 24%) in hexane. The products are sensitive to hydrolysis, particularly in the case of the less hindered N,N-dimethyl derivative. N-(Phenylseleno)morpholine $\underline{32}$, exhibits greater stability.[145] N-(Phenylseleno)phthalimide, $\underline{8}$, is conveniently prepared from PhSeCl and potassium phthalimide (Scheme 1).[25]

Kobayashi and Hiraoka[303,304] employed selenenamides derived from 7β-aminocephalosporins as key intermediates in the introduction of a 7α-methoxy substituent as shown in Scheme 13. The products, $\underline{58}$, were easily N-acylated with phenoxyacetyl chloride in a subsequent step, even in the absence of an added base, and proved superior to sulfenamide analogs in this respect. Application of the method to the penicillin series was less successful as β-lactam cleavage competed significantly during the introduction of the methoxyl group.

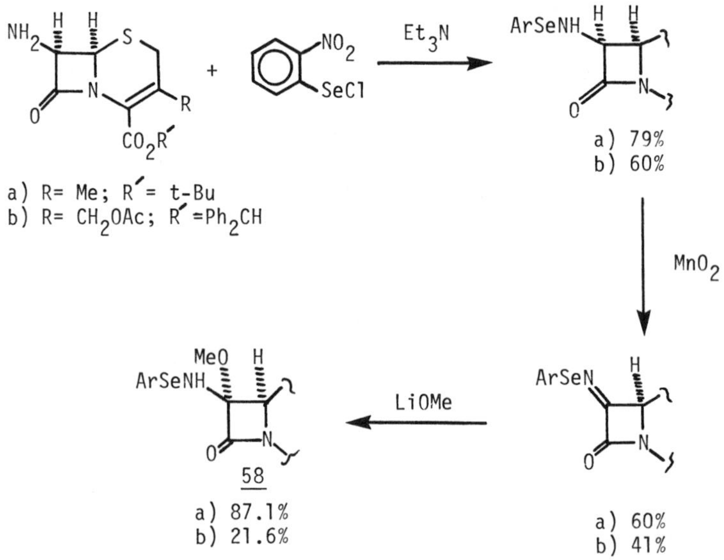

a) R= Me; R´= t-Bu
b) R= CH₂OAc; R´=Ph₂CH

SCHEME 13.

Two examples of intramolecular selenenamide formation are provided in Eqs. (133) and (134). The conversion of an imine to the selenenamides, $\underline{59}$, has

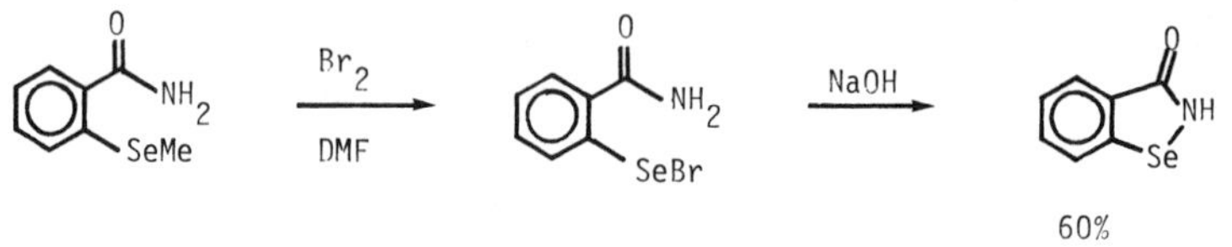

60%

(133) (Ref. 305)

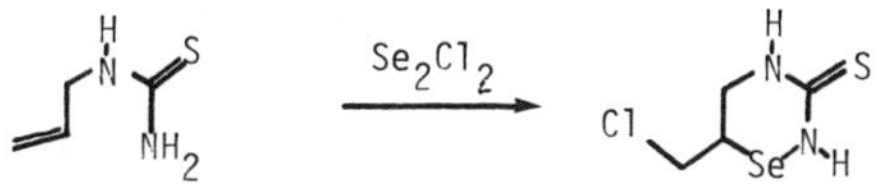

$$\text{(134) (Ref. 306)}$$

been reported[307] with ArSeCl and triethylamine [Eq. (135)]. The selenoxide, 61, was prepared when benzeneseleninyl chloride, 60, was used instead of its selenenyl counterpart.

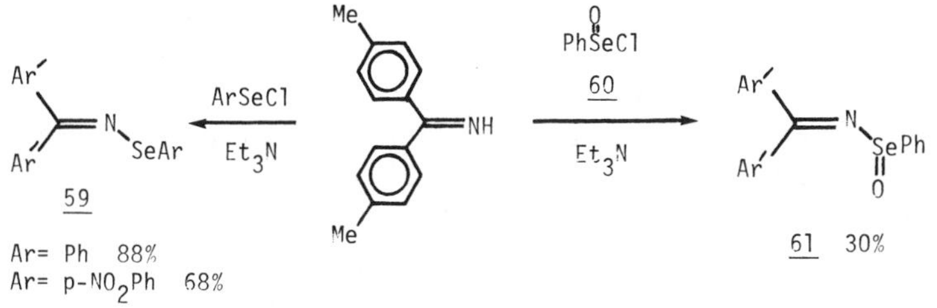

$$\text{(135)}$$

Ar= Ph 88%
Ar= p-NO$_2$Ph 68%

61   30%

Selenenamides are formed when a selenenic acid is generated in the presence of an amine [Eq. (136)]. This has been exploited by Reich and co-workers[11,18] in improving the efficiency of olefin formation from selenoxide eliminations. The addition of an unhindered secondary amine to the reaction mixture effectively scavenges the selenenic acid produced as a by-product during the elimination [see Eq. (5)] and so suppresses side reactions caused by it.

$$\text{PhSeOH} + \text{HNR}_2 \rightleftharpoons \text{PhSeNR}_2 + \text{H}_2\text{O} \tag{136}$$

11

Diazenes and PhSeOH, formed simultaneously in the oxidation of hydrazines or hydrazides with benzeneseleninic acid, 3, react further according to Eq. (137).[34,308,309] These and related oxidations are described in more detail in a number of the other chapters. Benzhydrazide and PhSeCl in pyridine form the corresponding selenoester [Eq. (138)],[309] probably via the same N-(phenylseleno)diazene intermediate, 62, postulated in the previous reaction.

$$\text{RN=NH} + \text{PhSeOH} \xrightarrow{-\text{H}_2\text{O}} [\text{RN=NSePh}] \xrightarrow{-\text{N}_2} \text{RSePh} \tag{137}$$

62

R = Ar, R-$\overset{\text{O}}{\overset{\|}{\text{C}}}$ or RSO$_2$

$$\tag{138}$$

47%

As in the case of phenols, variously substituted anilines[17,291,299,300,310,311] undergo chiefly aromatic selenenylation when treated with selenenyl halides or related electrophiles,[302] although selenenamide formation has occasionally been observed as well.[291,299] Selenenamides may be initially formed but rearrange rapidly to selenides [Eq. (139)].[312]

$$
\underset{\text{NH}_2}{R}\text{—} + \text{ArSeX} \xrightarrow{-HX} \underset{\text{NHSeAr}}{R}\text{—} \xrightarrow{HX} \text{ArSe—}\underset{\text{NH}_3^+}{R}\; X^-
$$

(139)

The reactions of *N*-chloroamines or amides with various selenium electrophiles have been studied by Derkach and co-workers[313-318] and generally produce halogenated Se (IV) amides [e.g., Eq. (140)].[313]

$$
\underset{\text{PhCNHCl}}{\overset{O}{\|}} + \text{ArSeCl} \longrightarrow \underset{\underset{Cl}{|}}{\overset{O}{\underset{\|}{\text{PhCNHSeAr}}}}\overset{Cl}{\overset{|}{}}
$$

(140)

5.3. With Sulfur and Selenium Nucleophiles

Selenenic electrophiles react readily with thiols and hydrogen sulfide, with anions such as thiocyanate, sulfinate, dithiocarbamate, thiosulfate, and thiosulfonate ions, as well as with most of their selenium counterparts. Ambident nucleophiles such as RSO_2^- attack exclusively through the softer sulfur atom. Early work in this area has been thoroughly reviewed by Klayman[1a] and will not be covered here.

Aromatic thiols react with *o*-nitrophenyl selenocyanate, 63, to afford the selenenyl sulfides, 64, in unspecified yields [Eq. (141)].[319] Pentafluorobenzene-thiol behaves anomalously and converts the selenocyanate to the corresponding diselenide. Surprisingly, the *p*-nitro derivative, 65, produces the diselenide, 66, as the chief product with all of the thiols studied [Eq. (141)].[319] The thiol in Eq. (142) was transformed to the selenenyl sulfide with PhSeBr and cyclized with excess reagent or upon oxidation.[320] Selenosulfates (seleno Bunte salts) 67 and cysteine generate the unsymmetrical selenenyl sulfides, 68, which can disproportionate to the diselenide and thiosulfate [Eq. (143)].[321] The relative amounts of the three products depend on the nature of the group R and the precise conditions.

$$
\underset{63}{\underset{\text{SeCN}}{\overset{NO_2}{}}} + \text{ArSH} \longrightarrow \underset{64}{\underset{\text{SeSAr}}{\overset{NO_2}{}}}
$$

(141a)

$$NO_2-\langle\bigcirc\rangle-SeCN \; + \; ArSH \; \longrightarrow \; NO_2-\langle\bigcirc\rangle-SeSe-\langle\bigcirc\rangle-NO_2$$

     65   66   (141b)

(142)

$$RSeSO_3^- \; + \; CySH \; \longrightarrow \; RSeSCy \; \longrightarrow \; RSeSeR \; + \; CySSO_3^-$$

    67   68   (143)

Selenenyl halides react with sodium benzenesulfinate,[28] sodium benzenethiosulfonate,[28] and potassium p-chloroselenobenzoate[322] to produce selenosulfonates, 69 [Eq. (144)], selenenyl thiosulfonates, 70 [Eq. (144)], and the diselenoperester, 71 [Eq. (145)], respectively.

$$PhSO_2Se-\langle\bigcirc\rangle-R \; \xleftarrow{PhSO_2^-Na^+} \; R-\langle\bigcirc\rangle-SeBr \; \xrightarrow{PhSO_2S^-Na^+} \; R-\langle\bigcirc\rangle-SeSSO_2Ph$$

69   80-85%   R= H, Me, Cl   70   70-75%

(144)

$$Cl-\langle\bigcirc\rangle-\overset{O}{\overset{\|}{C}}Se^-K^+ \; + \; PhSeCl \; \longrightarrow \; Cl-\langle\bigcirc\rangle-\overset{O}{\overset{\|}{C}}SeSePh$$

                                              (145)

                                            71   64%

Selenosulfonates are also formed when sulfinic acids react with PhSeOH [Eq. (146)], in turn produced *in situ* by the reduction of $PhSeO_2H$ with another equivalent of the sulfinic acid [cf. Eq. (10)].[35,36]

$$RSO_2H \; + \; PhSeOH \; \longrightarrow \; RSO_2SePh \; + \; H_2O$$

                                                  (146)

The sulfonium salt, 72, was prepared from PhSeBr and dimethyl sulfide in the presence of silver tetrafluoroborate [Eq. (147)].[302] It is itself an interesting selenenic electrophile as it performs aromatic selenenylations of aniline, phenols,

$$Me_2S \; + \; PhSeBr \; + \; AgBF_4 \; \longrightarrow \; Me_2\overset{+}{S}-SePh \; BF_4^- \; + \; AgBr$$

                                                     (147)

                                         72

and anisoles (*vide supra*). It also reacts with cyclohexene to produce the unsaturated sulfonium species, 74, after oxidation of the intermediate, 73, with hydrogen peroxide [Eq. (148)].[302]

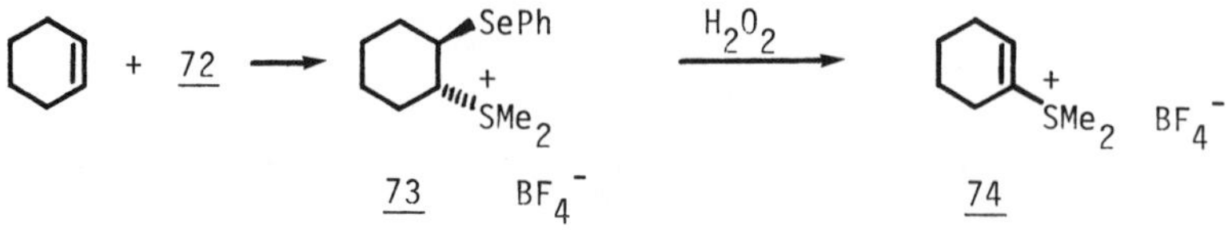

$$(148)$$

5.4. With Phosphorus Nucleophiles

The first example of a reaction between a trivalent phosphorus nucleophile and a selenium electrophile was reported in 1857 by Cahours and Hofmann,[323] who observed the formation of trialkylphosphine selenides from the parent phosphines and elemental selenium [Eq. (149)]. Analogous reactions of phosphites, aminophosphines, phosphorochloridites, and other trivalent phosphorus compounds have since been encountered and reviewed.[1b] A more recent application involves the synthesis of several biologically interesting aziridinyl and/or spin-labeled phosphine selenides such as 75–77.[324]

$$R_3P + Se \longrightarrow R_3P=Se$$

R = Me, Et

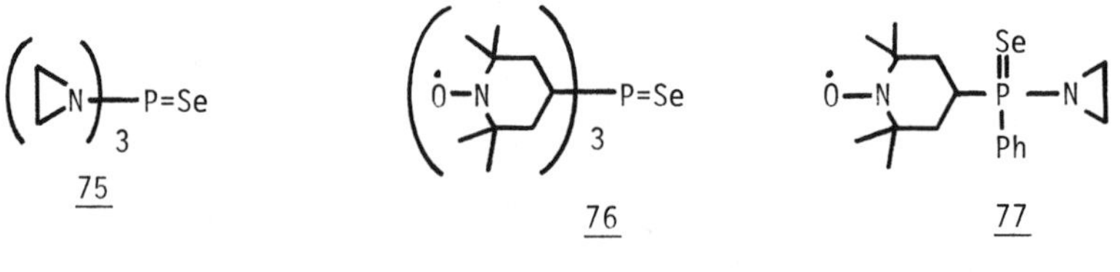

$$(149)$$

Selenium electrophiles other than the free element react with phosphorus nucleophiles.[1b] Recent examples include the deselenization of diselenides with phosphines [Eq. (150)][325] or aminophosphines [Eq. (151)],[d,326] and the conversion of diselenides to selenophosphonates via Eq. (152).[327]

$$RCSeSeCR + Ph_3P \longrightarrow RCSeCR + Ph_3P=Se \qquad (150)$$

77–98%

$$MeSeSeMe + (Et_2N)_3P \longrightarrow MeSeMe + (Et_2N)_3P=Se \qquad (151)$$

---

[d] Free-radical deselenizations of diselenides with phosphines are also known (see Chapter 7).

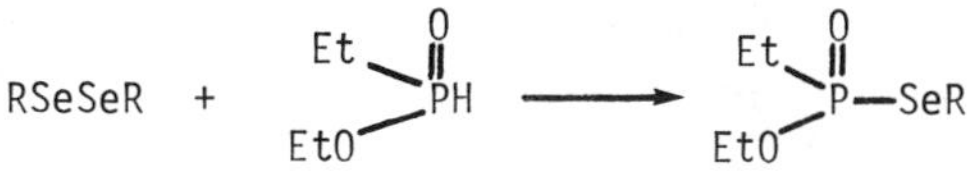

$$\text{RSeSeR} + \underset{\text{EtO}}{\overset{\text{Et}}{>}}\overset{\overset{\text{O}}{\parallel}}{\text{PH}} \longrightarrow \underset{\text{EtO}}{\overset{\text{Et}}{>}}\overset{\overset{\text{O}}{\parallel}}{\text{P—SeR}} \qquad (152)$$

$$\begin{aligned} R &= Me \quad 83.2\% \\ R &= Et \quad 76\% \end{aligned}$$

The reaction of tri-*n*-butylphosphine with electrophilic aryl selenocyanates has been exploited by Grieco et al.[328] in a convenient and efficient synthesis of aryl selenides from primary or secondary alcohols. A plausible mechanism is shown in Scheme 14, where initial displacement of CN⁻ from ArSeCN generates the selenophosphonium salt, 78. Next, attack by the alcohol produces an alkoxyphosphonium species, 79, from which displacement of tri-*n*-butyl-phosphine oxide by ArSe⁻ affords the product selenide. Inversion of configuration occurs with chiral secondary alcohols,[329] as expected from an S_N2-type displacement in the last step. The method provides a one pot alternative to the conversion of the alcohol to the selenide via the treatment of its mesylate or tosylate with ArSe⁻. The product selenides afford olefins by selenoxide elimination, and so the technique permits the overall dehydration of the original alcohol. An especially useful application involves the synthesis of *exo*-methylene compounds from cyclic substrates bearing hydroxymethylene substituents

$$\text{ArSeCN} + \text{n-Bu}_3\text{P} \longrightarrow \overset{+}{\text{ArSe-P(n-Bu)}_3} \quad {}^-\text{CN}$$
$$\underset{78}{}$$

$$\underset{78}{} + \text{ROH} \longrightarrow \overset{+}{\text{RO-P(n-Bu)}_3} \quad {}^-\text{SeAr}$$
$$\underset{79}{}$$

$$\underset{79}{} + {}^-\text{SeAr} \longrightarrow \text{RSeAr} + \text{n-Bu}_3\text{P=O}$$

SCHEME 14.

[Eq. (153)].[330–342] The arrangements that accompany many of the more traditional dehydration techniques can thus be avoided.

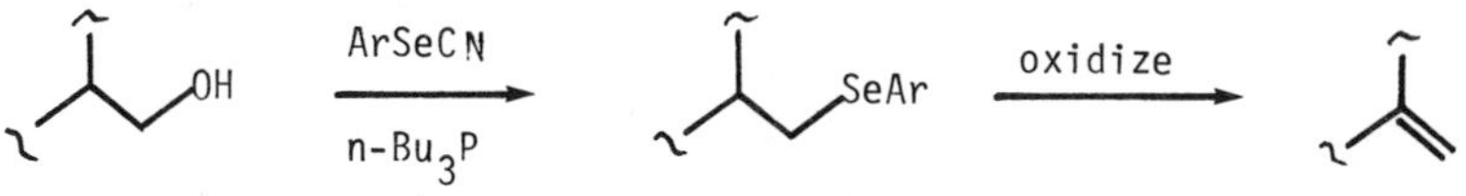

$$\text{(OH)} \xrightarrow[\text{n-Bu}_3\text{P}]{\text{ArSeCN}} \text{(SeAr)} \xrightarrow{\text{oxidize}} \text{(=CH}_2\text{)} \qquad (153)$$

Primary alcohols can be selectively functionalized in the presence of secondary,[174,338,341,343–345] tertiary,[334,337] or other more sterically hindered[159,346] primary hydroxyl functions. Diols can usually,[347–350] but not always,[159] be converted to bisselenides with excess reagent. *o*-Nitrophenyl selenocyanate, 63, has been the most widely employed reagent as it is stable and crystalline and provides selenoxides that eliminate with particular facility.[18,328,351] Phenyl selenocyanate[174,328,329,352] and the *p*-nitro derivative, 65,[353–355] have also been successfully utilized, but the reaction fails or gives lower yields with diselenides[328,341] or with PhSeCl.[328] Triphenylphosphine has been used in place of the tri-*n*-butyl derivative,[356] but is generally not recommended.[341] The reaction is typically carried out at or below room temperature in THF or pyridine.[328] Several other examples of the conversion of alcohols to olefins by this method have been reported.[357–366] Illustrative examples are shown in Eqs. (154)–(158).

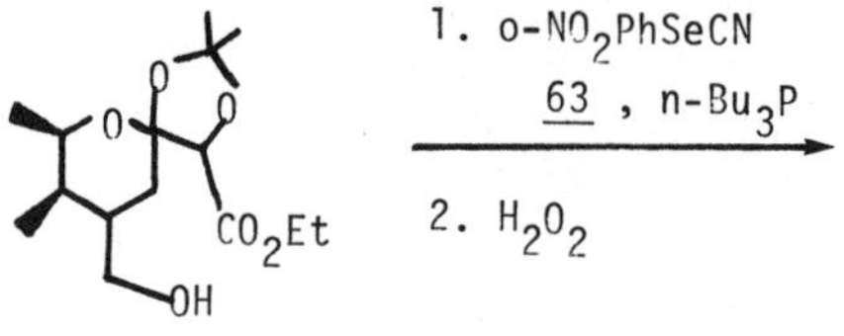

(154) (Ref. 330)

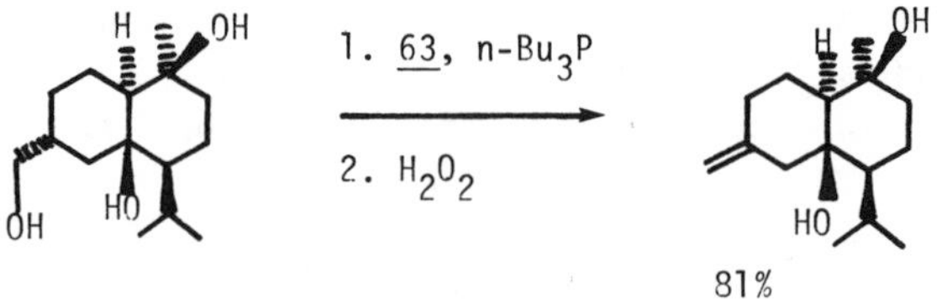

$$83\%$$

(155) (Ref. 337)

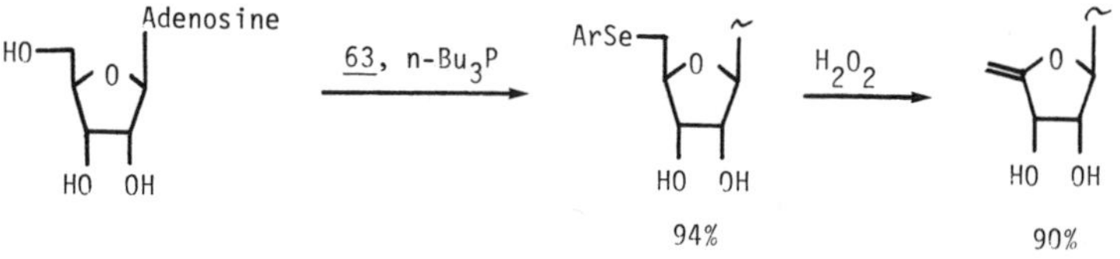

$$81\%$$

(156) (Ref. 341)

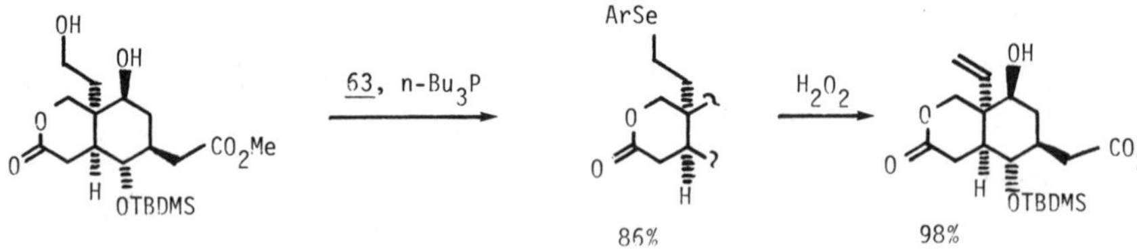

$$94\%$$ $$90\%$$

(157) (Ref. 345)

$$86\%$$ $$98\%$$

Several variations of the original method of Grieco et al.[328] have appeared. Clive et al.[353] developed an efficient 1,3-transposition of allylic alcohols based upon their conversion to selenides, followed by oxidation and [2,3]sigmatropic

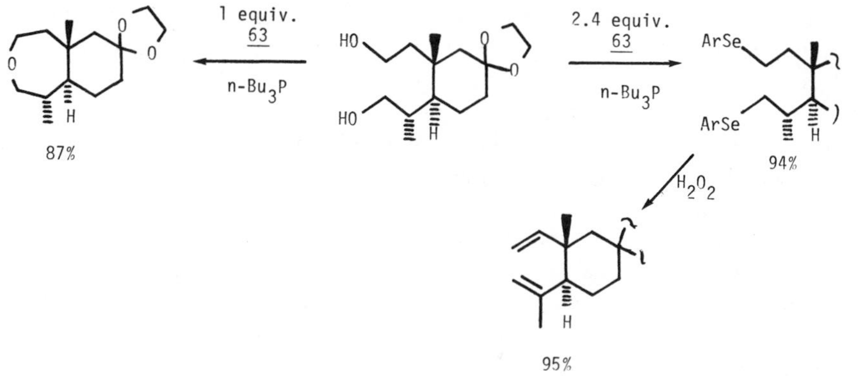

(158) (Ref. 347)

rearrangement [Eq. (159)]. A similar sequence has been employed by Kametani et al.[367,368] in the synthesis of linalyloxides from geraniol.

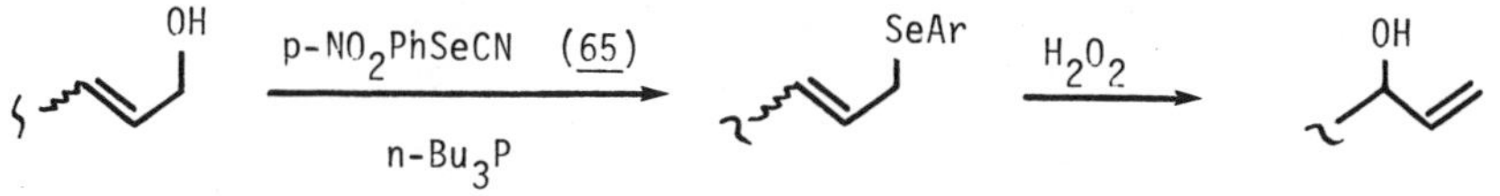

(159)

Sevrin and Krief[329] transformed secondary alcohols to bromides with overall retention of configuration via brominolysis of the corresponding phenyl selenide intermediates [Eq. (160)].

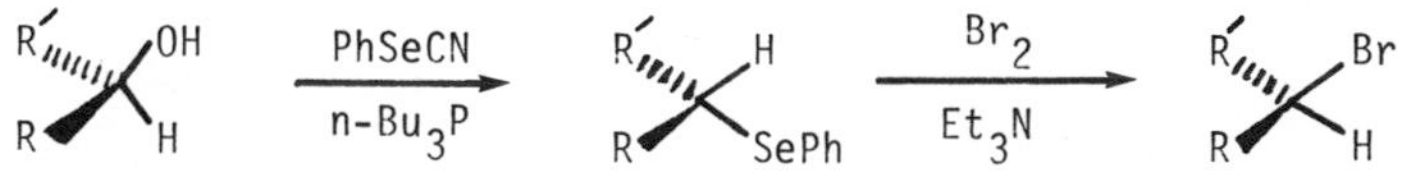

(160)

Carboxylic acids afford selenoesters when treated with aryl selenocyanates and tri-n-butylphosphine in methylene chloride.[369] Phenyl selenocyanate is superior to its o-nitro derivative for this purpose and phenyl thiocyanate produces the corresponding thioesters [Eq. (161)].

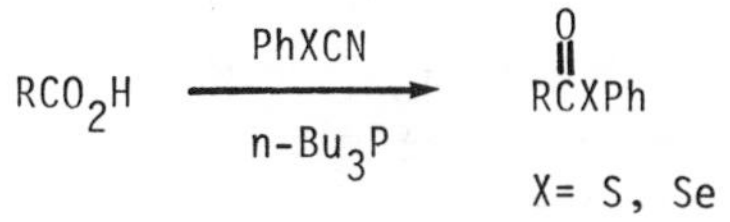

(161)

More recently, Grieco et al.[370] employed N-(phenylseleno)phthalimide, **8**, instead of ArSeCN for the conversion of primary alcohols to phenyl selenides and carboxylic acids to selenoesters [Eq. (162)]. Ley and co-workers[371–373] have applied the reaction with alcohols to the preparation of intermediates required for the synthesis of clerodane insect antifeedants [e.q., Eq. (163)]. With excess reagent, **8**, the steroidal alcohol, **80**, produced the imide, **81**, as the chief product, instead of the expected selenide [Eq. (164)].[374] If the mechanism for selenide formation with **8** resembles that with ArSeCN (Scheme 14), then it is likely that PhSe⁻ reacts preferentially with reagent **8** (or with **78**) rather than with the alkoxyphosphonium species, **79**, when the latter is derived from a secondary alcohol. The products are thus PhSeSePh and phthalimide anion, which ultimately attacks **79** to produce the observed product, **81**. When carboxylic acids react with **8** and tri-*n*-butylphosphine in the presence of amines, the corresponding amides are produced in high yields [Eq. (165)].[370] Presumably, the amine (instead of PhSe⁻) attacks a carboxyphosphonium intermediate analogous to **79** in this case.

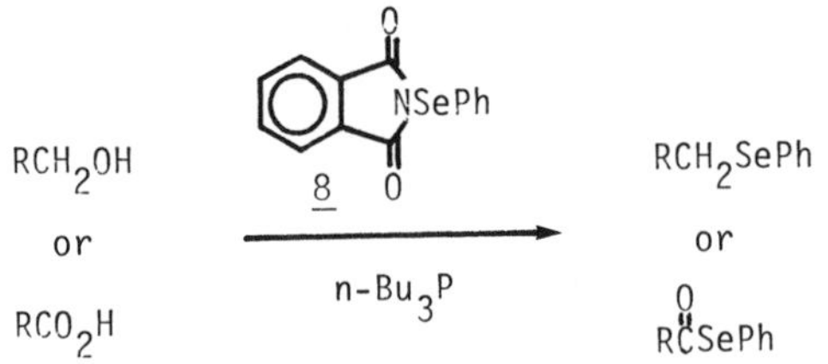

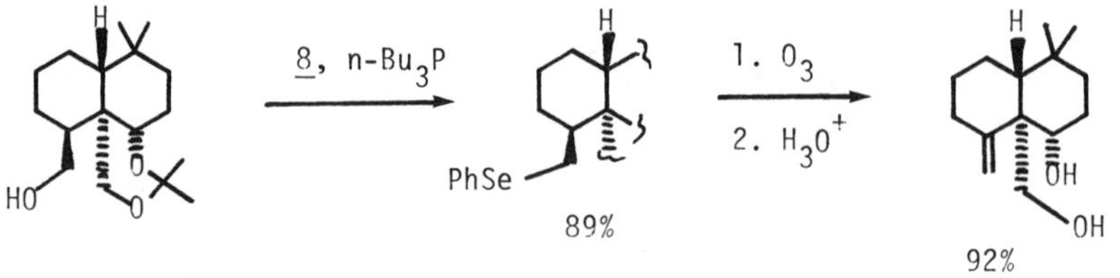

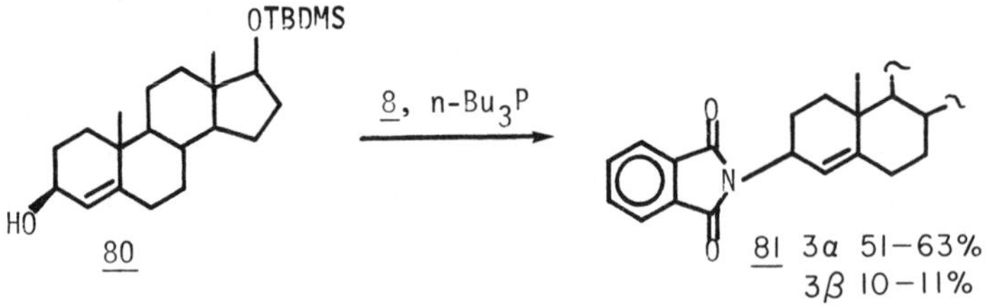

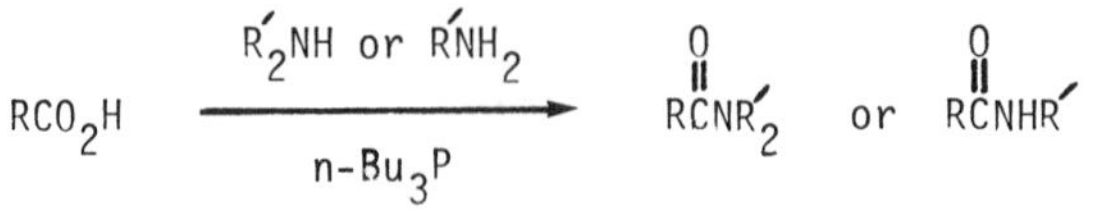

The cyanoselenenylation of aldehydes with PhSeCN or its *o*-nitro derivative, **63**, and tri-*n*-butylphosphine affords selenocyanohydrins[375] via Eq. (166). The products can in turn be converted to α,β-unsaturated nitriles by oxidation–elimination, or to α-metalated species that can be alkylated or made to undergo

Michael additions. Cyclohexenone produced a rearranged product [Eq. (167)].[375]

$$RCHO \xrightarrow[\text{n-Bu}_3\text{P}]{\text{ArSeCN}} RCH \overset{CN}{\underset{SeAr}{<}} \qquad (166)$$

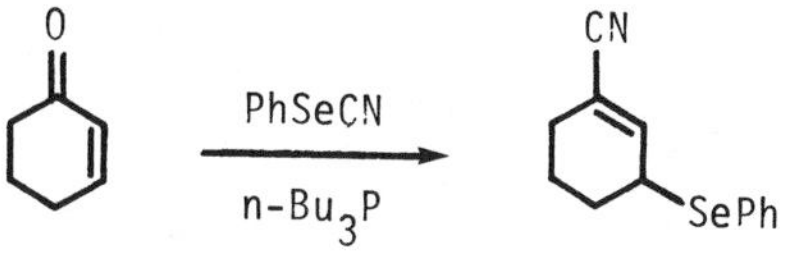

50%

$$(167)$$

The adduct, <u>82</u>, obtained from triphenylphosphine and selenocyanogen, has been used to prepare primary alkyl and benzylic selenocyanates from the corresponding alcohols [Eq. (168)].[376] Secondary alcohols gave mixtures of seleno- and isoselenocyanates while tertiary alcohols failed to react. An isolated example of the preparation of a selenophosphate from a phosphate[377] is shown in Eq. (169).

$$RCH_2OH + Ph_3P(SeCN)_2 \longrightarrow RCH_2SeCN \qquad (168)$$

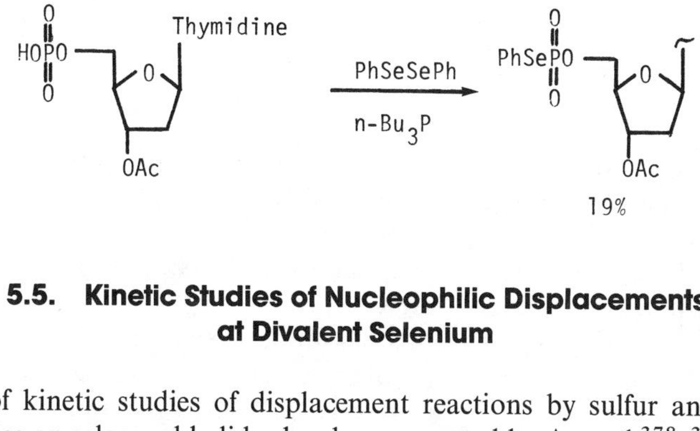

<u>82</u>

19%

$$(169)$$

5.5. Kinetic Studies of Nucleophilic Displacements at Divalent Selenium

A series of kinetic studies of displacement reactions by sulfur and selenium nucleophiles on selenenyl halides has been reported by Austad.[378–381] The rate constants for the reactions of o-NO$_2$PhSeBr, <u>83</u>, with an assortment of nucleophiles are shown in Table 1.[378] Second-order kinetics (first order in each reactant) were generally followed closely. There was little correlation between the proton basicities of the nucleophiles and their relative reaction rates, a finding that is not unusual for reactions between soft species of the type studied. However, a linear relationship was observed between the logarithms of the second-order rate constants and the oxidation potentials associated with the processes:

$$2\,Nu^- \rightleftharpoons Nu\!-\!Nu + 2e^-$$

TABLE 1. (Ref.[378])
SECOND-ORDER RATE CONSTANTS FOR THE REACTION OF
o-NO$_2$PhSeBr, 83, WITH NUCLEOPHILES IN METHANOL AT 25°C

Nucleophile	k (M^{-1} s^{-1})	Nucleophile	k (M^{-1} s^{-1})
PhS$^-$	7000	RSO$_2$S$^-$	
		R = Me	2.82
$\overset{S}{\overset{\|}{R_2NCS^-}}$		R = Ph	3.25
R$_2$N = piperidyl	4200	$\overset{S}{\overset{\|}{R_2PS^-}}$	
R$_2$N = Me$_2$N	2667	R = Me	350
		R = OMe	14.5
$\overset{X}{\overset{\|}{NH_2CNH_2}}$		R = OEt	35.3
X = S	53.3	$\overset{O}{\overset{\|}{(EtO)_2PX^-}}$	
X = Se	2625	X = S	1.97
		X = Se	160
CN$^-$	994		
		$\overset{O}{\overset{\|}{(i\text{-}PrO)_2PX^-}}$	
XCN$^-$		X = S	4.57
X = S	0.133	X = Se	222
X = Se	1.73		
S$_2$O$_3^{-2}$	125	$\overset{O}{\overset{\|}{(MeO)_2PSe^-}}$	120
PhSO$_2^-$	2.05		

Table 1 indicates that the rate constants for selenium nucleophiles are at least one order of magnitude greater than those for their sulfur analogs. Furthermore, o-NO$_2$PhSeCl, 84, and the selenenyl bromide, 83, display comparable reactivity, but exhibit reaction rates that are ca. 1000 times faster than those of the corresponding sulfenyl chloride.[380]

The reactions of most nucleophiles with selenenyl halides proceed via an S$_N$2-type transition state, 85.[380] However, unusual rate enhancements observed with bidentate nucleophiles such as R$_2$NCS$_2^-$ suggest a transition state, 86, where both of the dithiocarbamate sulfur atoms are coordinated to selenium.[380]

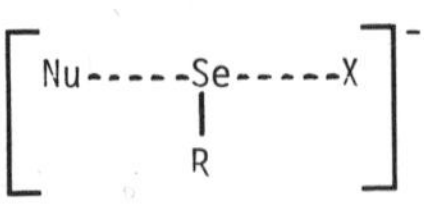

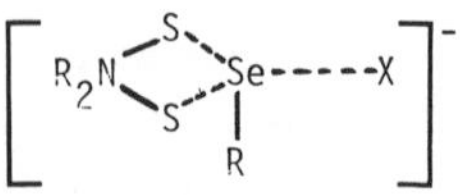

The effect of the o-nitro group in the reactions of selenenyl chloride, 84, has been the subject of several investigations. This compound is considerably more inert towards nucleophilic substitution than is PhSeCl.[379] Moreover, the o-nitro derivative is extremely sensitive to solvent effects in its reactions with the thiocyanate ion. For instance, this process occurs ca. 1.5×10^6 times faster in acetonitrile than in methanol.[380] These phenomena are consistent with structure 87 in which intramolecular coordination exists between the nitro oxygen and the selenium atom, an arrangement that has also been observed in the crystalline state.[382]

87

The nitro group thus impedes the approach of an attacking nucleophile from the required direction. The effect is greatly enhanced in protic solvents such as methanol, in which solvation of the nitro group increases its steric bulk. Kinetic evidence also indicates that the coordination between O and Se breaks down in solution when the leaving group is changed from halide to SCN^-, $ArSO_2^-$, or $(NH_2)_2C{=}S$.[379] The resulting free rotation around the aryl–selenium linkage lessens the blocking effect of the o-nitro group and increases the susceptibility of the leaving group towards displacement.

The transition states encountered in nucleophilic substitutions of selenenyl halides appear to be moderately stabilized by electron-donating substituents on either the nucleophile or the electrophile. Thus, the reaction of selenenyl bromide, 83, with p-substituted benzenesulfinate anions gave a Hammett plot with $\rho = -1.0$,[379] while the reactions of variously p-substituted derivatives of 83 with $PhSO_2^-$ or with thiourea produced $\rho = -0.34$ and -1.2, respectively.[381]

The reactions of bis-(alkylthio) selenides, 88, with thiolate anions and other nucleophiles (CN^-, SO_3^{-2}, PhLi, piperidine) were investigated by Kice and Slebocka-Tilk.[383] Compounds 88 have been implicated as intermediates in the assimilation of inorganic selenium compounds by living organisms and their chemical behavior is therefore of special significance.[1c] Such compounds present a nucleophile with a choice of sites for attack: S or Se [Eq. (170)]. Normally, divalent selenium represents a more electrophilic center than does divalent sulfur. In this case the better departing ability of $RSSe^-$ versus RS^-, however, constitutes an opposing effect. The authors determined that attack at selenium is favored when R is bulky (e.g., R = t-Bu), while less hindered systems (e.g., R = n-Bu) react faster at sulfur. Bis-(alkylthio) selenides are several orders of magnitude more reactive towards nucleophiles than are disulfides. In some cases, electron transfer processes may occur instead of normal S_N2 displacement.

$$
\begin{array}{c}
\nearrow \ RS\text{-}Nu \ + \ ^-SeSR \\
RSSeSR \ + \ Nu^- \\
88 \qquad\qquad \searrow \ RSSe\text{-}Nu \ + \ ^-SR
\end{array}
\qquad (170)
$$

A comparison of the reaction rates of $PhSO_2SePh$ and $PhSO_2SPh$ with cyanide ion in 90% acetonitrile–water at 25°C was performed by Gancarz and Kice,[36] who determined that the selenium compound reacts 70,000 times faster! The reactions of three selenosulfonates, 10 ($ArSO_2SePh$; Ar = Ph, p-ClPh, p-MePh) with cyanide ion showed only modest sensitivity to substituent effects and produced a Hammett plot with $\rho = +0.6$.

6. ADDITIONS OF SELENENIC ELECTROPHILES TO OLEFINS

Many selenenic electrophiles form 1,2 adducts with olefins according to Eq. (171). Selenenyl chlorides, bromides, and pseudohalides produce β-chloro-, bromo-, or various other β-functionalized selenides, respectively. When selenenyl halides react with olefins in the presence of other nucleophiles, the latter are sometimes incorporated in the β position instead of the halide atom. Cyclization occurs when a nucleophilic functionality exists at some other site in the unsaturated substrate, as described separately in Chapter 2. The use of divalent selenium electrophiles thus permits the elaboration of olefins to products with a variety of useful functionalities.

$$\text{(171)}$$

6.1. Additions of Selenenyl Halides

6.1.1. GENERAL CONSIDERATIONS

A review by Schmid and Garratt[384] contains references to earlier work as well as a compilation of kinetic data for the addition of PhSeCl to 44 different olefins, and a detailed discussion of mechanistic aspects of such processes. The reactions of selenenyl halides bear a superficial resemblance to these of sulfenyl halides. However, differences in regioselectivity, reactivity, and the nature of the intermediates are significant.

Selenenyl chlorides and bromides react very rapidly with olefins at or even below room temperature and rates are generally measured by stopped flow methods. A selenenyl iodide has been postulated as an intermediate in the reactions of dienes with PhSeSePh and iodine in acetonitrile.[385] The use of selenenyl fluorides as reagents remains undocumented. Common solvents include hexane, benzene, THF, methylene chloride, chloroform, and carbon tetrachloride. Polar solvents such as acetonitrile, acetic acid, and methanol have also been employed but can result in the competing formation of solvent incorporated products, particularly in the case of methanol.[290] The additions are generally second order overall, first order in each reactant.[384]

Both polar and steric effects are important in determining reaction rates. A study of the addition of PhSeCl to ethylene and its various methyl substituted derivatives in methylene chloride at 25°C was performed by Schmid and Garratt.[386] The effect of methyl substitution is noncumulative; propene displayed a relative reaction rate 8.8 times greater than that of ethylene, whereas di-, tri-, and tetramethylethylenes exhibited intermediate relative rates of 2.1–6.8 times that of the parent olefin. Steric effects thus appear to outweigh polar ones in the more heavily substituted substrates. This differs from analogous reactions of sulfenyl halides that are less susceptible to steric effects and show cumulative rate increases with additional methyl substitution.[386']

6.1.2. INTERMEDIATES AND STEREOCHEMISTRY

The 1,2 additions of selenenyl halides to olefins are generally highly *anti* stereospecific. For instance, cyclohexene produces the trans adducts, 89, when treated with PhSeCl or PhSeBr [Eq. (172)],[12] while (E)- and (Z)-1-phenylpropene afford the erythro and threo adducts, respectively, with 2,4-dinitrobenzeneselenenyl chloride [Eqs. (173) and (174)].[387] Product 89 (X = Cl) displays a somewhat unexpected preference for the diaxial conformation.[388,389]

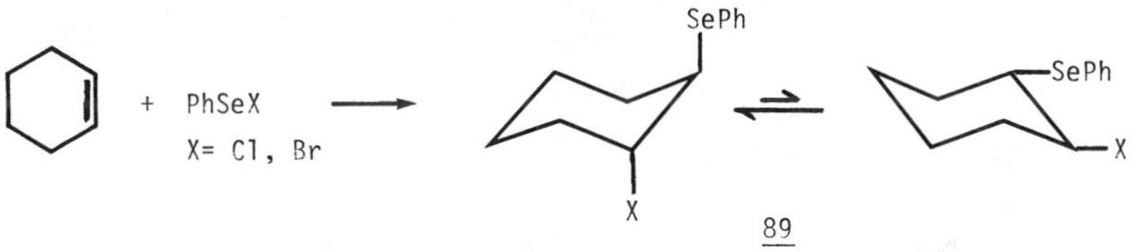

$$ (172) $$

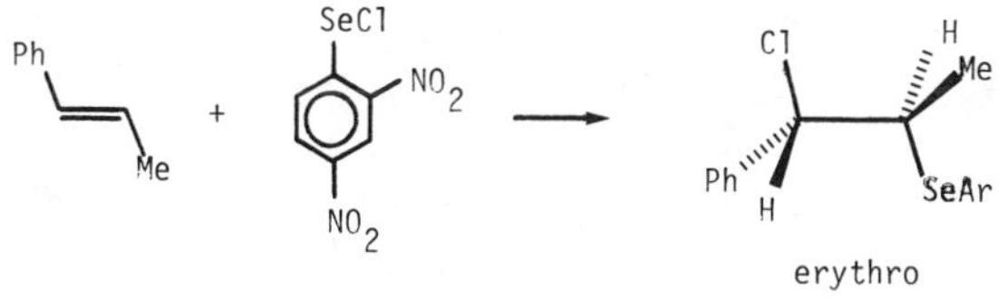

$$ (173) $$

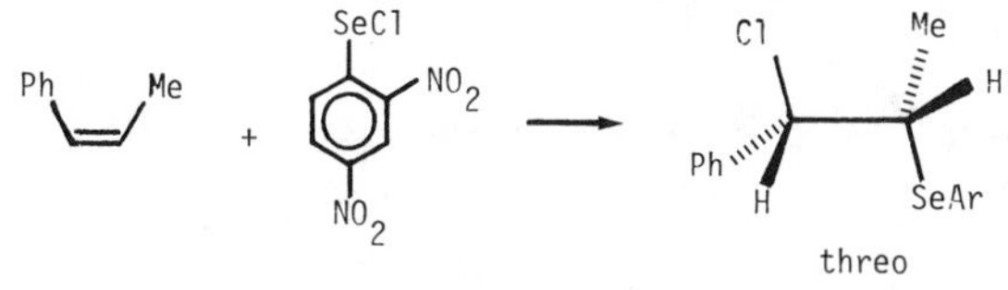

$$ (174) $$

The observed stereospecificity is consistent with the existence of selenium bridged intermediates. Based on stereochemical, kinetic, and other evidence, Schmid and Garratt[384] proposed a mechanism in which an episelenurane intermediate, 90, is initially produced and subsequently dissociates to a seleniranium ion, 91 (sometimes referred to as an episelenonium ion) and a halide ion. These may exist as an intimate, solvent separated, or fully dissociated ion pair, depending on factors such as the solvent polarity and the nature of the substrate. Ring opening of such species by halide ion or other nucleophiles then forms the observed products of *anti* addition. In some special cases such as arylpropenes containing strongly electron-donating substituents on the aryl moiety, the open carbonium ion, 92, is lower in energy than the seleniranium ion and its resulting formation is accompanied by a concomitant loss of stereospecificity. These processes are summarized in Scheme 15.

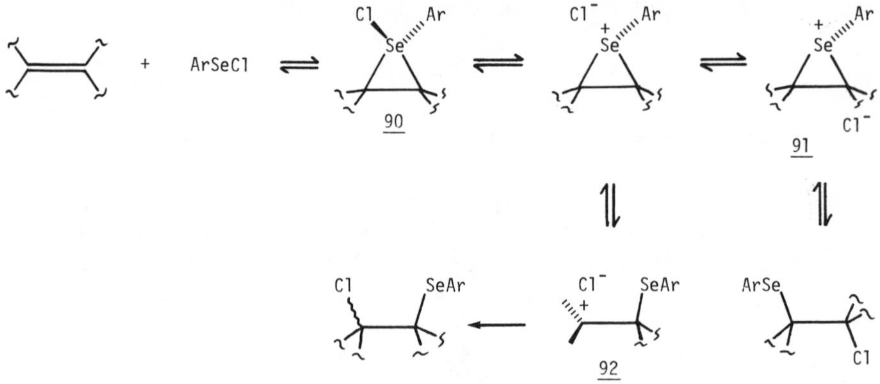

SCHEME 15.

The isolation of several episelenuranes of varying stability has been reported.[384,390] In the case of the reaction of p-tolueneselenenyl chloride with ethylene, Reich and Trend[391] failed to observe the corresponding episelenurane and noted only the formation of the 1,2 adduct in the presence of excess olefin, as well as its further chlorination by the selenenyl chloride when the ethylene was added slowly [Eq. (175)]. This of course does not rule out the existence of the episelenurane as a short-lived intermediate. Tetracoordinate selenium species have also been postulated as intermediates in the reactions of β-bromo selenides with soft nucleophiles such as selenocyanate ion.[392–394]

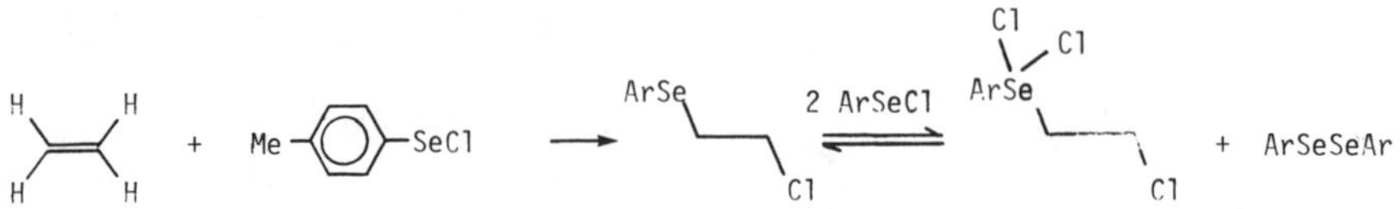

$$(175)$$

Schmid and Garratt[395] and Remion and Krief[396] characterized several seleniranium hexafluorophosphates and hexafluoroantimonates. The products were obtained by treating olefins with $ArSe^+ PF_6^-$ [395] or $ArSe^+ SbF_6^-$ [395,396] and from the reactions of β-chloroethyl selenides with $AgPF_6$ or $AgSbF_6$.[395] The parent olefins can be regenerated with triethylamine (Scheme 16).[396]

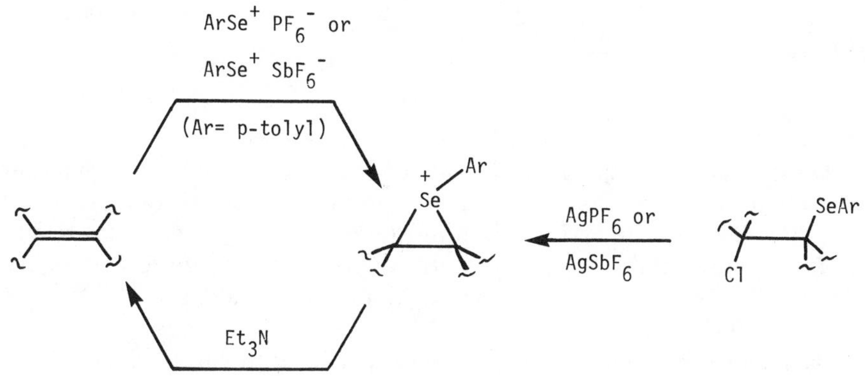

$ArSe^+ \ PF_6^- \ \text{or}$

$ArSe^+ \ SbF_6^-$

(Ar = p-tolyl)

$AgPF_6$ or $AgSbF_6$

Et_3N

SeAr

Cl

SCHEME 16.

6.1.3. REGIOCHEMISTRY

The control of regiochemistry in the 1,2 additions to unsymmetrical olefins is often poor and imposes limitations on the synthetic utility of such processes. Further complications stem from differences in regioisomer distributions obtained under conditions favoring either kinetic or thermodynamic control. Several recent investigations have provided insight into this area and permit product ratios to be predicted and sometimes regulated.

Monoalkyl olefins produce chiefly Markovnikov products at or near room temperature.[397–399] However, Raucher[400,401] demonstrated that anti-Markovnikov adducts are initially formed at $-78°C$ in THF or at $0°C$ in carbon tetrachloride, but isomerize readily to their Markovnikov regioisomers at room temperature. Polar solvents such as acetonitrile facilitate the isomerization [e.g., Eq. (176)].

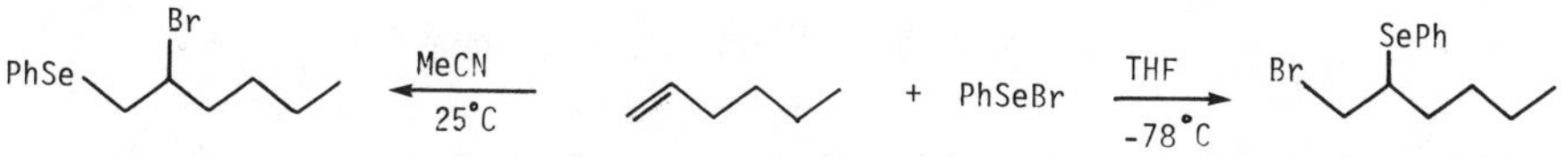

PhSe, Br

$\xrightarrow[25°C]{MeCN}$

+ PhSeBr

$\xrightarrow[-78°C]{THF}$

Br, SePh

$$(176)$$

Similarly, several groups have reported Markovnikov additions to 1,1-disubstituted olefins[290,386,397] at room temperature, whereas Ho and Kolt[402] observed anti-Markovnikov adducts at $-70°C$, followed by their subsequent isomerization upon warming [e.g., Eq. (177)]. At low temperatures attack by the chloride ion occurs at the less substituted carbon atom of the seleniranium ion even though this center is less capable of supporting positive charge. Hence, under kinetically controlled conditions steric factors outweigh polar effects. The thermodynamically favored Markovnikov regioisomers presumably arise through equilibration of the initial adducts via the corresponding seleniranium ions at higher temperatures.

(177)

1,2-Disubstituted olefins generally afford mixtures containing significant amounts of both regioisomers unless one group is considerably more bulky than the other, or is an aryl group. Anti-Markovnikov adducts are favored in the former case, while exclusively Markovnikov products are formed in the latter.[290,384] Trisubstituted olefins give Markovnikov adducts,[290,384,386,398] even at $-50°C$.[398]

The regiochemistry of addition of selenenyl halides to vinyl halides has not been thoroughly investigated, but a few examples are known. 1,1-Difluoroethylene reacts with PhSeCl or PhSeBr to afford only the 2,2,2-trihaloethyl selenide isomer [Eq. (178)],[403] while 1-chlorocyclohexene and related compounds produce mixtures of regioisomers [Eq. (179)].[404]

$$X = Cl \quad 91\%$$
$$X = Br \quad 90\%$$

(178)

(179)

$$75 : 25$$

6.1.4. BICYCLIC AND POLYCYCLIC SYSTEMS

Apart from consideration of stereoisomerism (*syn* versus *anti*) and regioisomerism (Markovnikov versus anti-Markovnikov) in the additions of selenenyl halides to olefins, it is also important to recognize that electrophilic attack may be favored upon one of the two faces of the π system. The term "configurational selectivity" has been employed by Garratt and Kabo[290] in this context. In general, the seleniranium ion is formed on the less sterically congested side of the double bond. Thus, the cis-fused bicyclic olefin, 93, reacts on the less hindered *exo* face[290] to produce chiefly the two regioisomers, 94 and 95 [Eq. (180)]. Similarly, 96 and 98, afford the products of *exo* attack upon the cyclobutene moiety to give 97 and 99, respectively, under the conditions shown [Eqs. (181) and (182)].[405] Norbornene, 100, is also *exo* selective [Eq. (183)] while norbornadiene, 102, is, surprisingly, *endo* selective [Eq. (184)].[290] Rearrangement products are formed in methanol (see Section 6.1.7).

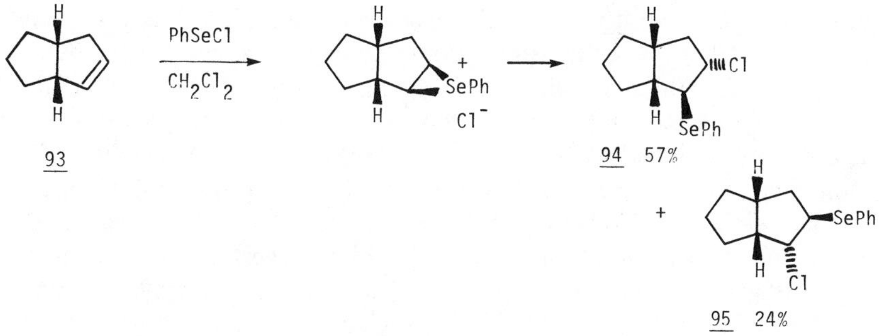

93

94 57%

95 24%

(180)

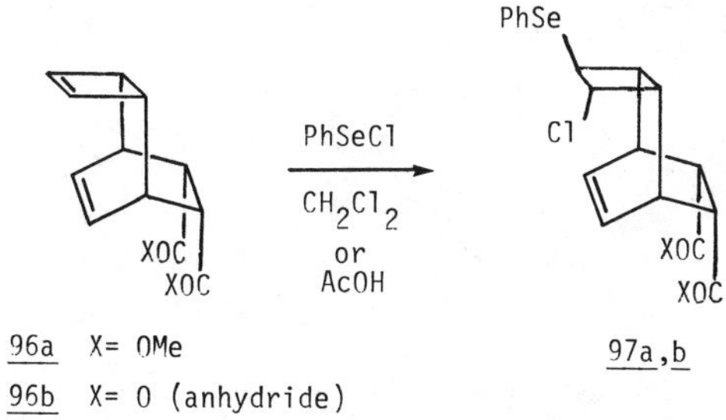

96a X= OMe
96b X= O (anhydride)

97a,b

(181)

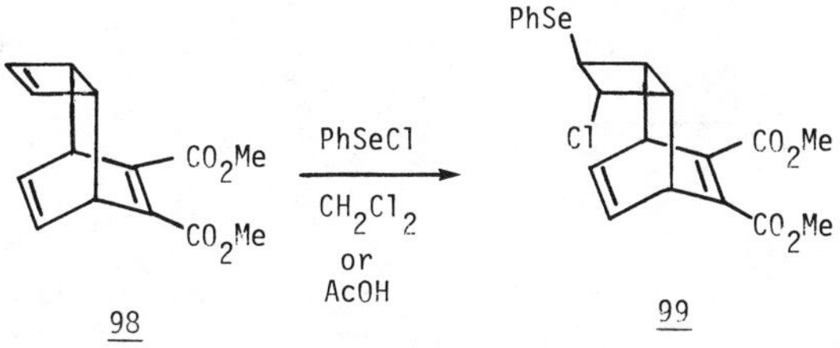

98

99

(182)

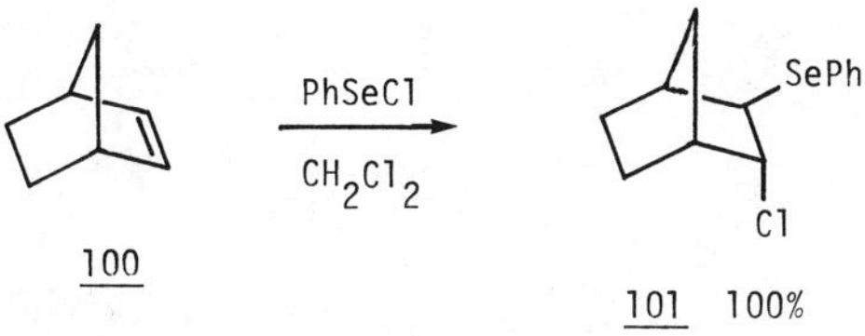

100

101 100%

(183)

102

103 major

104 minor

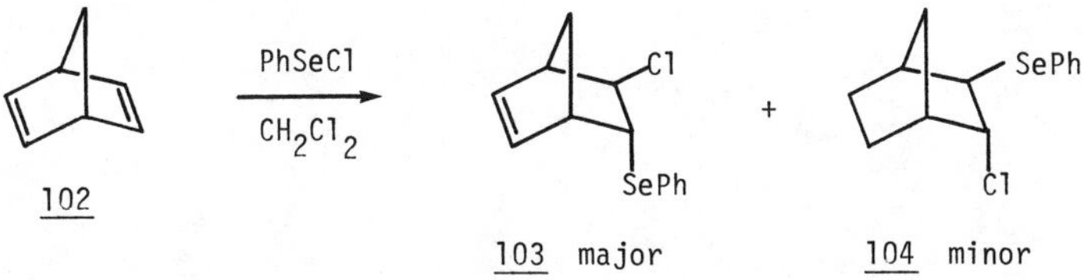

(184)

5-Norbornen-2-one, 105, forms the *exo* seleniranium ion, 106, which undergoes stereo- and regiospecific ring opening by *endo* attack of the halide ion at
C-6 to give 107 as the sole product [Eq. (185)].[406] The bicyclic keto olefin, 108,
displays similar stereo- and regiospecificity, but lower configurational selectivity
to afford both adducts 109 and 110 [Eq. (186)].[406] In contrast, the 2-chloro,
2-cyano, and 2-methoxy derivatives, 111 and 112, incorporate halide ion exclusively at C-5 [Eqs. (187) and (188)]. Carrupt and Vogel[406] rationalized the
difference in regiochemistry of these systems by suggesting that the carbonyl
group in 105 and 108 stabilizes a positive charge at C-6 through hyperconjuga-
tion, while field effects from 2-chloro, cyano, or methoxy substituents have the
opposite effect and are reinforced by steric hindrance to *endo* attack at C-6.
Consequently, the halide ion attacks preferentially at C-6 in Eqs. (185) and (186)
and at C-5 in Eqs. (187) and (188).

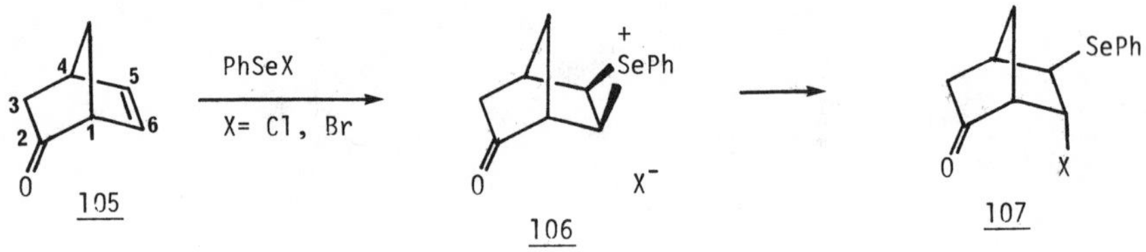

(185)

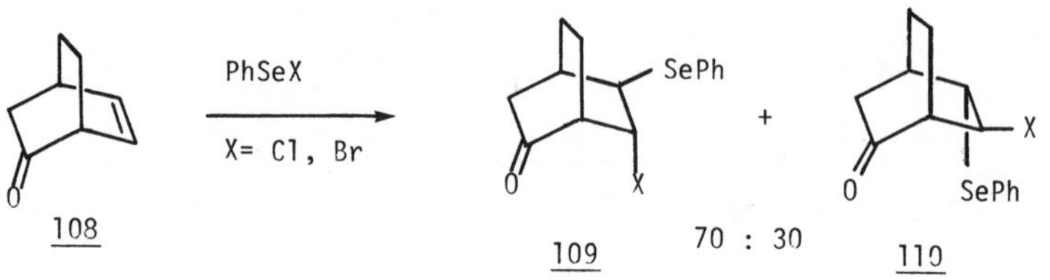

(186)

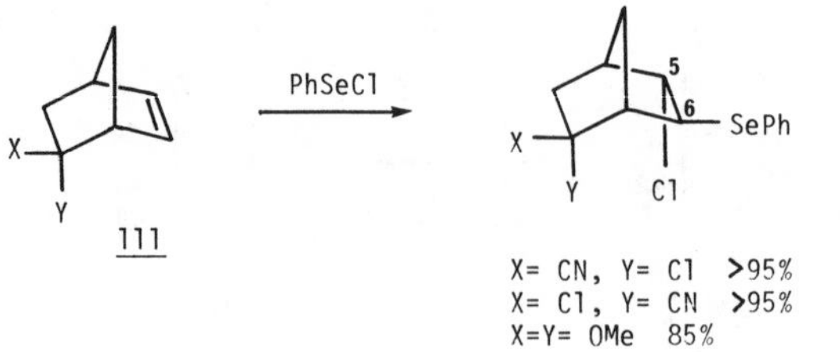

X= CN, Y= Cl >95%
X= Cl, Y= CN >95%
X=Y= OMe   85%

(187)

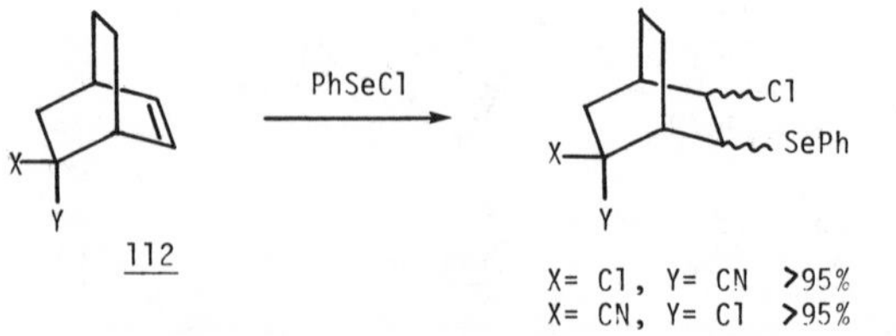

X= Cl, Y= CN >95%
X= CN, Y= Cl >95%

(188)

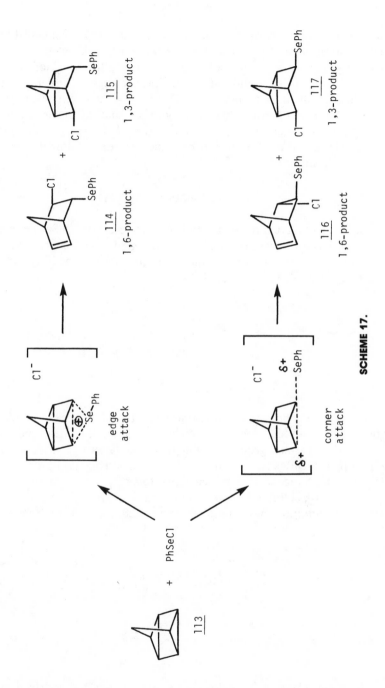

SCHEME 17.

The strained three-membered rings of quadricyclane, 113, react with PhSeCl to yield addition products. Garratt and co-workers[407] have identified four products, 114–117, in nonpolar solvents such as methylene chloride. Their formation arises from electrophilic attack at either an edge or corner of quadri-cyclane, which results in *endo* and *exo* orientations of the phenylseleno group, respectively (Scheme 17). In either case, both 1,3 and 1,6 incorporation of chloride ion is observed, with the 1,6 products 114 and 116 favored in nonpolar solvents. 1,3 Addition with solvent incorporation dominates in acetic acid, acetonitrile, and methanol.

6.1.5. ALLYLIC ALCOHOLS AND ACETATES

Liotta et al.[408] investigated the regio- and stereochemistry of PhSeCl additions to allylic alcohols and acetates, and formulated guidelines for predicting the results. In an acyclic system, the acetates react more regioselectively than the corresponding alcohols [e.g., Eq. (189)].

$$R= H \quad 7 : 3$$
$$R= Ac \quad 20 : 1$$

(189)

In cyclic allyl alcohols [e.g., 2-cyclohexenol, Eq. (190)] seleniranium ion formation occurs with high configurational selectivity on the side of the double bond *syn* to the hydroxyl group, providing that the latter is conformationally free to occupy a pseudoaxial position. This is the consequence of a stabilizing interaction between the oxygen and selenium atoms. After ring flip of the inter-mediate 118 to 119, chloride ion attacks preferentially from the axial direction to complete the addition. A single stereoisomer, 120, was thus produced.

(190)

The preceding scheme breaks down in systems where the hydroxyl group cannot occupy a pseudoaxial position or where steric hindrance prevents axial ring opening by the chloride ion. In contrast, an acetoxy substituent can effec-tively direct seleniranium ion formation *syn* to itself even from an equatorial position. This point is illustrated in Eqs. (191) and (192) where the equatorial acetate forms a unique stereoisomer while the corresponding alcohol furnishes a mixture of products.

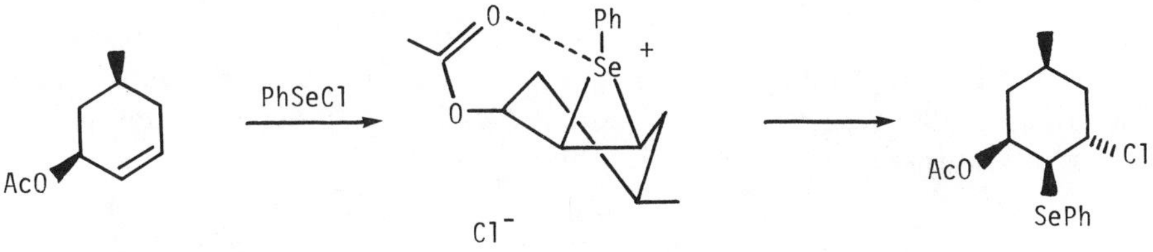

(191)

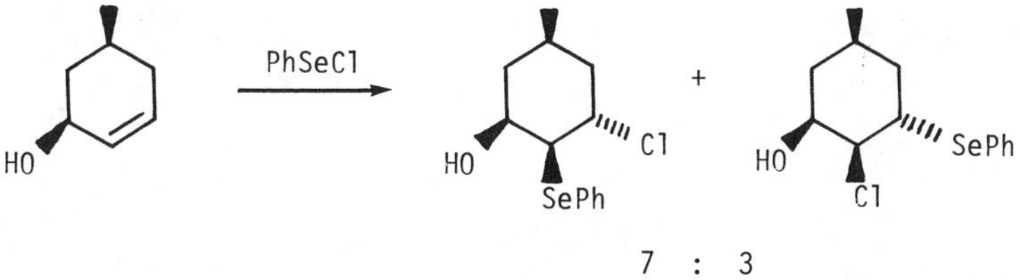

7 : 3

(192)

The application of such additions to enone transpositions is described in Section 6.1.8. Similar orienting effects have also been noted in the additions of other selenenic electrophiles to allylic substrates [see Eqs. (239) and (241)].

6.1.6. DIENES AND ALLENES

Acyclic dienes containing both terminal and internal double bonds react selectively at the terminal position [Eqs. (193) and (194)].[399] This is consistent with the previously mentioned observation that propene is more reactive than di-, tri-, or tetramethylethylene.[386] Similarly, addition of PhSeCl occurs at the less substituted double bond of the cyclic diene, 121, [Eq. (195)],[290] and exclusively to the isolated olefin instead of to the deactivated double bond of the quinone

$$Me(CH_2)_6CH=CH(CH_2)_8CH=CH_2 \xrightarrow[R= Me, Ph]{RSeBr} Me(CH_2)_6CH=CH(CH_2)_8\overset{\overset{\displaystyle Br}{|}}{C}HCH_2SeR$$

90%

(193)

$$Me_2C=CH(CH_2)_2\overset{\overset{\displaystyle Me}{|}}{C}HCH_2CH=CH_2 \xrightarrow[R= Me, Ph]{RSeBr} Me_2C=CH(CH_2)_2\overset{\overset{\displaystyle Me}{|}}{C}HCH_2\overset{\overset{\displaystyle Br}{|}}{C}HCH_2SeR$$

90%

(194)

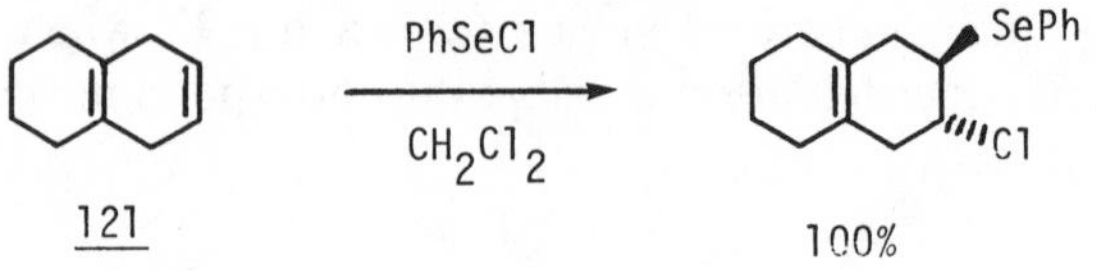

$$(195)$$

moiety of 122 [Eq. (196)].[398] Conjugated 1-silyldienes undergo Markovnikov addition to the monosubstituted double bond instead of 1,4 addition, or 1,2-addition to the vinylsilane moiety [Eq. (197)].[409]

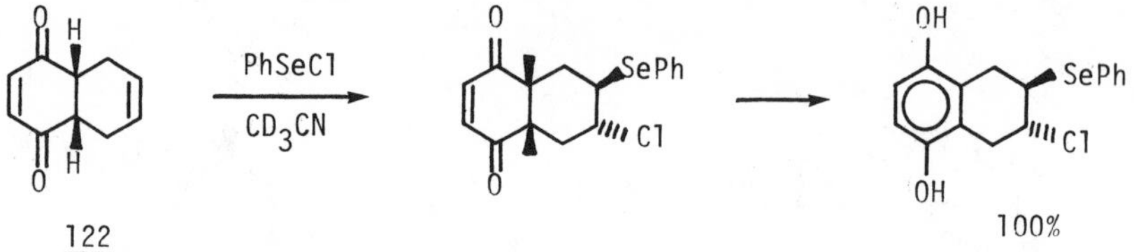

$$(196)$$

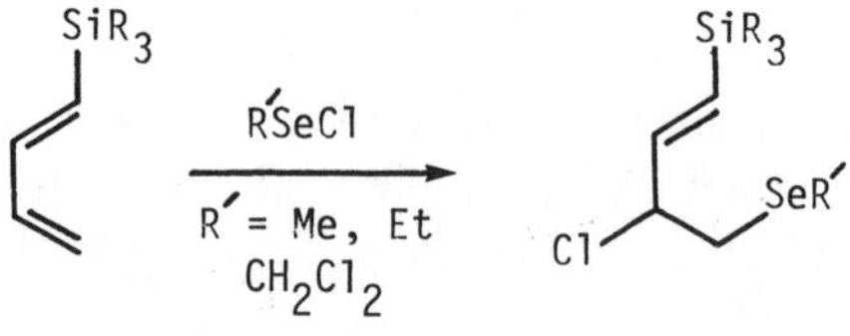

$$(197)$$

The additions of selenenyl halides to allenes have been studied by Schmid and Garratt and their co-workes[410–412] as well as by Halazy and Hevesi.[413] Like olefins, allenes show second-order kinetics, first order in both reactants. However, unlike olefins, allenes show cumulative rate increases when methyl substituents are introduced.[410] The additions are regiospecific with the RSe group bonding to the central allene carbon atom. In unsymmetrical allenes, either of the two π bonds can be attacked from either of two directions. Hence, considerations of chemo- and configurational selectivity arise. In general, the π bond with the bulkiest and most electron-donating substituents is attacked preferentially and the adduct with the (Z)-configuration at the remaining double bond dominates. The product distribution from allene, 123, and PhSeCl provides an illustrative example [Eq. (198)].[411]

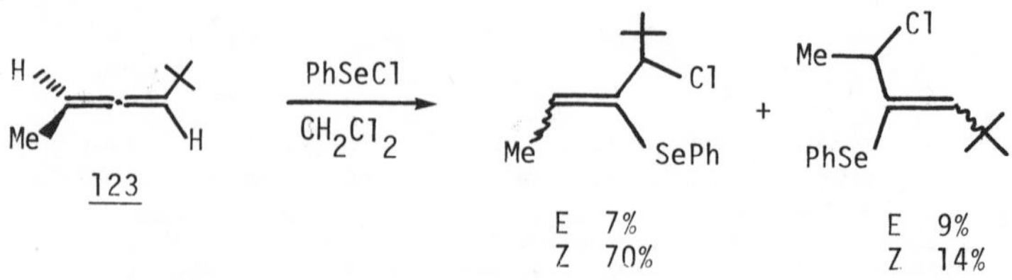

E   7%   E   9%
Z   70%   Z   14%

$$(198)$$

These reactions are believed to proceed via alkylidene seleniranium ions, possibly preceded by formation of their episelenurane counterparts.[412] Examination of the seleniranium ions, 124 and 125, leading to (E) and (Z) products, respectively, reveals that nucleophilic attack by halide ion is less hindered by the substituent R in the case of 125. Consequently the (Z) product is favored. An exceptional situation arises with mesityleneselenenyl bromide, where steric interactions between the bulky aryl group and the substituent R are sufficiently severe to destabilize 125 and result in (E) selectivity [Eq. (199)].[412] Steric and electronic effects are much less pronounced in the analogous additions of sulfenyl halides to allenes.[414]

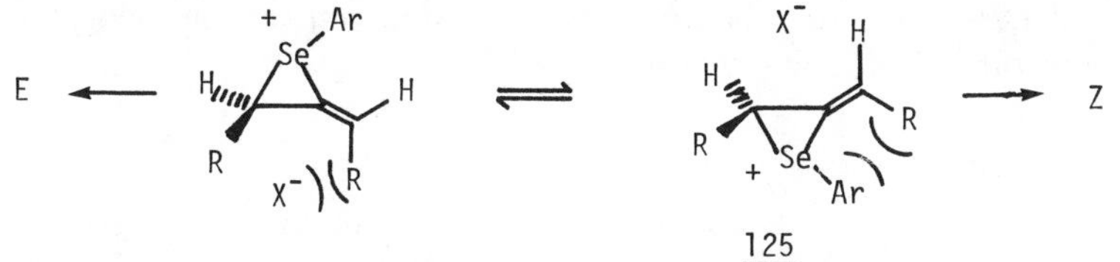

$$ (199) $$

The initial addition products are capable of undergoing $(E) \rightleftharpoons (Z)$ isomerization[413] as well as other rearrangements described in the next section.

6.1.7. REARRANGEMENTS

Apart from the simple equilibration of regioisomers, skeletal rearrangements, migrations of the phenylseleno group or the halogen atom, and prototropic shifts have also been observed during the addition of selenenyl halides to olefins and allenes. Camphene and α-pinene produce complex mixtures of products when treated with PhSeCl, presumably because of facile Wagner–Meerwein rearrangements.[398] β-Pinene, 126, affords the ring-opened limonene selenide, 127, as the principal product, along with a smaller amount of the selenide, 128 [Eq. (200)].[415]

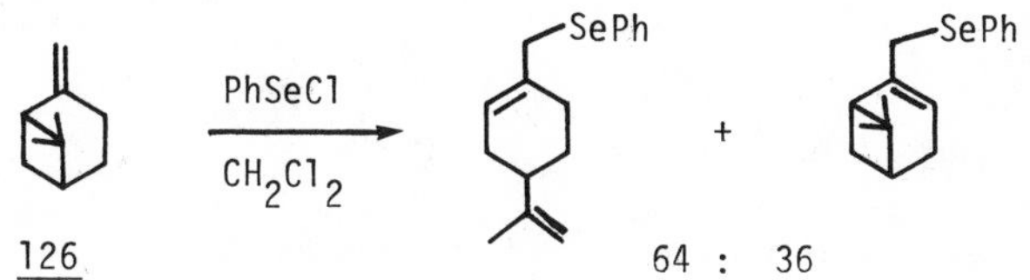

$$ (200) $$

Adamantylideneadamantane, 129, forms the dichlorohomoadamantane derivative, 130, nearly quantitatively when treated with two molar equivalents of PhSeCl [Eq. (201)].[416]

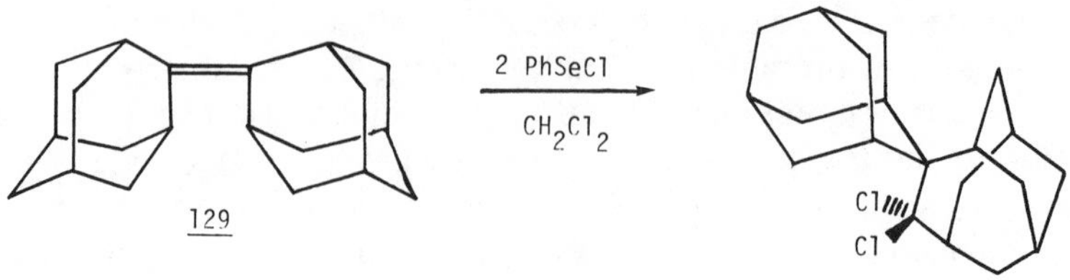

129 → 130 97% (201)

In contrast to its behavior in methylene chloride [cf. Eq. (183)], norbornene, 100, reacts with PhSeCl in methanol to give a mixture of products including the normal 1,2 adduct, 101, the products of *syn* and *anti* solvent incorporation and the rearranged products, 131 and 132.[290]

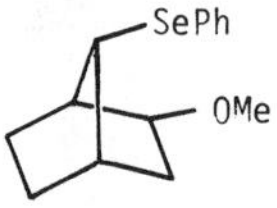

131

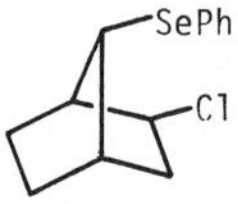

132

Norbornadiene [102; cf. Eq. (184)] forms the products of homoallylic rearrangement, 133–136, in methanol, including those with incorporated solvent, 135 and 136.[290] The normal 1,2 adducts, 103 and 104, that are obtained in methylene chloride are completely absent in methanol. Other olefins that are prone to homoallylic rearrangements in methanol include 137 and 138,[290] while transannular bond formation occurs in 96 and 98 under similar conditions.[405]

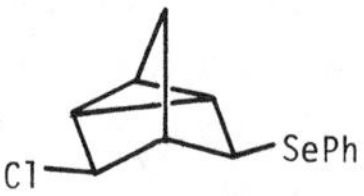

133

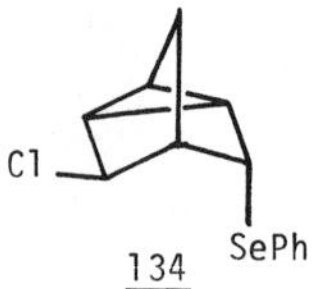

134

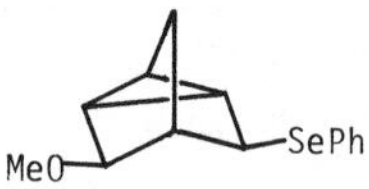

135

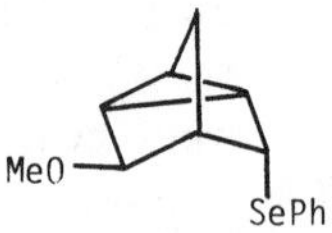

136

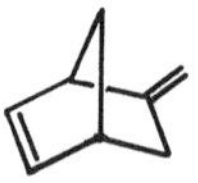

137

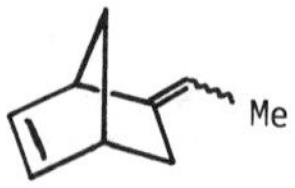

138

Migrations of the phenylseleno group have been observed by Itoh and co-workers[417,418] during the addition of PhSeCl to allylsilanes. The initial 1,2 adducts were formed at −78°C, but eliminated chlorotrimethylsilane in the presence of tin (II) chloride at 0°C or upon chromatography on Florisil.[e] The

[e] Florisil is a registered trademark of the Floridin Co.

resulting allyl selenides, 139, rearranged spontaneously to afford the less substituted isomers, 140 (Scheme 18; for further discussion of selenide rearrangements, see Chapter 8). A similar procedure employing PhSCl as the electrophile produced unrearranged sulfides.[418] The two reagents thus provide access to regioisomeric allylic alcohols via oxidation and [2,3] sigmatropic rearrangement of the corresponding selenoxides or sulfoxides.

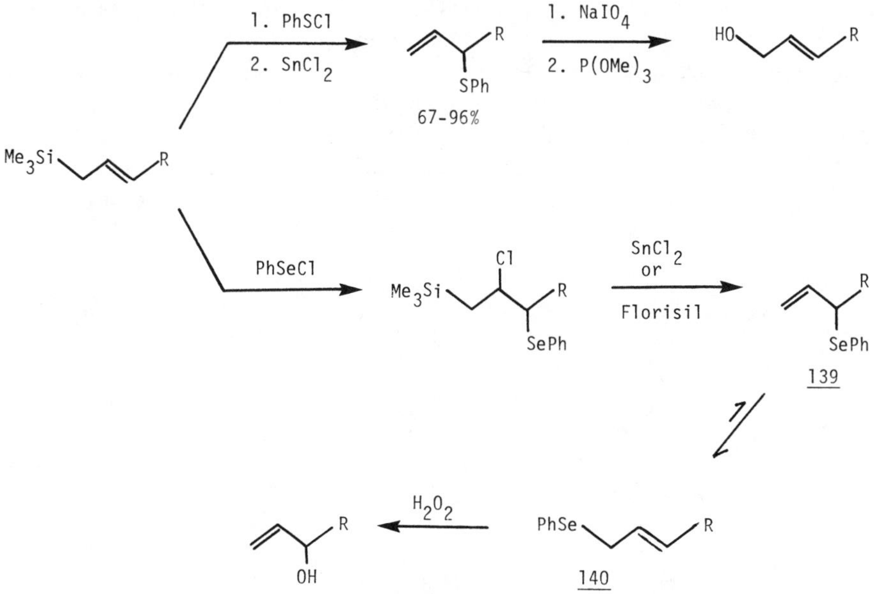

SCHEME 18.

Prototropic shifts [e.g., Eq. (202)] and halogen migrations from the more to the less substituted allylic position [e.g., Eq. (203)] have been observed during the addition of PhSeCl or PhSeBr to allenes.[413]

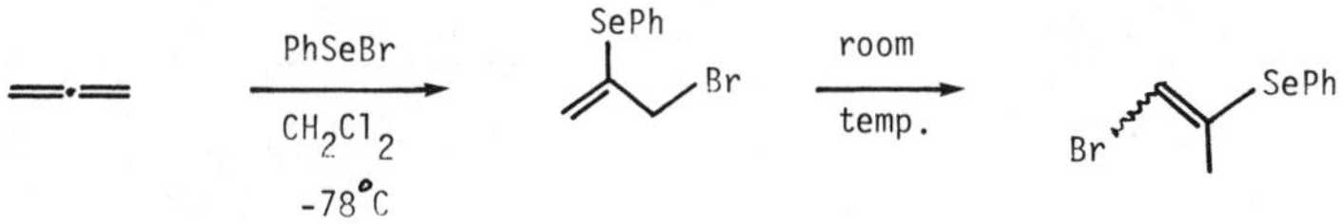

(202)

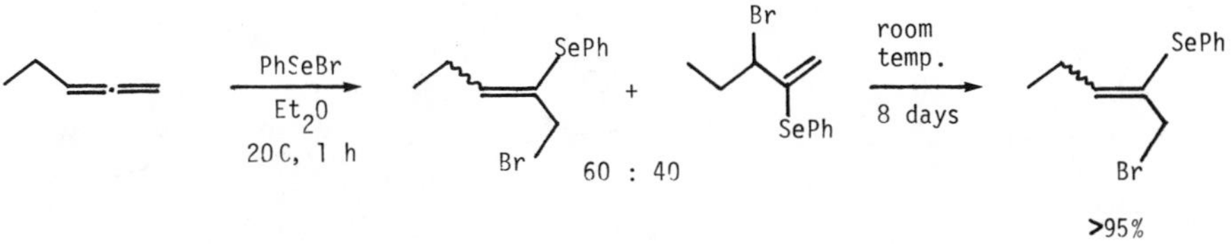

(203)

6.1.8.   SYNTHETIC APPLICATIONS

There are relatively few direct applications of the electrophilic additions of selenenyl halides to olefins. The adducts serve as precursors of vinyl selenides via dehydrohalogenation and of vinyl or allyl halides through selenoxide elimination. Specific examples of reactions of the former type are given in Eqs. (204)–(208).[282,401,419-421] Terminal olefins can be transformed into vinyl selenides with the phenylseleno group in either the 1 or 2 position by applying thermodynamic or kinetic control during the addition step [e.g., Eq. (206)].[401] 1-Vinyl selenides produce mixtures of (E) and (Z) isomers in ratios that vary with the base–solvent system employed. Raucher et al.[282] found that LDA–ether at 0°C provides the highest proportion of the (E) isomer of several systems studied [e.g., Eq. (207)]. Other stereoselective syntheses of vinyl selenides have also been developed by these authors (also see Scheme 12).[282] In one instance a β-chloro

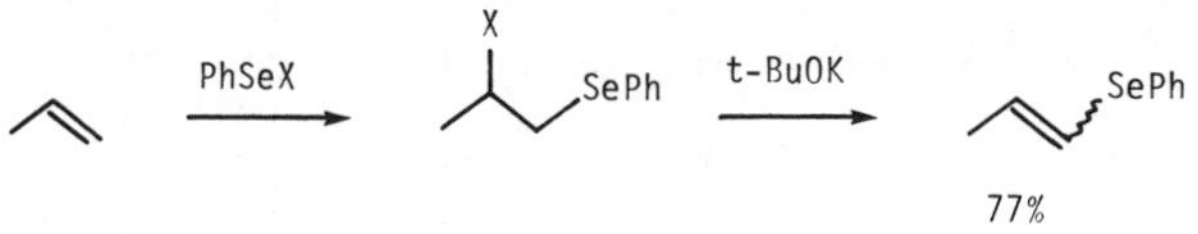

(204) (Ref. 419)

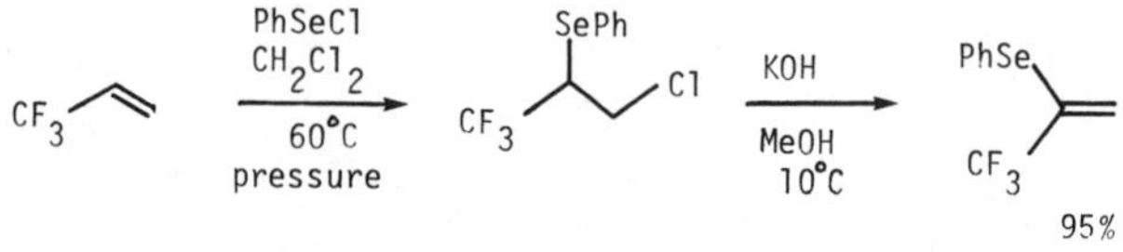

(205) (Ref. 420)

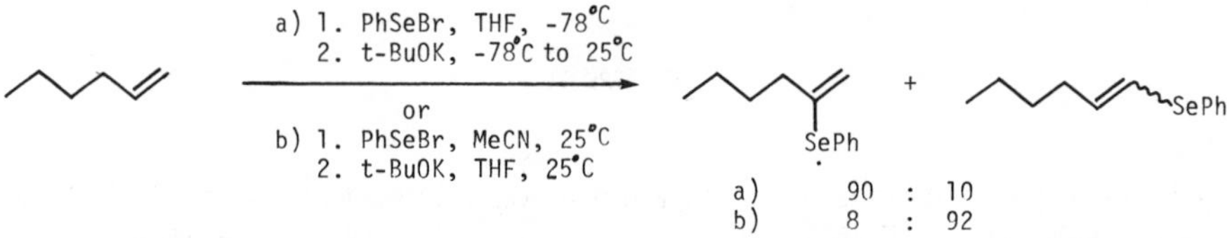

(206) (Ref. 401)

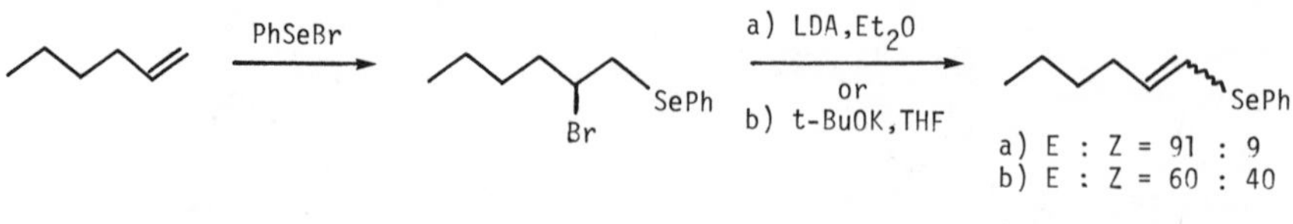

(207) (Ref. 282)

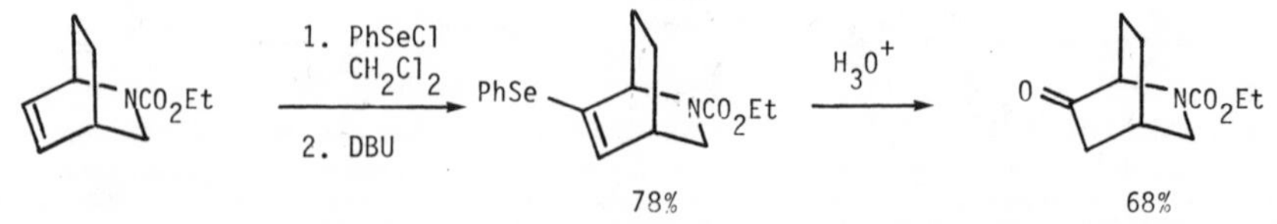

(208) (Ref. 421)

selenide adduct reverted to the original olefin when treated with base or with silica gel.[422]

Examples of selenoxide eliminations of β-halo selenides[400,402,423,424] are provided in Eqs. (209)–(214). Markovnikov adducts of terminal olefins produce only the corresponding vinyl halides upon elimination [e.g., Eq. (209)], but both allyl and vinyl halides can be formed from anti-Markovnikov adducts. The allyl isomers are generally preferred [Eqs. (210)–(212) and (214) where X = H], unless protons are unavailable for elimination from the required site, or the elimination is impeded by conformational [Eq. (213)] or electronic effects [Eq. (214) where X = OH].

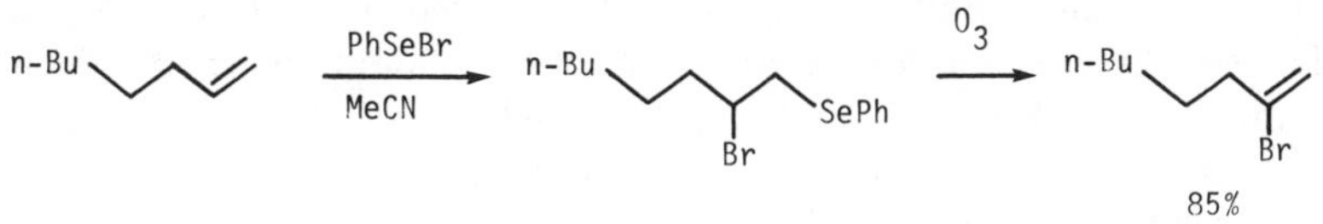

85%

(209) (Ref. 400)

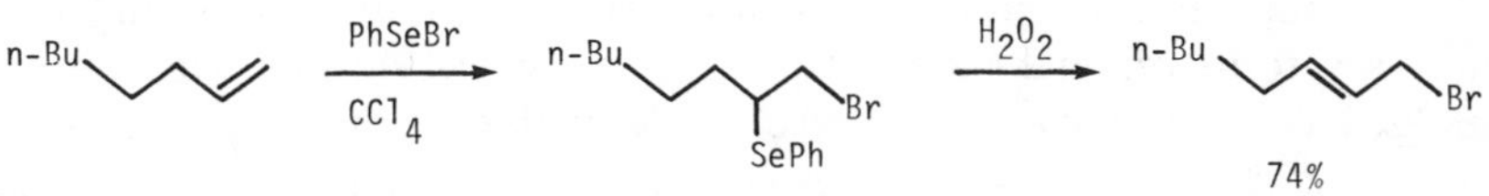

74%

(210) (Ref. 400)

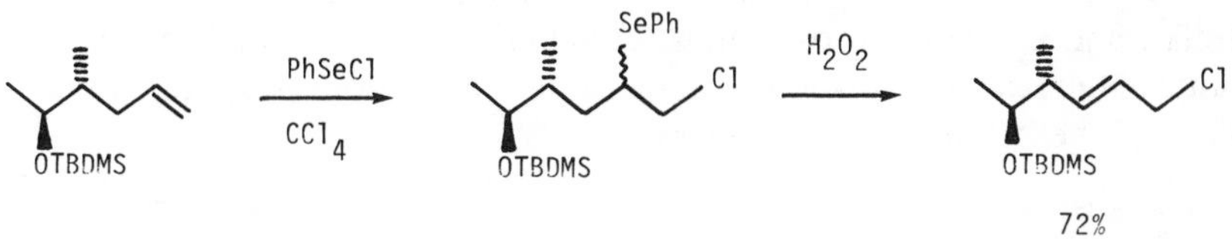

72%

(211) (Ref. 423)

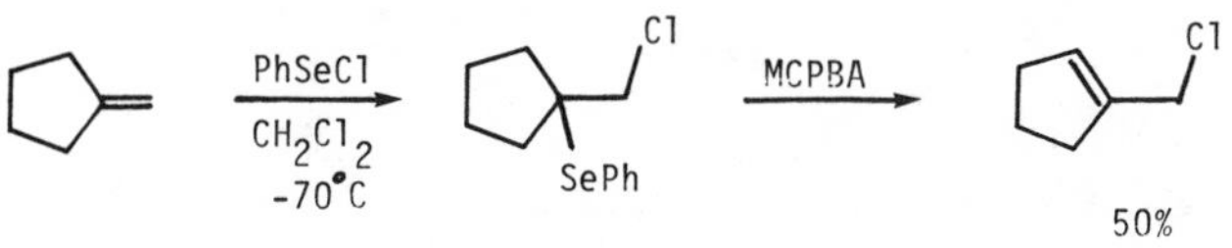

50%

(212) (Ref. 402)

71%   3%

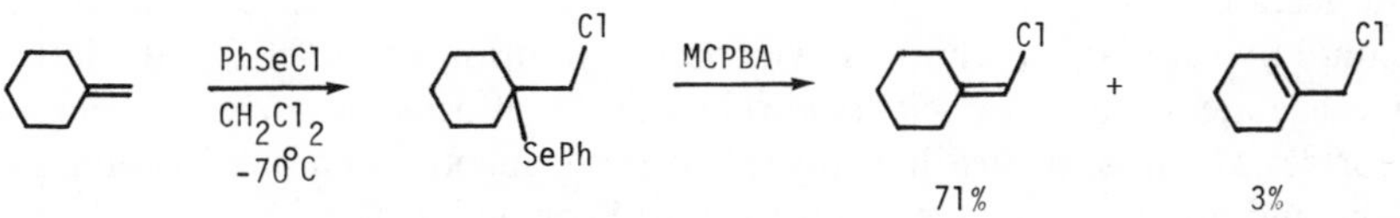

(213) (Ref. 402)

major
if X= H

major
if X= OH

minor

X= H, OH

(214) (Ref. 424)

Hori and Sharpless[425] obtained rearranged allyl chlorides together with smaller amounts of vinyl chlorides by treating olefins with a catalytic amount of a diaryl diselenide in the presence of N-chlorosuccinimide (NCS). A plausible mechanism involves the formation of ArSeCl, its addition to the olefin, and NCS mediated elimination of the β-chloro selenides, 142 and 143, as in the example in Scheme 19. β-Pinene, 126, anomalously afforded the less stable unrearranged allyl chloride, 144, as the principal product.[415] In this case N-(phenylseleno)succinimide, 141, assumes the role of the electrophile instead of PhSeCl, and produces the rearranged selenide, 128. Subsequent reaction of 128 with NCS furnishes the product 144 with further rearrangement of the double bond back to its original exocyclic position (Scheme 19). Allyl selenides other than 128 have been converted to allyl chlorides with NCS in a similar fashion.[415]

Raucher[426] prepared α-phenylseleno ketones from monosubstituted olefins by sequentially treating them with PhSeBr in dimethyl sulfoxide (DMSO), silver hexafluorophosphate, and triethylamine [Eq. (215)]. The reaction is believed to proceed via an alkoxysulfonium ion intermediate, 146, reminiscent of the Moffatt and Corey–Kim oxidations of alcohols. The use of silver tetrafluoroborate in the reaction of 1-decene under similar conditions produced the alcohol in lieu of the ketone as the major product.[427]

146

(215)

Liotta and co-workers[408,428] developed a novel enone transposition procedure based on their studies of the additions of PhSeCl to allylic alcohols and acetates (see Section 6.1.5). The initial enone is first reduced to the allyl alcohol by standard methods. Regiospecific addition of PhSeCl to the latter or to its acetate followed by selenoxide elimination affords the corresponding vinyl chloride, which is in turn hydrolyzed to the transposed enone. Examples of a cyclic and an acyclic system are shown in Scheme 20.

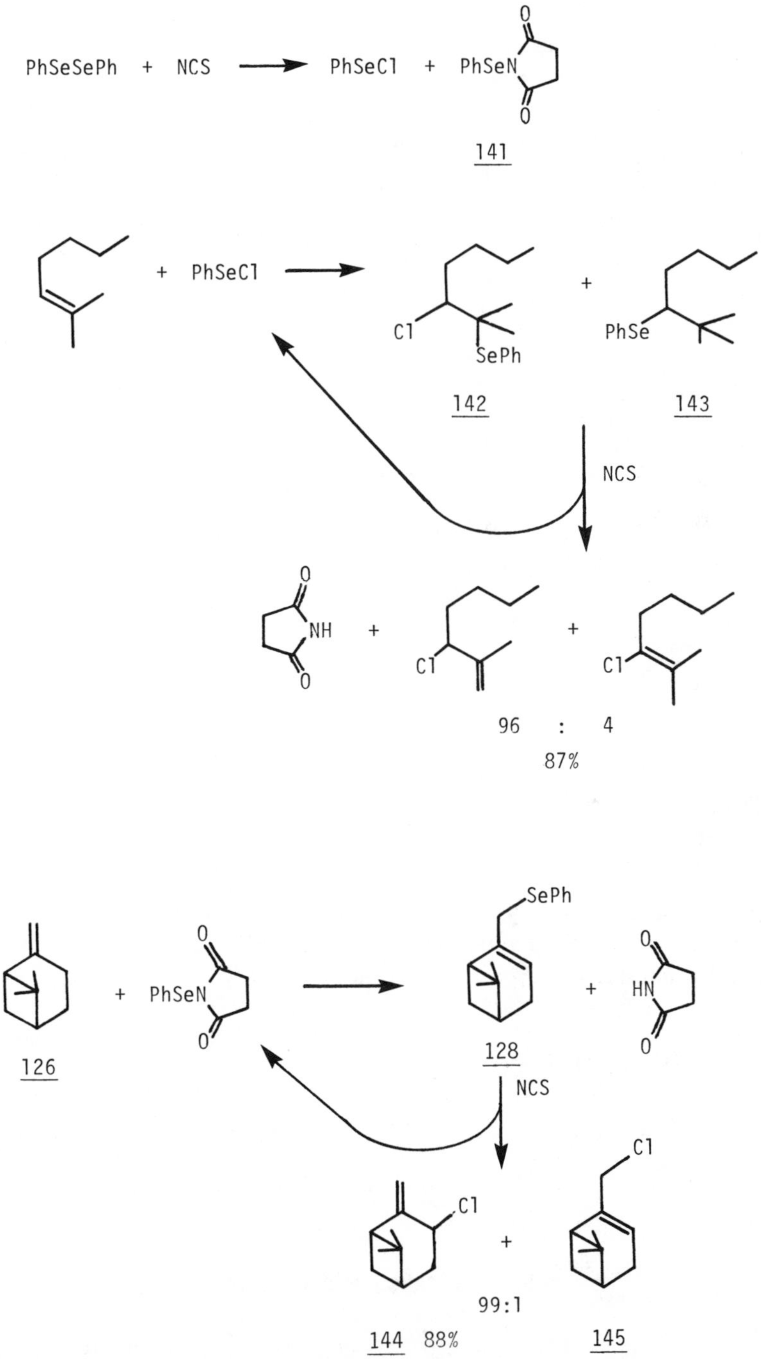
PhSeSePh + NCS
PhSeCl + PhSeN
O
O
141
+ PhSeCl
Cl
SePh
PhSe
142
143
NCS
NH
Cl
Cl
96 : 4
87%
+ PhSeN
126
SePh
HN
128
NCS
Cl
Cl
99:1
144 88%
145

SCHEME 19.

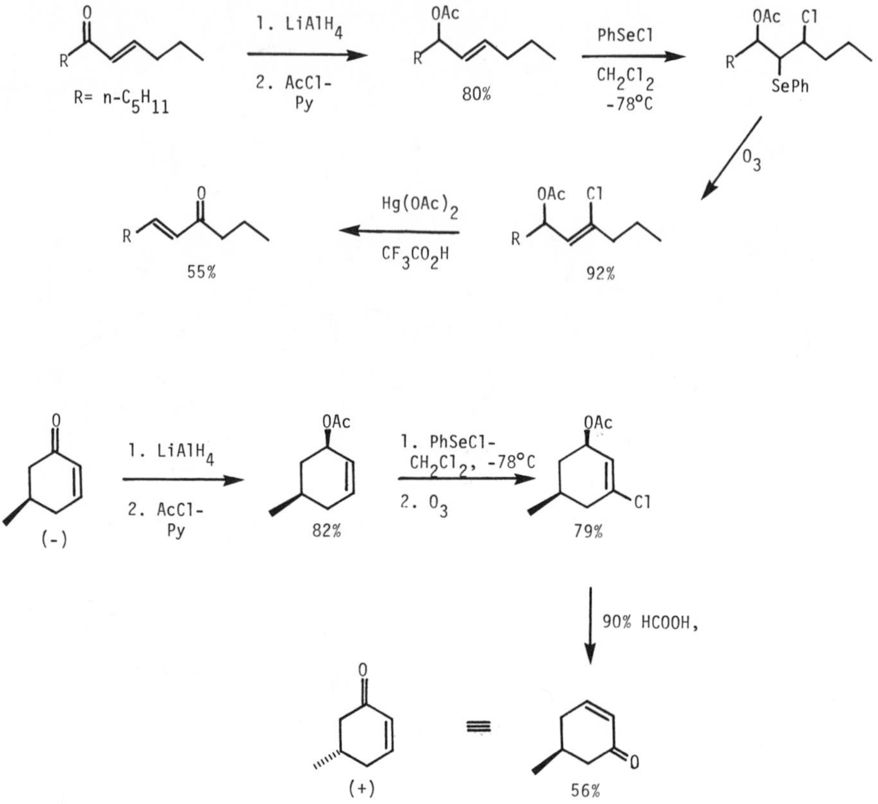

SCHEME 20.

β,γ-Unsaturated carboxylic acids containing a β-substituent decarboxylate when treated with PhSeCl under basic conditions [Eq. (216)].[429] Unsubstituted or γ-substituted analogs cyclize to lactones instead.

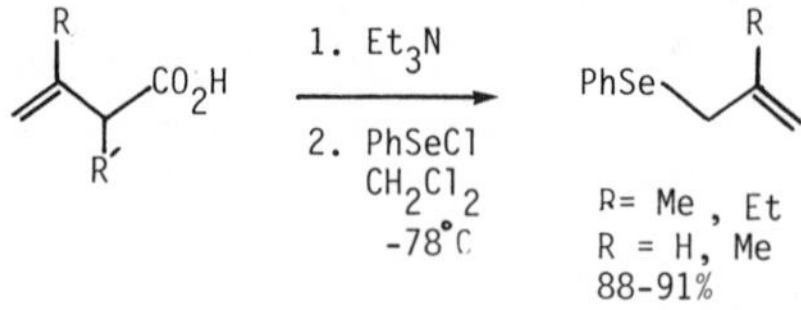

(216)

6.2. β-Oxyselenenylation

This term will be used to denote processes in which the reaction of an olefin with a selenenic electrophile produces a β-hydroxy-, alkoxy-, or acetoxy selenide.

6.2.1. VIA SOLVENT INCORPORATION

When the addition of a selenenyl halide to an olefin is performed in the presence of water, alcohols, or acetic acid, solvent incorporation is generally observed. In principle, this can transpire in either of three ways. First, the selenenyl halide can react directly with the solvent to form a new electrophilic species, which then adds to the olefin. This may be the case in acetoxyselenenylation reactions where the electrophile is the preformed selenenyl acetate, 6. Such processes are discussed separately in the next section. Second, the seleniranium ion formed from the selenenyl halide and olefin may be subjected to preferential attack by the solvent instead of by halide ion. Third, the β-chloro selenide may form initially in the usual manner and undergo subsequent solvolysis to the final product. In this instance the same seleniranium ion is formed as an intermediate due to neighboring group participation by the selenium atom in the solvolysis of the chloride.[430] All three mechanisms are expected to proceed stereospecifically, affording the products of *anti* addition (Scheme 21).

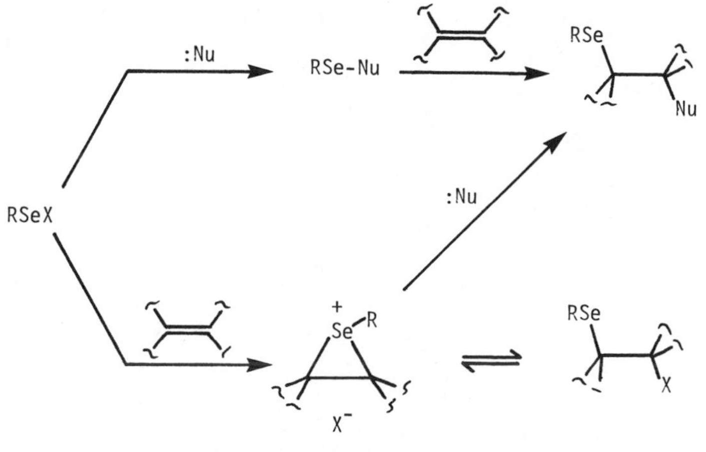

SCHEME 21.

When methanol is employed as the solvent and PhSeCl as the electrophile, the formation of the β-methoxy selenide is faster than the solvolysis of either the selenenyl halide[27,290] or the corresponding β-chloro selenide.[290,430] Thus, product formation arises chiefly through solvolysis of the initially formed seleniranium ion, although the competing formation of the β-chloro selenide may also be observed[290] under kinetically controlled conditions.

β-Alkoxy selenides derived from the alkoxyselenenylation of terminal alkenes are formed with predominantly Markovnikov orientation.[290,431] Internal olefins afford mixtures of regioisomers[290,432] and conjugated dienes form mainly the Markovnikov products of 1,2 addition.[433] Cyclic olefins produce trans adducts exclusively.[12,434] Examples and further transformations of the products to allyl ethers,[12,434] phenylselenomethyl ketones,[431] or vinyl selenides[433] are shown in Eqs. (217)–(219).

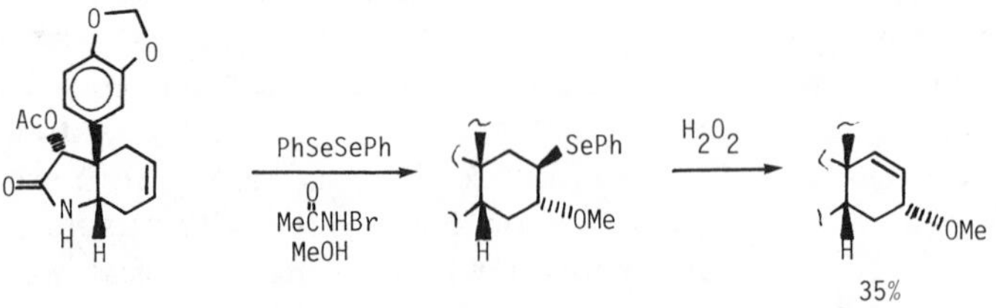

(217) (Ref. 434)

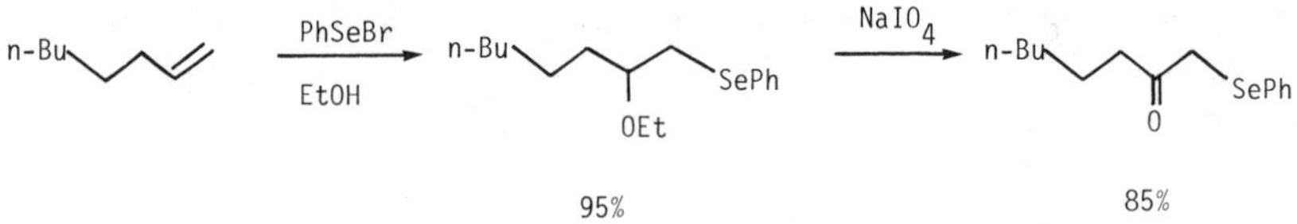

(218) (Ref. 431)

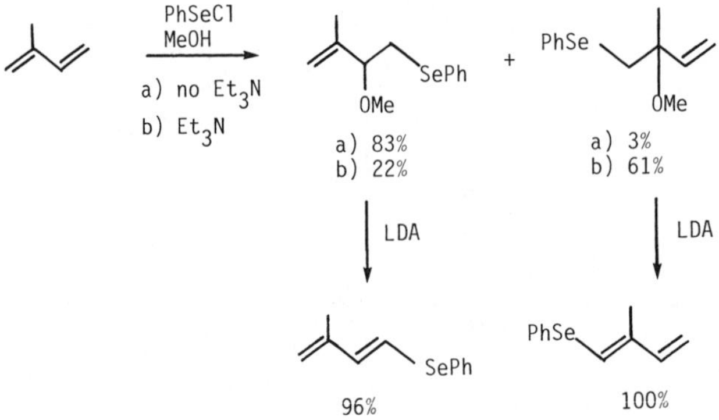

(219) (Ref. 433)

An alternative procedure for the preparation of β-alkoxy selenides involves the treatment of cyclic or acyclic olefins with PhSeCN in the alcohol in the presence of catalytic amounts of Cu(II) or Ni(II) halides [Eq. (220)].[435,436]

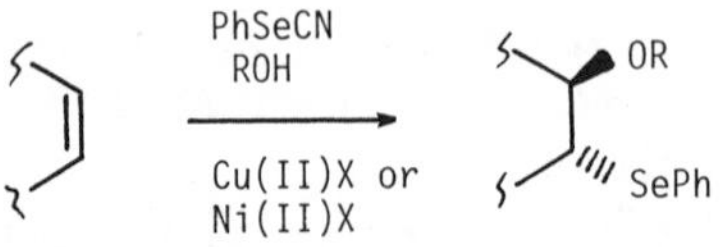

(220)

The preparation of β-hydroxy selenides has been reported from the reaction of olefins with PhSeCl,[229] N-(phenylseleno)phthalimide, 8,[25,437] or the succinimide derivative, 141,[25] in aqueous acetonitrile or methylene chloride, as well as from the hydrolysis of preformed β-halo selenides.[399] Examples are provided in Eqs. (221)–(224).

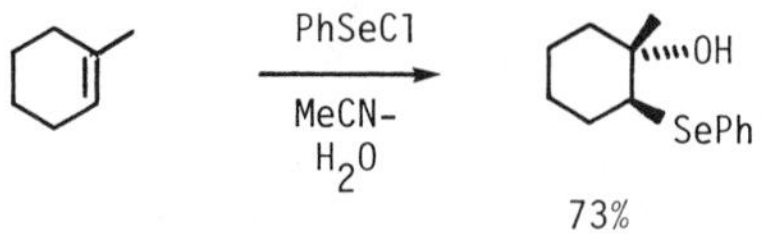

(221) (Ref. 229)

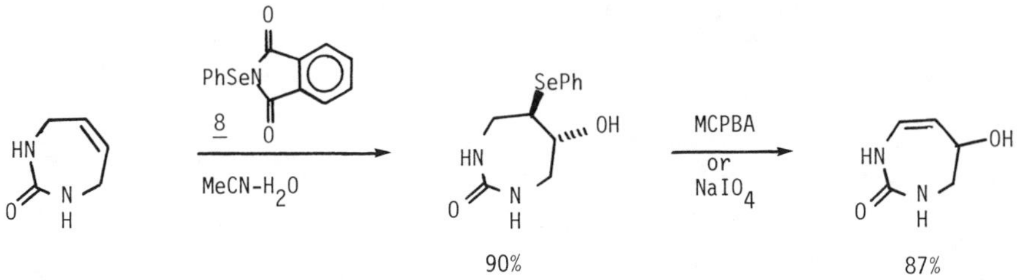

73%

90%   87%

(222) (Ref. 437)

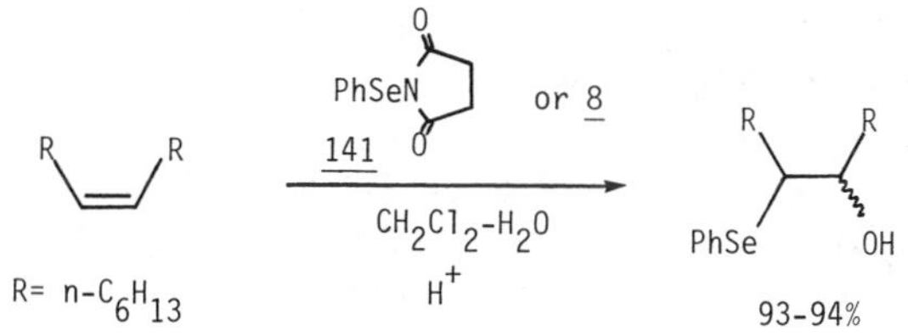

R= n-C$_6$H$_{13}$ 93-94%

(223) (Ref. 25)

R= n-C$_6$H$_{13}$ 68-74%

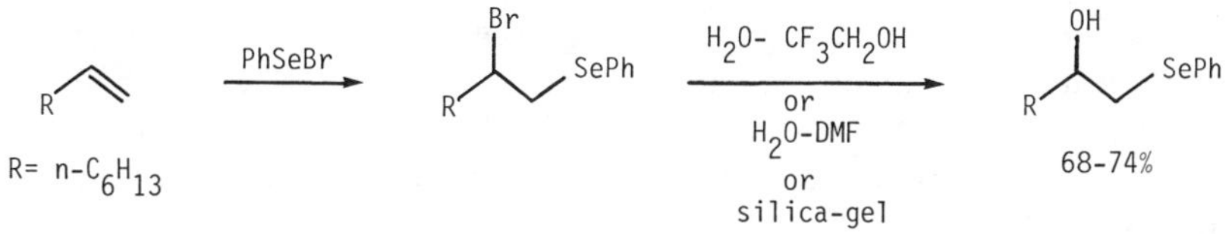

(224) (Ref. 399)

An electrochemical oxyselenenylation technique developed by Torii and co-workers[231,232,438] permits the efficient and highly regioselective (Markovnikov) preparation of β-hydroxy-, alkoxy-, or acetoxy selenides from olefins in the appropriate nucleophilic solvent [Eq. (225)]. As in the electrochemical selenenylation of ketones [see Eq. (24)], the anodic oxidation of PhSeSePh in the presence of a tetraethylammonium halide generates the corresponding selenenyl halide, which then functions as the electrophile. The further electrochemical oxidation of β-hydroxy- or methoxy selenides formed in this manner can be effected *in situ* to produce allyl alcohols or ethers, presumably via selenoxide intermediates.[438] The electrophilic PhSeOH produced in the second step can be efficiently recycled as it reacts readily with more of the starting olefin. This permits the use of only catalytic amounts of the diselenide.[232] Electrochemical oxyselenenylation has been employed in the synthesis of marmelolactone and rose oxide.[438]

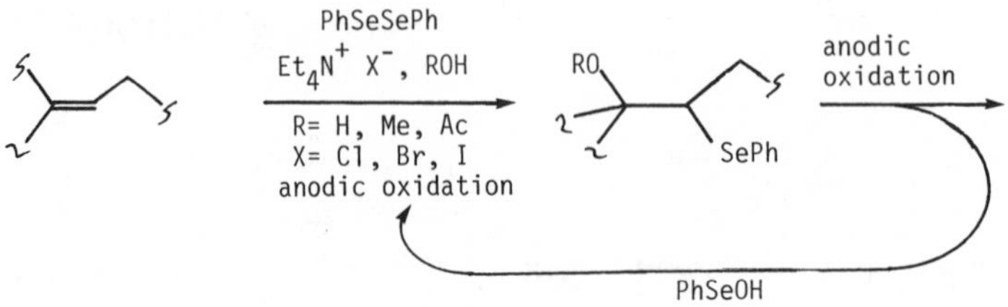

$$(225)$$

6.2.2. VIA SELENENYL ACETATES AND TRIFLUOROACETATES

Selenenyl acetates such as 6, prepared *in situ* from the selenenyl halide and silver acetate (Scheme 1),[12,21] add readily to cyclohexene to afford the *trans*-acetoxy selenides.[12,439] A less expensive alternative that does not require the use of the silver salt was reported by Sharpless and Lauer.[12] The olefin is treated with PhSeBr in acetic acid containing potassium acetate with comparable results [Eq. (226)]. Under these conditions it is less certain whether prior formation of PhSeOAc occurs, or whether the olefin suffers attack by PhSeBr followed by solvolysis. In most situations the question is of little concern, however, as Garratt and Schmid[387] have demonstrated that the same products are observed when (E)- or (Z)-1-phenylpropene react with either 2,4-dinitrobenzeneselenenyl acetate, or with the corresponding selenenyl chloride followed by acetic acid in a separate step. Acetoxyselenenylations of olefins generally proceed with high *anti* stereospecificity but often with poor regioselectivity. As with other β-oxygen-substituted selenides, oxidation and elimination provides the allylic acetates exclusively. Examples and applications are shown in Eqs. (226)–(231).

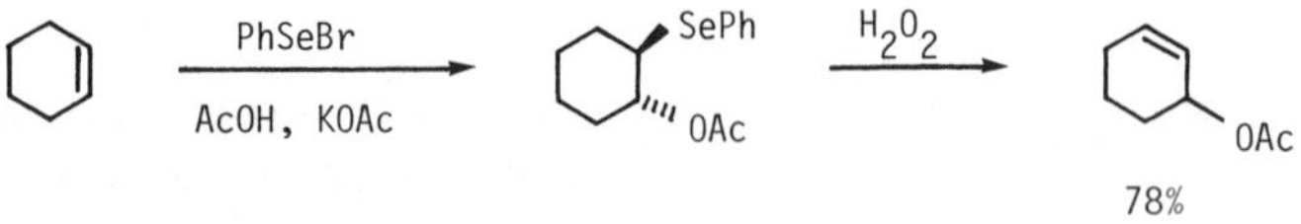

$$(226) \text{ (Ref. 12)}$$

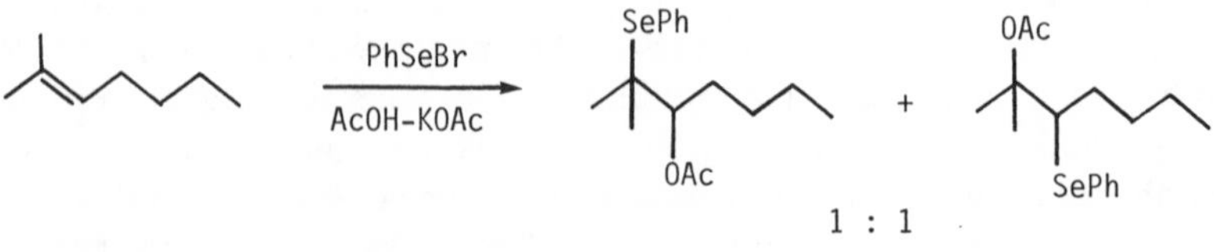

$$(227) \text{ (Ref. 12)}$$

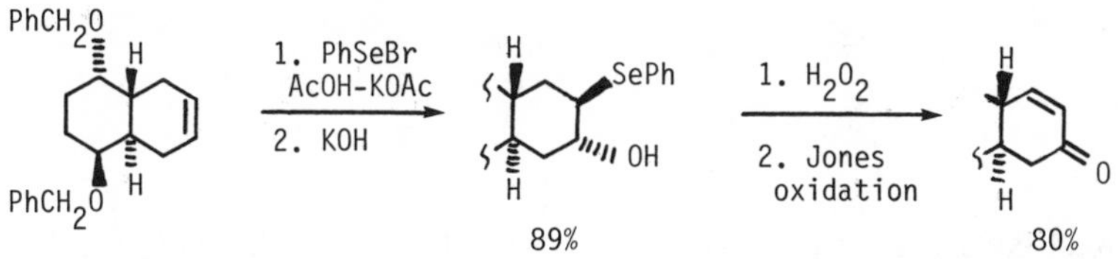

(228) (Refs. 440, 441)

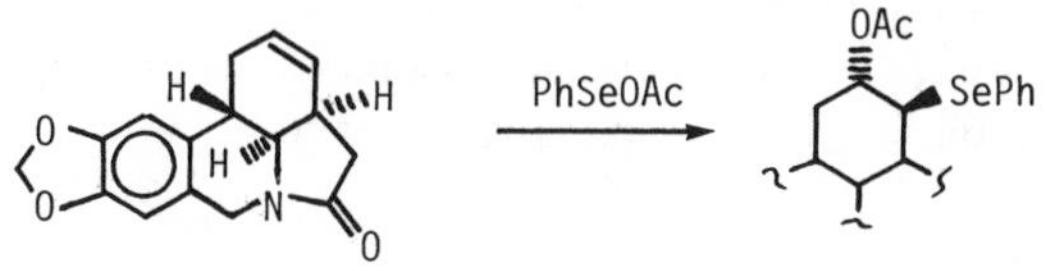

(229) (Ref. 442)

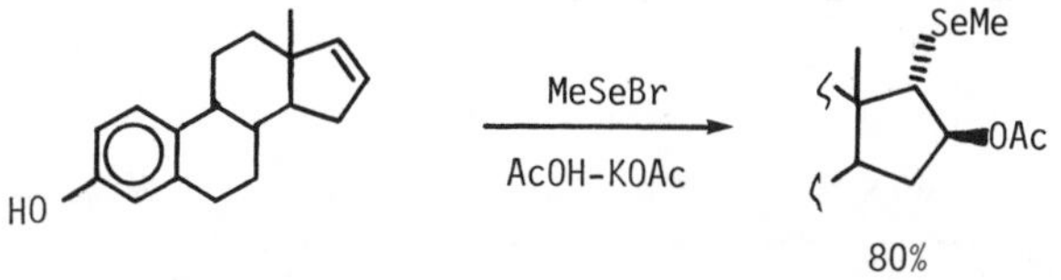

(230) (Ref. 228)

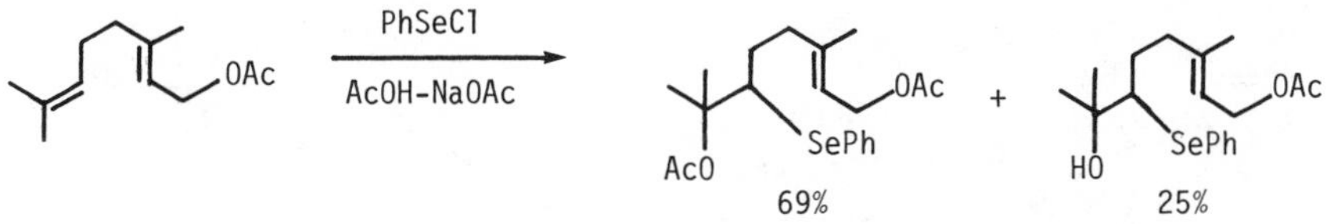

(231) (Refs. 443, 444)

Olefins can be acetoxyselenenylated with acetoxymethyl methyl selenide, 147, and hydrogen peroxide.[445] A plausible mechanism is based on the fragmentation of the selenoxide, 148, to formaldehyde and MeSeOAc, followed by 1,2 addition of the latter to the olefin [Eq. (232)]. Similar results were obtained by heating the olefin with excess dimethyl selenoxide in acetic acid–chloroform [Eq. (233)].[446] No added oxidant was required. The mechanism in the latter instance is unclear, although it is interesting to note that selenide, 147, is the expected Pummerer product of dimethyl selenoxide. Since 147 itself fails to react with olefins in the absence of oxidizing agents capable of converting it to selenoxide, 148,[445] a redox reaction between 147 and dimethyl selenoxide may occur to produce the required 148.

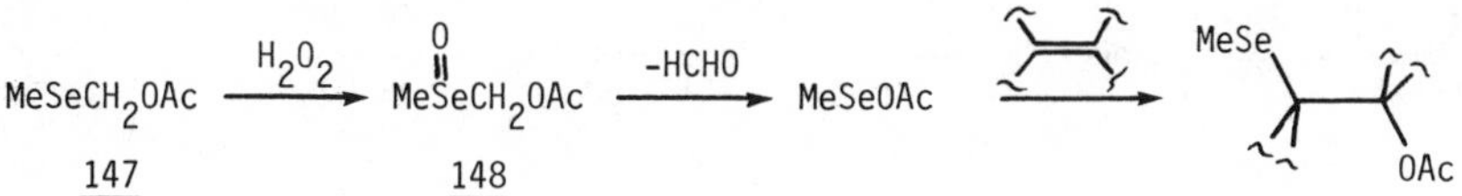

(232)

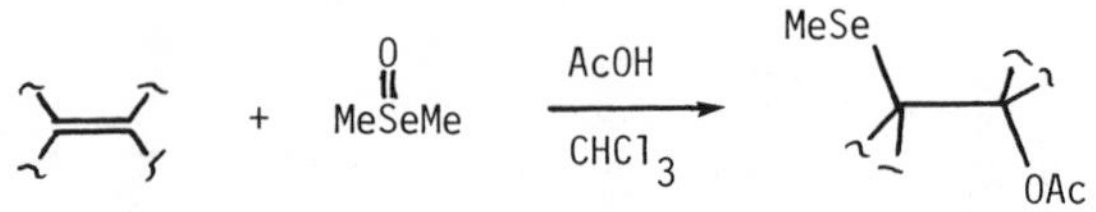

$$\text{(233)}$$

β-Acetoxy selenides are also formed when olefins are treated with benzene-seleninic acid, **3**, in acetic acid,[12,447] when cyclohexene reacts with PhSeNMe$_2$ in acetic anhydride,[29] or when olefins react with PhSeSePh in the presence of oxygen and a catalytic amount of copper (II) acetate, or with the diselenide and lead (IV) acetate (Scheme 22).[448] The selenenyl acetate, **6**, (PhSeOAc) is a probable intermediate in all of these transformations.

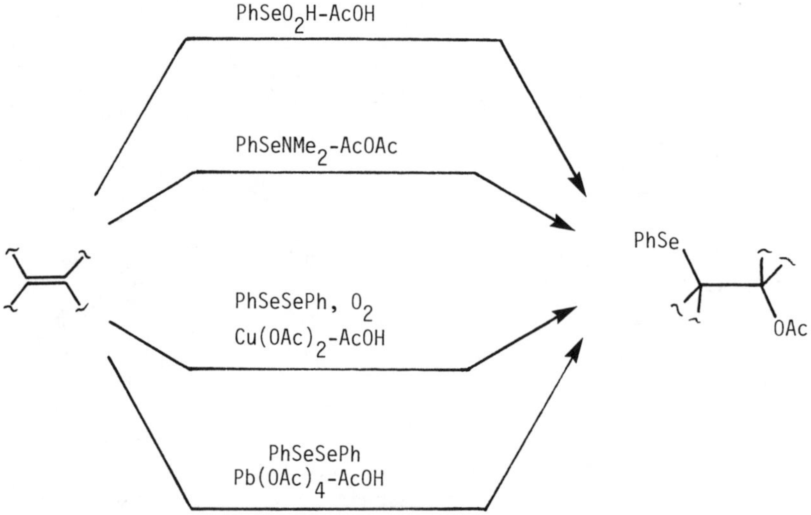

SCHEME 22.

Benzeneselenenyl trifluoroacetate, **7**, prepared according to Scheme 1, has been independently studied by Clive[23] and Reich.[24] It adds readily to olefins with a high degree of *anti* stereospecificity but with poor regioselectivity and so resembles the acetate derivative. The products are easily hydrolyzed to β-hydroxy selenides [Eq. (234)].

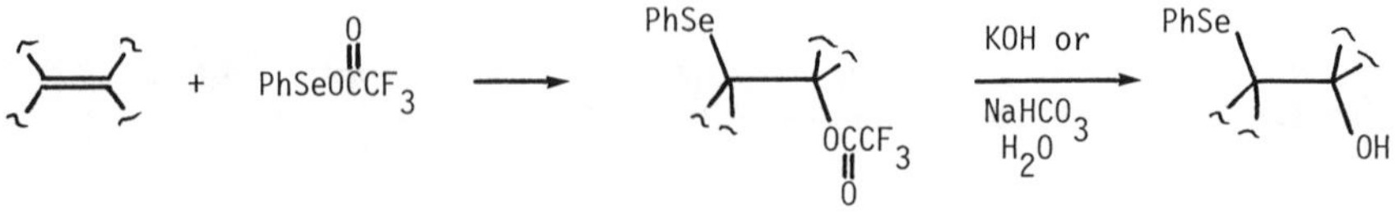

$$\text{(234)}$$

6.2.3. VIA SELENENIC ACIDS AND ANHYDRIDES

Unstable selenenic acids are formed as by-products in selenoxide eliminations [Eq. (5)], as well as from the comproportionation of diselenides with seleninic acids [i.e., the reverse of Eq. (4)], and from the reduction of seleninic acids with various reducing agents. When selenenic acids are generated in the presence of olefins, 1,2 additions occur to give β-hydroxy selenides [Eq. (235)].[f]

$$\text{(235)}$$

Reich and co-workers[18,449] provided an effective demonstration of this facet of selenenic acid behavior by observing the decomposition of selenoxide 149 to 151 and 152. The selenenic acid, 150, is first produced by selenoxide elimination, and reacts with the newly formed double bond to afford the two regio-isomeric products of intramolecular 1,2 addition [Eq. (236)].

$$\text{(236)}$$

Intermolecular additions, as well as redox processes of selenenic acids, are sometimes encountered as unwanted side reactions during the preparation of olefins by selenoxide elimination. Reich and co-workers[11,18] and Hori and Sharpless[19] reported detailed investigations of methods for the suppression of such reactions. Along with appropriate choices of solvent, oxidant, and arene-seleno group, the readdition of ArSeOH can be avoided by its further oxidation to $ArSeO_2H$ or by its removal with amines (see Section 5.2) or activated olefins such as vinyl acetate.[54] 1,4-Diazabicyclo[2.2.2]octane (DABCO) is particularly effective in suppressing both addition and redox processes of selenenic acids during the elimination of vinyl selenoxides to acetylenes and/or allenes.[450] Although electron deficient olefins such as enones do not add selenenic acids readily, House et al.[451] reported the formation of the β-hydroxy selenide, 155, from the selenoxide, 153. This anomalous reaction is believed to proceed via

---

[f] The reader will recall from Section 1 that the active electrophile may actually be another species such as PhSeOSePh instead of the selenenic acid PhSeOH in such processes.

conjugate instead of electrophilic addition to the bridgehead enone, 154, followed by intramolecular selenenylation of the resulting enolate [Eq. (237)].

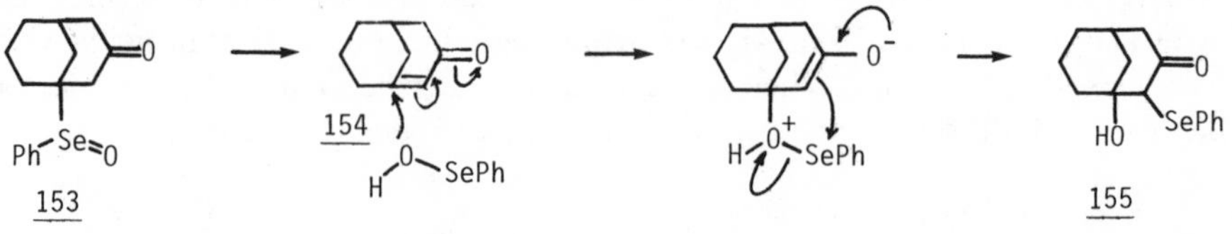

$$(237)$$

The equilibrium between a mixture of a diselenide and a seleninic acid and the corresponding selenenic acid [Eq. (4)] strongly favors the former species. However, in the presence of an olefin, the selenenic acid is continuously removed by 1,2 addition, thus driving the reaction in the direction of comproportionation. Hori and Sharpless[19] prepared several rearranged allyl alcohols from olefins by treating them with a mixture of PhSeSePh and enough hydrogen peroxide to produce the required amount of $PhSeO_2H$. The addition of the resulting selenenic acid (PhSeOH) to the olefins afforded β-hydroxy selenides, which were in turn converted to the desired products by selenoxide elimination [Eq. (238)]. The rates, yields, and regioselectivity (favoring Markovnikov orientation) of the addition step were enhanced by the presence of anhydrous magnesium sulfate to remove excess water.

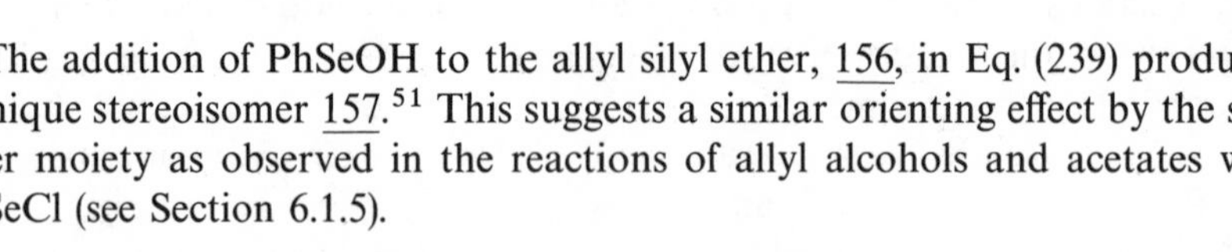

$$(238)$$

The addition of PhSeOH to the allyl silyl ether, 156, in Eq. (239) produced a unique stereoisomer 157.[51] This suggests a similar orienting effect by the silyl ether moiety as observed in the reactions of allyl alcohols and acetates with PhSeCl (see Section 6.1.5).

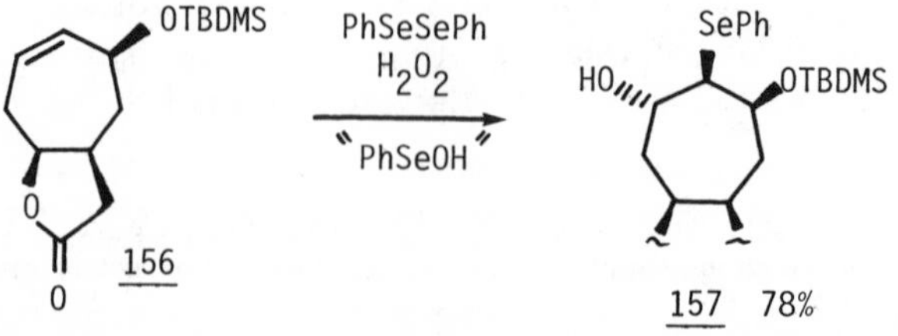

$$(239)$$

Krief and co-workers[452] generated MeSeOH and PhSeOH from the reduction of the corresponding seleninic acids with hypophosphorus acid in the presence of olefins. The resulting 1,2 additions were *anti* stereospecific; trisubstituted olefins gave only Markovnikov adducts while monosubstituted ones produced mixtures of regioisomers. The addition of PhSeOH to cyclohexene was also observed when benzeneseleninic acid was reduced with triphenylphosphine in the presence of the olefin [Eq. (240)].[309]

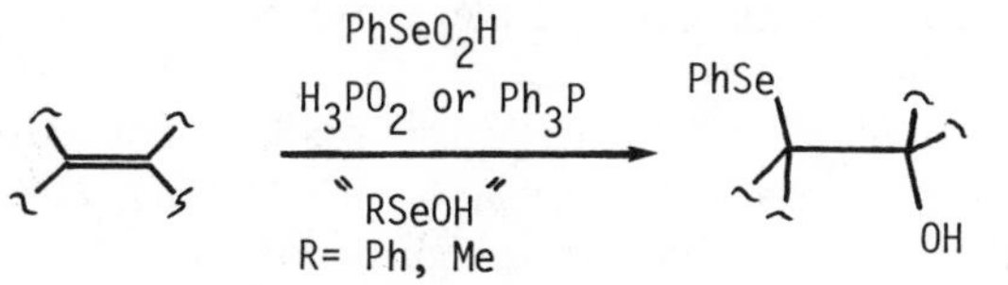

$$
\underset{\text{R= Ph, Me}}{\overset{\overset{\text{PhSeO}_2\text{H}}{\underset{\text{"RSeOH"}}{H_3PO_2 \text{ or } Ph_3P}}}{\longrightarrow}}
\tag{240}
$$

Kuwajima and co-workers[427,453–455] produced an electrophilic species believed to be the selenenic anhydride PhSeOSePh, 4, from the oxidation of PhSeSePh with *t*-butyl hydroperoxide or by the comproportionation of the diselenide with benzeneseleninic anhydride, 16 (Scheme 23). It is relevant to

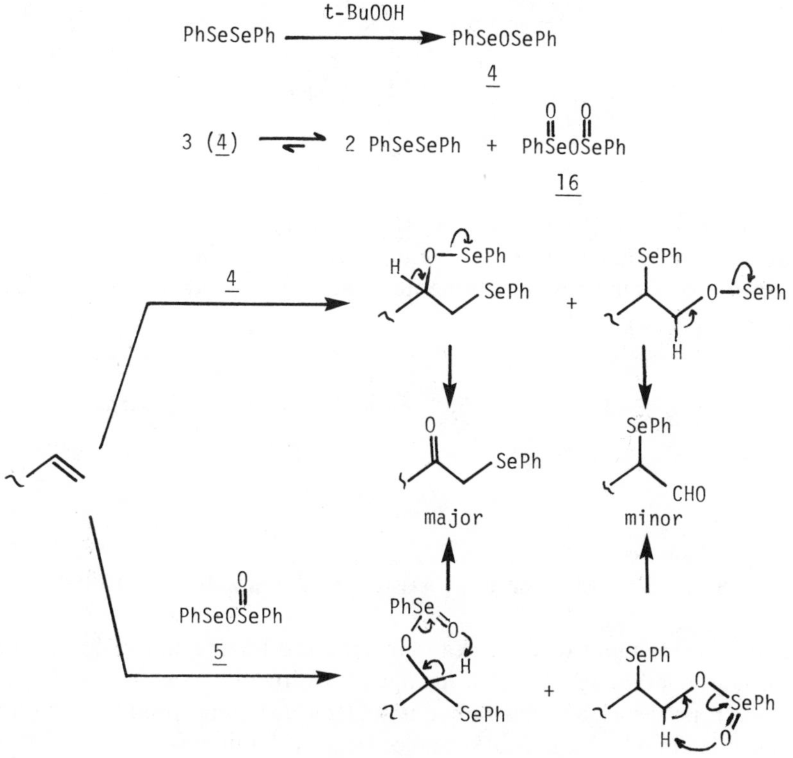

$$\text{PhSeSePh} \xrightarrow{\quad t\text{-BuOOH}\quad} \text{PhSeOSePh}$$
$$\underline{4}$$

$$3\ (\underline{4}) \rightleftharpoons 2\ \text{PhSeSePh} + \overset{\overset{O\ \ \ O}{\|\ \ \ \|}}{\text{PhSeOSePh}}$$
$$\underline{16}$$

SCHEME 23.

note that Gancarz and Kice[20] were unable to detect significant amounts of ArSeOSeAr (Ar = p-FPh) by ^{19}F nmr when bis-(p-fluorophenyl) diselenide was oxidized with t-butyl hydroperoxide. This indicates that, as in the case of PhSeOH, the equilibrium in Scheme 23 strongly favors the disproportionation of $\underline{4}$ and can only be driven in the opposite direction by the continuous removal of the selenenic anhydride. Olefins react with this putative electrophile to produce α-phenylseleno ketones and aldehydes (Scheme 23).[427,453] The formation of carbonyl-containing products would also be consistent with the intermediacy of a mixed selenenic–seleninic anhydride, $\underline{5}$ (Scheme 23) and such a possibility cannot be ruled out.[427]

Mixtures of the ketone and aldehyde are usually obtained from monosubstituted olefins, with the ketone strongly favored in DMSO solutions.[427,453] Allyl ethers and allyl silyl ethers react with high regioselectivity (except in the case of primary allyl derivatives), again probably because of a stabilizing interaction between the allylic oxygen and the selenium atoms [Eq. (241)].[453–455] Allyl acetates, benzoates, and pivaloates are less effective in controlling the regiochemistry.[453,455]

$$(241)$$

The formation of the electrophile, $\underline{158}$, has been proposed in the reaction of hexabutylstannoxane with PhSeBr [Eq. (242)].[427,456] This hypothetical species reacts with olefins in a similar manner to PhSeOSePh, producing α-phenylseleno ketones and aldehydes.

$$Bu_3SnOSnBu_3 \ + \ PhSeBr \ \longrightarrow \ ``Bu_3SnOSePh\text{''} \ + \ Bu_3SnBr \tag{242}$$
$$\underline{158}$$

6.3. β-Amido-, Amino-, Azido-, and Nitroselenenylation

Electrophilic selenium reactions can be employed for the conversion of olefins to selenides containing various β-nitrogen substituents. As with oxyselenenylations, such processes are generally *anti* stereospecific and produce mixtures of regioisomers in which the Markovnikov adduct dominates.

6.3.1. β-AMIDES

Toshimitsu and co-workers[457,458] demonstrated that adducts derived from PhSeCl and mono- or disubstituted olefins react further in aqueous acetonitrile containing a strong acid such as trifluoromethanesulfonic acid. Solvent incorporation of acetonitrile, followed by hydrolysis of the nitrile function affords the corresponding amides (Scheme 24). Benzonitrile, propionitrile, butyronitrile, and ethyl cyanoacetate may be used in place of acetonitrile.[458] Oxidation–elimination of the products provides allylic amides selectively,[458,459] while reduction with triphenyltin hydride affords saturated amides.[458] β-Amido selenoxides derived from cyclohexene, 159, are remarkably stable at room temperature because of intramolecular hydrogen bonding, but eliminate smoothly at 250°C.[458,459]

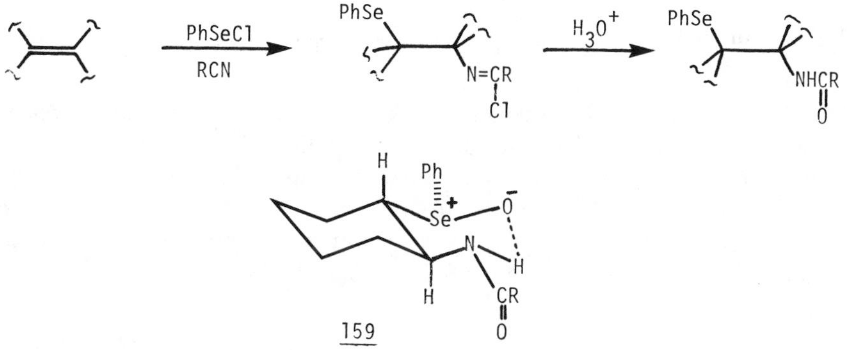

SCHEME 24.

Similar products are obtained from an electrochemical procedure in which a mixture of the olefin and PhSeSePh is subjected to anodic oxidation in acetonitrile [Eq. (243)].[460]

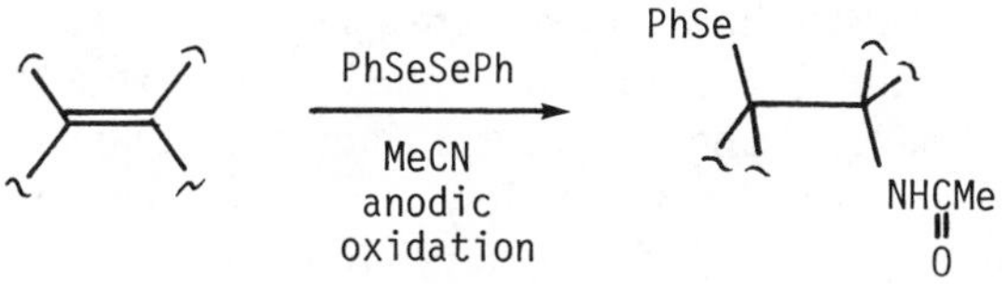

(243)

Barton et al.[461] converted olefins to β-sulfonamido selenides by treatment with PhSeSePh and chloramine T, followed by reductive work-up with sodium borohydride (Scheme 25). The intermediate, 160, is believed to act as the olefin attacking electrophile. (This process is fundamentally different from the Sharpless allylic amination reaction employing 17).[462]

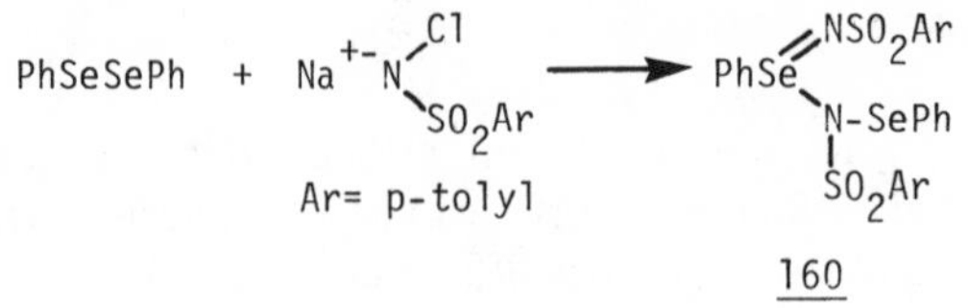
PhSeSePh
+
$Na^{+-}N$
Cl
SO_2Ar
Ar= p-tolyl
PhSe
NSO_2Ar
N-SePh
SO_2Ar
160

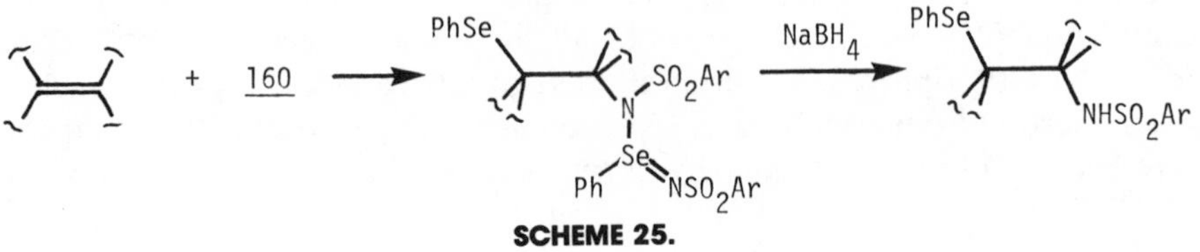
+
160
PhSe
SO_2Ar
N
Se
Ph
NSO_2Ar
$NaBH_4$
PhSe
$NHSO_2Ar$

SCHEME 25.

6.3.2. β-AMINES AND β-AZIDES

Although enones react with selenenamides via Michael additions as in Scheme 4, unactivated olefins are generally inert and the preparation of β-amino selenides by direct 1,2 addition fails. An exceptional intramolecular addition was, however, observed when selenenamide, 161 (prepared by trapping selenenic acid, 150, with diethylamine) was allowed to stand in the presence of excess diethylamine, thus producing adduct 162 in 17% yield [Eq. (244)].[18,449]

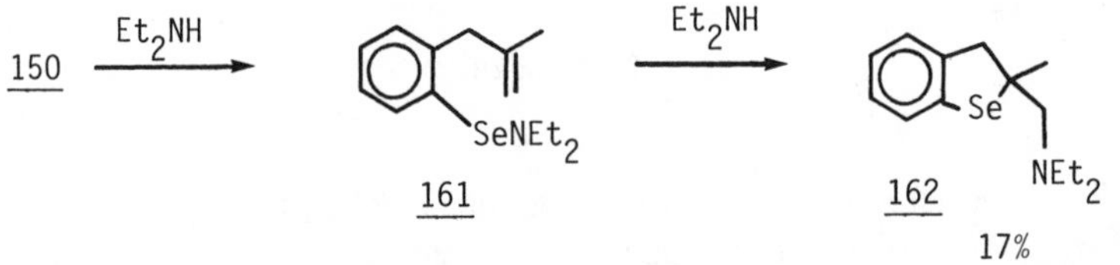
150
Et_2NH
$SeNEt_2$
161
Et_2NH
Se
NEt_2
162
17%

(244)

β-Azido selenides were prepared by Krief et al.[399] from the reaction of olefins with PhSeBr or MeSeBr, followed by sodium azide in trifluoroethanol or lithium azide in DMF (Scheme 26). Oxidation of the cyclohexene adduct produced a 60:40 mixture of the allyl and vinyl azides, indicating that the azido function is less effective than either the acetamido group or various oxygen substituents in directing elimination towards the allylic position. An overall aminoselenenylation was effected by reducing the β-azido selenide with lithium aluminum hydride (Scheme 26).

6.3.3. β-NITRO COMPOUNDS

Olefins furnish β-nitro selenides in modest yields when treated with PhSeBr followed by silver nitrite in methylene chloride–acetonitrile.[463] Yields are en-

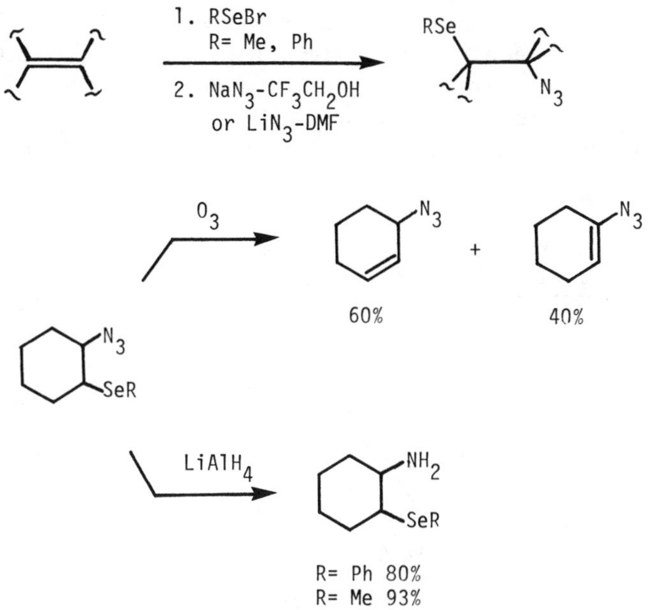

SCHEME 26.

hanced in the presence of mercuric chloride as the competing formation of β-hydroxy selenides is suppressed.[464] The adducts afford nitro olefins in excellent yield upon oxidation–elimination [Eq. (245)].

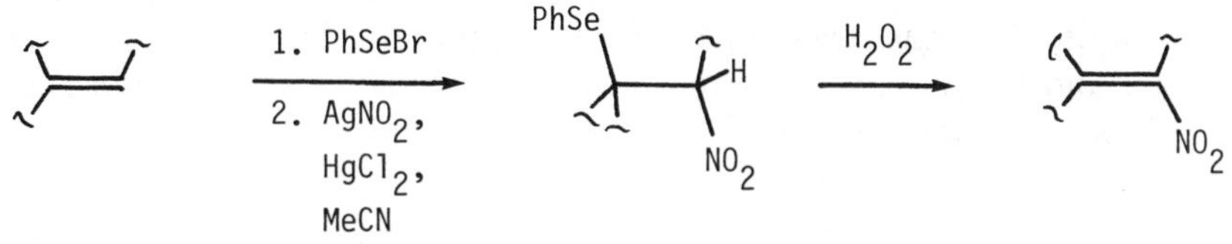

$$(245)$$

6.4. Reactions of Other Selenenic Electrophiles

6.4.1. SELENOSULFONATION

Selenosulfonates, 10 (Ar = Ph or p-MePh) are not sufficiently reactive to add spontaneously to olefins. However, Back and Collins[465,466] discovered that such additions are readily performed with mono- and disubstituted olefins in the presence of boron trifluoride etherate. The reactions are *anti* stereospecific and sulfones rather than sulfinate esters are produced [Eq. (246)]. The regiochemistry is predominantly Markovnikov and as such is opposite and complementary to that observed in free-radical selenosulfonation (see Chapter 7). The cyclohexene adduct, 163, exhibits the same somewhat surprising preference for the diaxial conformation[467] as was previously noted with the PhSeCl adduct, 89. The products of selenosulfonation can be converted to vinyl sulfones in excellent yield via their selenoxides [Eq. (246)]. It is worthy of note that the addition of selenosulfonates to olefins has no parallel in the chemistry of thiosulfonates.

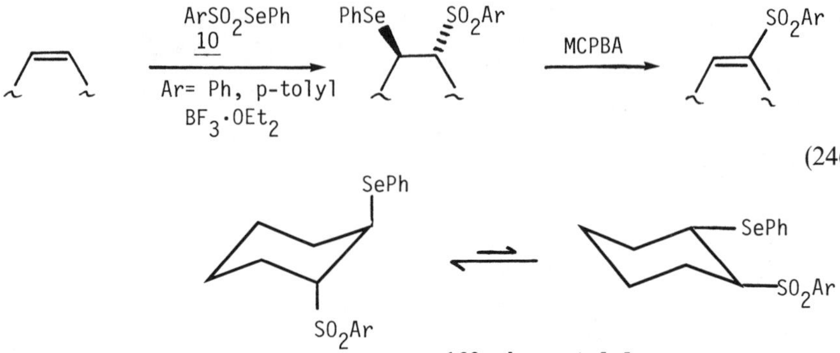

$$(246)$$

163   Ar= p-tolyl

6.4.2.   CYANOSELENENYLATION

As in the case of selenosulfonates, selenocyanate, 12 (PhSeCN), does not react with unactivated olefins in the absence of a Lewis acid catalyst. Tomoda et al.[468] obtained generally high yields of β-cyano selenides in the presence of tin (IV) chloride, aluminum trichloride, or boron trifluoride [Eq. (247)]. The reactions are *anti* stereospecific but the regiochemistry is difficult to predict and depends on the precise nature of the substrate.

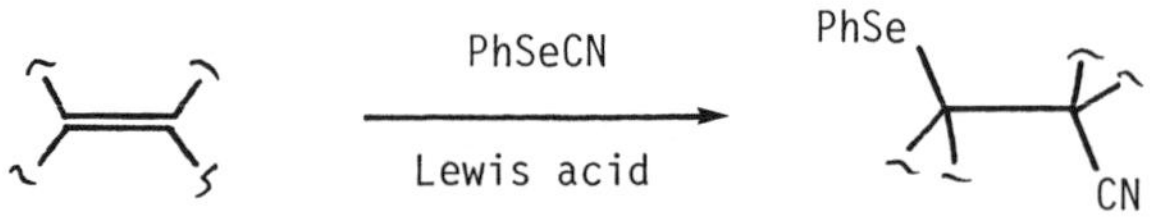

$$(247)$$

The more reactive enamines[469] and ketene acetals[470] undergo uncatalyzed 1,2 additions in ethanol solution without solvent incorporation [Eqs. (248) and (249)]. Both processes are completely regiospecific due to the activating effect of the nitrogen and oxygen substituents, respectively. Although enamines react

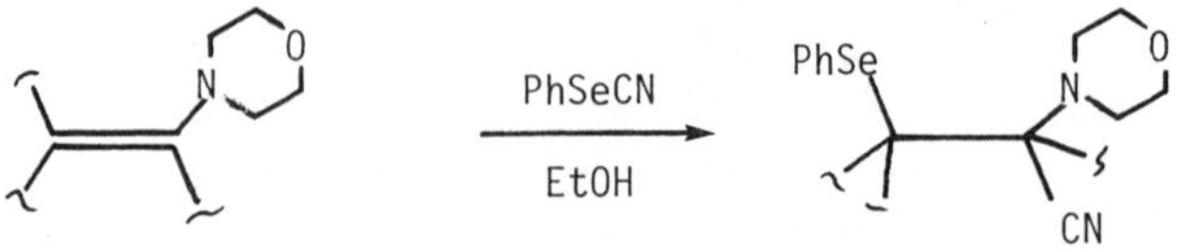

$$(248)$$

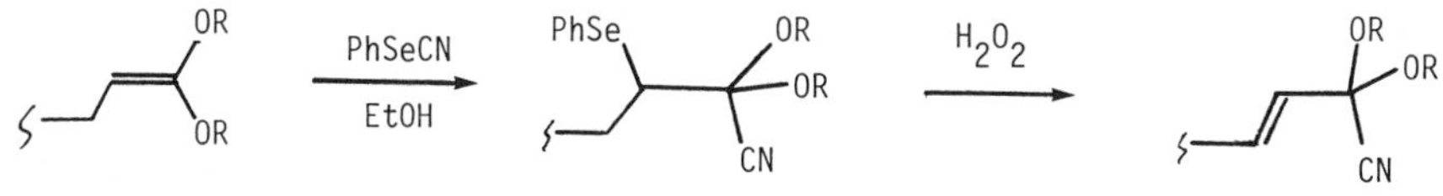

$$(249)$$

anti stereospecifically, ketene acetals do not, suggesting a high degree of carbonium ion character in the intermediates of the latter process. The reader will also recall that olefins are oxyselenenylated with PhSeCN in alcoholic media in the presence of Cu(II) or Ni(II) catalysts [see Eq. (220)].

6.4.3. THIOCYANOSELENENYLATION

Garratt et al.[26] found that selenenyl thiocyanate, 9 (PhSeSCN) adds rapidly to olefins without catalysis, producing β-thio- and/or isothiocyanates (Scheme 27). Under kinetic conditions in methylene chloride, monosubstituted olefins provide mainly anti-Markovnikov thiocyanates while 1,1-disubstituted olefins form all four possible products with the Markovnikov thiocyanate dominant. 1,2-Disubstituted olefins produce thiocyanates that readily isomerize to isothiocyanates at room temperature. More heavily substituted substrates afford chiefly isothiocyanates with Markovnikov orientation favored in the case of trisubstituted derivatives.

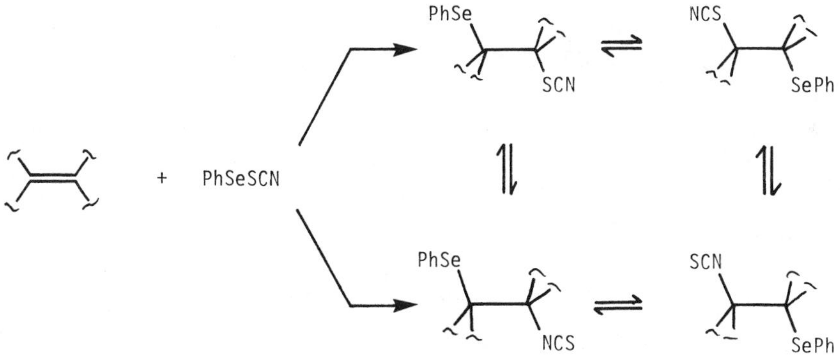

SCHEME 27.

The stereochemistry of addition is less predictable than with other selenenic electrophiles. Norbornene, 100, produces the trans adduct, 164, exclusively while diene, 96b, forms a 2:1 mixture of the trans and cis adducts, 165 and 166, respectively.[26] The reaction of PhSeSCN with (E)- and (Z)-1-phenylpropene is especially surprising as both olefins are reported to give the products of stereospecific *syn* addition.[471] Furthermore, the (E) isomer provides the isothiocyanate, 167, while its (Z) counterpart produces the thiocyanate, 168!

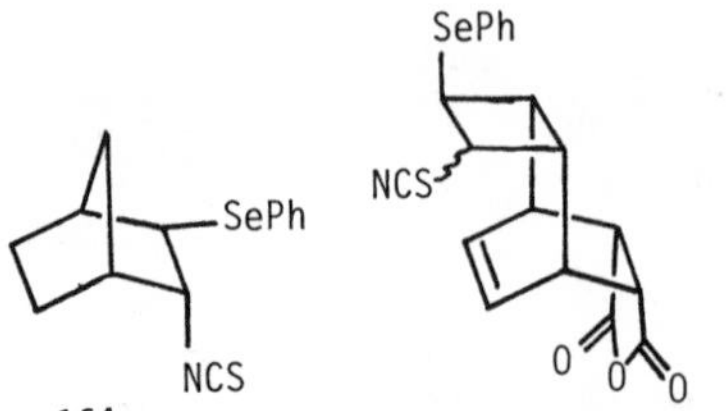

164   165 endo SCN
       166 exo SCN   167   168

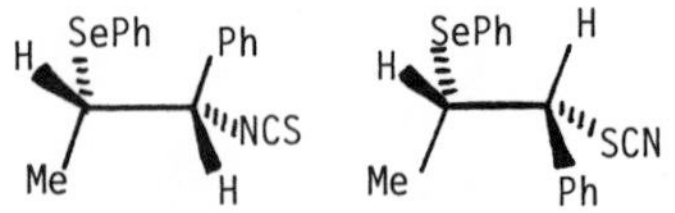

A study of the reaction rates of ring substituted styrenes revealed a greater sensitivity to the nature of the substituent in the additions of PhSeSCN ($\rho^+ = -3.78$)[27] than in the corresponding reactions with PhSeCl. This suggests that the intermediates in the additions of PhSeSCN have more carbonium ion character in the rate limiting transition state. Clearly the capricious chemo-, regio-, and stereoselectivity of these processes severely curtails their synthetic appeal.

6.4.4. REACTIONS OF SELENIUM MONOCHLORIDE

Selenium monochloride, 1, adds to excess ethylene with loss of selenium to afford the selenium analog, 169, of mustard gas [Eq. (250)].[472] When 1 is in excess, further chlorination of the product occurs to form 170 [Eq. (251)]. Propylene produces a mixture of regioisomers [Eq. (252)] while dienes undergo cyclization.[9] 1,1-Diarylethylenes are converted to fused benzoselenophenes, 172, probably via an initial 1,2 adduct, 171 [Eq. (253)].[473] Vicinal[474,475] and β-bisoximes,[476] as well as β-bishydrazones[477] form heterocycles as in the example in Eq. (254).[474]

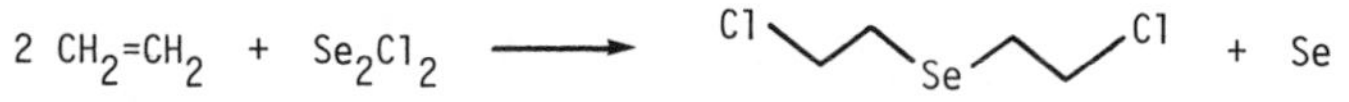

$$2\ CH_2\text{=}CH_2\ +\ Se_2Cl_2\ \longrightarrow\ \underset{169}{Cl\diagdown\diagup_{Se}\diagdown\diagup Cl}\ +\ Se \tag{250}$$

$$2\ CH_2\text{=}CH_2\ +\ 2\ Se_2Cl_2\ \longrightarrow\ \underset{170}{Cl\diagdown\diagup\underset{\underset{Cl}{|}}{\overset{\overset{Cl}{|}}{Se}}\diagdown\diagup Cl}\ +\ 3\ Se \tag{251}$$

$$2\ \diagup\!\!\diagdown\ +\ Se_2Cl_2\ \longrightarrow\ Cl\diagdown\diagup\!\!\diagdown_{Se}\diagup R\ +\ Cl\diagdown\diagup\!\!\diagdown_{Se}\diagup R\ +\ Se$$

$$1\ :\ 1$$

$$R=\ \diagup\!\!\!\!\diagdown\!\!-Cl\quad or\quad \diagup\!\!\!\!\diagdown\!\!-Cl \tag{252}$$

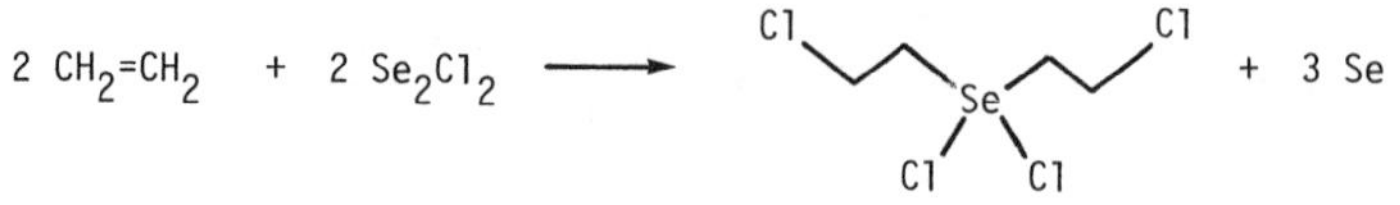

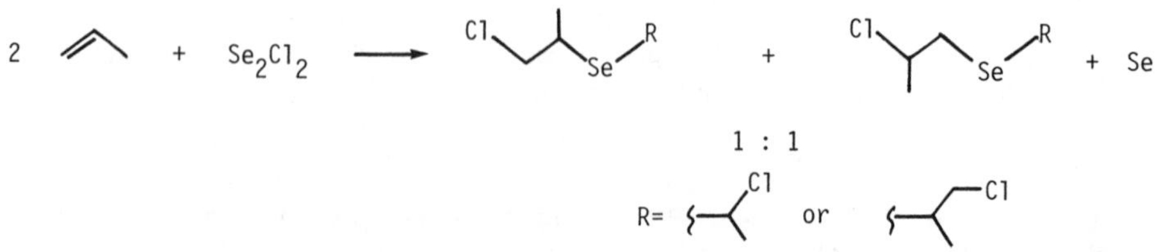

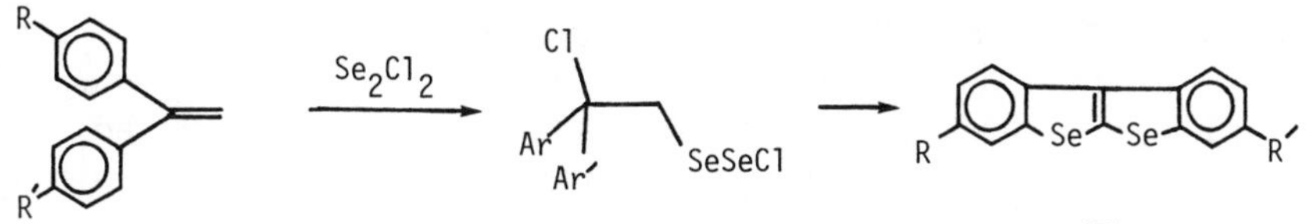

(253)

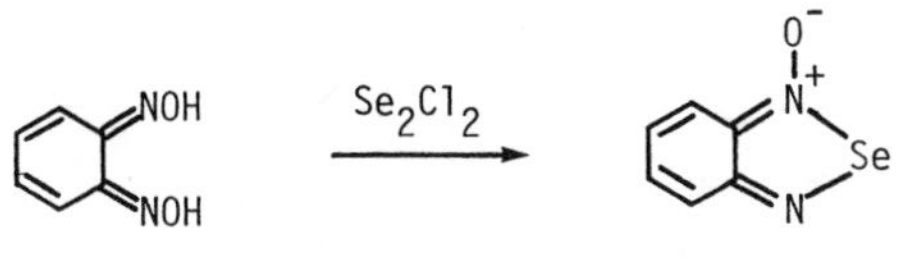

66%

(254)

7. ADDITIONS OF SELENENIC ELECTROPHILES TO ENOL ETHERS

Dihydropyran reacts with PhSeCl in methylene chloride to afford a mixture of cis and trans adducts (Scheme 28).[478] As in the case of the ketene acetals in Eq. (249), the lack of stereospecificity is attributed to carbonium ion character in the intermediate. In the presence of triethylamine and either water[479] or methanol,[480] dihydropyran gives oxyselenenylation products (Scheme 28).

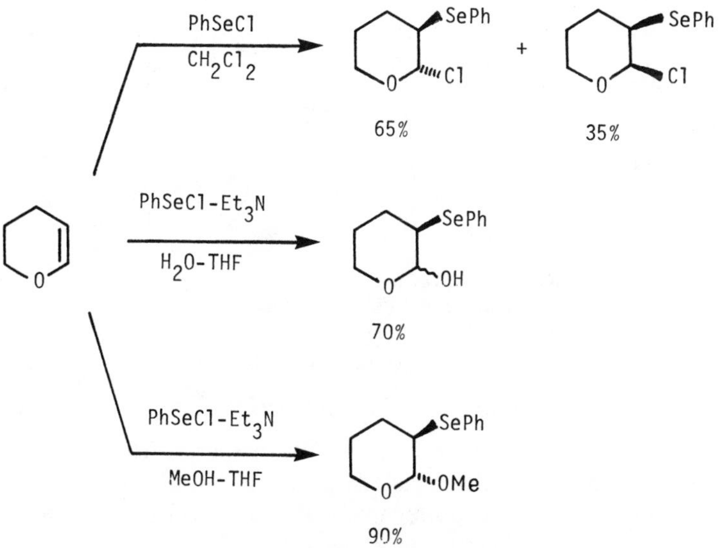

SCHEME 28.

These, and similar products derived from substituted dihydropyrans and alcohols other than methanol (e.g., Scheme 29), act as latent aldehydes, which can be further elaborated by means of Wittig reactions. Such compounds have served as synthons for the synthesis of pseudomonic acids.[479-481] Another application involves the α-glycosylation of carbohydrate alcohols[482] with 3,4,6-tri-O-benzyl-D-glucal and PhSeCl to provide 2′-deoxy-disaccharides after deselenization (e.g., Scheme 29).[481-482]

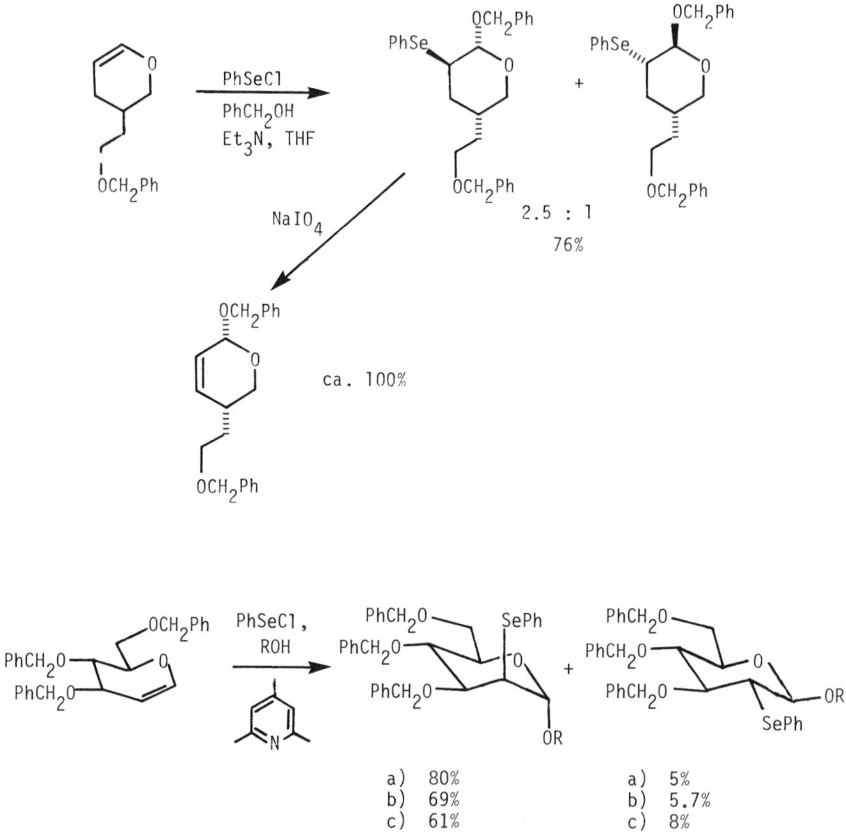

a)   ROH=   1,2:3,4-diisopropylidene-α-D-galactopyranose
b)   ROH=   1,2:5,6-diisopropylidene-α-D-glucofuranose
c)   ROH=   benzyl 2-acetamido-3,6-di-O-benzyl-2-deoxy-α-D-glucopyranoside

SCHEME 29.

α-Phenylseleno aldehydes are conveniently prepared from enol ethers by hydrolysis of their alkoxyselenenylation products. For instance, α-(phenylseleno)acetaldehyde is difficult to prepare by the direct selenenylation of acetaldehyde, but is obtained quantitatively via Eq. (255).[483,484] Related examples are provided in Eqs. (256)–(260).

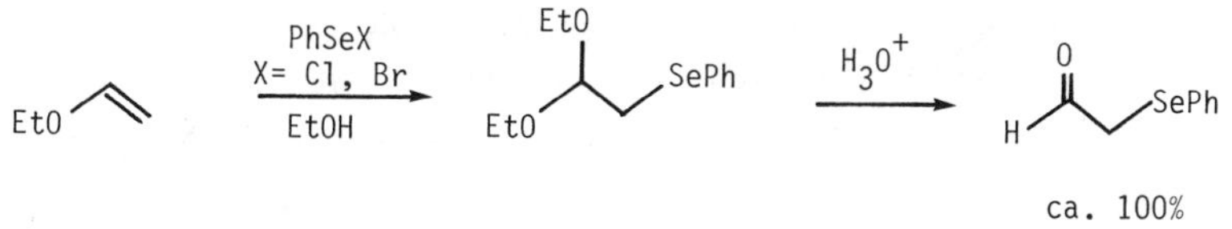

(255) (Refs. 483, 484)

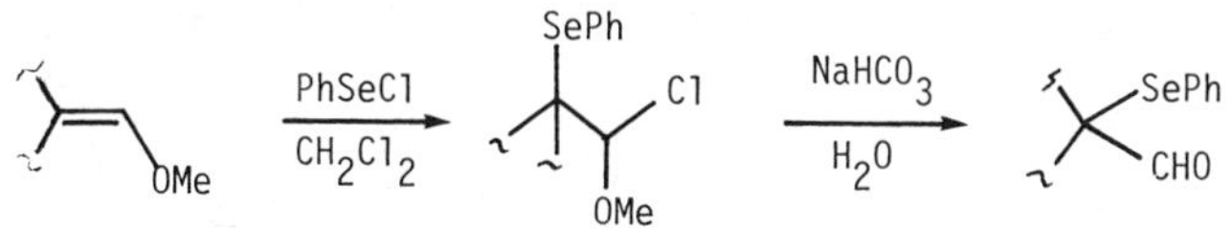

(256) (Ref. 485)

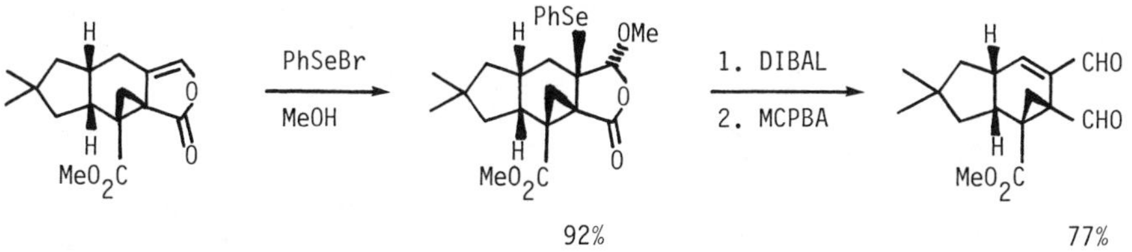

92%　　　　　　　　　77%

(257) (Ref. 486)

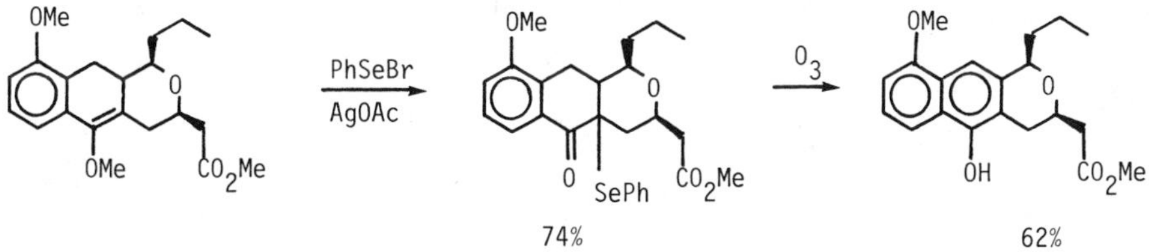

74%　　　　　　　　62%

(258) (Ref. 487)

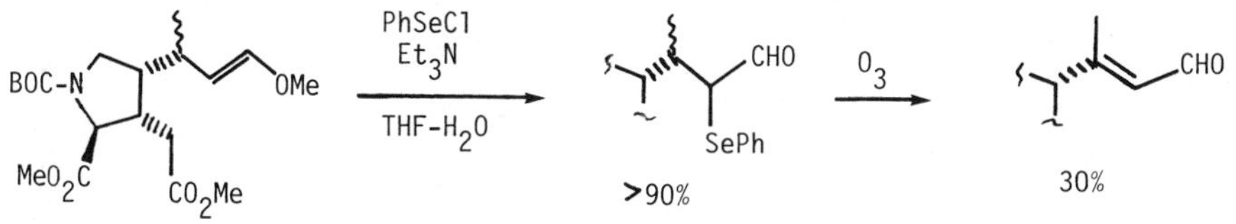

>90%　　　　　　　30%

(259) (Ref. 203)

73%

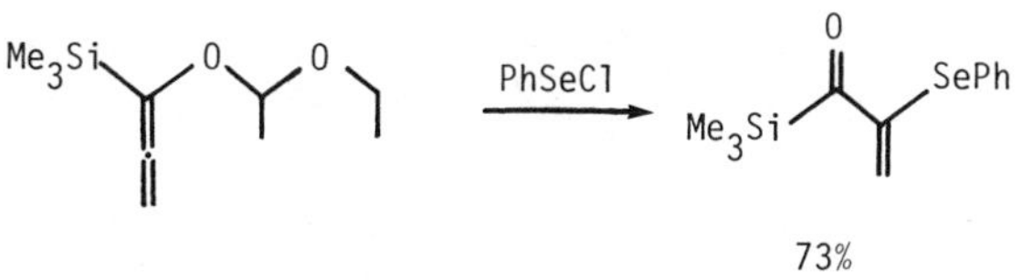

(260) (Ref. 287)

The alkoxyselenenylation of enol ethers with allylic alcohols provides ketene acetals, 173, after selenoxide elimination. Petrzilka[488,490] and Pitteloud and Petrzilka[489] effected Claisen rearrangements of such compounds to produce

γ,δ-unsaturated esters [Eq. (261)]. Applications of this elegant sequence include the synthesis of the 10 membered macrolide phoracantholide J (*cis*-174) [Eq. (262)][491] and the preparation of 175, an intermediate in guaianolide synthesis [Eq. (263)].[492]

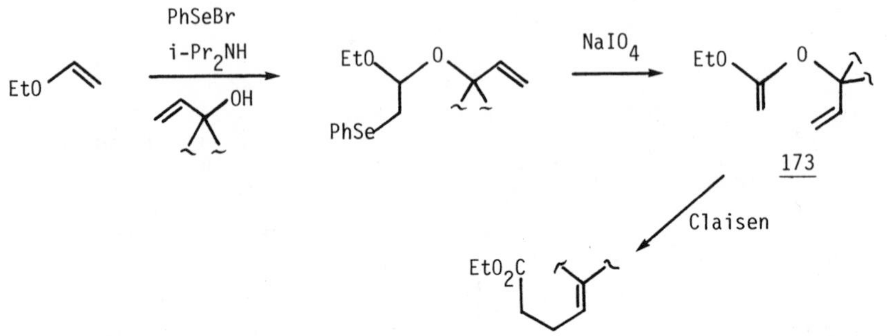

(261)

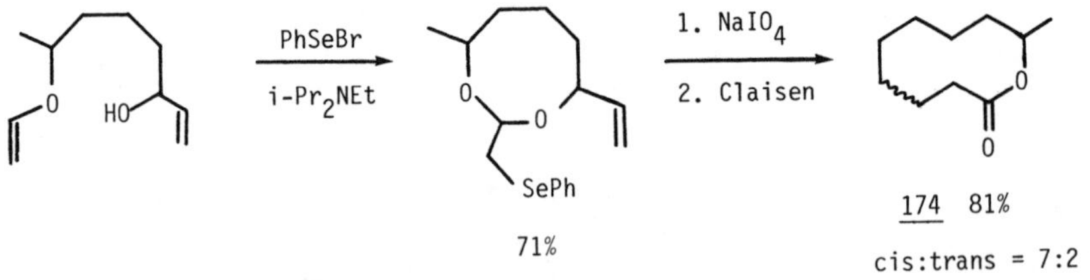

(262)

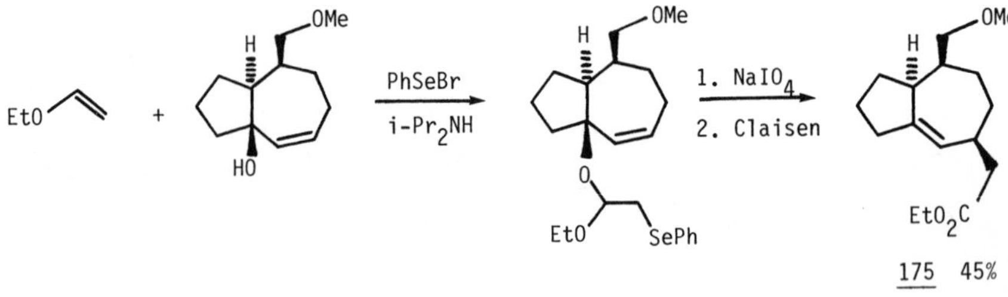

(263)

Ketals can be α-selenenylated by PhSeCl.[485,493] Presumably enol ethers are formed as intermediates and react further by intramolecular alkoxyselenenylation [e.g., Eqs. (264) and (265)].

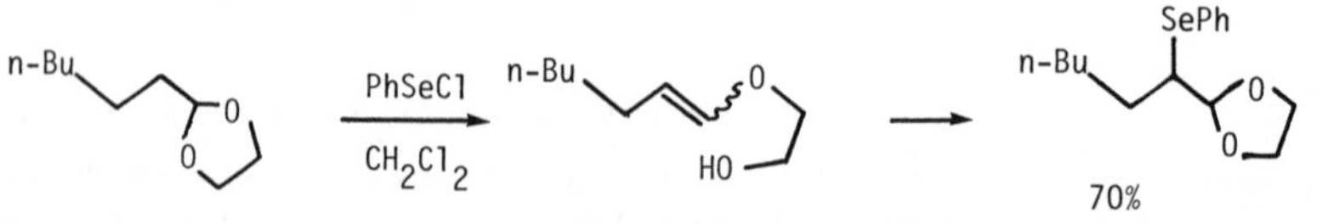

(264) (Ref. 485)

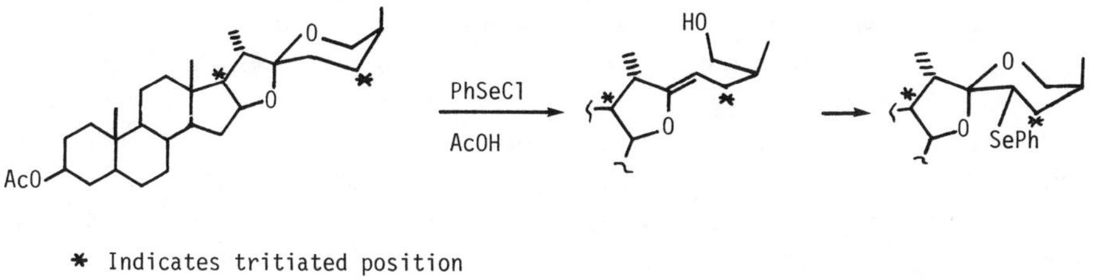

* Indicates tritiated position

(265) (Ref. 493)

8. ADDITIONS OF SELENENIC ELECTROPHILES TO ACETYLENES

The 1,2 additions of selenenic electrophiles to acetylenes have been less frequently studied than those to olefins. A brief review of the subject by Schmid[494] appeared in 1978. Early work by Kataeva et al.[495,496] and Kataev et al.[497] indicated that the stereochemistry of addition varies with the solvent. Surprisingly, *syn* additions of selenenyl halides to several acetylenes were observed in ethyl acetate while *anti* additions occurred in acetic acid or DMF [Eq. (266)].

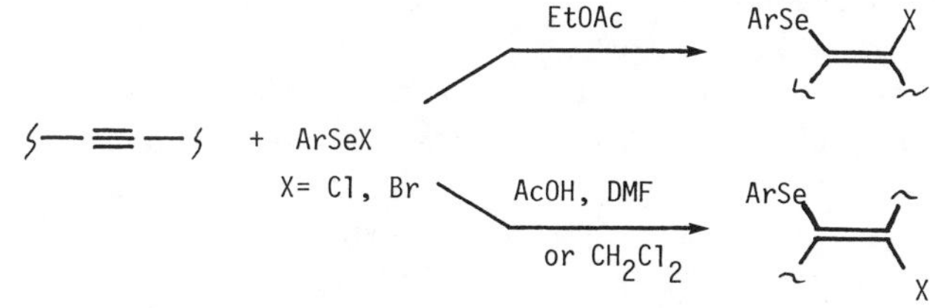

(266)

Unusual kinetic results were also obtained with PhSeCl and several acetylenes in dioxane, THF, and ethyl acetate.[498] The reactions were third order overall, first order in alkyne, and second order in PhSeCl. Subsequently, Schmid and Garratt[499] reported that the additions of PhSeCl to a series of acetylenes in methylene chloride are first order in each reactant and proceed exclusively by *anti* addition. Acetylenes react more slowly than olefins and the rates are increased by all of several alkyl substituents studied except for the *t*-butyl group. Polar effects thus outweigh steric effects in all but the most hindered substrates.

Selenirenium ions, 176, have been proposed as intermediates in such additions[395,499] and one such species (where R = Me and $X^- = SbF_6^-$) has been characterized spectroscopically.[395] It has been suggested that selenirenium ion formation is preceded by that of other intermediates containing tetracoordinated selenium.[499]

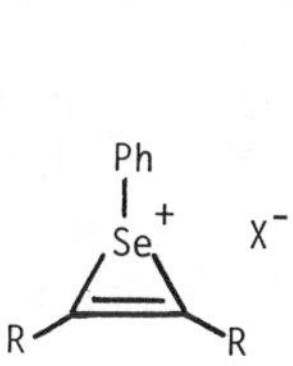

176

Regioselectivity is often poor with unsymmetrical acetylenes, although anti-Markovnikov products tend to dominate.[499] Propiolic acid derivatives and acetylenic ketones furnish adducts in which the arylseleno group is incorporated α to the carbonyl group.[500] The additions of PhSeCl to acetylenic alchohols have been independently studied by Filer et al.[501] and Garratt et al.[502] The former group reported that acetylenes such as 177 produce chiefly the anti-Markovnikov adducts, 178, in acetic acid at 24°C or in methylene chloride at −78°C [Eq. (267)]. The products were stable towards isomerization even at elevated temperatures.

$$R-\!\!\equiv\!\!-(CH_2)_nOH \;+\; PhSeCl \quad \xrightarrow[\substack{(-78°C)}]{\substack{AcOH\ (24°C) \\ or\ CH_2Cl_2}} \quad \underset{178}{\overset{R\diagup\diagdown SePh}{Cl\diagup\diagdown(CH_2)_nOH}} \quad (267)$$

177

R= H, Me; n= 2-4

In contrast to these results, Garratt et al.[502] found that propargyl alcohols, 179, produced anti-Markovnikov adducts, 180, in methylene chloride at ambient temperature, which then isomerized to their Markovnikov counterparts, 181 [Eq. (268)]. In some cases Markovnikov adducts were the only detectable products. These authors suggested that anti-Markovnikov orientation occurs under conditions of kinetic control due to the steric hindrance from the geminal substituents R' and R'' towards attack by the chloride ion. Subsequent isomerization produces the thermodynamically favored Markovnikov regioisomers. When this process takes place at a rate comparable to or faster than the addition, only the latter products are observed. The possibility of a stabilizing interaction between the hydroxyl group and the selenium atom in the formation of the selenirenium ion intermediate has been adumbrated,[502,503] but its precise role in determining the regiochemical outcome of such processes remains unclear. Cyclization was not observed by either group.

(268)

Electrochemical oxyselenenylation of 3-hydroxyalkynes is reminiscent of that of olefins [cf. Eq. (225)] and proceeds with the regiochemistry shown in Eq. (269). Concomitant dehydration affords α-arylseleno-α,β-unsaturated aldehydes or ketones.[503] Methyl ethers behave similarly to the free alcohols, but acetates give poorer yields.

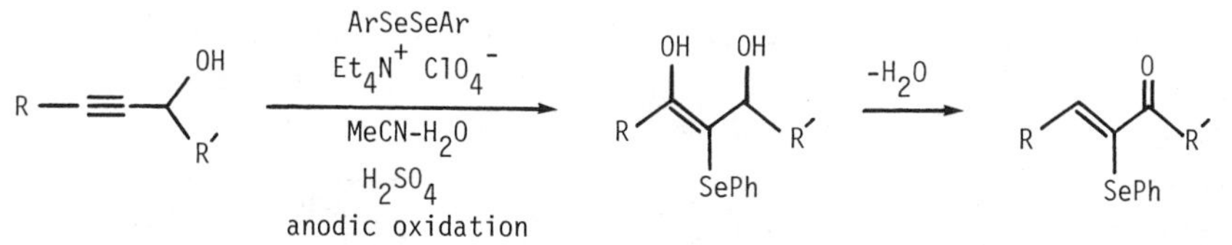

$$\text{(269)}$$

The adducts obtained from PhSeCl or PhSeBr and 1,4-dichloro-2-butyne are precursors of substituted dienes,[504] which in turn are synthetically useful in Diels–Alder reactions and related processes [Eq. (270)].[505]

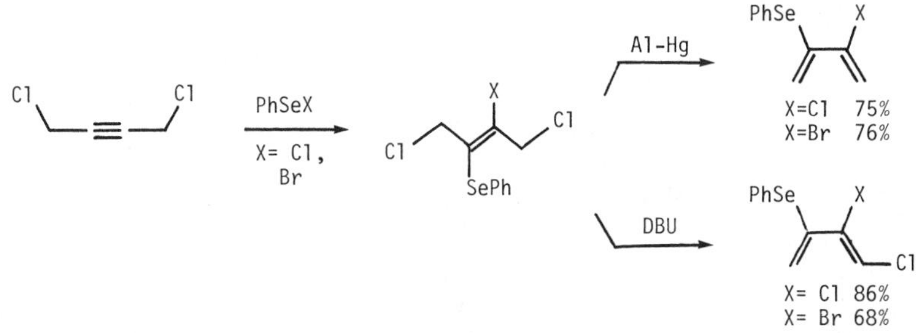

$$\text{(270)}$$

Lithium trialkylalkynylborates, <u>182</u>, react with PhSeCl with migration of an alkyl group from boron to carbon to afford compounds <u>183</u>. These undergo acid catalyzed hydrolysis to vinyl selenides,[506] or oxidative cleavage of the B—C bond to product α-phenylseleno ketones.[507] Since the latter compounds can be converted to enones via their selenoxides, the procedure permits the overall transformation of acetylenes to acyclic enones (Scheme 30).

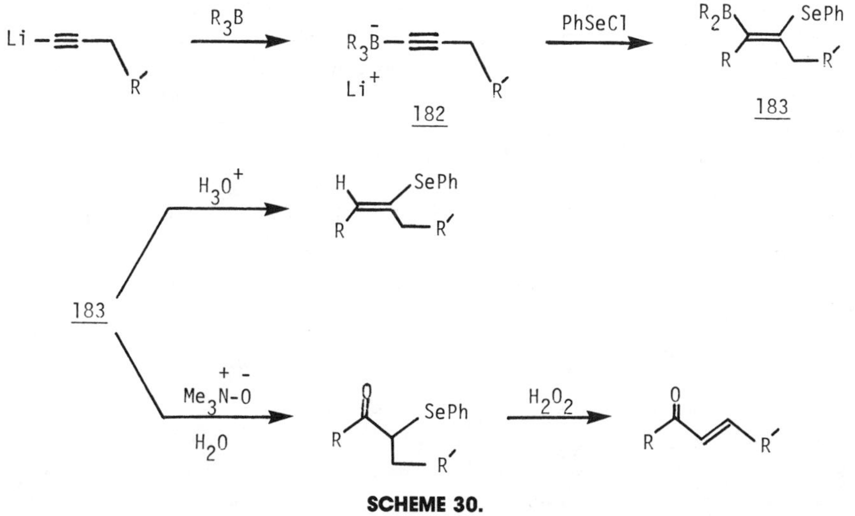

SCHEME 30.

Other selenenic electrophiles have also been investigated. Internal acetylenes react with PhSeCN in the presence of copper (II) chloride or bromide and triethylamine to afford excellent yields of β-halovinyl selenides. The additions are *anti* stereospecific, but unsymmetrical alkynes produce mixtures of regio-isomers [Eq. (271)].[508] On the other hand, terminal acetylenes react regiospecifically with anti-Markovnikov orientation, but not stereospecifically [Eq. (272)].[508]

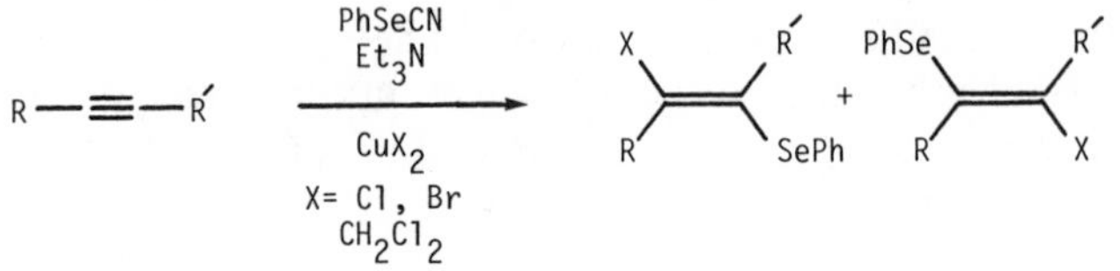

$$(271)$$

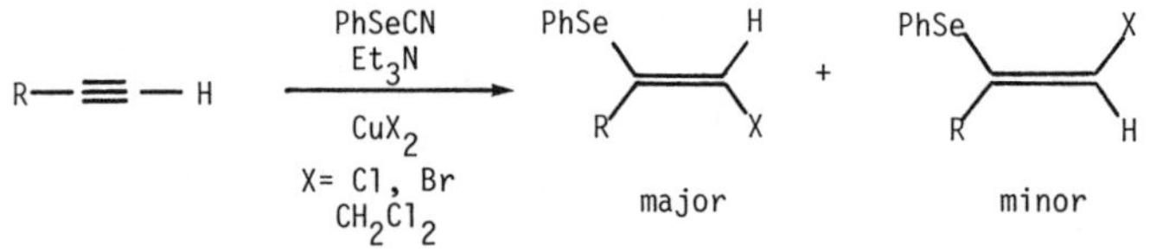

$$(272)$$

The addition of selenenyl trifluoroacetate, $\underline{7}$, to several acetylenes has been reported.[11,24] The adducts are easily hydrolyzed to α-phenylseleno ketones [e.g., Eq. (273)].

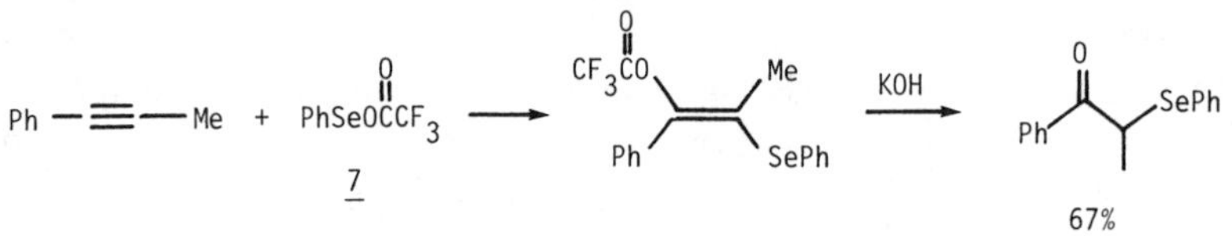

$$(273)$$

Selenenamides react with dimethyl acetylenedicarboxylate via an initial Michael addition followed by intramolecular selenenylation[509] in a manner resembling their reaction with enones (cf. Scheme 4). A single adduct is initially formed but equilibrates to an (E)–(Z) mixture [Eq. (274)].

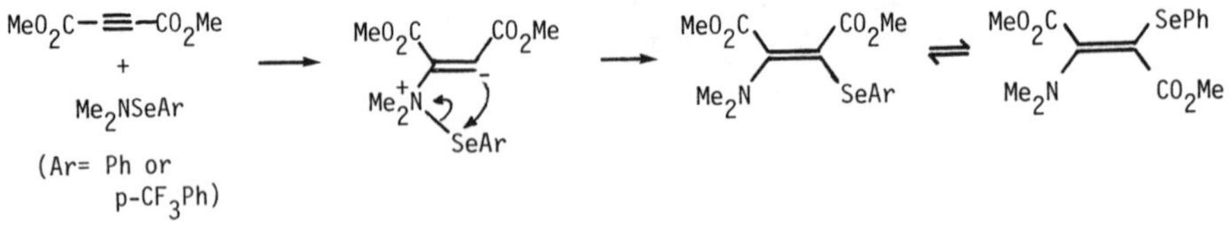

$$(274)$$

Acetylene produces the 1:1 adduct, $\underline{184}$, when treated with selenium monochloride [Eq. (275)].[510] In contrast to the analogous process with olefins [cf. Eqs. (250)–(252)], a 2:1 adduct is not formed in the presence of excess acetyl-

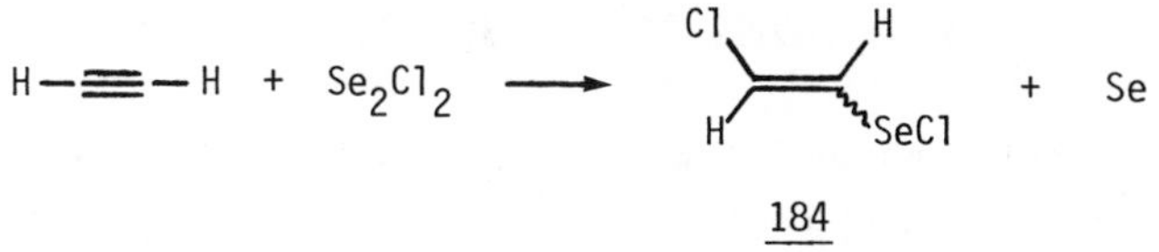

$$H-\!\!\equiv\!\!-H \ + \ Se_2Cl_2 \ \longrightarrow \ \underset{\underset{\textbf{184}}{}}{\overset{Cl\diagup\diagdown H}{\underset{H\diagup\diagdown SeCl}{}}} \ + \ Se \qquad (275)$$

ene. On the other hand, dialkylacetylenes afford bis-(β-chloroalkyl) diselenides in modest yields [Eq. (276)],[511] while phenylacetylenes are cyclized to benzo-selenophenes [Eq. (277)].[511]

$$2 \ \ R-\!\!\equiv\!\!-R \ + \ Se_2Cl_2 \ \longrightarrow \ \underset{R\diagup\diagdown Se-Se}{\overset{Cl\diagdown \quad R}{}} \underset{}{\overset{R\diagdown \quad Cl}{\diagup\diagdown R}} \qquad (276)$$

$$Ph-\!\!\equiv\!\!-R \ + \ Se_2Cl_2 \ \longrightarrow \ \underset{Se\diagdown R}{\overset{Cl}{\bigcirc\!\!\bigcirc}} \ + \ HCl \ + \ Se \quad (277)$$

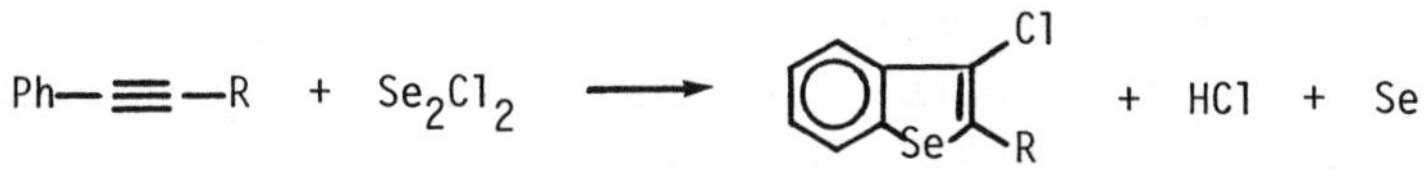

The additions of selenenyl chlorides to several enynes have been reported.[512–514] The trimethylsilyl derivative, 185, can react at either unsaturated site,[512,513] with chemoselectivity dependent on the nature of the solvent [Eq. (278)]. The t-butyl analog of 185 is attacked preferentially at the double bond[512] while the stannane, 188, undergoes an overall metathesis reaction [Eq. (279)].[514]

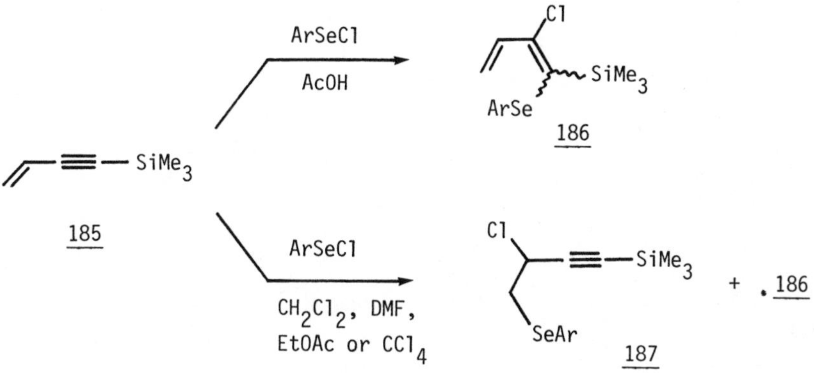

$$\text{185} \quad \xrightarrow[\text{AcOH}]{\text{ArSeCl}} \quad \text{186}$$
$$\xrightarrow[\substack{CH_2Cl_2, \ DMF, \\ EtOAc \ or \ CCl_4}]{ArSeCl} \quad \text{187} \ + \ \text{186}$$

$$(278)$$

$$\underset{\textbf{188}}{\diagup\!\!\equiv\!\!-SnMe_3} \ \xrightarrow[CH_2Cl_2]{PhSeCl} \ \underset{45\%}{\diagup\!\!\equiv\!\!-SePh} \ + \ \underset{91\%}{Me_3SnCl}$$

$$(279)$$

9. REACTIONS OF Se (IV) ELECTROPHILES

Olefins react with selenium tetrachloride, **2**, and tetrabromide, **13**, to produce 2:1 adducts according to Eq. (280).[515–520] The addition is believed to proceed in stepwise fashion with high *anti* stereospecificity[515] but often with poor regioselectivity.

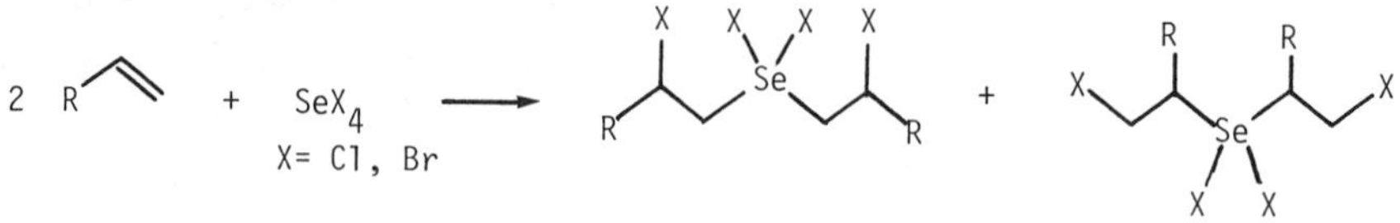

$$2 \ R \diagdown + \ SeX_4 \quad (X= Cl, \ Br) \longrightarrow \quad + \quad $$

(280)

Norbornadiene, **102**,[521] and other dienes[522,523] produce 1:1 adducts such as **189** [Eq. (281)] while phenylacetylenes afford benzoselenophenes [Eq. (282)].[524] Loss of halogen from the initial product is sometimes observed and is facilitated by refluxing the adduct in acetone.[522] Selenium tetrachloride chlorinates norbornene, **100**, in lieu of forming an addition product.[525] It attacks anisole and ethoxybenzene at the para position to form diaryl selenium dichlorides (Eq. (283)]526 and undergoes metathesis reactions with tetraphenyllead [Eq. (284)].[526] Polyketones, **190**, cyclize via Eq. (285).[527]

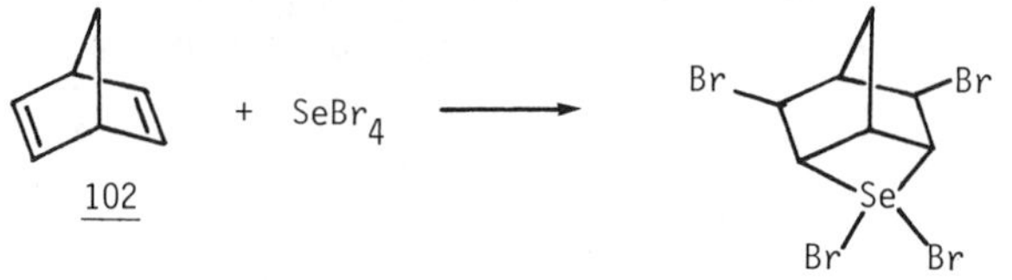

$$\underset{102}{} + \ SeBr_4 \longrightarrow \underset{189}{}$$

(281)

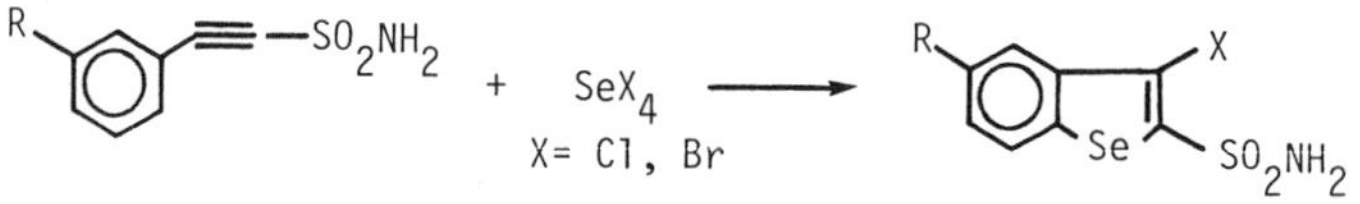

$$R \diagdown \equiv-SO_2NH_2 + \underset{X= \ Cl, \ Br}{SeX_4} \longrightarrow $$

(282)

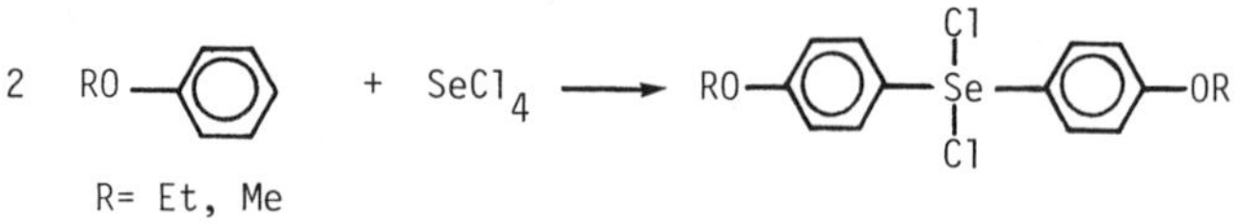

$$2 \ RO \diagdown + SeCl_4 \longrightarrow RO- \underset{\overset{Cl}{|}}{\overset{Cl}{\underset{|}{Se}}} -OR$$
R= Et, Me

(283)

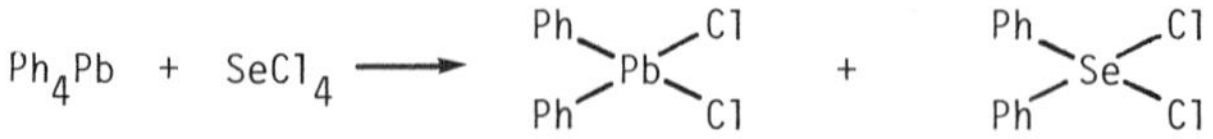

$$Ph_4Pb + SeCl_4 \longrightarrow \underset{Ph}{\overset{Ph}{}}Pb\overset{Cl}{\underset{Cl}{}} + \underset{Ph}{\overset{Ph}{}}Se\overset{Cl}{\underset{Cl}{}}$$

(284)

$$\underset{190}{} \rightleftharpoons \quad + \ SeCl_4 \longrightarrow $$

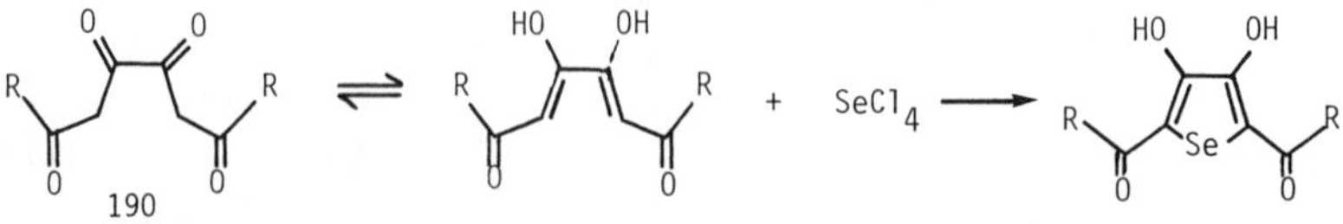

(285)

The electrophilic additions of methylselenium trichloride ($\underline{15}$, R = Me) in its dimeric β-form were investigated by Garratt and Schmid.[528-530] As in the case of selenenyl halides, olefins react *anti* stereospecifically unless they contain strongly electron-donating substituents capable of stabilizing a carbonium ion intermediate.[530] Stereospecificity has been attributed to bridged perseleniranium ion intermediates, $\underline{191}$ [Eq. (286)].[530] Aryl substituted olefins produce Markovnikov adducts whereas alkyl derivatives tend to react nonregioselectively.[529,530]

$$\underline{191} \tag{286}$$

Further reactions of the adducts have been reported. Thus, compound $\underline{192}$, obtained from isobutylene and MeSeCl$_3$ under thermodynamic conditions, reacts further through the loss of chlorine or via dehydrochlorination [Eq. (287)].[529]

$$\tag{287}$$

The addition of 2,4-dinitrophenylselenium trichloride [$\underline{15}$, R = 2,4-(NO$_2$)$_2$Ph] to olefins forms the same products, $\underline{193}$, expected from the addition of the corresponding selenenyl halide, together with the vicinal dichlorides, $\underline{194}$ (Scheme 31).[387,531] Lawson and Kharasch[531] originally attributed this to prior disproportionation of the selenium trichloride to ArSeCl and chlorine, followed by the separate addition of these species to the olefin, as in Scheme 31. More recently, however, Garratt and Schmid[387] noted that (*E*)- and (*Z*)-1-phenylpropene react faster with ArSeCl$_3$ than with ArSeCl, suggesting a different mechanism. A detailed investigation revealed that the two isomeric olefins produce *anti* adducts, $\underline{195}$ and $\underline{196}$, respectively, along with their regioisomers (Scheme 32). Subsequent stereospecific decomposition of $\underline{195}$ and $\underline{196}$ to the dichlorides, $\underline{197}$ and $\underline{198}$, liberates the selenenyl chloride, which then reacts with excess olefin in the normal manner to produce the observed β-chloro selenides.

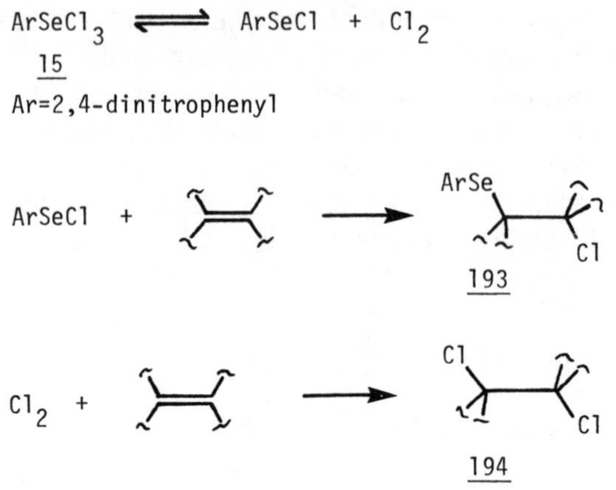
ArSeCl3
ArSeCl + Cl2
15
Ar=2,4-dinitrophenyl
ArSeCl +
ArSe
Cl
193
Cl2 +
Cl
Cl
194

SCHEME 31.

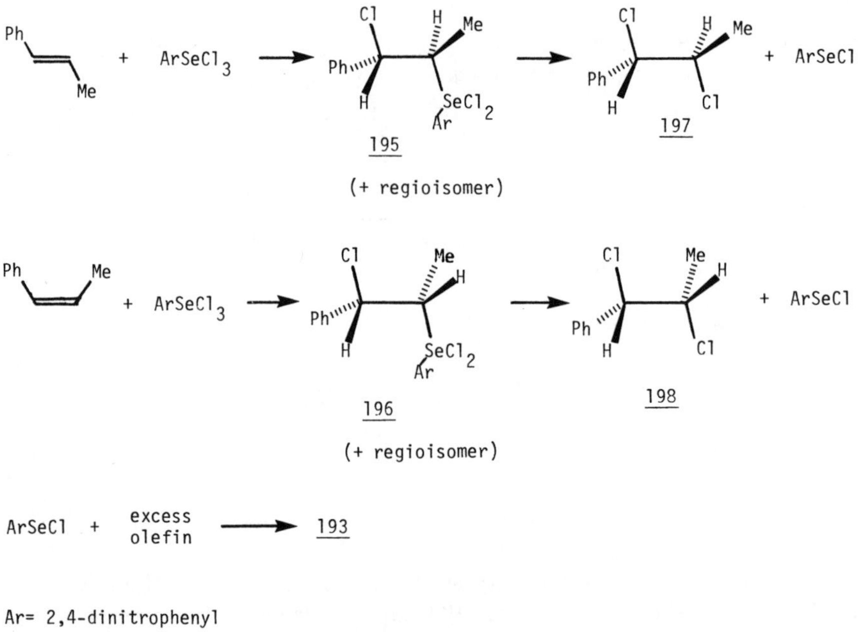
Ph
Me
+ ArSeCl3
Cl
H
Me
Ph
H
SeCl2
Ar
195
(+ regioisomer)
Cl
H
Me
Ph
H
Cl
197
+ ArSeCl
Ph
Me
+ ArSeCl3
Cl
Me
H
Ph
H
SeCl2
Ar
196
(+ regioisomer)
Cl
Me
H
Ph
H
Cl
198
+ ArSeCl
ArSeCl + excess olefin
193
Ar= 2,4-dinitrophenyl

SCHEME 32.

Terminal acetylenes form 1,2 adducts with β-$MeSeCl_3$ in a nonstereospecific and nonregioselective fashion. Disubstituted derivatives react *anti* stereospecifically [Eq. (288)].[528]

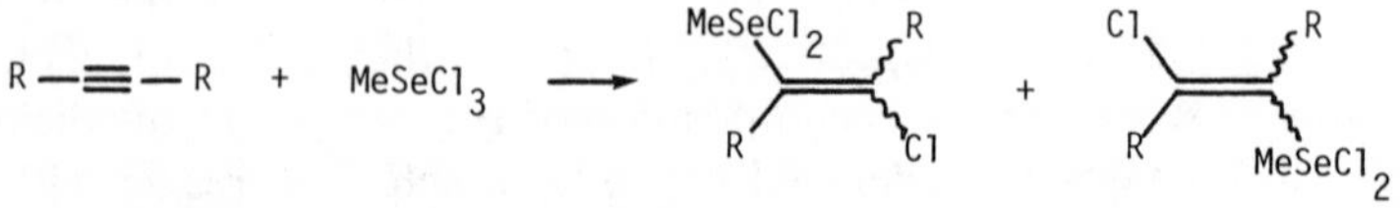
R—≡—R + MeSeCl3
MeSeCl2
R
R
Cl
+
Cl
R
R
MeSeCl2

(288)

Phenylselenium trichloride (15, R = Ph) reacts with hexamethyldisilazane or its *N*-chloro derivative to produce 199 [Eq. (289)].[318]

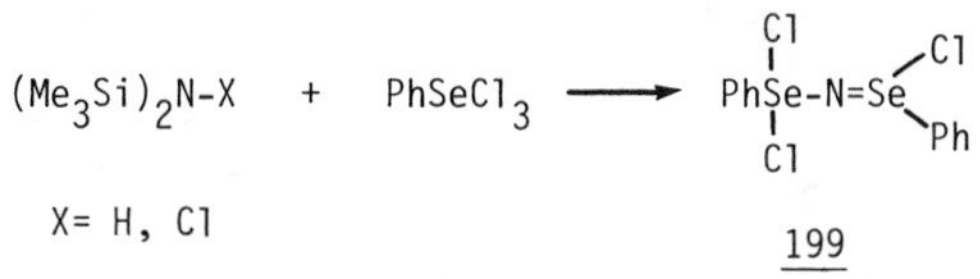
(Me3Si)2N-X
+
PhSeCl3
X= H, Cl
Cl
Cl
PhSe-N=Se
Cl
Ph
199

(289)

Selenium oxychloride, 14, and benzeneseleninyl chloride, 60, are other examples of Se (IV) electrophiles; however, their use in organic chemistry has been rather limited. The oxychloride may be used instead of Se_2Cl_2 in the transformation in Eq. (253). It also adds to acetylene to give an isolable 1:1 adduct 200 which, as in the case of the Se_2Cl_2 adduct 184, does not react further with excess alkyne [Eq. (290)].[510] Certain diamines are converted to selenadiazoles and related heterocycles[532,533] upon treatment with 14.

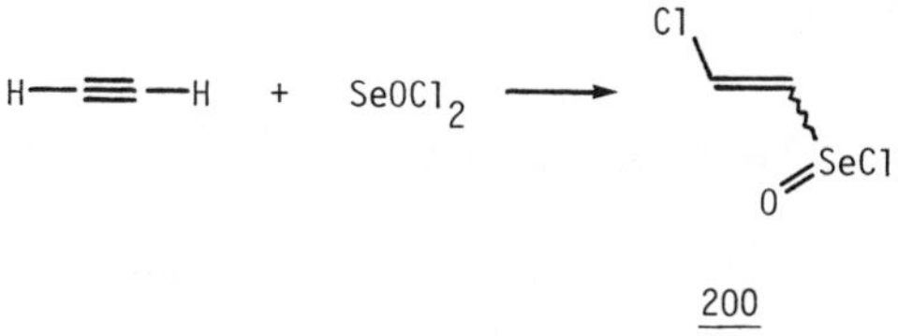
H—≡—H
+
SeOCl2
Cl
SeCl
O
200

(290)

The seleninyl chloride, 60, reacts with ketone[11] or ester enolates[534] to produce α-selenoxides in one step [Eqs. (291) and (292)]. However, the difficulties in handling this unstable and moisture sensitive reagent have rendered it less attractive than the conventional two-step approach of selenenylation and oxidation.

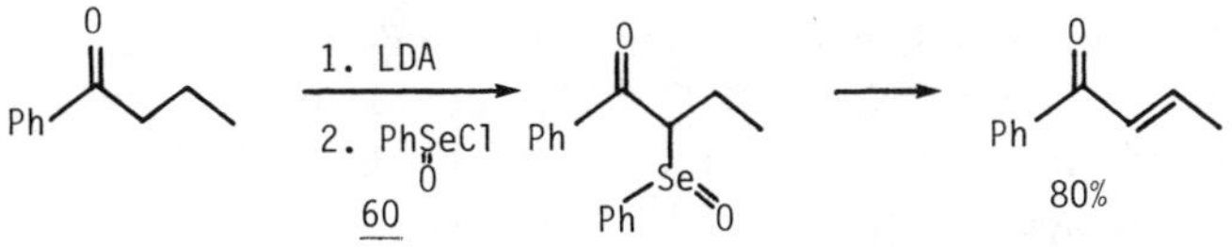
O
Ph
1. LDA
2. PhSeCl
O
60
O
Ph
Se
Ph
O
O
Ph
80%

(291)

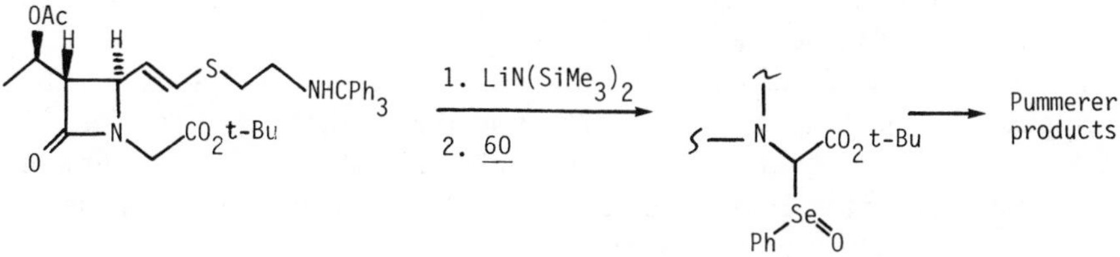
OAc
H
H
S
NHCPh3
N
CO2t-Bu
O
1. LiN(SiMe3)2
2. 60
N
CO2t-Bu
Se
Ph
O
Pummerer
products

(292)

The reagent has also been employed in the conversion of a limited variety of secondary amines to ketones [Eq. (293)],[535] and primary amines[535] or aldoximes[536] to nitriles [Eq. (294)]. The latter transformation can also be effected with PhSeCl or various sulfur electrophiles.[536]

$$R_2CHNH_2 \xrightarrow{60} R_2C=NH \xrightarrow{H_2O} R_2C=O \tag{293}$$

$$RCH_2NH_2 \xrightarrow{60} RC\equiv N \xleftarrow[\substack{PhSeCl \\ Et_3N}]{60 \ or} RCH=NOH \tag{294}$$

10. ADDENDUM

Organoselenium chemistry continues to proliferate rapidly. Approximately 100 articles relevant to the subject of this chapter have appeared between early 1983 and mid-1984. These are briefly surveyed in the following update.

10.1. α-Selenenylation of Carbonyl Compounds

The selenenylation of ketones followed by selenoxide elimination remains a popular method for their transformation to enones. New examples of both the direct method[537–539] and the variation employing enolates[540–548] have appeared. Applications include synthetic approaches to coriamyrtin,[537] descarboxyquadrone,[538] paniculides,[539] steroids,[540–542] tricothecanes,[542] acrylodan,[543] hexaquinane analogs,[544] and β-elemenone.[545] A key step in a synthesis of periplanone-B[549] involved the conversion of a ketone to an α-diketone by selenenylation, oxidation, and a Pummerer reaction.

The dehydrogenations of several esters,[550,551] including a precursor of giberellin A_{38} methyl ester,[552] have been effected by selenenylation–oxidation. Lactones have been similarly transformed to α,β-unsaturated products containing endo-[553–558] or exocyclic[556,559,560] double bonds. The protocol has been exploited in the synthesis of telekin,[559] pinnatifidin,[559] yomogin,[560] dihydroactinidiolide,[553] macrolides,[554] steganacin,[555] and steroids.[556]

A dehydronicotine product was prepared[561] from a pyrrolidone precursor by selenenylation in the presence of excess base (cf. Scheme 7), carbonyl reduction with diborane, and selenoxide elimination.

Recent reports of the selenenylation of β-dicarbonyl compounds include examples of a diketone,[562] a keto lactone in the synthesis of otonecine,[563] a keto ester in the synthesis of a thromboxane A_2 analog,[564] dimethyl methylmalonate,[565] and several keto lactams,[566–568] including uridine derivatives[566] and cytochalasan precursors.[567]

10.2. Selenenylation of Other Functional Groups

The conversion of an enol acetate to an α-phenylseleno aldehyde with benzeneselenenyl trifluoroacetate has been employed in nucleoside synthesis.[569] Two examples of the selenenylation of enol silyl ethers with PhSeBr have been reported.[570,571] The 2-selenenylation of a 3,4-disubstituted indole with PhSeCl has been observed in low yield.[572] The reaction of nitromethane with PhSeBr and sodium ethoxide affords nitro(phenylseleno)methane, which can be converted to 1-nitro-1-phenylseleno-1-alkenes.[573] The α-selenenylations of several phosphoranes[574–576] and a phosphonate[577] have been described.

10.3. Reactions of Se (II) Electrophiles with Acetylenic, Vinylic, Allenic, and Benzylic Carbanions

The selenenylation of a lithioenamine with PhSeBr,[578] as well as of 3-lithiothiophene[579] and the dilithio derivative of 2-methyl-1,3-butenyne[580] with elemental selenium have been reported. Further studies of the processes depicted in Eq. (126) have appeared in a full paper.[581]

10.4. Reactions of Se (II) Electrophiles with Heteroatom Nucleophiles

9-O-Methylclavulanic acid, 201, undergoes decarboxylation to afford the 3-selenide when treated with p-ClPhSeBr and N-methylmorpholine. The mixed anhydride, 202, is a possible intermediate [Eq. (295)].[582]

Ketone hydrazones[583] or their N,N-dimagnesio derivatives[584] react with Se_2Br_2 or Se_2Cl_2, respectively, in the presence of tertiary amines to afford selenoketones [Eq. (296)]. A new, convenient preparation of Se_2Br_2 has been developed in conjunction with this work[583]. Hydrazones also react with PhSeBr and t-butyltetramethylguanidine to provide vinyl selenides[585] [Eq. (296)], while N-arylbenzamides produce heterocycles, 203, when treated with PhSeSePh and N-chlorosuccinimide [Eq. (297)].[586] The corresponding sulfur heterocycles were also prepared.

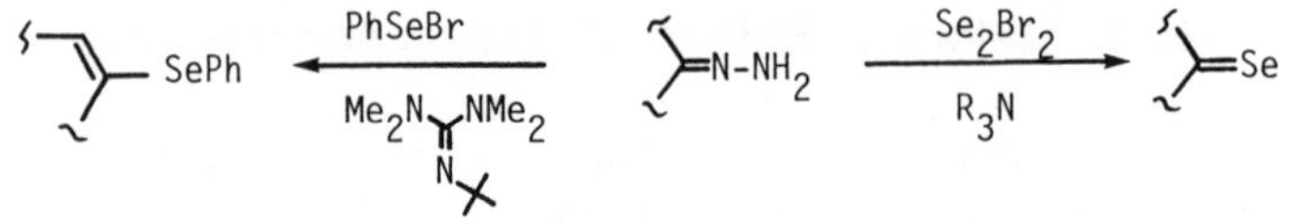

(296)

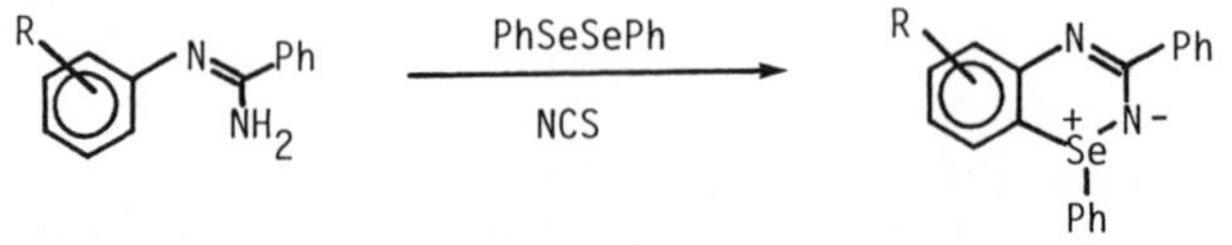

(297)

203

N-(Phenylseleno)morpholine, 32, has been used in the further synthetic transformations of aldehydes via their α-selenenylated derivatives [cf. Eqs. (48) and (49)].[587,588]

The transformation of cyclic dithioketals to ring expanded dihydro-1,4-dithiins and 1,4-dithiepins has been effected with PhSeCl,[589] according to Eq. (298).

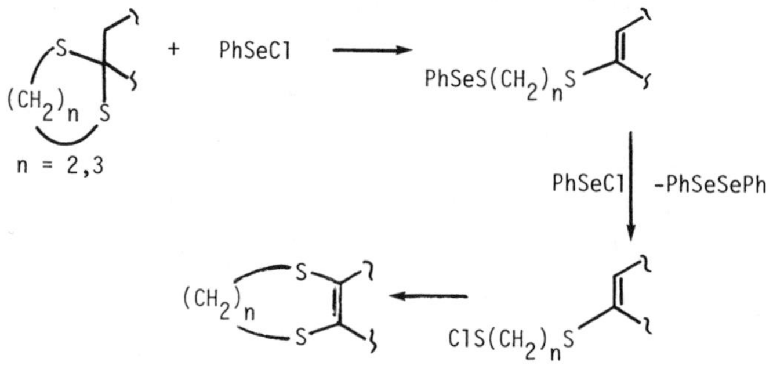

(298)

The dehydration of primary alcohols with *o*-NO$_2$PhSeCN and tri-*n*-butylphosphine followed by oxidation [cf. Eq. (153)] has been employed in the syntheses of carpesiolin,[590] quadrone,[591] nucleosides,[592] eldanolide,[593] β-elemol,[594] macrolides,[595] melitensin,[596] asteltoxin,[597,598] β-chamigrene,[599] megaphone,[600] coleon A,[601] and other products.[602–605] The deoxygenation of a primary alcohol via selenide formation in the preceding manner followed by reductive deselenization was exploited in a recent synthesis of dihydrocorynantheol.[606] Two new examples of the conversion of secondary alcohols to selenides have also been reported.[607,608]

A detailed kinetic study under conditions of varying pH of the hydrolysis[609] and thiolysis[610] of *o*-nitro- and *o*-benzoylbenzeneselenenic anhydrides, as well as a similar study of the thiolysis of the corresponding selenenic acids, have been performed by Kice et al.[610] The ortho substituent plays a key role in these mechanisms, which are complex and do not proceed by a simple S$_N$2 displacement at selenium.

10.5. Additions of Selenenic Electrophiles to Olefins

An investigation of substituent effects in the reactions of ArSeCl with (*E*)- and (*Z*)-1-phenylpropene, and in those of PhSeCl with 1-arylpropenes, suggests that C—Se bond formation lags behind Se—Cl bond cleavage in the rate limiting transition state.[611,612] These processes are roughly equal in the case of the corresponding sulfur systems.[612]

The additions of PhSeCl of PhSeBr to a series of acrylates and related compounds, followed by dehydrohalogenation [cf. Eqs. (204)–(208)], afforded the corresponding α-(phenylseleno)acrylates.[613] The products are of interest as captodative olefins. Oxidation of the β-chloroselenide obtained from oxacephem, 204, and PhSeCl produced the endocyclic olefin, 205, in good yield [Eq. (299)].[614]

$$\tag{299}$$

β-Chloro selenides obtained from olefins and PhSeCl have been further transformed to vicinal cis-dichlorides[615] or to β-allyl-substituted selenides[616] by treatment with PhSeCl (or Cl$_2$) and tetrabutylammonium chloride, or with allyltrimethylsilane and zinc dibromide, respectively. Displacement of chloride from the β-chloro selenides may also be similarly effected with enol silyl ethers and zinc bromide.[616]

The addition of PhSeCl to isoprene furnishes the products of both 1,2 and 1,4 addition, whose ratio is temperature dependent.[617] Further synthetically useful transformations of the products have been described.[617] The adducts obtained from allenes and PhSeBr react with electron rich aromatic compounds (furan, thiophene, *N*-methylpyrrole, and 1,3,5-trimethoxybenzene) in the presence of silver perchlorate to afford electrophilic substitution products in good yield.[618]

The 1-hydroxycylcopropyl substituted olefin, 206, underwent ring expansion in the presence of PhSeBr and triethylamine [Eq. (300)].[619]

$$\tag{300}$$

Several examples of oxyselenenylations of olefins with N-(phenylseleno) phthalimide, 8,–methanol[620] or with PhSeCl-methanol[621] have been reported. Applications of acetoxyselenenylation in the synthesis of ingenanes[622] and lycorine[623] have appeared. The reactions of olefins with PhSeCl and silver crotonate afford the corresponding β-acyloxy selenides, which undergo free-radical cyclization when treated with triphenyltin hydride.[624] Further studies of the electrochemical oxidation of olefins to allylic ethers and alcohols in the presence of catalytic amounts of PhSeSePh [cf. Eq. (225)] have been reported.[625] The utility of norbornadiene as a scavenger for the PhSeOH produced as a by-product during a difficult selenoxide elimination has been demonstrated.[626]

Cyclohexene underwent amidoselenenylation (cf. Scheme 24) when present during the anodic oxidation of PhSeSePh in acetonitrile.[627] The addition of PhSeSCN to olefins (cf. Scheme 27) was found to strongly favor the isothiocyanate products when a mixture of PhSeCl and mercury (II) thiocyanate was employed as the selenenylating reagent.[628] A procedure for the oxidative elimination of the products to vinyl isothiocyanates was developed.[628]

10.6. Additions of Selenenic Electrophiles to Enol Ethers

New examples of the conversion of an enol ether to an α-seleno aldehyde with PhSeCl,[629] and of the oxyselenenylation of a dihydropyran[630] in a synthesis of methyl desosaminide have been reported.

10.7. Additions of Selenenic Electrophiles to Acetylenes

Further investigations of the additions of PhSeCl to 1,4-dichloro-2-butyne and of the synthetic applications of the products [cf. Eq. (270)] have appeared[631] in a full paper.

10.8. Reactions of Se (IV) Electrophiles

The additions of selenium tetrachloride and tetrabromide to allyl halides[632] produce mixtures of regioisomers, as indicated previously in Eq. (280). Kinetic studies of the additions of the tetrachloride to vinyl- and allylphthalimide have been performed.[633] These processes are first order in the selenium halide and second order in the alkene. The reaction of anisole (and other electron rich arenes) with PhSeCl and silver hexafluorophosphate produced the aromatic selenenylation product, 207 [Eq. (301)], along with PhSeSePh and the selenonium salt, 208.[634] The latter product is probably formed via the Se (IV) intermediate, 209.

(301)

209

REFERENCES

1. *Organic Selenium Compounds: Their Chemistry and Biology*, D. L. Klayman and W. H. H. Günther, Eds., Wiley-Interscience, New York, 1973. (a) D. L. Klayman, Chap. 4; (b) J. Michalski and A. Markowska, Chap. 10.; (c) J. L. Martin, Chap. 13B.

2. K. B. Sharpless, K. M. Gordon, R. F. Lauer, D. W. Patrick, S. P. Singer, and M. W. Young, *Chem. Scr.*, **8**, 9 (1975).

3. D. L. J. Clive, *Tetrahedron*, **34**, 1049 (1978).

4. D. L. J. Clive, *Aldrichimica Acta*, **11**, 43 (1978).

5. H. J. Reich, in *Oxidation in Organic Chemistry*, W. S. Trahanovsky, Ed., Part C, Academic, New York, 1978, Chap. 1.

6. H. J. Reich, *Acc. Chem. Res.*, **12**, 22 (1979).

7. D. M. Yost and C. E. Kircher, *J. Am. Chem. Soc.*, **52**, 4680 (1930).

8. H. Brintzinger, K. Pfannstiel, and H. Vogel, *Z. Anorg. Allg. Chem.*, **256**, 75 (1948).

9. F. Lautenschlaeger, *J. Org. Chem.*, **34**, 4002 (1969).

10. H. J. Reich, M. L. Cohen, and P. S. Clark, *Org. Synth.*, **59**, 141 (1979).

11. H. J. Reich, J. M. Renga, and I. L. Reich, *J. Am. Chem. Soc.*, **97**, 5434 (1975).

12. K. B. Sharpless and R. F. Lauer, *J. Org. Chem.*, **39**, 429 (1974).

13. K. B. Sharpless, R. F. Lauer, and A. Y. Teranishi, *J. Am. Chem. Soc.*, **95**, 6137 (1973).

14. H. J. Reich, W. W. Willis, Jr., and S. Wollowitz, *Tetrahedron Lett.*, **23**, 3319 (1982).

15. J. L. Kice, F. McAfee, and H. Slebocka-Tilk, *Tetrahedron Lett.*, **23**, 3323 (1982).

16. H. J. Reich, C. A. Hoeger, and W. W. Willis, Jr., *J. Am. Chem. Soc.*, **104**, 2936 (1982).

17. O. Behaghel and H. Siebert, *Ber.*, **66**, 708 (1933).

18. H. J. Reich, S. Wollowitz, J. E. Trend, F. Chow, and D. F. Wendelborn, *J. Org. Chem.*, **43**, 1697 (1978).

19. T. Hori and K. B. Sharpless, *J. Org. Chem.*, **43**, 1689 (1978).

20. R. A. Gancarz and J. L. Kice, *Tetrahedron Lett.*, **22**, 1661 (1981).

21. O. Behaghel and W. Müller, *Ber.*, **68**, 1540 (1935).

22. D. L. J. Clive, *J. Chem. Soc. Chem. Commun.*, 695 (1973).

23. D. L. J. Clive, *J. Chem. Soc. Chem. Commun.*, 100 (1974).

24. H. J. Reich, *J. Org. Chem.*, **39**, 428 (1974).

25. K. C. Nicolaou, D. A. Claremon, W. E. Barnette, and S. P. Seitz, *J. Am. Chem. Soc.*, **101**, 3704 (1979).

26. D. G. Garratt, M. D. Ryan, and M. Ujjainwalla, *Can. J. Chem.*, **57**, 2145 (1979).

27. W. J. E. Parr and R. C. Crafts, *Tetrahedron Lett.*, **22**, 1371 (1981).

28. T. Austad, *Acta Chem. Scand., Ser. A*, **30**, 479 (1976).

29. H. J. Reich and J. M. Renga, *J. Org. Chem.*, **40**, 3313 (1975).

30. H. Bauer, *Ber.*, **46**, 92 (1913).

31. O. Behaghel and H. Siebert, *Ber.*, **65**, 812 (1932).

32. S. Tomoda, Y. Takeuchi, and Y. Nomura, *Chem Lett.*, 1069 (1981).

33. T. G. Back and R. G. Kerr, *J. Organomet. Chem.*, **286**, 171 (1985).

34. T. G. Back and S. Collins, *Tetrahedron Lett.*, **21**, 2213 (1980).

35. R. A. Gancarz and J. L. Kice, *Tetrahedron Lett.*, **21**, 1697 (1980).

36. R. A. Gancarz and J. L. Kice, *J. Org. Chem.*, **46**, 4899 (1981).

37. N. Rabjohn, *Org. React.*, **24**, 261 (1976).

38. J. L. Huguet, *Adv. Chem. Ser.*, **76**, 345 (1967).

39. D. N. Jones, D. Mundy, and R. D. Whitehouse, *J. Chem. Soc. Chem. Commun.*, 86 (1970).

40. R. Walter and J. Roy, *J. Org. Chem.*, **36**, 2561 (1971).

41. K. B. Sharpless, M. W. Young, and R. F. Lauer, *Tetrahedron Lett.*, 1979 (1973).

42. K. B. Sharpless and R. F. Lauer, *J. Am. Chem. Soc.*, **95**, 2697 (1973).

43. H. J. Reich, I. L. Reich, and J. M. Renga, *J. Am. Chem. Soc.*, **95**, 5813 (1973).

44. H. Rheinboldt and M. Perrier, *Bull. Soc. Chim. Fr.*, **17**, 759 (1950).

45. H. Berner, G. Schulz, G. Fischer, and H. Reinshagen, *Monatsh. Chem.*, **109**, 557 (1978).

46. W. J. E. Parr, *J. Chem. Soc. Perkin Trans. 1*, 3002 (1981).

47. G. Quinkert, H. Englert, F. Cech, A. Stegk, E. Haupt, D. Leibfritz, and D. Rehm, *Chem. Ber.*, **112**, 310 (1979).

48. I. Iijima, K. C. Rice, and J. V. Silverton, *Heterocycles*, **6**, 1157 (1977).

49. J. Reden, M. F. Reich, K. C. Rice, A. E. Jacobson, A. Brossi, R. A. Streaty, and W. A. Klee, *J. Med. Chem.*, **22**, 256 (1979).

50. C. H. Heathcock, C. M. Tice, and T. C. Germroth, *J. Am. Chem. Soc.*, **104**, 6081 (1982).

51. J. A. Marshall and R. D. Royce, Jr., *J. Org. Chem.*, **47**, 693 (1982).

52. P. A. Grieco, T. Oguri, S. Burke, E. Rodriguez, G. T. DeTitta, and S. Fortier, *J. Org. Chem.*, **43**, 4552 (1978).

53. T. Cynkowski and M. Kocór, *Rocz. Chem.*, **50**, 257 (1976).

54. P. E. Eaton, G. D. Andrews, E.-P. Krebs, and A. Kunai, *J. Org. Chem.*, **44**, 2824 (1979).

55. P. Callant, R. Ongena, and M. Vandewalle, *Tetrahedron*, **37**, 2085 (1981).

56. M. Miyano, J. N. Smith, and C. R. Dorn, *Tetrahedron*, **38**, 3447 (1982).

57. P. Kok, P. DeClercq, and M. Vandewalle, *Bull. Soc. Chim. Belg.*, **87**, 615 (1978).

58. W. Gramlich and H. Plieninger, *Tetrahedron Lett.*, 475 (1978).

59. W. Gramlich and H. Plieninger, *Chem. Ber.*, **112**, 1550 (1979).

60. W. Gramlich and H. Plieninger, *Chem. Ber.*, **112**, 1571 (1979).

61. H. Plieninger and W. Gramlich, *Chem. Ber.*, **111**, 1944 (1978).

62. Y. K. Yee and A. G. Schultz, *J. Org. Chem.*, **44**, 719 (1979).

63. R. B. Boar, *J. Chem. Soc. Perkin Trans. 1*, 1275 (1975).

64. D. Caine and H. Deutsch, *J. Am. Chem. Soc.*, **100**, 8030 (1978).

65. J. P. Konopelski, C. Djerassi, and J. P. Raynaud, *J. Med. Chem.*, **23**, 722 (1980).

66. S. Danishefsky, K. Vaughan, R. Gadwood, and K. Tsuzuki, *J. Am. Chem. Soc.*, **103**, 4136 (1981).

67. S. Danishefsky, K. Vaughan, R. C. Gadwood, and K. Tsuzuki, *J. Am. Chem. Soc.*, **102**, 4262 (1980).

68. P. A. Grieco, G. F. Majetich, and Y. Ohfune, *J. Am. Chem. Soc.*, **104**, 4226 (1982).

69. P. A. Grieco, Y. Ohfune, and G. Majetich, *J. Am. Chem. Soc.*, **99**, 7393 (1977).

70. M. R. Roberts and R. H. Schlessinger, *J. Am. Chem. Soc.*, **103**, 724 (1981).

71. J. E. Baldwin and P. L. M. Beckwith, *J. Chem. Soc. Chem. Commun.*, 279 (1983).

72. J. H. Zaidi and A. J. Waring, *J. Chem. Soc. Chem. Commun.*, 618 (1980).

73. C. Giordano, *Gazz. Chim. Ital.*, **105**, 1265 (1975).

74. N. Miyoshi, T. Yamamoto, N. Kambe, S. Murai, and N. Sonoda, *Tetrahedron Lett.*, **23**, 4813 (1982).

75. S. Torii, K. Uneyama, and K. Handa, *Tetrahedron Lett.*, **21**, 1863 (1980).

76. R. Michels, M. Kato, and W. Heitz, *Makromol. Chem.*, **177**, 2311 (1976).

77. A. Toshimitsu, H. Owada, S. Uemura, and M. Okano, *Tetrahedron Lett.*, **23**, 2105 (1982).

78. H. J. Reich, J. M. Renga, and I. L. Reich, *J. Org. Chem.*, **39**, 2133 (1974).

79. M. D. Taylor, G. Minaskanian, K. N. Winzenberg, P. Santone, and A. B. Smith, III, *J. Org. Chem.*, **47**, 3960 (1982).

80. H. O. House, P. C. Gaa, J. H. C. Lee, and D. Van Derveer, *J. Org. Chem.*, **48**, 1670 (1983).

81. J. M. Luteijn and A. de Groot, *Tetrahedron Lett.*, **22**, 789 (1981).

82. T. Ishida and K. Wada, *J. Chem. Soc. Chem. Commun.*, 337 (1977).

83. T. Ishida and K. Wada, *J. Chem Soc. Perkin Trans. 1*, 323 (1979).

84. D. L. Snitman, R. J. Himmelsbach, and D. S. Watt, *J. Org. Chem.*, **43**, 4758 (1978).

85. C. Romming and P. E. Hansen, *Acta Chem. Scand., Sect. A*, **33**, 265 (1979).

86. J. E. McMurry and M. G. Silvestri, *J. Org. Chem.*, **41**, 3953 (1976).

87. A. Murai, A. Abiko, M. Ono, N. Katsui, and T. Masamune, *Chem. Lett.*, 1209 (1978).

88. A. Murai, A. Abiko, M. Ono, and T. Masamune, *Bull. Chem. Soc. Jpn.*, **55**, 1191 (1982).

89. M. Hirayama, S. Fukatsu, and N. Ikekawa, *J. Chem. Soc., Perkin Trans. 1*, 88 (1981).

90. G. Mehta, A. V. Reddy, A. N. Murthy, and D. S. Reddy, *J. Chem. Soc., Chem. Commun.*, 540 (1982).

91. R. E. Ireland, P. Beslin, R. Giger, U. Hengartner, H. A. Kirst, and H. Maag, *J. Org. Chem.*, **42**, 1267 (1977).

92. G. Stork and S. Raucher, *J. Am. Chem. Soc.*, **98**, 1583 (1976).

93. S. Pennanen, *Acta Chem. Scand. Ser. B*, **34**, 261 (1980).

94. L. A. Paquette and Y.-K. Han, *J. Org. Chem.*, **44**, 4014 (1979).

95. L. A. Paquette and Y.-K. Han, *J. Am. Chem. Soc.*, **103**, 1835 (1981).

96. G. D. Annis and L. A. Paquette, *J. Am. Chem. Soc.*, **104**, 4504 (1982).

97. Z. Lidert and C. W. Rees, *J. Chem. Soc., Chem. Commun.*, 317 (1983).

98. J. S. Swenton, R. M. Blankenship, and R. Sanitra, *J. Am. Chem. Soc.*, **97**, 4941 (1975).

99. A. B. Smith, III, P. A. Levenberg, P. J. Jerris, R. M. Scarborough, Jr., and P. M. Wovkulich, *J. Am. Chem. Soc.*, **103**, 1501 (1981).

100. D. Caine, A. A. Boucugnani, and W. R. Pennington, *J. Org. Chem.*, **41**, 3632 (1976).

101. D. Caine and A. S. Frobese, *Tetrahedron Lett.*, 3107 (1977).

102. D. Caine, H. Deutsch, and J. T. Gupton, III, *J. Org. Chem.*, **43**, 343 (1978).

103. R. E. Zipkin, N. R. Natale, I. M. Taffer, and R. O. Hutchins, *Synthesis*, 1035 (1980).

104. T. A. Hase and P. Kukkola, *Synth. Commun.*, **10**, 451 (1980).

105. J. Blumbach, D. A. Hammond, and D. A. Whiting, *Tetrahedron Lett.*, **23**, 3949 (1982).

106. L.-F. Tietze, G. V. Kiedrowski, and B. Berger, *Tetrahedron Lett.*, **23**, 51 (1982).

107. G. Quinket, G. Dürner, E. Kleiner, F. Adam, E. Haupt, and D. Leibfritz, *Chem. Ber.*, **113**, 2227 (1980).

108. P. A. Grieco, M. Nishizawa, S. D. Burke, and N. Marinovic, *J. Am. Chem. Soc.*, **98**, 1612 (1976).

109. P. A. Grieco, M. Nishizawa, T. Oguri, S. D. Burke, and N. Marinovic, *J. Am. Chem. Soc.*, **99**, 5773 (1977).

110. M. Nishizawa, P. A. Grieco, S. D. Burke, and W. Metz, *J. Chem. Soc., Chem. Commun.*, 76 (1978).

111. L. Laitem, P. Thibaut, and L. Christiaens, *J. Heterocycl. Chem.*, **13**, 469 (1976).

112. J. P. H. Verheyden, A. C. Richardson, R. S. Bhatt, B. D. Grant, W. L. Fitch, and J. G. Moffatt, *Pure Appl. Chem.*, **50**, 1363 (1978).

113. T. Wakamatsu, K. Akasaka, and Y. Ban, *J. Org. Chem.*, **44**, 2008 (1979).

114. C. J. Sih, D. Massuda, P. Corey, R. D. Gleim, and F. Suzuki, *Tetrahedron Lett.*, 1285 (1979).

115. H. Reinshagen and A. Stütz, *Monatsh. Chem.*, **110**, 567 (1979).

116. H. E. Zimmerman and R. J. Pasteris, *J. Org. Chem.*, **45**, 4864 (1980).

117. M. T. Edgar, G. Barth, and C. Djerassi, *J. Org. Chem.*, **45**, 2680 (1980).

118. R. L. Cargill, D. F. Bushey, J. R. Dalton, R. S. Prasad, R. D. Dyer, and J. Bordner, *J. Org. Chem.*, **46**, 3389 (1981).

119. T. -Y. Luh and K. L. Lei, *J. Chem. Soc. Chem. Commun.*, 214 (1981).

120. H. H. Seltzman, S. D. Wyrick, and C. G. Pitt, *J. Labelled Compds. Radiopharm.*, **18**, 1365 (1981).

121. S. A. Sadek, W. V. Kessler, S. M. Shaw, J. N. Anderson, and G. C. Wolf, *J. Med. Chem.*, **25**, 1488 (1982).

122. W. H. Rastetter, R. B. Nachbar, Jr., S. Russo-Rodriguez, R. V. Wattley, W. G. Thilly, B. M. Andon, W. L. Jorgensen, and M. Ibrahim, *J. Org. Chem.*, **47**, 4873 (1982).

123. S. H. Bertz, G. Rihs, and R. B. Woodward, *Tetrahedron*, **38**, 63 (1982).

124. S. Danishefsky, S. Chackalamannil, and B.-J. Uang, *J. Org. Chem.*, **47**, 2231 (1982).

125. G. M. Rubottom and H. D. Juve, Jr., *J. Org. Chem.*, **48**, 422 (1983).

126. S. Sakane, Y. Matsumura, Y. Yamamura, Y. Ishida, K. Maruoka, and H. Yamamoto, *J. Am. Chem. Soc.*, **105**, 672 (1983).

127. S. H. Korzeniowski, D. P. Vanderbilt, and L. B. Hendry, *Org. Prep. Proced. Int.*, **8**, 81 (1976).

128. P. A. Grieco and M. Nishizawa, *J. Chem. Soc. Chem. Commun.*, 582 (1976).

129. F. C. Brown, D. G. Morris, and A. M. Murray, *Tetrahedron*, **34**, 1845 (1978).

130. F. C. Brown, R. K. Fraser, R. W. Jemison, D. G. Morris, A. M. Murray, and J. D. Stephen, *Aust. J. Chem.*, **31**, 695 (1978).

131. E. Piers and E. H. Ruediger, *J. Org. Chem.*, **45**, 1725 (1980).

132. K. Nagao, I. Yoshimura, M. Chiba, and S. W. Kim, *Chem. Pharm. Bull.*, **31**, 114 (1983).

133. D. Liotta, G. Zima, C. Barnum, and M. Saindane, *Tetrahedron Lett.*, **21**, 3643 (1980).

134. P. A. Chaloner and A. B. Holmes, *J. Chem. Soc., Perkin Trans. 1*, 1838 (1976).

135. A. Pelter, R. S. Ward, D. Ohlendorf, and D. H. J. Ashdown, *Tetrahedron*, **35**, 531 (1979).

136. J. Schwartz and Y. Hayasi, *Tetrahedron Lett.*, **21**, 1497 (1980).

137. H. E. Zimmerman and M. C. Hovey, *J. Org. Chem.*, **44**, 2331 (1979).

138. W. Oppolzer and K. Bättig, *Helv. Chim. Acta*, **64**, 2489 (1981).

139. R. K. Hill and L. A. Renbaum, *Tetrahedron*, **38**, 1959 (1982).

140. G. Zima and D. Liotta, *Synth. Commun.*, **9**, 697 (1979).

141. D. Liotta, C. S. Barnum, and M. Saindane, *J. Org. Chem.*, **46**, 4301 (1981).

142. G. Zima, C. S. Barnum, and D. Liotta, *J. Org. Chem.*, **45**, 2736 (1980).

143. S. V. Ley and A. J. Whittle, *Tetrahedron Lett.*, **22**, 3301 (1981).

144. M. Jefson and J. Meinwald, *Tetrahedron Lett.*, **22**, 3561 (1981).

145. C. Paulmier and P. Lerouge, *Tetrahedron Lett.*, **23**, 1557 (1982).

146. N. Petragnani, J. V. Comasseto, R. Rodrigues, and T. J. Brocksom, *J. Organomet. Chem.*, **124**, 1 (1977).

147. D. Liotta, M. Saindane, and D. Brothers, *J. Org. Chem.*, **47**, 1598 (1982).

148. A. B. Smith, III, and R. E. Richmond, *J. Am. Chem. Soc.*, **105**, 575 (1983).

149. W. Oppolzer and K. Thirring, *J. Am. Chem. Soc.*, **104**, 4978 (1982).

150. E. Pfaff and H. Plieninger, *Chem. Ber.*, **115**, 1967 (1982).

151. M. J. Pearson, *J. Chem. Soc. Perkin Trans. 1*, 2544 (1981).

152. M. Hayashi, H. Miyake, H. Wakatsuka, and S. Iguchi, Japan. Kokai, 78149956, 1978; *Chem. Abstr.*, **90**, 151678m (1979).

153. S. P. Singer and K. B. Sharpless, *J. Org. Chem.*, **43**, 1448 (1978).

154. J. Tsuji, K. Masaoka, T. Takahashi, A. Suzuki, and N. Miyaura, *Bull. Chem. Soc. Jpn.*, **50**, 2507 (1977).

155. G. Ohloff, W. Giersch, K. H. Schulte-Elte, and C. Vial, *Helv. Chim. Acta*, **59**, 1140 (1976).

156. L. Lombardo, L. N. Mander, and J. V. Turner, *J. Am. Chem. Soc.*, **102**, 6626 (1980).

157. P. A. Grieco and M. Miyashita, *J. Org. Chem.*, **39**, 120 (1974).

158. P. A. Grieco, C. S. Pogonowski, and S. Burke, *J. Org. Chem.*, **40**, 542 (1975).

159. P. A. Grieco and M. Nishizawa, *J. Org. Chem.*, **42**, 1717 (1977).

160. K. Yamakawa, K. Nishitani, and A. Yamamoto, *Chem. Lett.*, 177 (1976).

161. K. Yamakawa and T. Tominaga, Japan. Kokai 7809759, 1978; *Chem. Abstr.*, **89**, 24563z (1978).

162. K. Yamakawa and T. Tominaga, Japan. Kokai 7809758, 1978; *Chem. Abstr.*, **89**, 24564a (1978).

163. N. Ikota and B. Ganem, *J. Org. Chem.*, **43**, 1607 (1978).

164. T. R. Hoye and M. J. Kurth, *J. Org. Chem.*, **43**, 3693 (1978).

165. A. R. Battersby, A. L. Gutman, C. J. R. Fookes, H. Günther, and H. Simon, *J. Chem. Soc. Chem. Commun.*, 645 (1981).

166. K. Yamakawa, K. Nishitani, and T. Tominaga, *Tetrahedron Lett.*, 2829 (1975).

167. S. W. Rollinson, R. A. Amos, and J. A. Katzenellenbogen, *J. Am. Chem. Soc.*, **103**, 4114 (1981).

168. J. N. Marx and P. J. Dobrowolski, *Tetrahedron Lett.*, **23**, 4457 (1982).

169. Y. Fujimoto, H. Miura, T. Shimizu, and T. Tatsuno, *Tetrahedron Lett.*, **21**, 3409 (1980).

170. M. Watanabe and A. Yoshikoshi, *Chem. Lett.*, 1315 (1980).

171. M. Ando, K. Tajima, and K. Takase, *J. Org. Chem.*, **48**, 1210 (1983).

172. T. Kawamata, S. Inayama, and K. Sata, *Chem. Pharm. Bull.*, **28**, 277 (1980).

173. M. J. Darmon and G. B. Schuster, *J. Org. Chem.*, **47**, 4658 (1982).

174. P. A. Grieco, J. Inanaga, N.-H. Lin, and T. Yanami, *J. Am. Chem. Soc.*, **104**, 5781 (1982).

175. T. J. Brocksom, N. Petragnani, and R. Rodrigues, *J. Org. Chem.*, **39**, 2114 (1974).

176. D. Buddhsukh and P. Magnus, *J. Chem. Soc. Chem. Commun.*, 952 (1975).

177. D. A. Chass, D. Buddhasukh, and P. D. Magnus, *J. Org. Chem.*, **43**, 1750 (1978).

178. P. J. Kocienski, G. Cernigliaro, and G. Feldstein, *J. Org. Chem.*, **42**, 353 (1977).

179. T. A. Hase and R. Kivikari, *Acta Chem. Scand. Ser. B*, **33**, 589 (1979).

180. M. Hayashi, H. Miyake, H. Wakatsuka, and S. Iguchi, Japan. Kokai, 78149954, 1978; *Chem. Abstr.*, **90**, 151676j (1979).

181. M. Hayashi, Y. Arai, H. Wakatsuka, M. Kawamura, Y. Konishi, T. Tsuda, and K. Matsumoto, *J. Med. Chem.*, **23**, 525 (1980).

182. S. Escher, W. Giersch, and G. Ohloff, *Helv. Chim. Acta*, **64**, 943 (1981).

183. M. Fetizon, M.-T. Montaufier, and J. Rens, *J. Chem. Res. (S)*, 9 (1982).

184. S. Danishefsky, M. Hirama, K. Gombatz, T. Harayama, E. Berman, and P. F. Schuda, *J. Am. Chem. Soc.*, **101**, 7020 (1979).

185. S. Danishefsky, M. Hirama, K. Gombatz, T. Harayama, E. Berman, and P. Schuda, *J. Am. Chem. Soc.*, **100**, 6536 (1978).

186. D. J. Robins and S. Sakdarat, *J. Chem. Soc. Perkin Trans. 1*, 1734 (1979).

187. Y. Terao, N. Imai, K. Achiwa, and M. Sekiya, *Chem. Pharm. Bull.*, **30**, 3167 (1982).

188. S. R. Wilson, L. R. Phillips, Y. Pelister, and J. C. Huffman, *J. Am. Chem. Soc.*, **101**, 7373 (1979).

189. Y. Tobe, K. Kakiuchi, Y. Odaira, T. Hosaki, Y. Kai, and N. Kasai, *J. Am. Chem. Soc.*, **105**, 1376 (1983).

190. J. A. Marshall and R. H. Ellison, *J. Am. Chem. Soc.*, **98**, 4312 (1976).

191. J. C. Sih and D. R. Graber, *J. Org. Chem.*, **43**, 3798 (1978).

192. J. M. Luteijn, M. van Doorn, and A. de Groot, *Tetrahedron Lett.*, **21**, 4127 (1980).

193. R. S. Lott, E. G. Breitholle, and C. H. Stammer, *J. Org. Chem.*, **45**, 1151 (1980).

194. L. A. Paquette, E. Farkas, and R. Galemmo, *J. Org. Chem.*, **46**, 5434 (1981).

195. W. R. Roush and A. P. Spada, *Tetrahedron Lett.*, **23**, 3773 (1982).

196. J. P. Vigneron, R. Méric, M. Larcheveque, A. Debal, G. Kunesch, P. Zagatti, and M. Gallois, *Tetrahedron Lett.*, **23**, 5051 (1982).

197. F. Rouessac, H. Zamarlik, and N. Gnonlonfoun, *Tetrahedron Lett.*, **24**, 2247 (1983).

198. T. Wakamatsu, K. Akasaka, and Y. Ban, *Tetrahedron Lett.*, 2755 (1977).

199. C. A. Wilson II and T. A. Bryson, *J. Org. Chem.*, **40**, 800 (1975).

200. T.-Y. Luh, W. H. So, and S. W. Tam, *J. Organomet. Chem.*, **218**, 261 (1981).

201. P. A. Zoretic and P. Soja, *J. Org. Chem.*, **41**, 3587 (1976).

202. J. Ficini, A. Guingant, and J. D'Angelo, *J. Am. Chem. Soc.*, **101**, 1318 (1979).

203. Y. Ohfune and M. Tomita, *J. Am. Chem. Soc.*, **104**, 3511 (1982).

204. M. H. Benn and H. Rüeger, unpublished results.

205. J. Lévy, J.-Y. Laronze, J. Laronze and J. Le Men, *Tetrahedron Lett.*, 1579 (1978).

206. K. K. Mahalanabis, M. Mumtaz, and V. Snieckus, *Tetrahedron Lett.*, **23**, 3971 (1982).

207. J. A. Oakleaf, M. T. Thomas, A. Wu, and V. Snieckus, *Tetrahedron Lett.*, 1645 (1978).

208. M. Majewski, G. B. Mpango, M. T. Thomas, A. Wu, and V. Snieckus, *J. Org. Chem.*, **46**, 2029 (1981).

209. J. M. Renga and H. J. Reich, *Org. Synth.*, **59**, 58 (1980).

210. J. Bruhn, H. Heimgartner, and H. Schmid, *Helv. Chim. Acta*, **62**, 2630 (1979).

211. J. N. Marx and G. Minaskanian, *Tetrahedron Lett.*, 4175 (1979).

212. J. N. Marx and G. Minaskanian, *J. Org. Chem.*, **47**, 3306 (1982).

213. D. L. Snitman, M.-Y. Tsai, and D. S. Watt, *Synth. Commun.*, **8**, 195 (1978).

214. T. Kametani, S. A. Surgenor, and K. Fukumoto, *Heterocycles*, **14**, 303 (1980).

215. S. Ohuchida, N. Hamanaka, and M. Hayashi, *Tetrahedron Lett.*, **22**, 5301 (1981).

216. Y. Tamura, A. Wada, S. Okuyama, S. Fukumori, Y. Hayashi, N. Gohda, and Y. Kita, *Chem. Pharm. Bull.*, **29**, 1312 (1981).

217. R. L. Sobczak, M. E. Osborn, and L. A. Paquette, *J. Org. Chem.*, **44**, 4886, (1979).

218. G. Nicollier, M. Rebetez, R. Tabacchi, H. Gerlach, and A. Thalmann, *Helv. Chim. Acta*, **61**, 2899 (1978).

219. R. F. C. Brown, F. W. Eastwood, and G. L. McMullen, *Aust. J. Chem.*, **30**, 179 (1977).

220. R. Bloch and P. Orvane, *Synth. Commun.*, **11**, 913 (1981).

221. D. Liotta, C. Barnum, R. Puleo, G. Zima, C. Bayer, and H. S. Kezar, III, *J. Org. Chem.*, **46**, 2920 (1981).

222. S. C. Welch, C. P. Hagan, D. H. White, W. P. Fleming, and J. W. Trotter, *J. Am. Chem. Soc.*, **99**, 549 (1977).

223. A. G. Schultz, *J. Org. Chem.*, **40**, 3466 (1975).

224. D. Liotta, M. Saindane, C. Barnum, H. Ensley, and P. Balakrishnan, *Tetrahedron Lett.*, **22**, 3043 (1981).

225. J. Tsuji, K. Masaoka, and T. Takahashi, *Tetrahedron Lett.*, 2267 (1977).

226. G. M. Ksander, J. E. McMurry, and M. Johnson, *J. Org. Chem.*, **42**, 1180 (1977).

227. S. J. Falcone and M. E. Munk, *Synth. Commun.*, **9**, 719 (1979).

228. T. Arunachalam and E. Caspi, *J. Org. Chem.*, **46**, 3415 (1981).

229. A. Toshimitsu, T. Aoai, H. Owada, S. Uemura, and M. Okano, *J. Chem. Soc. Chem. Commun.*, 412 (1980).

230. F. G. Bordwell, J. E. Bares, J. E. Bartmess, G. E. Drucker, J. Gerhold, G. J. McCollum, M. Van Der Puy, N. R. Vanier, and W. S. Matthews, *J. Org. Chem.*, **42**, 326 (1977).

231. S. Torii, K. Uneyama, and M. Ono, *Tetrahedron Lett.*, **21**, 2741 (1980).

232. S. Torii, K. Uneyama, M. Ono, and T. Bannou, *J. Am. Chem. Soc.*, **103**, 4606 (1981).

233. I. Ryu, S. Murai, I. Niwa, and N. Sonoda, *Synthesis*, 874 (1977).

234. S. Danishefsky and C. F. Yan, *Synth. Commun.*, **8**, 211 (1978).

235. S. Danishefsky, R. Zamboni, M. Kahn, and S. J. Etheredge, *J. Am. Chem. Soc.*, **102**, 2097 (1980).

236. S. Danishefsky, R. Zamboni, M. Kahn, and S. J. Etheredge, *J. Am. Chem. Soc.*, **103**, 3460 (1981).

237. G. D. Crouse and L. A. Paquette, *J. Org. Chem.*, **46**, 4272 (1981).

238. S. Danishefsky, C. F. Yan, and P. M. McCurry Jr., *J. Org. Chem.*, **42**, 1819 (1977).

239. S. Danishefsky, C. F. Yan, R. K. Singh, R. B. Gammill, P. M. McCurry Jr., N. Fritsch, and J. Clardy, *J. Am. Chem. Soc.*, **101**, 7001 (1979).

240. J. Hooz and J. Oudenes, *Synth. Commun.*, **10**, 667 (1980).

241. D. R. Williams and K. Nishitani, *Tetrahedron Lett.*, **21**, 4417 (1980).

242. D. R. Williams, B. A. Barner, K. Nishitani, and J. G. Phillips, *J. Am. Chem. Soc.*, **104**, 4708 (1982).

243. E. E. Knaus, T. A. Ondrus, and C. S. Giam, *J. Heterocycl. Chem.*, **13**, 789 (1976).

244. T. G. Back and N. Ibrahim, *Tetrahedron Lett.*, 4931 (1979).

245. T. G. Back, N. Ibrahim, and D. J. McPhee, *J. Org. Chem.*, **47**, 3283 (1982).

246. K. Anzai, *J. Heterocycl. Chem.*, **16**, 567 (1979).

247. D. N. Brattesani and C. H. Heathcock, *Tetrahedron Lett.*, 2279 (1974).

248. D. N. Brattesani and C. H. Heathcock, *J. Org. Chem.*, **40**, 2165 (1975).

249. H. E. Zimmerman and D. R. Diehl, *J. Am. Chem. Soc.*, **101**, 1841 (1979).

250. T. Sakakibara, I. Takai, E. Ohara, and R. Sudoh, *J. Chem. Soc. Chem. Commun.*, 261 (1981).

251. T. Sakakibara, S. Ikuta, and R. Sudoh, *Synthesis*, 261 (1982).

252. V. I. Erashko, O. M. Sazonova, A. A. Tishaninova, and A. A. Fainzil'berg, *Izv. Akad. Nauk SSSR, Ser. Khim.*, 161 (1982).

253. D. Seebach and D. Enders, *J. Med. Chem.*, **17**, 1225 (1974).

254. R. Kupper and C. J. Michejda, *J. Org. Chem.*, **44**, 2326 (1979).

255. N. Petragnani and M. de Moura Campos, *Chem. Ind.*, 1461 (1964).

256. N. Petragnani, R. Rodrigues, and J. V. Comasseto, *J. Organomet. Chem.*, **114**, 281 (1976).

257. G. Saleh, T. Minami, Y. Ohshiro, and T. Agawa, *Chem. Ber.*, **112**, 355 (1979).

258. J. V. Comasseto and N. Petragnani, *J. Organomet. Chem.*, **152**, 295 (1978).

259. W. A. Kleschick and C. H. Heathcock, *J. Org. Chem.*, **43**, 1256 (1978).

260. M. Mikolajczyk, S. Grzejszczak, and K. Korbacz, *Tetrahedron Lett.*, **22**, 3097 (1981).

261. B. Harirchian and P. Magnus, *J. Chem. Soc. Chem. Commun.*, 522 (1977).

262. K. Fuji, M. Ueda, K. Sumi, and E. Fujita, *Tetrahedron Lett.*, **22**, 2005 (1981).

263. S. Raucher and G. A. Koolpe, *J. Org. Chem.*, **43**, 4252 (1978).

264. H. J. Reich, W. W. Willis Jr., and P. D. Clark, *J. Org. Chem.*, **46**, 2775 (1981).

265. J. M. Renga, unpublished doctoral dissertation, University of Wisconsin, Madison, 1975.

266. E. Gipstein, C. G. Willson, and H. S. Sachdev, *J. Org. Chem.*, **45**, 1486 (1980).

267. T. G. Back, *J. Org. Chem.*, **46**, 5443 (1981).

268. M. Isobe, Y. Ichikawa, and T. Goto, *Tetrahedron Lett.*, **22**, 4287 (1981).

269. T. Kauffmann, R. Kriegesmann, B. Altepeter, and F. Steinseifer, *Chem. Ber.*, **115**, 1810 (1982).

270. D. J. Buckley, S. Kulkowit, and A. McKervey, *J. Chem. Soc., Chem. Commun.*, 506 (1980).

271. P. J. Giddings, D. I. John, and E. J. Thomas, *Tetrahedron Lett.*, **21**, 399 (1980).

272. P. J. Giddings, D. I. John, E. J. Thomas, and D. J. Williams, *J. Chem. Soc., Perkin Trans. 2*, 2757 (1982).

273. N. Petragnani and G. Schill, *Chem. Ber.*, **103**, 2271 (1970).

274. R. Pellicciari, M. Curini, P. Ceccherelli, and R. Fringuelli, *J. Chem. Soc. Chem. Commun.*, 440 (1979).

275. T. G. Back and R. G. Kerr, *Tetrahedron Lett.*, **23**, 3241 (1982).

276. P. J. Giddings, D. I. John, and E. J. Thomas, *Tetrahedron Lett.*, **21**, 395 (1980).

277. M. Taboury, *Bull. Soc. Chim. Fr.*, **29**, 761 (1903).

278. L. Brandsma, H. Wijers, and J. F. Arens, *Rec. Trav. Chim. Pays-Bas*, **81**, 583 (1962).

279. L. Brandsma, H. E. Wijers, and C. Jonker, *Rec. Trav. Chim. Pays-Bas*, **83**, 208 (1964).

280. R. Mayer and A. K. Müller, *Z. Chem.*, **4**, 384 (1964).

281. Y. A. Boiko, B. S. Kupin, and A. A. Petrov, *Zh. Org. Khim.*, **4**, 1355 (1968).

282. S. Raucher, M. R. Hansen, and M. A. Colter, *J. Org. Chem.*, **43**, 4885 (1978).

283. J. V. Comasseto, J. T. B. Ferreira, and N. Petragnani, *J. Organomet. Chem.*, **216**, 287 (1981).

284. S. Tomoda, Y. Takeuchi, and Y. Nomura, *Chem. Lett.*, 253 (1982).

285. T. Hayama, S. Tomoda, Y. Takeuchi, and Y. Nomura, *Chem. Lett.*, 1249 (1982).

286. Y. L. Gol'dfarb, V. P. Lifvinov, and V. Y. Mortikov, *Khim. Geterotsikl. Soedin.*, 898 (1979).

287. H. J. Reich and M. J. Kelly, *J. Am. Chem. Soc.*, **104**, 1119 (1982).

288. A. Haces, E. M. G. A. van Kruchten and W. H. Okamura, *Tetrahedron Lett.*, **23**, 2707 (1982).

289. F. M. Hauser, R. P. Rhee, and S. Prasanna, *Synthesis*, 72 (1980).

290. D. G. Garratt and A. Kabo, *Can. J. Chem.*, **58**, 1030 (1980).

291. W. S. Cook and R. A. Donia, *J. Am. Chem. Soc.*, **73**, 2275 (1951).

292. G. Hölzle and W. Jenny, *Helv. Chim. Acta*, **41**, 331 (1958).

293. W. Jenny, *Helv. Chim. Acta*, **35**, 1591 (1952).

294. H. J. Reich, I. L. Reich, and S. Wollowitz, *J. Am. Chem. Soc.*, **100**, 5981 (1978).

295. H. J. Reich and S. Wollowitz, *J. Am. Chem. Soc.*, **104**, 7051 (1982).

296. K. B. Sharpless and R. F. Lauer, *J. Am. Chem. Soc.*, **94**, 7154 (1972).

297. M. Shimizu and I. Kuwajima, *Tetrahedron Lett.*, 2801 (1979).

298. I. Kuwajima, M. Shimizu, and H. Urabe, *J. Org. Chem.*, **47**, 837 (1982).

299. L. R. M. Pitombo, *Chem. Ber.*, **92**, 745 (1959).

300. F. O. Ayorinde, *Tetrahedron Lett.*, **24**, 2077 (1983).

301. A. M. Kuliev, M. A. Shakhgel'diev, and S. M. Shikhaliev, *Dokl. Akad. Nauk Az. SSR*, **37**, 36 (1981).

302. P. G. Gassman, A. Miura, and T. Miura, *J. Org. Chem.*, **47**, 951 (1982).

303. T. Kobayashi and T. Hiraoaka, *Bull. Chem. Soc. Jpn.*, **52**, 3366 (1979).

304. T. Hiraoka and T. Kobayashi, Japan. Kokai, 7955592, 1979; *Chem. Abst.*, **91**, 175372d (1979).

305. R. Weber and M. Renson, *Bull. Soc. R. Sci. Liege*, **48**, 146 (1979).

306. M. Apostolescu, *Bull. Inst. Politeh. Iasi*, **20**, 9 (1974).

307. F. A. Davis and E. W. Kluger, *J. Am. Chem. Soc.*, **98**, 302 (1976).

308. T. G. Back and S. Collins, *Tetrahedron Lett.*, 2661 (1979).

309. T. G. Back, S. Collins, and R. G. Kerr, *J. Org. Chem.*, **46**, 1564 (1981).

310. V. I. Naddaka, I. I. Logacheva, and V. S. Yur'eva, *Zh. Org. Khim.*, **12**, 2606 (1976).

311. T. G. Back and R. G. Kerr, *Can. J. Chem.*, **60**, 2711 (1982).

312. P. T. Southwell-Keely, I. L. Johnstone, and E. R. Cole, *Phosphorus and Sulfur*, **1**, 261 (1976).

313. N. Y. Derkach, T. V. Lyapina, and E. S. Levchenko, *Zh. Org. Khim.*, **10**, 139 (1974).

314. N. Y. Derkach, N. A. Pasmurtseva, T. V. Lyapina, and E. S. Levchenko, *Zh. Org. Khim.*, **10**, 1873 (1974).

315. N. Y. Derkach, T. V. Lyapina, and E. S. Levchenko, *Zh. Org. Khim.*, **14**, 280 (1978).

316. N. Y. Derkach, T. V. Lyapina, and E. S. Levchenko, *Zh. Org. Khim.*, **16**, 33 (1980).

317. N. Y. Derkach and T. V. Lyapina, *Zh. Org. Khim.*, **17**, 529 (1981).

318. N. Y. Derkach, T. V. Lyapina, and E. S. Levchenko, *Zh. Org. Khim.*, **17**, 622 (1981).

319. E. R. Clark and M. A. S. Al-Turaihi, *J. Organomet. Chem.*, **134**, 181 (1977).

320. J. E. Baldwin, S. B. Haber, and J. Kitchin, *J. Chem. Soc. Chem. Commun.*, 790 (1973).

321. A. Scarf, E. R. Cole, and P. T. Southwell-Keely, *Phosphorus and Sulfur*, **3**, 285 (1977).

322. H. Ishihara and Y. Hirabayashi, *Chem. Lett.*, 203 (1976).

323. A. Cahours and A. W. Hofmann, *Ann. Chem.*, **104**, 1 (1857).

324. G. Sosnovsky and M. Konieczny, *Synthesis*, 583 (1978).

325. H. Ishihara, S. Sato, and Y. Hirabayashi, *Bull. Chem. Soc. Jpn.*, **50**, 3007 (1977).

326. R. H. Mitchell, *Can. J. Chem*, **54**, 238 (1976).

327. I. A. Nuretdinov, D. N. Sadkova, and E. V. Bayandina, *Izv. Akad. Nauk SSSR, Ser. Khim.*, 2635 (1977).

328. P. A. Grieco, S. Gilman, and M. Nishizawa, *J. Org. Chem.*, **41**, 1485 (1976).

329. M. Sevrin and A. Krief, *J. Chem. Soc., Chem. Commun.*, 656 (1980).

330. J. Meinwald, M. Adams, and A. J. Duggan, *Heterocycles*, **7**, 989 (1977).

331. J. Meinwald, *Pure Appl. Chem.*, **49**, 1275 (1977).

332. M. A. Adams, A. J. Duggan, J. Smolanoff, and J. Meinwald, *J. Am. Chem. Soc.*, **101**, 5364 (1979).

333. W. C. Still, *J. Am. Chem. Soc.*, **101**, 2493 (1979).

334. J. R. Williams and J. F. Callahan, *J. Org. Chem.*, **45**, 4479 (1980).

335. W. Oppolzer, K. Bättig, and T. Hudlicky, *Helv. Chim. Acta*, **62**, 1493 (1979).

336. W. Oppolzer, K. Bättig, and T. Hudlicky, *Tetrahedron*, **37**, 4359 (1981).

337. P. A. Wender and J. C. Hubbs, *J. Org. Chem.*, **45**, 365 (1980).

338. W. R, Roush and T. E. D'Ambra, *J. Org. Chem.*, **46**, 5045 (1981).

339. W. R. Roush, H. R. Gillis, and A. I. Ko, *J. Am. Chem. Soc.*, **104**, 2269 (1982).

340. R. L. Snowden, *Tetrahedron Lett.*, **22**, 101 (1981).

341. H. Takaku, T. Nomoto, and K. Kimura, *Chem. Lett.*, 1221 (1981).

342. K. Kano, K. Hayashi, and H. Mitsuhashi, *Chem. Pharm. Bull.*, **30**, 1198 (1982).

343. P. A. Grieco, T. Takigawa, and W. J. Schillinger, *J. Org. Chem.*, **45**, 2247 (1980).

344. S. Raucher, J. E. Burks, Jr., K.-J. Hwang, and D. P. Svedberg, *J. Am. Chem. Soc.*, **103**, 1853 (1981).

345. J. A. Marshall and G. A. Flynn, *J. Org. Chem.*, **44**, 1391 (1979).

346. S. Takano, M. Takahashi, and K. Ogasawara, *J. Am. Chem. Soc.*, **102**, 4282 (1980).

347. G. Majetich, P. A. Grieco, and M. Nishizawa, *J. Org. Chem.*, **42**, 2327 (1977).

348. D. J. Hart, *J. Org. Chem.*, **46**, 3576 (1981).

349. D. J. Hart and K. Kanai, *J. Am. Chem. Soc.*, **105**, 1255 (1983).

350. P. Müller and M. Rey, *Helv. Chim. Acta*, **65**, 1191 (1982).

351. K. B. Sharpless and M. W. Young, *J. Org. Chem.*, **40**, 947 (1975).

352. D. L. J. Clive, V. Farina, A. Singh, C. K. Wong, W. A. Kiel, and S. M. Menchen, *J. Org. Chem.*, **45**, 2120 (1980).

353. D. L. J. Clive, G. Chittattu, N. J. Curtis, and S. M. Menchen, *J. Chem. Soc. Chem. Commun.*, 770 (1978).

354. W. A. Ayer and R. H. McCaskill, *Can. J. Chem.*, **59**, 2150 (1981).

355. U. Schmidt, A. Lieberknecht, H. Griesser, and H. Bökens, *Tetrahedron Lett.*, **23**, 4911 (1982).

356. C. R. Hutchinson, K. C. Mattes, M. Nakane, J. J. Partridge, and M. R. Uskoković, *Helv. Chim. Acta*, **61**, 1221 (1978).

357. T. Kametani, H. Matsumoto, H. Nemoto, and K. Fukumoto, *J. Am. Chem. Soc.*, **100**, 6218 (1978).

358. T. R. Hoye and M. J. Kurth, *J. Org. Chem.*, **44**, 3461 (1979).

359. P. A. Grieco, E. Williams, H. Tanaka, and S. Gilman, *J. Org. Chem.*, **45**, 3537 (1980).

360. L. E. Friedrich and P. Y.-S. Lam, *J. Org. Chem.*, **46**, 306 (1981).

361. D. J. Morgans, Jr., *Tetrahedron Lett.*, **22**, 3721 (1981).

362. S. Takano, M. Takahashi, S. Hatakeyama, and K. Ogasawara, *J. Chem. Soc. Chem. Commun.*, 556 (1979).

363. S. Takano, K. Shibuya, M. Takahashi, S. Hatakeyama, and K. Ogasawara, *Heterocycles*, **16**, 1125 (1981).

364. L. Calabi, B. Danieli, G. Lesma, and G. Palmisano, *Tetrahedron Lett.*, **23**, 2139 (1982).

365. Y. Ohfune, K. Takaki, H. Kameoka, and T. Sakai, *Chem. Lett.*, 209 (1982).

366. B. B. Snider and G. B. Phillips, *J. Org. Chem.*, **48**, 464 (1983).

367. T. Kametani, H. Nemoto, and K. Fukumoto, *Heterocycles*, **6**, 1365 (1977).

368. T. Kametani, H. Nemoto, and K. Fukumoto, *Biorg. Chem.*, **7**, 215 (1978).

369. P. A. Grieco, Y. Yokoyama, and E. Williams, *J. Org. Chem.*, **43**, 1283 (1978).

370. P. A. Grieco, J. Y. Jaw, D. A. Claremon, and K. C. Nicolaou, *J. Org. Chem.*, **46**, 1215 (1981).

371. S. V. Ley, N. S. Simpkins, and A. J. Whittle, *J. Chem. Soc. Chem. Commun.*, 1001 (1981).

372. S. V. Ley, D. Neuhaus, N. S. Simpkins, and A. J. Whittle, *J. Chem. Soc. Perkin Trans. 1*, 2157 (1982).

373. S. V. Ley, N. S. Simpkins, and A. J. Whittle, *J. Chem. Soc. Chem. Commun.*, 503 (1983).

374. T. G. Back and D. J. McPhee, *J. Org. Chem.*, **49**, 3842 (1984).

375. P. A. Grieco and Y. Yokoyama, *J. Am. Chem. Soc.*, **99**, 5210 (1977).

376. Y. Tamura, M. Adachi, T. Kawasaki, and Y. Kita, *Tetrahedron Lett.*, 2251 (1979).

377. M. Sekine and T. Hata, *Chem. Lett.*, 801 (1979).

378. T. Austad, *Acta Chem. Scand. Ser. A*, **29**, 895 (1975).

379. T. Austad, *Acta Chem. Scand. Ser. A*, **30**, 579 (1976).

380. T. Austad, *Acta Chem. Scand. Ser. A*, **31**, 93 (1977).

381. T. Austad, *Acta Chem. Scand. Ser. A*, **31**, 227 (1977).

382. R. Eriksen and S. Hauge, *Acta Chem. Scand.*, **26**, 3153 (1972).

383. J. L. Kice and H. Slebocka-Tilk, *J. Am. Chem. Soc.*, **104**, 7123 (1982).

384. G. H. Schmid and D. G. Garratt, in *The Chemistry of Double-Bonded Functional Groups. Supplement A, Part 2*, S. Patai, Ed., Wiley-Interscience, London, 1977, Chap. 9.

385. A. Toshimitsu, S. Uemura, and M. Okano, *J. Chem. Soc. Chem. Commun.*, 87 (1982).

386. G. H. Schmid and D. G. Garratt, *Tetrahedron*, **34**, 2869 (1978).

387. D. G. Garratt and G. H. Schmid, *Can. J. Chem.*, **52**, 3599 (1974).

388. N. S. Zefirov, L. G.. Gurvich, A. S. Shashkov, and V. A. Smit, *Zh. Org. Khim.*, **10**, 1786 (1974).

389. N. S. Zefirov, L. G. Gurvich, A. S. Shashkov, M. Z. Krimer, and E. A. Vorob'eva, *Tetrahedron*, **32**, 1211 (1976).

390. D. G. Garratt and G. H. Schmid, *Can. J. Chem.*, **52**, 1027 (1974).

391. H. J. Reich and J. E. Trend, *Can J. Chem.*, **53**, 1922 (1975).

392. B. Lindgren, *Tetrahedron Lett.*, 4347 (1974).

393. B. Lindgren, *Acta Chem. Scand. Ser. B*, **30**, 941 (1976).

394. B. Lindgren, *Acta Chem. Scand. Ser. B*, **31**, 1 (1977).

395. G. H. Schmid and D. G. Garratt, *Tetrahedron Lett.*, 3991 (1975).

396. J. Rémion and A. Krief, *Tetrahedron Lett.*, 3743 (1976).

397. E. G. Kataev, T. G. Mannafov, E. G. Berdnikov, and O. A. Komarovskaya, *Zh. Org. Khim.*, **9**, 1983 (1973).

398. D. Liotta and G. Zima, *Tetrahedron Lett.*, 4977 (1978).

399. J. N. Denis, J. Vicens, and A. Krief, *Tetrahedron Lett.*, 2697 (1979).

400. S. Raucher, *Tetrahedron Lett.*, 3909 (1977).

401. S. Raucher, *J. Org. Chem.*, **42**, 2950 (1977).

402. P.-T. Ho and R. J. Kolt, *Can. J. Chem.*, **60**, 663 (1982).

403. A. E. Feiring, *J. Org. Chem.*, **45**, 1958 (1980).

404. I. L. Reich and H. J. Reich, *J. Org. Chem.*, **46**, 3721 (1981).

405. D. G. Garratt, M. D. Ryan, and A. Kabo, *Can. J. Chem.*, **58**, 2329 (1980).

406. P.-A. Carrupt and P. Vogel, *Tetrahedron Lett.*, **23**, 2563 (1982).

407. P. L. Beaulieu, V. M. Morisset, and D. G. Garratt, *Can. J. Chem.*, **58**, 1005 (1980).

408. D. Liotta, G. Zima, and M. Saindane, *J. Org. Chem.*, **47**, 1258 (1982).

409. V. I. Koshutin, *Zh. Obshch. Khim.*, **43**, 2221 (1973).

410. G. H. Schmid, D. G. Garratt, and P. L. Beaulieu, *Chem. Scr.*, **15**, 128 (1980).

411. D. G. Garratt, P. L. Beaulieu, and M. D. Ryan, *Tetrahedron*, **36**, 1507 (1980).

412. D. G. Garratt, P. L. Beaulieu, V. M. Morisset, and M. Ujjainwalla, *Can. J. Chem.*, **58**, 2745 (1980).

413. S. Halazy and L. Hevesi, *Tetrahedron Lett.*, **24**, 2689 (1983).

414. D. G. Garratt and P. L. Beaulieu, *Can. J. Chem.*, **58**, 2737 (1980).

415. T. Hori and K. B. Sharpless, *J. Org. Chem.*, **44**, 4208 (1979).

416. D. G. Garratt, *Tetrahedron Lett.*, 1915 (1978).

417. H. Nishiyama, K. Itagaki, K. Sakuta, and K. Itoh, *Tetrahedron Lett.*, **22**, 5285 (1981).

418. H. Nishiyama, S. Narimatsu, and K. Itoh, *Tetrahedron Lett.*, **22**, 5289 (1981).

419. E. G. Kataev, T. G. Mannafov, A. B. Remizov, and O. A. Komarovskaya, *Zh. Org. Khim.*, **11**, 2322 (1975).

420. A. E. Feiring, *J. Org. Chem.*, **45**, 1962 (1980).

421. G. R. Krow and D. A. Shaw, *Synth. Commun.*, **12**, 313 (1982).

422. J. C. Sih and D. R. Graber, *J. Org. Chem.*, **47**, 4919 (1982).

423. J.-M. Beau, S. Aburaki, J.-R. Pougny, and P. Sinaÿ, *J. Am. Chem. Soc.*, **105**, 621 (1983).

424. M. Miyano, *J. Org. Chem.*, **46**, 1846 (1981).

425. T. Hori and K. B. Sharpless, *J. Org. Chem.*, **44**, 4204 (1979).

426. S. Raucher, *Tetrahedron Lett.*, 2261 (1978).

427. M. Shimizu and I. Kuwajima, *Bull. Chem. Soc. Jpn.*, **54**, 3100 (1981).

428. D. Liotta and G. Zima, *J. Org. Chem.*, **45**, 2551 (1980).

429. D. Goldsmith, D. Liotta, C. Lee, and G. Zima, *Tetrahedron Lett.*, 4801 (1979).

430. S. P. McManus and D. H. Lam, *J. Org. Chem.*, **43**, 650 (1978).

431. T. Takahashi, H. Nagashima, and J. Tsuji, *Tetrahedron Lett.*, 799 (1978).

432. G. W. Francis and T. Tande, *J. Chromatogr.*, **150**, 139 (1978).

433. A. Toshimitsu, S. Uemura, and M. Okano, *J. Chem. Soc. Chem. Commun.*, 965 (1982).

434. K. Isobe, J. Taga, and Y. Tsuda, *Tetrahedron Lett.*, 2331 (1976).

435. A. Toshimitsu, S. Uemura, and M. Okano, *J. Chem. Soc. Chem. Commun.*, 166 (1977).

436. A. Toshimitsu, T. Aoai, S. Uemura, and M. Okano, *J. Org. Chem.*, **45**, 1953 (1980).

437. P. S. Liu, V. E. Marques, J. A. Kelley, and J. S. Driscoll, *J. Org. Chem.*, **45**, 5225 (1980).

438. S. Torii, K. Uneyama, and M. Ono, *Tetrahedron Lett.*, **21**, 2653 (1980).

439. W. Jenny, *Helv. Chim. Acta*, **36**, 1278 (1953).

440. N.-Y. Wang. C.-T. Hsu, and C. J. Sih, *J. Am. Chem. Soc.*, **103**, 6538 (1981).

441. C.-T. Hsu, N.-Y. Wang, L. H. Latimer, and C. J. Sih, *J. Am. Chem. Soc.*, **105**, 593 (1983).

442. B. Umezawa, O. Hoshino, S. Sawaki, S. Satoh, H. Numao, H. Sashida, K. Mori, and Y. Iitaka, *Tennen Yuki Kagobutsu Toronkai Koen Yoshishu*, **22**, 581 (1979); via *Chem. Abstr.*, **93**, 114787n (1980).

443. T. Kametani, K. Suzuki, H. Kurobe, and H. Nemoto, *J. Chem. Soc. Chem. Commun.*, 1128 (1979).

444. T. Kametani, K. Suzuki, H. Kurobe, and H. Nemoto, *Chem. Pharm. Bull.*, **29**, 105 (1981).

445. N. Miyoshi, S. Murai, and N. Sonoda, *Tetrahedron Lett.*, 851 (1977).

446. N. Miyoshi, S. Furui, S. Murai, and N. Sonoda, *J. Chem. Soc., Chem. Commun.*, 293 (1975).

447. N. Miyoshi, Y. Takai, S, Murai, and N. Sonoda, *Bull. Chem. Soc. Jpn.*, **51**, 1265 (1978).

448. N. Miyoshi, Y. Ohno, K. Kondo, S. Murai, and N. Sonoda, *Chem. Lett.*, 1309 (1979).

449. H. J. Reich and J. E. Trend, *J. Org. Chem.*, **41**, 2503 (1976).

450. H. J. Reich and W. W. Willis, Jr., *J. Am. Chem. Soc.*, **102**, 5967 (1980).

451. H. O. House, W. A. Kleschick, and E. J. Zaiko, *J. Org. Chem.*, **43**, 3653 (1978).

452. D. Labar, A. Krief, and L. Hevesi, *Tetrahedron Lett.*, 3967 (1978).

453. M. Shimizu, R. Takeda, and I. Kuwajima, *Tetrahedron Lett.*, 419 (1979).

454. M. Shimizu, R. Takeda, and I. Kuwajima, *Tetrahedron Lett.*, 3461 (1979).

455. M. Shimizu, R. Takeda, and I. Kuwajima, *Bull. Chem. Soc. Jpn.*, **54**, 3510 (1981).

456. I. Kuwajima and M. Shimizu, *Tetrahedron Lett.*, 1277 (1978).

457. A. Toshimitsu, T. Aoai, S. Uemura, and M. Okano. *J. Chem. Soc., Chem. Commun.*, 1041 (1980).

458. A. Toshimitsu, T. Aoai, H. Owada, S. Uemura, and M. Okano, *J. Org. Chem.*, **46**, 4727 (1981).

459. A. Toshimitsu, H. Owada, T. Aoai, S. Uemura, and M. Okano *J. Chem. Soc. Chem. Commun.*, 546 (1981).

460. A. Bewick, D. E. Coe, G. B. Fuller, and J. M. Mellor, *Tetrahedron Lett.*, **21**, 3827 (1980).

461. D. H. R. Barton, M. R. Britten-Kelly, and D. Ferreira, *J. Chem. Soc., Perkin Trans. 1*, 1090 or 1682 (1978).

462. K. B. Sharpless, T. Hori, L. K. Truesdale, and C. O. Dietrich, *J. Am. Chem. Soc.*, **98**, 269 (1976).

463. T. Hayama, S. Tomoda, Y. Takeuchi, and Y. Nomura, *Chem. Lett.*, 1109 (1982).

464. T. Hayama, S. Tomoda, Y. Takeuchi, and Y. Nomura, *Tetrahedron Lett.*, **23**, 4733 (1982).

465. T. G. Back and S. Collins, *Tetraheron Lett.*, **21**, 2215 (1980).

466. T. G. Back and S. Collins, *J. Org. Chem.*, **46**, 3249 (1981).

467. N. S. Zefirov, T. G. Back, and S. Collins, *Zh. Org. Khim.*, **17**, 2014 (1981).

468. S. Tomoda, Y. Takeuchi, and Y. Nomura, *J. Chem. Soc. Chem. Commun.*, 871 (1982).

469. S. Tomoda, Y. Takeuchi, and Y. Nomura, *Tetrahedron Lett.*, **23**, 1361 (1982).

470. S. Tomoda, Y. Takeuchi, and Y. Nomura, *Chem. Lett.*, 1733 (1982).

471. D. G. Garratt, *Can. J. Chem.*, **57**, 2180 (1979).

472. C. E. Boord and F. F. Cope, *J. Am. Chem. Soc.*, **44**, 395 (1922).

473. D. Elmaleh, S. Patai, and Z. Rappoport, *Isr. J. Chem.*, **9**, 155 (1971).

474. C. L. Pedersen, *J. Chem. Soc. Chem. Commun.*, 704 (1974).

475. C. L. Pedersen, *Acta Chem. Scand. Ser. B.*, **30**, 675 (1976).

476. M. Perrier and J. Vialle, *Bull. Soc., Chem. Fr.*, 199 (1979).

477. M. Perrier and J. Vialle, *Bull. Soc. Chim. Fr.*, 205 (1979).

478. D. G. Garratt, *Can. J. Chem.*, **56**, 2184 (1978).

479. R. A. Raphael, J. H. A. Stibbard, and R. Tidbury, *Tetrahedron Lett.*, **23**, 2407 (1982).

480. A. P. Kozikowski, K. L. Sorgi, and R. J. Schmiesing, *J. Chem. Soc. Chem. Commun.*, 477 (1980).

481. A. P. Kozikowski, R. J. Schmiesing, and K. L. Sorgi, *J. Am. Chem. Soc.*, **102**, 6577 (1980).

482. G. Jaurand, J.-M. Beau and P. Sinaÿ, *J. Chem. Soc., Chem. Commun.*, 572 (1981).

483. R. Baudat and M. Petrzilka, *Helv. Chim. Acta*, **62**, 1406 (1979).

484. D. L. J. Clive, C. G. Russell, and S. C. Suri, *J. Org. Chem.*, **47**, 1632 (1982).

485. K. C. Nicolaou, R. L. Magolda, and W. J. Sipio, *Synthesis*, 982 (1979).

486. R. K. Boeckman, Jr. and S. S. Ko, *J. Am. Chem. Soc.*, **102**, 7146 (1980).

487. M. F. Semmelhack and A. Zask, *J. Am. Chem. Soc.*, **105**, 2034 (1983).

488. M. Petrzilka, *Helv. Chim Acta*, **61**, 2286 (1978).

489. R. Pitteloud and M. Petrzilka, *Helv. Chim. Acta*, **62**, 1319 (1979).

490. M. Petrzilka, U.S. Patent 4,234,741, 1980; *Chem. Abstr.*, **94**, 156341p (1981).

488. M. Petrzilka, *Helv. Chim. Acta*, **61**, 2286 (1978).

492. P. Metz and H.-J. Schäfer, *Tetrahedron Lett.*, **23**, 4067 (1982).

493. A. G. González, C. Betancor, C. G. Francisco, R. Hernández, J. A. Salazar, and E. Suárez, *Tetrahedron Lett.*, 2959 (1977).

494. G. H. Schmid, in *The Chemistry of the Carbon-Carbon Triple Bond*, S. Patai, Ed., Wiley, London, 1978, Chap. 8.

495. L. M. Kataeva, E. G. Kataev, and T. G. Mannafov, *Zh. Strukt. Khim.*, **7**, 226 (1966).

496. L. M. Kataeva, E. G. Kataev, and T. G. Mannafov, *Zh. Strukt. Khim.*, **10**, 830 (1969).

497. E. G. Kataev, T. G. Mannafov, and Y. Y. Samitov, *Zh. Org. Khim.*, **11**, 2324 (1975).

498. E. G. Kataev, T. G. Mannafov, and M. K. Mannanov, *Kinet. Katal.*, **9**, 1161 (1968).

499. G. H. Schmid and D. G. Garratt, *Chem. Scr.*, **10**, 76 (1976).

500. E. G. Kataev, T. G. Mannafov, and O. O. Saidov, *Zh. Org. Khim.*, **7**, 2229 (1971).

501. C. N. Filer, D. Ahern, R. Fazio, and E. J. Shelton, *J. Org. Chem.*, **45**, 1313 (1980).

502. D. G. Garratt, P. L. Beaulieu, and V. M. Morisset, *Can. J. Chem.*, **59**, 927 (1981).

503. K. Uneyama, K. Takano, and S. Torii, *Tetrahedron Lett.*, **23**, 1161 (1982).

504. A. J. Bridges and J. W. Fischer, *Tetrahedron Lett.*, **24**, 445 (1983).

505. A. J. Bridges and J. W. Fischer, *Tetrahedron Lett.*, **24**, 447 (1983).

506. J. Hooz and R. Mortimer, *Tetrahedron Lett.*, 805 (1976).

507. J. Hooz and R. D. Mortimer, *Can. J. Chem.*, **56**, 2786 (1978).

508. S. Tomoda, Y. Takeuchi, and Y. Nomura, *Chem. Lett.*, 1715 (1981).

509. H. J. Reich, J. M. Renga, and J. E. Trend, *Tetrahedron Lett.*, 2217 (1976).

510. C. D. Hurd and O. Fancher, *Int. J. Sulfur Chem., Part A*, **1**, 18 (1971).

511. W. Ried and G. Sell, *Synthesis*, 447 (1976).

512. M. D. Stadnichuk, V. A. Ryazantsev, and A. A. Petrov, *Zh. Obshch. Khim.*, **49**, 956 (1979).

513. V. A. Ryazantsev, M. D. Stadnichuk, and A. A. Petrov, *Zh. Obshch. Khim.*, **50**, 1301 (1980).

514. V. A. Ryazantsev, M. D. Stadnichuk, and A. A. Petrov, *Zh. Obshch. Khim.*, **50**, 694 (1980).

515. D. G. Garratt, M. Ujjainwalla, and G. H. Schmid, *J. Org. Chem.*, **45**, 1206 (1980).

516. E. S. Mamedov, R. S. Salakhova, T. M. Gadzhily, and T. N. Shakhtakhtinskii, *Dokl. Akad. Nauk Az. SSR*, **34**, 38 (1978).

517. Y. V. Migalina, S. V. Gallabobik, I. I. Ershova, and V. I. Staninets, *Zh. Obshch. Khim.*, **52**, 1559 (1982).

518. Y. V. Migalina, S. V. Gallabobik, V. G. Lendel, and V. I. Staninets, *Zh. Obshch. Khim.*, **52**, 1563 (1982).

519. Y. V. Migalina, S. V. Gallabobik, S. M. Khripak, and V. I. Staninets, *Khim. Geterotsikl. Soedin*, 911 (1982).

520. Y. V. Migalina, S. V. Gallabobik, and V. I. Staninets, *Zh. Org. Khim.*, **18**, 2609 (1982).

521. Y. V. Migalina, V. G. Lendel, I. M. Balog, and V. I. Staninets, *Ukr. Khim. Zh.*, **47**, 1293 (1981).

522. Y. V. Migalina, V. I. Staninets, V. G. Lendel, A. S. Koz'min and N. S. Zefirov, *Khim. Geterotsikl. Soedin.*, 1633 (1977).

523. Y. V. Migalina, V. G. Lendel, A. S. Koz'min, and N. S. Zefirov, *Khim. Geterotsikl. Soedin.*, 708 (1978).

524. Y. V. Migalina, S. V. Gallabobik, V. G. Lendel, and V. I. Staninets, *Khim. Geterotsikl. Soedin.*, 1283 (1981).

525. S. Uemura, A. Onoe, and M. Okano, *Bull. Chem. Soc. Jpn.*, **48**, 3702 (1975).

526. E. R. Clark and M. A. Al-Turaihi, *J. Organomet. Chem.*, **96**, 251 (1975).

527. K. Balenović, A. Deljac, B. Gaspert, and Z. Stefanac, *Monatsh. Chem.*, **98**, 1344 (1967).

528. D. G. Garratt and G. H. Schmid, *Chem. Scr.*, **11**, 170 (1977).

529. D. G. Garratt and G. H. Schmid, *J. Org. Chem.*, **42**, 1776 (1977).

530. D. G. Garratt and G. H. Schmid, *Chem. Scr.*, **15**, 132 (1980).

531. D. D. Lawson, and N. Kharasch, *J. Org. Chem.*, **24**, 857 (1959).

532. A. P. Komin and M. Carmack, *J. Heterocycl. Chem.*, **13**, 13 (1976).

533. M. L. Kaplan, R. C. Haddon, F. C. Schilling, J. H. Marshall, and F. B. Bramwell, *J. Am. Chem. Soc.*, **101**, 3306 (1979).

534. M. Shiozaki and T. Hiraoka, *Tetrahedron*, **23**, 3457 (1982).

535. M. R. Czarny, *Synth. Commun.*, **6**, 285 (1976).

536. G. Sosnovsky and J. A. Krogh, *Z. Naturforsch., Ser. B*, **34**, 511 (1979).

537. K. Tanaka, F. Uchiyama, T. Ikeda, and Y. Inubushi, *Chem. Pharm. Bull.*, **31**, 1958 (1983).

538. K. Kakiuchi, T. Nakao, M. Takeda, Y. Tobe, and Y. Odaira, *Tetrahedron Lett.*, **25**, 557 (1984).

539. R. Baker, C. L. Gibson, C. J. Swain, and D. J. Tapolczay, *J. Chem. Soc. Chem. Commun.*, 619 (1984).

540. D. F. Crowe, P. H. Christie, J. I. DeGraw, A. N. Fujiwara, E. Grange, P. Lim, M. Tanabe, T. Cairns, and G. Skelly, *Tetrahedron*, **39**, 3083 (1983).

541. P. Wieland and J. Kalvoda, *Tetrahedron Lett.*, **24**, 5603 (1983).

542. M. E. Jung and G. L. Hatfield, *Tetrahedron Lett.*, **24**, 2931 (1983).

543. F. G. Prendergast, M. Meyer, G. L. Carlson, S. Iida, and J. D. Potter, *J. Biol. Chem.*, **258**, 7541 (1983).

544. G. Mehta and M. S. Nair, *J. Chem. Soc. Chem. Commun.*, 439 (1983).

545. T. Sato, Y. Gotoh, M. Watanabe, and T. Fujisawa, *Chem. Lett.*, 1533 (1983).

546. T. A. Lyle, H. B. Mereyala, A. Pascual, and B. Frei, *Helv. Chim. Acta*, **67**, 774 (1984).

547. A. K. Bhattacharya and B. Miller, *J. Org. Chem.*, **48**, 2412 (1983).

548. W. J. Cummins, M. G. B. Drew, J. Mann, and E. B. Walsh, *J. Chem. Soc., Perkin Trans. 1*, 167 (1983).

549. S. L. Schreiber and C. Santini, *J. Am. Chem. Soc.*, **106**, 4038 (1984).

550. Y. Tobe, T. Iseki, K. Kakiuchi, and Y. Odaira, *Tetrahedron Lett.*, **25**, 3895 (1984).

551. K. Furuta, A. Misumi, A. Mori, N. Ikeda, and H. Yamamoto, *Tetrahedron Lett.*, **25**, 669 (1984).

552. L. Lombardo and L. N. Mander, *J. Org. Chem.*, **48**, 2298 (1983).

553. T. K. Chakraborty and S. Chandrasekaran, *Tetrahedron Lett.*, **25**, 2895 (1984).

554. T. Ohta, M. Sunagawa, K. Nishimaki, and S. Nozoe, *Heterocycles*, **20**, 1567 (1983).

555. K. Tomioka, T. Ishiguro, Y. Iitaka, and K. Koga, *Tetrahedron*, **40**, 1303 (1984).

556. M. Kocór, M. M. Kabat, J. Wicha, and W. Peczyńska-Czoch, *Steroids*, **41**, 55 (1983).

557. K. Suga, S. Watanabe, T. Fujita, and S. Inoki, *Yukagaku*, **32**, 447 (1983); via *Chem. Abstr.*, **99**, 212388b (1983).

558. K. S. Bhat and A. S. Rao, *Ind. J. Chem. Sec. B*, **22B**, 360 (1983).

559. K. Yamakawa, K. Nishitani, and A. Murakami, *Chem. Pharm. Bull.*, **31**, 3411 (1983).

560. K. Yamakawa, K. Nishitani, A. Murakami, and A. Yamamoto, *Chem. Pharm. Bull.*, **31**, 3397 (1983).

561. C. G. Chavdarian, *J. Org. Chem.*, **48**, 1529 (1983).

562. P. E. Eaton and W. H. Bunnelle, *Tetrahedron Lett.*, **25**, 23 (1984).

563. H. Niwa, Y. Uosaki, and K. Yamada, *Tetrahedron Lett.*, **24**, 5731 (1983).

564. S. Ohuchida, N. Hamanaka, and M. Hayashi, *Tetrahedron*, **39**, 4269 (1983).

565. S. Raucher and R. F. Lawrence, *Tetrahedron Lett.*, **24**, 2927 (1983).

566. H. Hayakawa, H. Tanaka, and T. Miyasaka, *Chem. Pharm. Bull.*, **30**, 4589 (1982).

567. S. A. Harkin and E. J. Thomas, *Tetrahedron Lett.*, **24**, 5535 (1983).

568. M. L. Durrant and E. J. Thomas, *J. Chem. Soc. Perkin Trans. 1*, 901 (1984).

569. M. Arita, K. Adachi, Y. Ito, H. Sawai, and M. Ohno, *J. Am. Chem. Soc.*, **105**, 4049 (1983).

570. J. Salaün and Y. Almirantis, *Tetrahedron*, **39**, 2421 (1983).

571. R. McCague, C. J. Moody, and C. W. Rees, *J. Chem. Soc. Perkin Trans. 1*, 2399 (1983).

572. A. P. Kozikowski and M. N. Greco, *J. Org. Chem.*, **49**, 2310 (1984).

573. T. Sakakibara, M. D. Manandhar, and Y. Ishido, *Synthesis*, 920 (1983).

574. T. Minami, H. Sako, T. Ikehira, T. Hanamoto, and I. Hirao, *J. Org. Chem.*, **48**, 2569 (1983).

575. A. L. Braga, J. V. Comasseto, and N. Petragnani, *Synthesis*, 240 (1984).

576. H. Schmidbaur, C. Zybill, C. Krüger and H.-J. Kraus, *Chem. Ber.*, **116**, 1955 (1983).

577. B. B. Snider and G. B. Phillips, *J. Org. Chem.*, **48**, 3685 (1983).

578. L. Duhamel, J.-M. Poirier, and N. Tedga, *J. Chem. Research* (S), 222 (1983).

579. V. P. Litvinov, I. A. Dzhumaev, and B. M. Zolotarev, *Izv. Akad. Nauk SSSR, Ser. Khim.*, 2105 (1983).

580. W. Kulik, H. D. Verkruijsse, R. L. P. de Jong, H. Hommes, and L. Brandsma, *Tetrahedron Lett.*, **24**, 2203 (1983).

581. H. J. Reich, M. J. Kelly, R. E. Olson, and R. C. Holtan, *Tetrahedron*, **39**, 949 (1983).

582. G. Brooks and E. Hunt, *J. Chem. Soc. Perkin Trans. 1*, 2513 (1983).

583. F. S. Guziec, Jr., and C. A. Moustakis, *J. Org. Chem.*, **49**, 189 (1984).

584. R. Okazaki, A. Ishii, and N. Inamoto, *J. Chem. Soc. Chem. Commun.*, 1429 (1983).

585. D. H. R. Barton, G. Bashiardes, and J.-L. Fourrey, *Tetrahedron Lett.*, **25**, 1287 (1984).

586. T. L. Gilchrist, C. W. Rees, and D. Vaughan, *J. Chem. Soc. Perkin Trans. 1*, 49 (1983).

587. P. Lerouge and C. Paulmier, *Tetrahedron Lett.*, **25**, 1983 (1984).

588. P. Lerouge and C. Paulmier, *Tetrahedron Lett.*, **25**, 1987 (1984).

589. C. G. Francisco, R. Freire, R. Hernández, J. A. Salazar, and E. Suarez, *Tetrahedron Lett.*, **25**, 1621 (1984).

590. K. Nagao, M. Chiba, and S.-W. Kim, *Chem. Pharm. Bull.*, **31**, 414 (1983).

591. J. M. Dewanckele, F. Zutterman, and M. Vandewalle, *Tetrahedron*, **39**, 3235 (1983).

592. C. Boullais, N. Zylber, J. Zylber, J. Guilhem, and A. Gaudemer, *Tetrahedron*, **39**, 759 (1983).

593. T. Uematsu, T. Umemura, and K. Mori, *Agric. Biol. Chem.*, **47**, 597 (1983).

594. J. P. Kutney and A. K. Singh, *Can. J. Chem.*, **61**, 1111 (1983).

595. I. Paterson, *Tetrahedron Lett.*, **24**, 1311 (1983).

596. M. Arnó, B. García, J. R. Pedro, and E. Seoane, *Tetrahedron Lett.*, **24**, 1741 (1983).

597. S. L. Schreiber and K. Satake, *J. Am. Chem. Soc.*, **105**, 6723 (1983).

598. S. L. Schreiber and K. Satake, *J. Am. Chem. Soc.*, **106**, 4186 (1984).

599. R. E. Ireland, W. C. Dow, J. D. Godfrey, and S. Thaisrivongs, *J. Org. Chem.*, **49**, 1001 (1984).

600. T. R. Hoye and M. J. Kurth, *Tetrahedron Lett.*, **24**, 4769 (1983).

601. T. Matsumoto, S. Imai, T. Hirata, Y. Fukuda, T. Yamaguchi, and K. Inoue, *Bull. Chem. Soc. Jpn.*, **56**, 3471 (1983).

602. P. G. Gassman and S. M. Bonser, *Tetrahedron Lett.*, **24**, 3431 (1983).

603. G. Stork and K. S. Atwal, *Tetrahedron Lett.*, **24**, 3819 (1983).

604. R. L. Funk, L. H. M. Horcher II, J. U. Daggett, and M. M. Hansen, *J. Org. Chem.*, **48**, 2632 (1983).

605. T. Matsumoto, S. Imai, K. Ondo, H. Katoaka, and K. Kato, *Bull. Chem. Soc. Jpn.*, **56**, 2985 (1983).

606. B. Danieli, G. Lesma, G. Palmisano, and S. Tollari, *J. Chem. Soc. Perkin Trans. 1*, 1237 (1984).

607. K. L. Chasey, L. A. Paquette, and J. F. Blount, *J. Org. Chem.*, **47**, 5262 (1982).

608. U. Schmidt, A. Lieberknecht, H. Bökens, and H. Griesser, *J. Org. Chem.*, **48**, 2680 (1983).

609. J. L. Kice, F. McAfee, and H. Slebocka-Tilk, *J. Org. Chem.*, **49**, 3100 (1984).

610. J. L. Kice, F. McAfee, and H. Slebocka-Tilk, *J. Org. Chem.*, **49**, 3106 (1984).

611. G. H. Schmid and D. G. Garratt, *J. Org. Chem.*, **48**, 4169 (1983).

612. G. H. Schmid and D. G. Garratt, *Tetrahedron Lett.*, **24**, 5299 (1983).

613. Z. Janousek, S. Piettre, F. Gorissen-Hervens and H. G. Viehe, *J. Organomet. Chem.*, **250**, 197 (1983).

614. T. Aoki, T. Konoike, H. Itani, T. Tsuji, M. Yoshioka, and W. Nagata, *Tetrahedron*, **39**, 2515 (1983).

615. A. M. Morella and A. D. Ward, *Tetrahedron Lett.*, **25**, 1197 (1984).

616. R. P. Alexander and I. Paterson, *Tetrahedron Lett.*, **24**, 5911 (1983).

617. R. S. Brown, S. C. Eyley, and P. J. Parsons, *J. Chem. Soc. Chem. Commun.*, 438 (1984).

618. S. Halazy and L. Hevesi, *J. Org. Chem.*, **48**, 5242 (1983).

619. B. M. Trost, J. M. Balkovec, and M. K.-T. Mao, *J. Am. Chem. Soc.*, **105**, 6755 (1983).

620. A. B. Smith, III, B. H. Toder, R. E. Richmond, and S. J. Branca, *J. Am. Chem. Soc.*, **106**, 4001 (1984).

621. M. Perrier and F. Rouessac, *C. R. Acad. Sci. Ser. 2*, **295**, 729 (1982).

622. L. A. Paquette, T. J. Nitz, R. J. Ross, and J. P. Springer, *J. Am. Chem. Soc.*, **106**, 1446 (1984).

623. B. Umezawa, O. Hoshino, S. Sawaki, H. Sashida, K. Mori, Y. Hamada, K. Kotera, and Y. Iitaka, *Tetrahedron*, **40**, 1783 (1984).

624. D. L. J. Clive, and P. L. Beaulieu, *J. Chem. Soc. Chem. Commun.*, 307 (1983).

625. K. Uneyama, M. Ono, and S. Torii, *Phosphorus and Sulfur*, **16**, 35 (1983).

626. B. M. Trost and D. M. T. Chan, *J. Org. Chem.*, **48**, 3346 (1983).

627. A. Kunai, J. Harada, J. Izumi, H. Tachihara, and K. Sasaki, *Electrochim. Acta*, **28**, 1361 (1983).

628. A. Toshimitsu, S. Uemura, M. Okano, and N. Watanabe, *J. Org. Chem.*, **48**, 5246 (1983).

629. Suntry Ltd., Japan Kokai 8365270, 1983; *Chem. Abstr.*, **100**, 23009g (1984).

630. G. Bérubé, E. Luce, and K. Jankowski, *Bull. Soc. Chim. Fr., Part 2*, 109 (1983).

631. A. J. Bridges and J. W. Fischer, *J. Org. Chem.*, **49**, 2954 (1984).

632. E. S. Mamedov, S. B. Kurbanov, and R. D. Mishiev, *Dokl. Akad. Nauk Az. SSR*, **39**, 42 (1983).

633. R. S. Salakhova, N. F. Musaeva, M. S. Salakhov, E. S. Mamedov, and T. N. Shakhtakhtinskii, *Dokl. Akad. Nauk Az. SSR*, **39**, 37 (1983).

634. G. Lindgren and G. H. Schmid, *Chem. Scr.*, **23**, 98 (1984).

2

Organoselenium-Based Ring Closure Reactions

K. C. NICOLAOU,* N. A. PETASIS,* and D. A. CLAREMON†

*Department of Chemistry, University of Pennsylvania,
Philadelphia, Pennsylvania
†Merck Sharp and Dohme Research Laboratories,
West Point, Pennsylvania

CONTENTS

1. INTRODUCTION

Heterocycles and carbocycles feature heavily in the structures of natural products and other synthetic targets. This has made the construction of rings one of the most essential tasks in organic synthesis. As a consequence, a plethora of ring-forming reactions have been developed. Many of these reactions are restricted to certain types or sizes of rings, while others are more general.

Among the most useful ring-forming reactions are those based on an intra-molecular reaction between a nucleophilic end (Nu) and an electrophilic end

(C—X), Eq. (1). The efficiency of these ring closures depends on the nature of Nu and X, as well as the size of the ring formed.

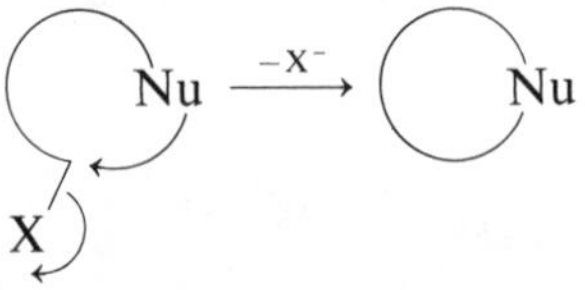

$$\text{(1)}$$

Olefinic double bonds have been utilized in these cyclizations to generate the electrophilic site, by addition of various electrophiles (E^+) to form an intermediate *onium* species that is intramolecularly captured by the nucleophilic end (Nu), Eq. (2).[1] The internal nucleophile (Nu) can be   OH, CO_2H, SR, NHR, or an olefin. Common electrophiles (E^+) that induce such cyclizations are the following: H^+,[2] X^+ (Br^+ or I^+),[3] Pb (IV),[4] Hg (II),[5] Tl (I) or Tl (III),[6] and PhS^+.[7]

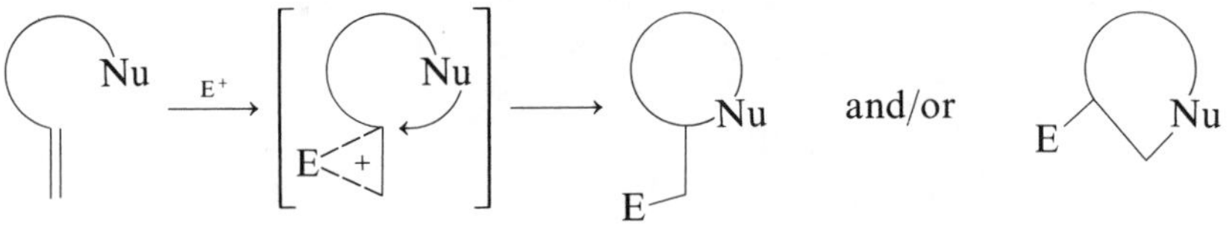

$$\text{(2)}$$

The renaissance in organoselenium chemistry during the past decade[8–10,25] has led to the recognition of some novel and powerful electrophilic organoselenium reagents such as PhSeCl and PhSeBr. These reagents were found to add readily to olefinic bonds giving β-substituted phenylselenide adducts, which could be subjected to oxidation–selenoxide *syn*-elimination, forming selenium-free products, e.g., Eq. (3).[11]

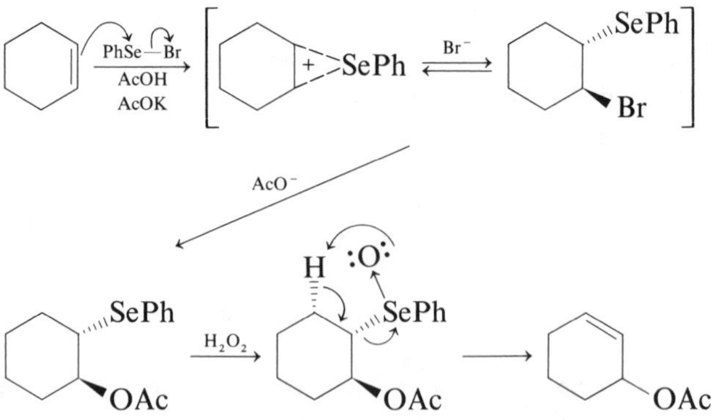

$$\text{(3)}$$

Utilizations of these organoselenium electrophiles to induce cyclizations of the type shown in Eq. (2) were explored by Nicolaou, Clive, Ley, Kametani, Sharpless, and others.[12,25] Using this approach it is possible to synthesize *O*-, *S*-, and *N*-heterocycles as well as lactones and carbocycles. The term *cyclofunctionalization* was introduced by Clive et al.[13] to describe this process. The various organoselenium induced ring closures will be the subject of this chapter.

2. PHENYLSELENO-ETHERIFICATION REACTIONS

Cyclic ethers have been synthesized traditionally through cyclizations of olefinic alcohols induced by halogens[3] or other electrophiles. Many of these reactions, however, lack generality and require rather drastic conditions. Probably the most useful example is the halo-etherification reaction that has been applied to the synthesis of cyclic vinyl ethers. This was illustrated nicely by the conversion of prostaglandin $F_{2\alpha}$ methyl ester, $\underline{1}$, to prostacyclin methyl ester, $\underline{2}$, shown in Eq. (4).[14]

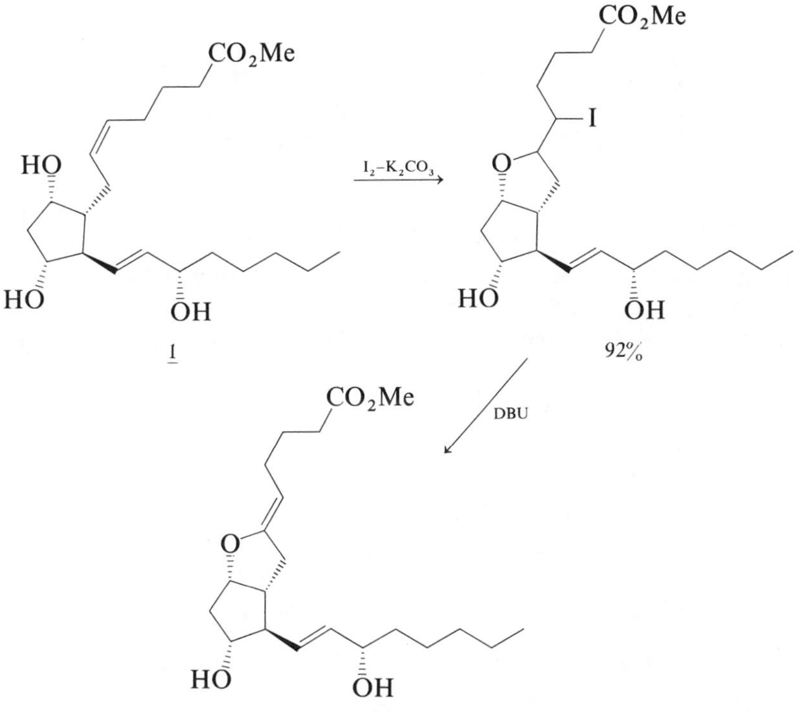

$$(4)$$

The use of PhSeCl as an electrophile in a cyclo-etherification reaction was independently examined by Nicolaou and co-workers[15,16] and Clive and co-workers.[13,17] This process, which turned out to be a mild and efficient operation, was termed *phenylseleno-etherification* and is exemplified in Eq. (5).[15,16] The reaction is believed to proceed via an intermediate such as $\underline{4}$, and gives the trans adduct, $\underline{5}$. In order to neutralize the liberated HCl and avoid any side reactions resulting from it, a base such as Et_3N or anhydrous K_2CO_3 can be added to the reaction mixture. The resulting cyclic phenylseleno ether, $\underline{5}$, is a stable intermediate and can be converted to cyclic allylic ether, $\underline{7}$, by oxidation to the selenoxide, $\underline{6}$, followed by *syn* elimination, Eq. (6).[15,16] For the oxidation of selenides to selenoxides, several oxidizing agents can be used, including: H_2O_2, O_3, peracids, $NaIO_4$, or chloramine T.

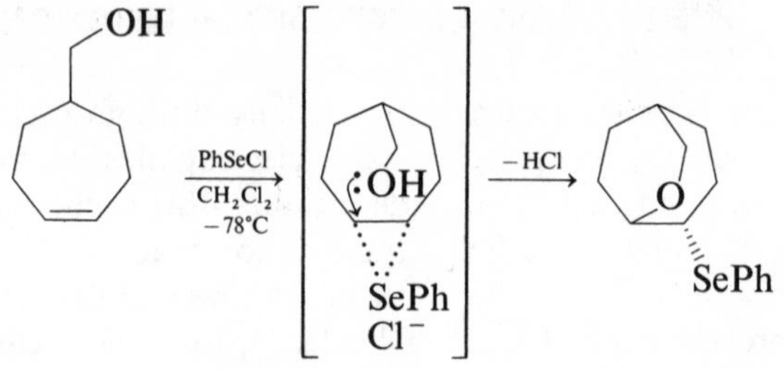

OH

$$\xrightarrow[\substack{CH_2Cl_2 \\ -78°C}]{PhSeCl}$$

$\Big[$:OH   SePh   Cl⁻ $\Big]$

$\xrightarrow{-HCl}$

O   SePh

(5)

<u>3</u>   <u>4</u>   <u>5</u>   95%

O   SePh

$$\xrightarrow[0 \rightarrow 25°C]{H_2O_2}$$

O   H   Se   O   Ph

$\longrightarrow$

O

(6)

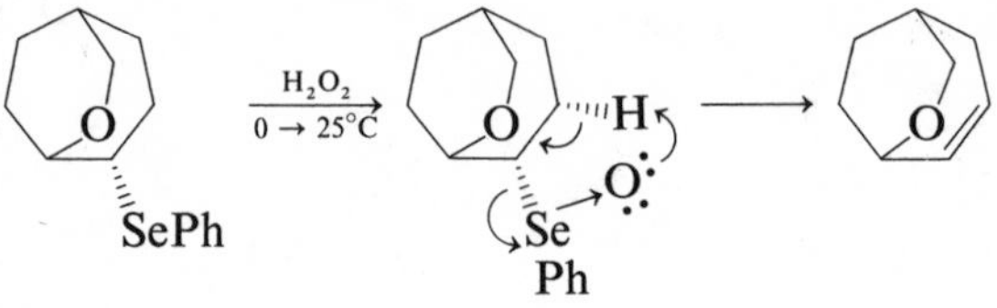

<u>5</u>   <u>6</u>   <u>7</u>   87%

The mild conditions for the introduction and elimination of the phenylseleno group, and the elimination away from oxygen, makes this methodology synthetically useful and complimentary to the halo-etherification reaction. Thus,

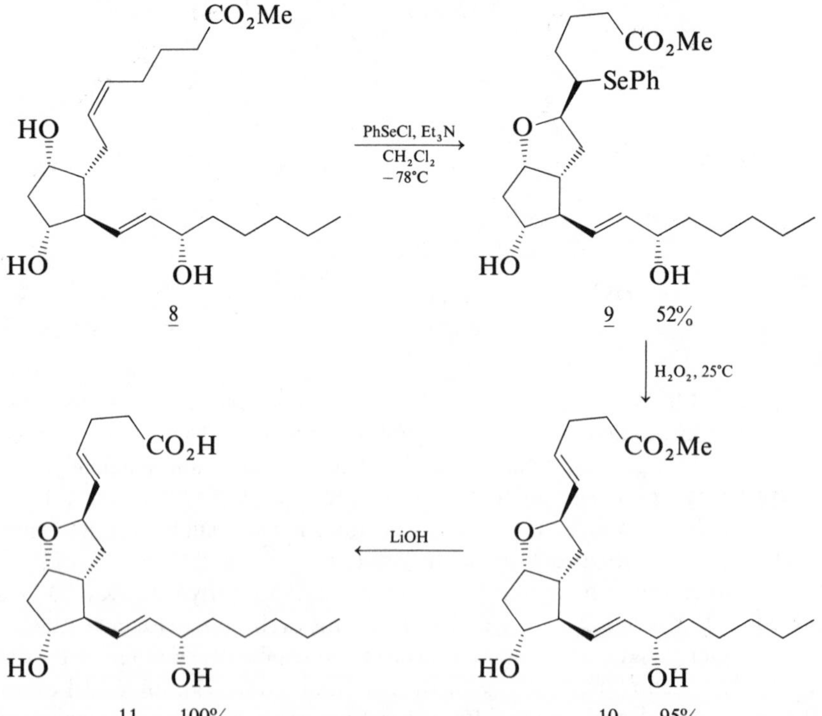

CO₂Me

HO   HO   OH

<u>8</u>

$$\xrightarrow[\substack{CH_2Cl_2 \\ -78°C}]{PhSeCl, Et_3N}$$

CO₂Me   SePh

O   HO   OH

<u>9</u>   52%

$\Big\downarrow$ H₂O₂, 25°C   (7)

CO₂Me

O   HO   OH

<u>10</u>   95%

$\xleftarrow{LiOH}$

CO₂H

O   HO   OH

<u>11</u>   100%

it was used by Nicolaou et al.[18] and by Corey et al.[19] in the synthesis of the prostacyclin analog, $\underline{11}$, from prostaglandin $F_{2\alpha}$ methyl ester, $\underline{8}$, Eq. (7).[18]

Some more examples that illustrate the versatility and efficiency of this methodology for the synthesis of cyclic allylic ethers are given in Table 1.

Joullie and co-workers[16,20] have employed this reaction in the synthesis of the muscarine analog, $\underline{12}$, Eq. (8).

This methodology was also used by Kraus and Taschner in their approach to the quassinoid skeleton, shown in Eq. (9).[21]

TABLE 1
SYNTHESIS OF CYCLIC ALLYLIC ETHERS

Substrate	Phenylseleno Ether Yield (%)	Cyclic Allylic Ether Yield (%)	Reference
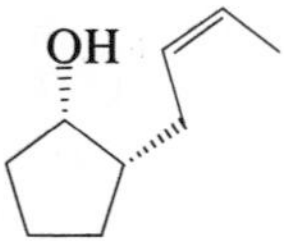	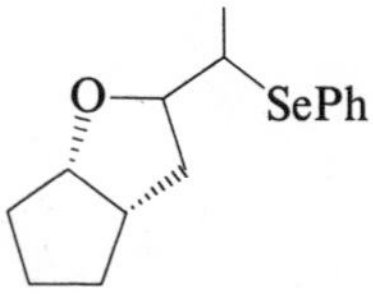 (83)	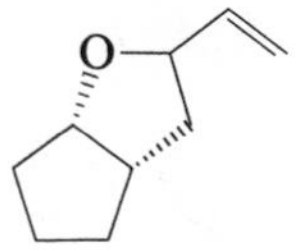 (84)	16
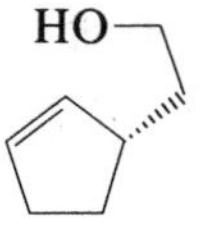	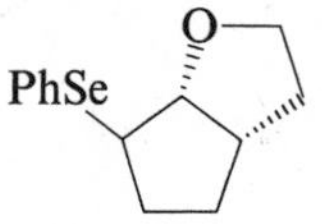 (92)	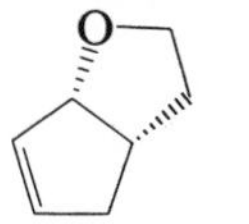 (82)	16, 17
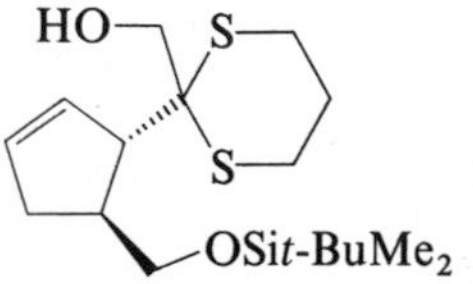	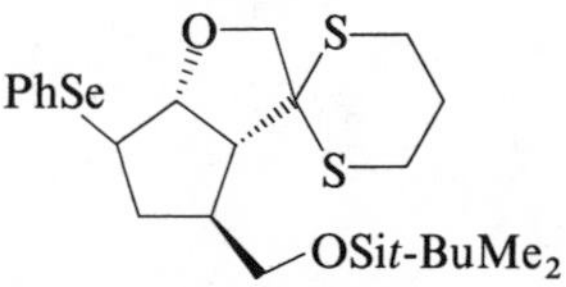 (86)	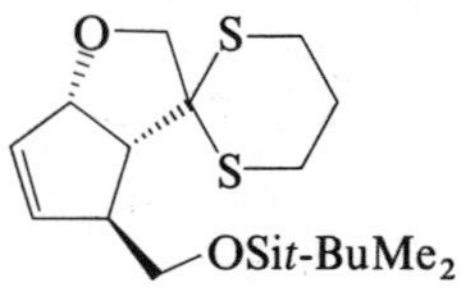 (75)	15, 16
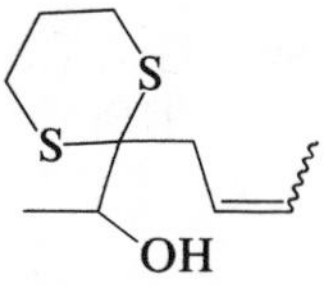	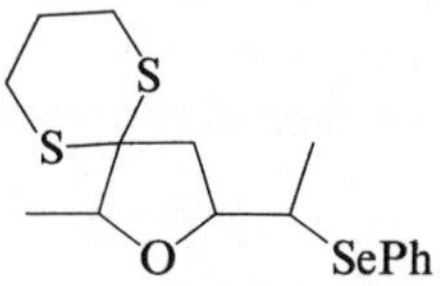 (83)	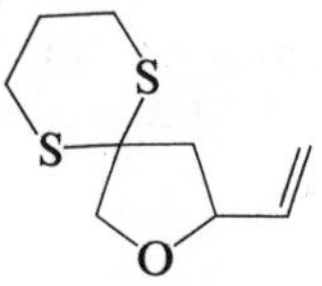 (83)	16
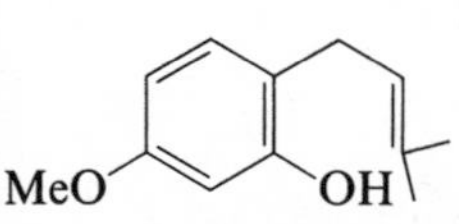	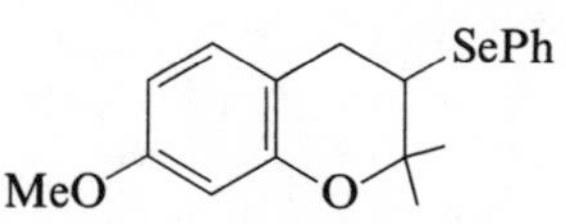 (79)	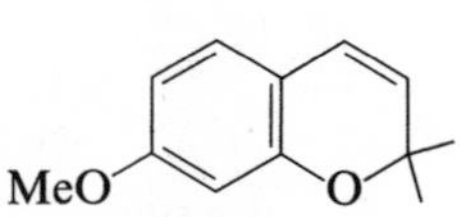 (54)	13

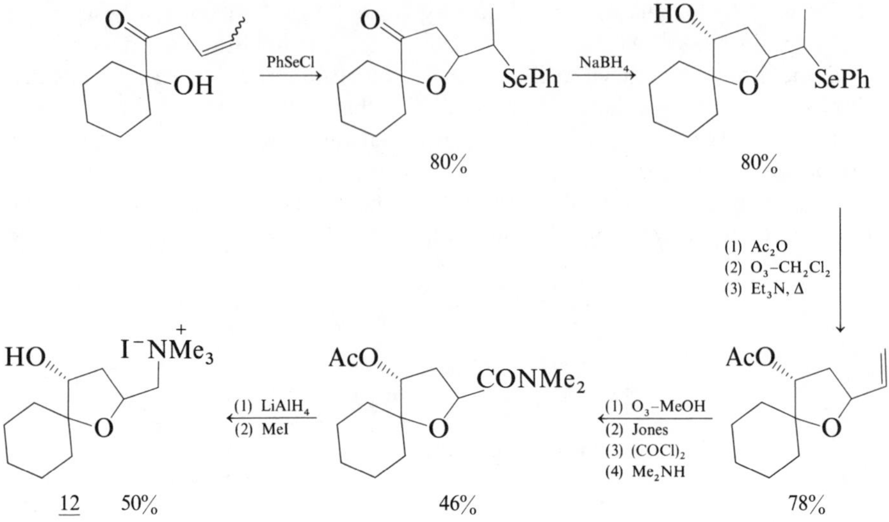

$$(8)$$

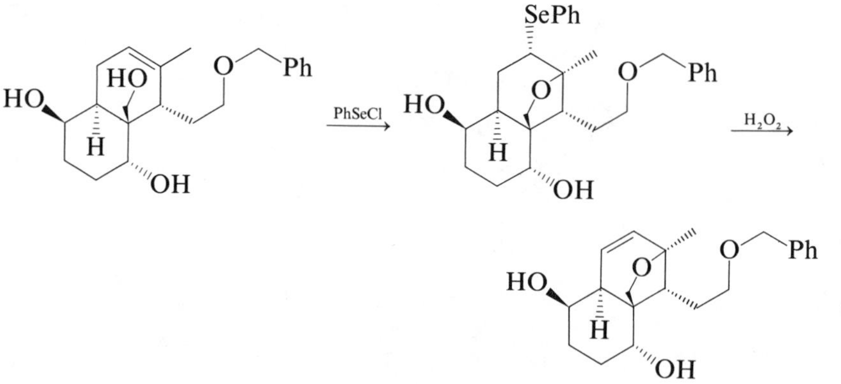

$$(9)$$

Another useful transformation of cyclic phenylselenoethers, e.g., <u>5</u>, is the reductive cleavage of the phenylselenogroup (PhSe) to form saturated cyclic ethers, <u>14</u>.[15,16] This conversion can be carried out with Raney Ni, Eq. (10).[15]

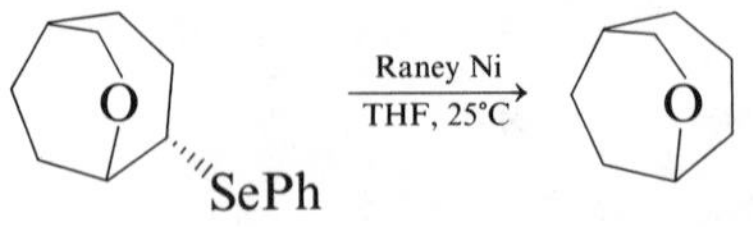

$$(10)$$

Alternatively, the PhSe group can be cleaved with tri-n-butyltin hydride (n-Bu$_3$SnH) in the presence of catalytic amounts of azobisisobutyronitrile (AIBN) as a radical initiator,[16] or with triphenyltin hydride (Ph_3SnH).[23] The tin hydride methods are particularly useful with substrates having other reducible groups, such as olefins or sulfur, but the required high temperatures make them inapplicable to thermally labile molecules. An example is given in Eq. (11).[17,22]

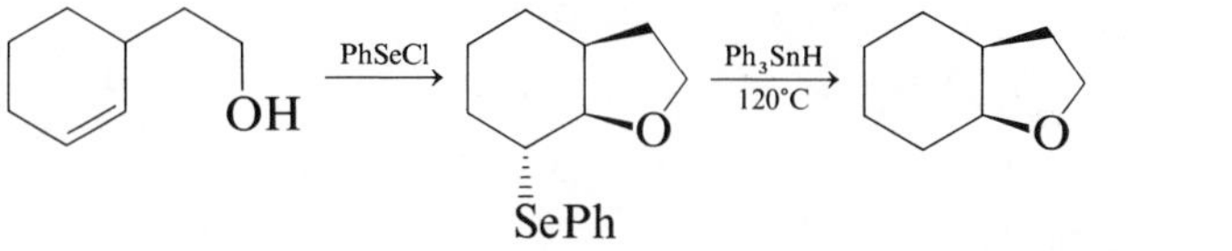

$$\text{(11)}$$

Applications of the preceding methodology to the synthesis of the prostacyclin analog $\underline{15}$[18] and the eleutherin analog $\underline{16}$[23] are shown in Eqs. (12) and (13), respectively.

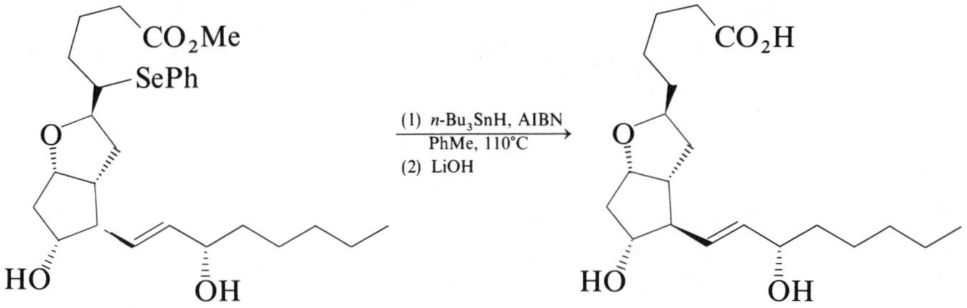

$$\text{(12)}$$

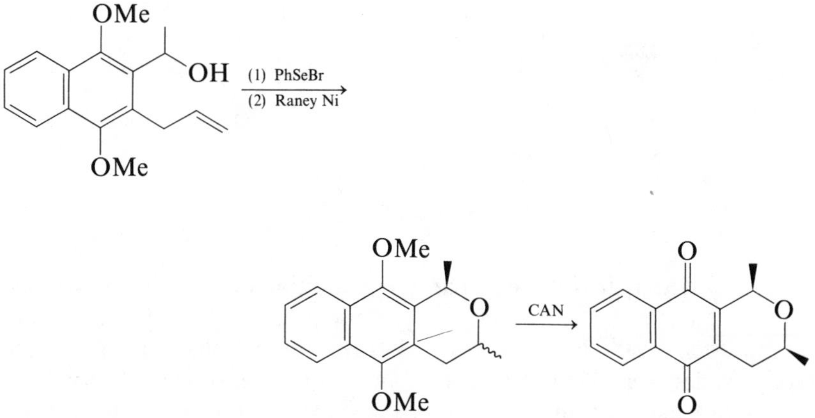

$$\text{(13)}$$

$$CAN = Ce(NH_4)(NO_3)_4$$

The required geometrical relationship between the reacting OH and olefin components in the ring closure offers a chemical tool for the determination of stereochemistry as exemplified in Eq. (14),[24] or for the differentiation of two similar OH groups, as shown in Eqs. (15).[25] Furthermore, it provides a means of structural assignment of olefins, as indicated in Eq. (16).[26]

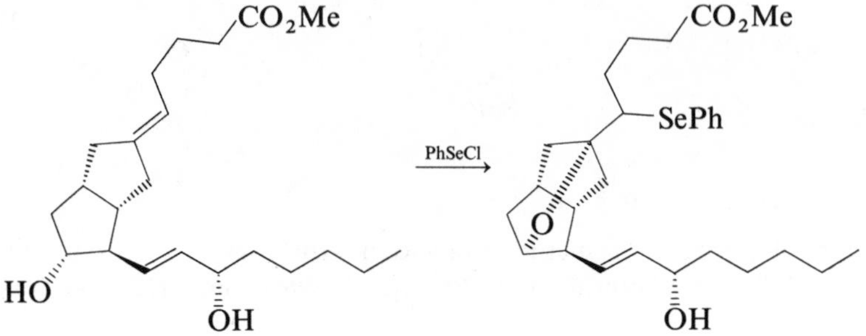

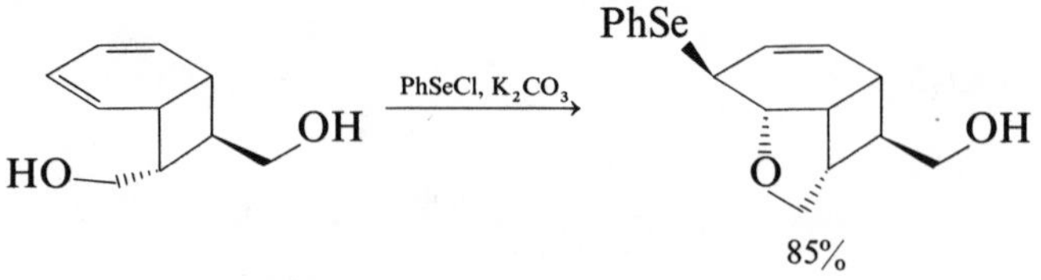

In situ generated hydroxyl groups, as in the addition of alcohols to carbonyl compounds, can also participate in phenylseleno-etherification type reactions. Current and Sharpless[27] have applied such a reaction for the synthesis of 2,6-dideoxyglycosides, e.g., $\underline{17}$, Eq. (17). Ley and Lygo[28] have cleverly used the intramolecular version of this reaction for the synthesis of spiroketals, e.g., (18) Eq. (18). The electrophilic organoselenium reagent used in this reaction is the recently introduced N-phenylselenophthalimide (N-PSP).[29]

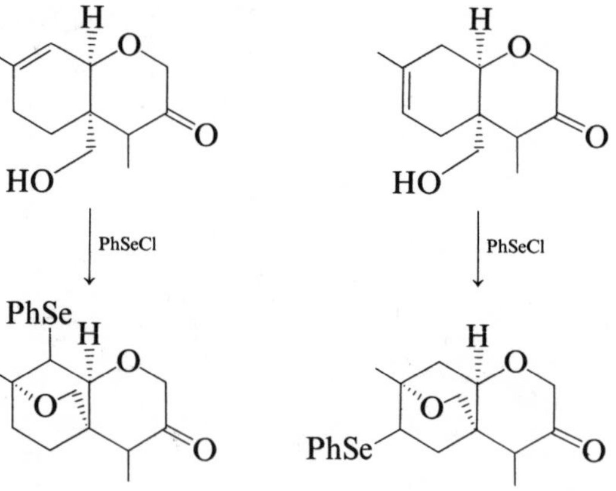

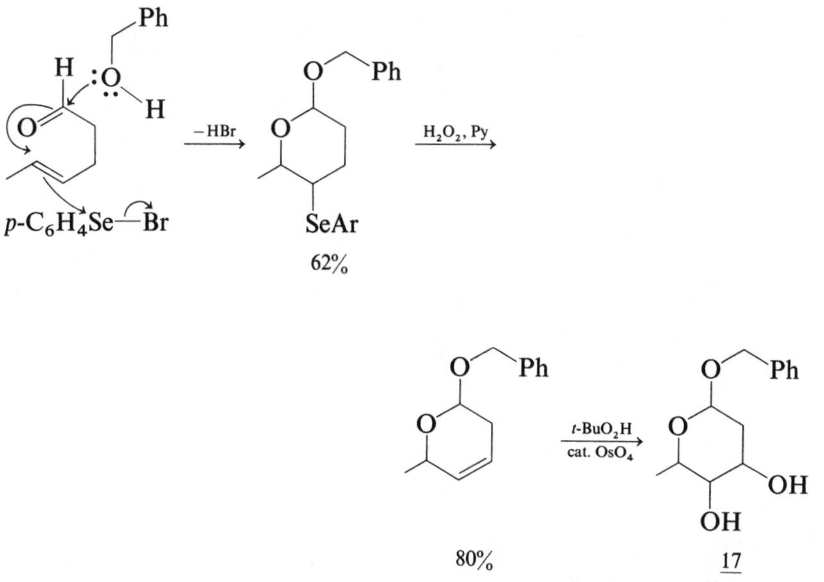

Ph

H :O H

p-C$_6$H$_4$Se—Br

$-$HBr

O—Ph

SeAr

62%

H$_2$O$_2$, Py

O—Ph

O

t-BuO$_2$H
cat. OsO$_4$

O—Ph

OH

OH

17

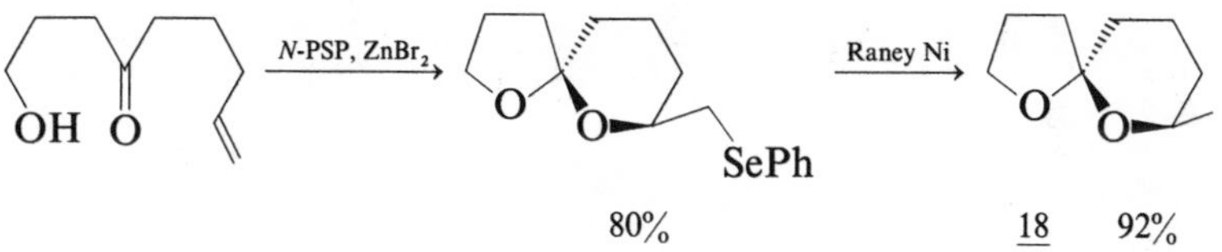

(17)

80%

OH O

N-PSP, ZnBr$_2$

O O

SePh

Raney Ni

O O

18   92%

(18)

80%

N-PSP =

O

N-SePh

O

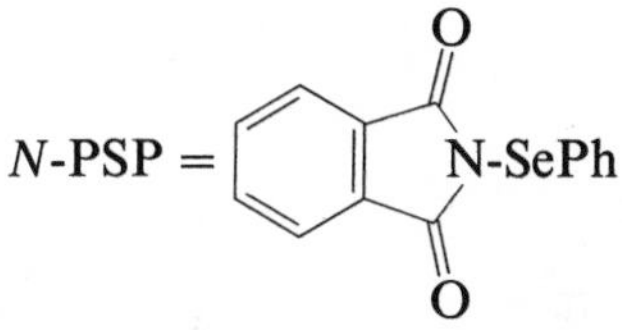

Another type of OH that was found to take part in a phenylseleno-etherifi-cation reaction is the enolic OH of β-dicarbonyl systems. Ley and co-workers[30,33] have found a variety of conditions that are effective for this transformation, which results in the formation of cyclic phenylselenovinyl ethers. Reductive elimination of the PhSe group from these adducts with a tin hydride reagent, gives the corresponding selenium-free products. Table 2 shows some of these cyclic vinyl ethers.

β-Hydroxyselenides, resulting from the addition of PhSeOH across an ole-finic bond, can undergo phenylseleno etherification with another olefinic bond present in the molecule leading to an overall conversion of a diene to a cyclic ether bearing two PhSe groups.[34] These PhSe groups can then be eliminated oxidatively or reductively to give unsaturated or saturated cyclic ethers, respec-tively. This process is illustrated with 1,5-cyclooctadiene, 19, Eq. (19).[34]

TABLE 2
SYNTHESIS OF CYCLIC VINYL ETHERS

Substrate	Conditions	Phenylseleno Ether Yield (%)	Cyclic Vinyl Ether Yield (%)	References
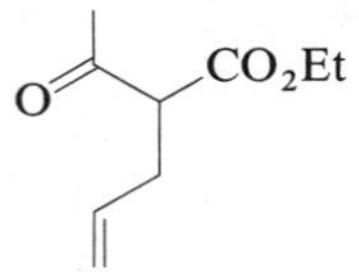	N-PSP, ZnI_2 CH_2Cl_2	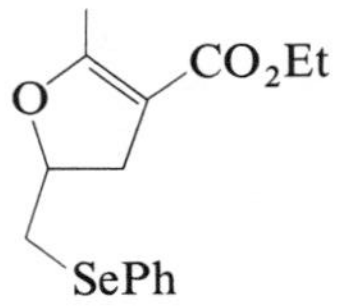 (76)	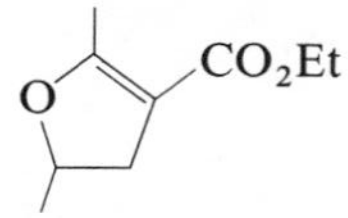 (92)	30
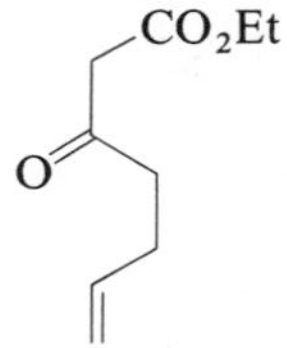	N-PSP, ZnI_2 CH_2Cl_2	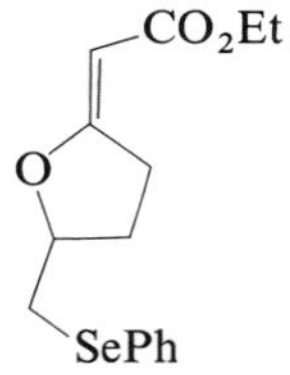 (82)	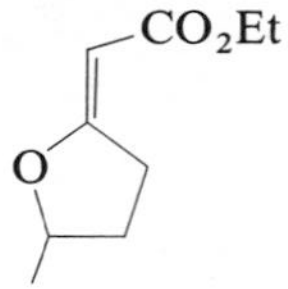	30
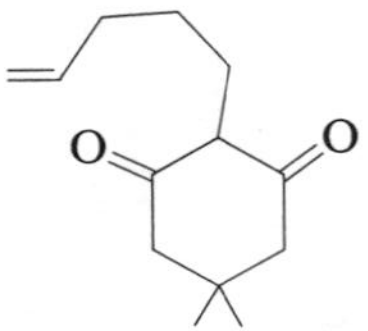	N-PSP, TsOH CH_2Cl_2	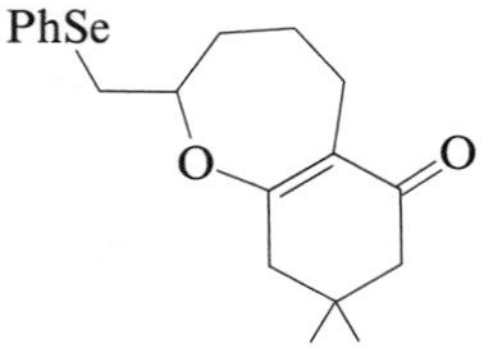 (49)	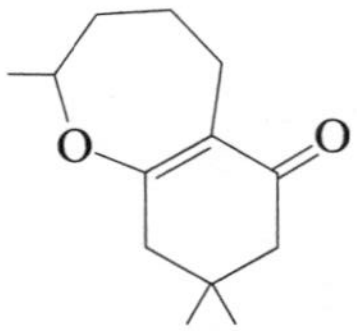 (91)	30

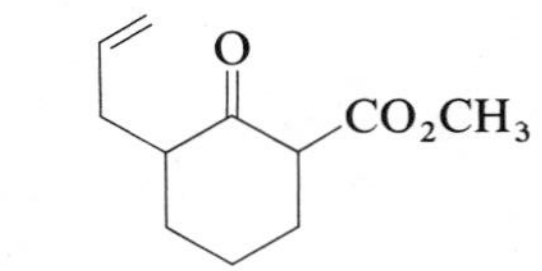	PhSePF$_6$ CH$_2$Cl$_2$	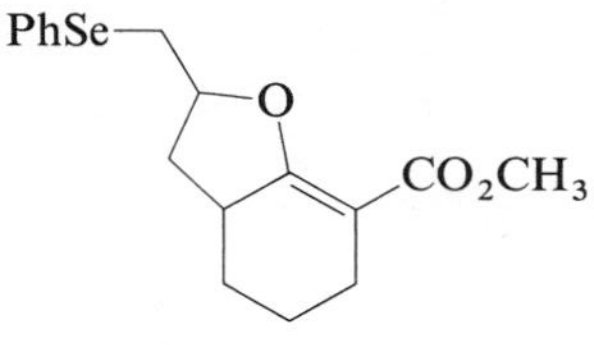 (70)	31
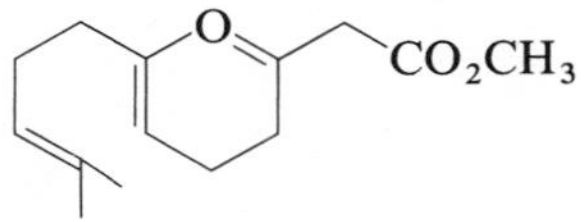	PhSeSbF$_6$ CH$_2$Cl$_2$	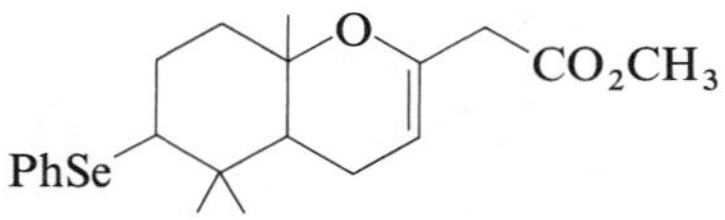 (58)	31
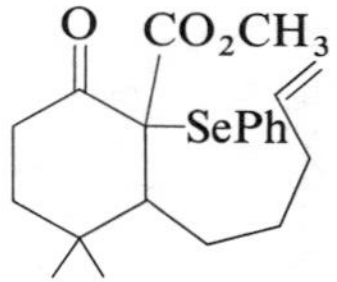	cat. SnCl$_4$ CH$_2$Cl$_2$	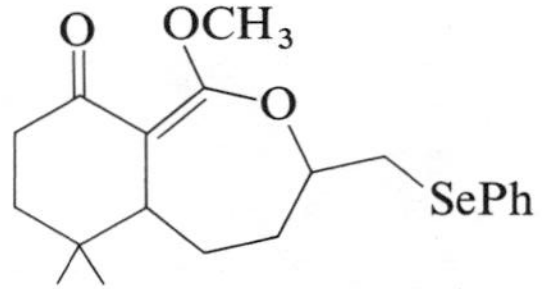 (84)	32
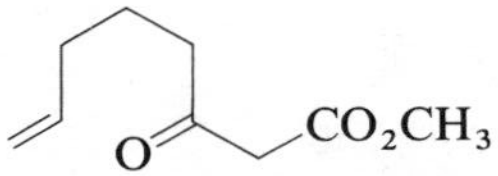	N-PSP cat. SnCl$_4$ CH$_2$Cl$_2$	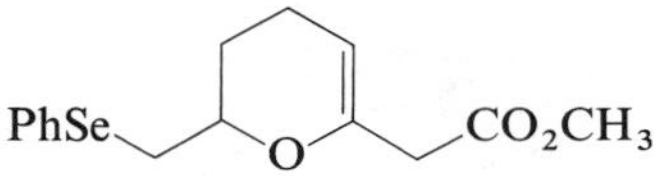 (84) 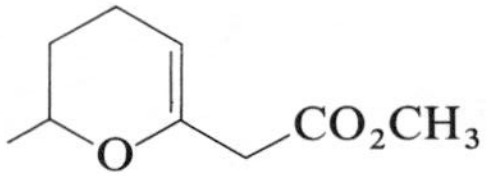 (64)	33

CO$_2$CH$_3$

PhSe

OCH$_3$

SePh

TABLE 3
SYNTHESIS OF CYCLIC ETHERS FROM DIOLEFINS

Diolefin	Conditions	Diphenylseleno Ether Yield (%)	Cyclic Ether Yield (%)	References
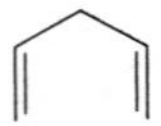	N-PSP–H_2O	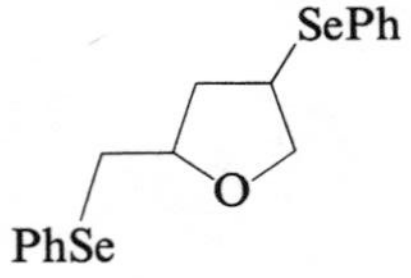 (55)		29
	PhSeSePh–H_2O_2	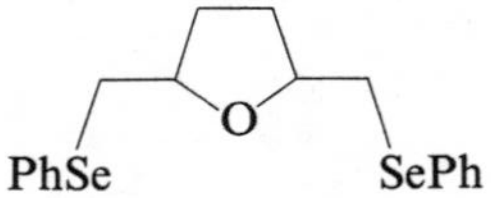 (64)		34
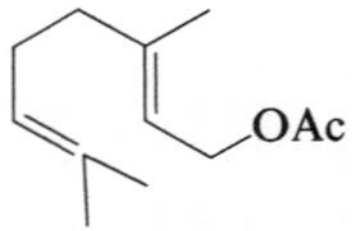	PhSeSePh–H_2O_2	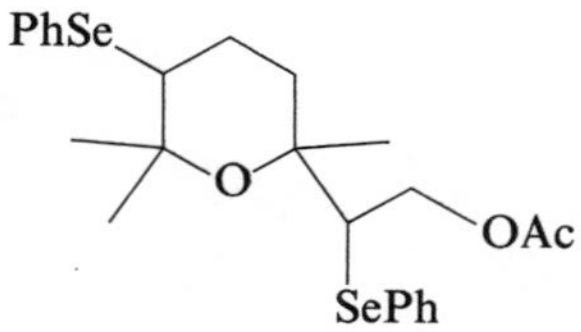 (58)		34
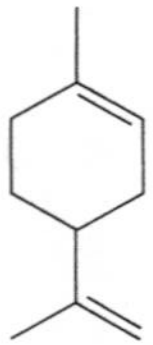	PhSeSePh–H_2O_2	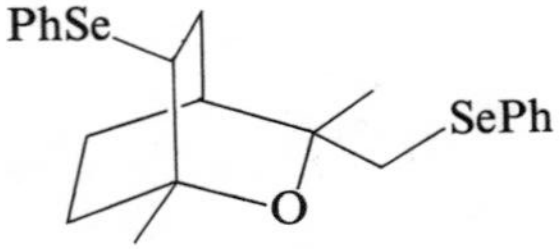 (47)	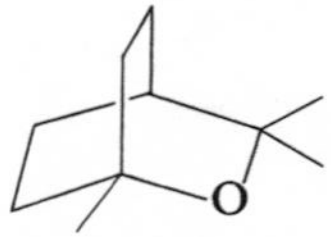 (80)	34

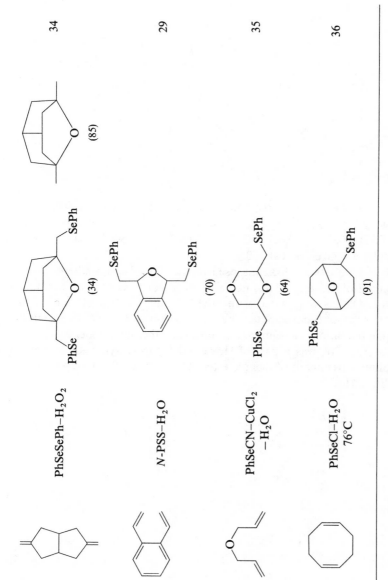

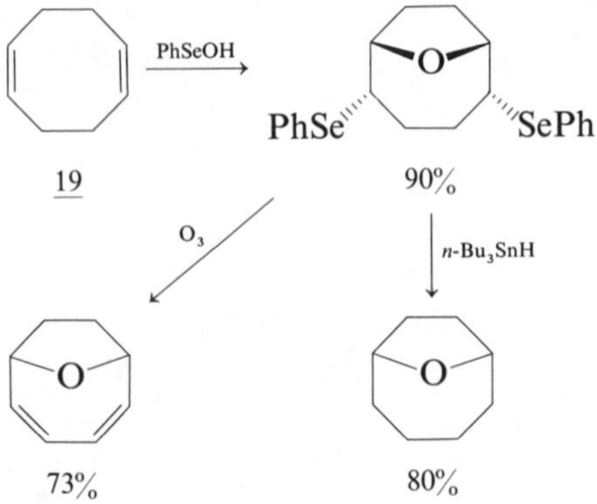

$$(19)$$

Several sources of PhSeOH were found to be applicable in reaction (19). They are the following: PhSeSePh–H_2O_2,[34] (N-PSP)–H_2O,[29] N-Phenylseleno succinimide (N-PSS)–H_2O,[29] PhSeCN–$CuCl_2$–H_2O[35] and PhSeCl–H_2O.[36] Some typical examples are given in Table 3.

The most common type of olefins used in the phenylseleno-etherification reaction are the isolated ones. This, however, does not constitute a limitation of the reaction. Olefins conjugated with other olefins [e.g., Eq. (15)] or with carbonyl groups can also participate. Thus, cyclization of 20 gave the selenide, 21, which upon oxidation–*syn* elimination furnished the butenolide, 22, (20).[37] This sequence, Eq. (20), was a part of Hoye and Caruso's synthesis of ancistrofuran.[37] Similar treatment of dienoic ester, 23, produced the 1,4-cyclization adduct, 24, Eq. (21).[37]

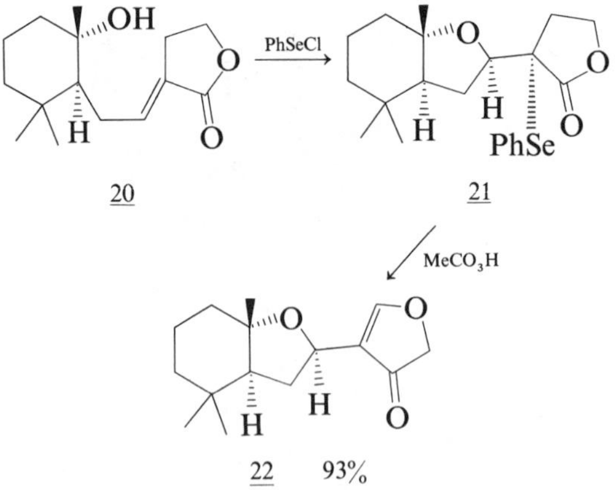

$$(20)$$

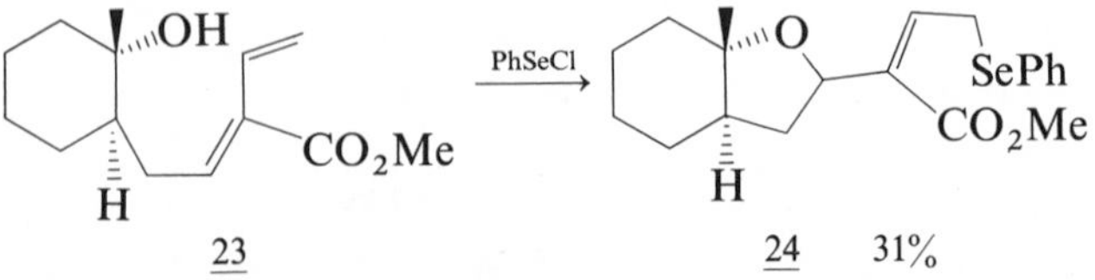

$$(21)$$

More recently, Kane and Mann[38] subjected a ribose derivative, <u>25</u>, to this reaction and found that the cyclization proceeded with high stereoselectivity giving only the selenide, <u>26</u>, and none of its α isomer, Eq. (22).[38]

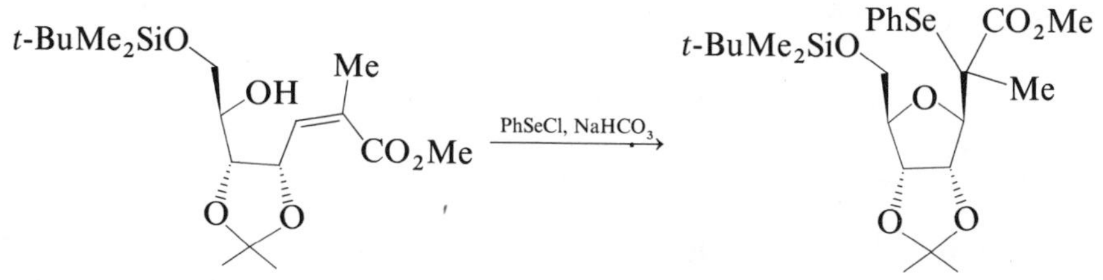

$$\tag{22}$$

3. PHENYLSELENO LACTONIZATION REACTIONS

Lactones are not only found in the structures of many natural products, but they also constitute very useful synthetic intermediates, especially for the stereo-controlled building of complex molecules. The cyclization of hydroxy acids (lactonization) mediated by a variety of reagents is one of the most effective routes to lactones, particularly useful for macrolide synthesis.[39] Other important precursors of lactones are the acyclic olefinic carboxylic acids, <u>27</u>, which are cyclized with the aid of an electrophile (E^+), in the general fashion shown in Eq. (23).

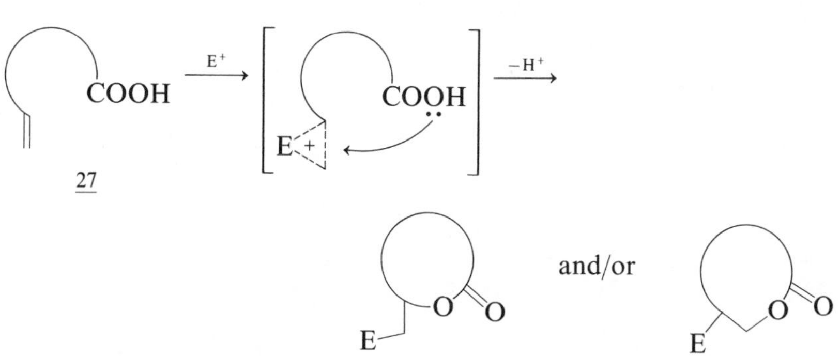

$$\tag{23}$$

Among the most widely used electrophiles in this transformation are the halogens (bromide or iodine). A typical example of this halo-lactonization reaction[3] is shown in Eq. (24), taken from the classic synthesis of prostaglandins by Corey et al.[40]

The preceding lactonization procedures, however, suffer from several limitations, including rather undesirable conditions and incompatibility with other

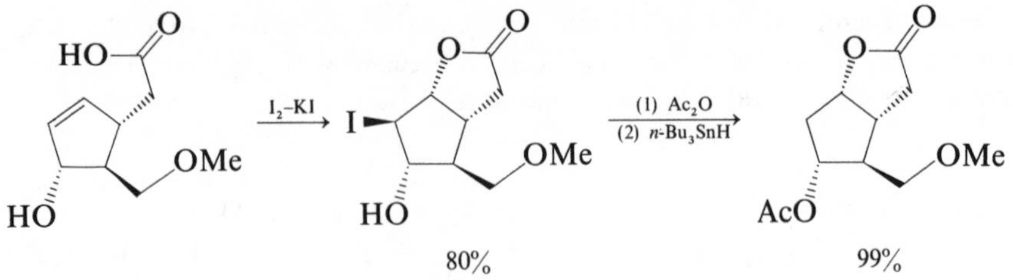

$$(24)$$

functionalities. A much milder method for the lactonization of olefinic acids, based on electrophilic organoselenium reagents was developed by Nicolaou and co-workers[41,42] and by Clive and co-workers.[43,44] The reaction was termed *phenylseleno lactonization,* and is illustrated with 28, Eq. (25).[41,42] As indicated, this reaction proceeds similarly to the phenylseleno etherification discussed previously in Section 2. The reaction works in the absence or presence of base, but is advantageous to perform the carboxylate salt with triethylamine (Et$_3$N), pyridine (Py) or anhydrous potassium carbonate (K$_2$CO$_3$). The stereochemistry of the selenolactones, such as 29, is again trans, as confirmed by X-ray crystallography.[42] The selenolactones, 29, can be further elaborated oxidatively to the unsaturated lactones, 30, or reductively to their saturated derivatives, 31, Eq. (25).

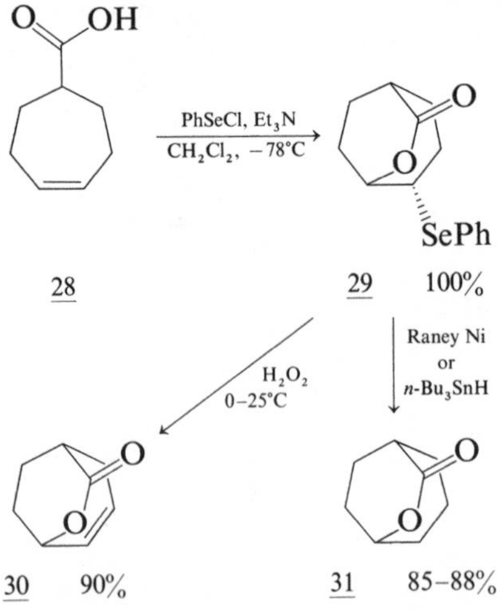

$$(25)$$

An important feature of the phenylseleno-lactonization reaction is that it forms preferably the five-membered rings over the four- or six-membered rings, the six- over the seven- and the seven- over the eight-membered rings. The versatility and synthetic utility of this reaction is demonstrated by the examples given in Table 4.

TABLE 4
SYNTHESIS OF LACTONES

Substrate	Selenolactone Yield (%)	Unsaturated Lactone Yield (%)	Saturated Lactone Yield (%)	References
(structure)	PhSe (30–35)			42, 45
(structure)	PhSe (95)	(93)		45
(structure)	PhSe (80–100)			29, 42, 43
(structure)	PhSe (74)			42
(structure)	PhSe (70)			42

TABLE 4 (*Continued*)
SYNTHESIS OF LACTONES

Substrate	Selenolactone Yield (%)	Unsaturated Lactone Yield (%)	Saturated Lactone Yield (%)	References
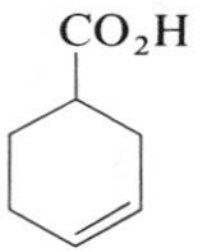 CO_2H	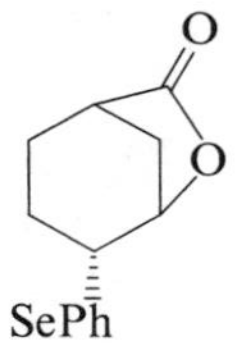 (90) SePh	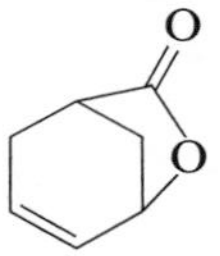 (81)	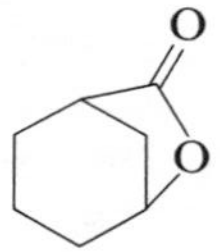 (80)	41, 42
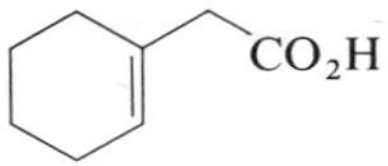 CO_2H	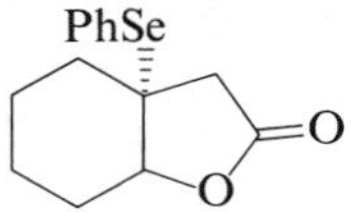 (91) PhSe	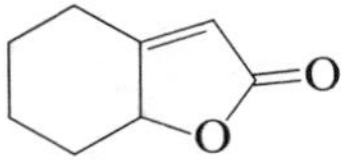 (87)	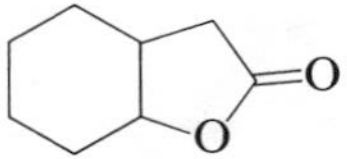 (84–85)	41, 42
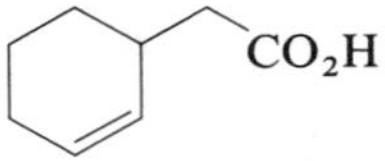 CO_2H	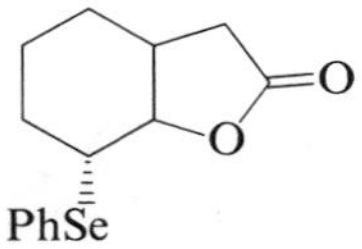 (82–93) PhSe	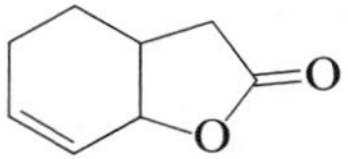 (92)	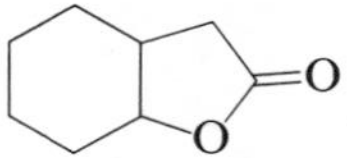 (76)	41, 42
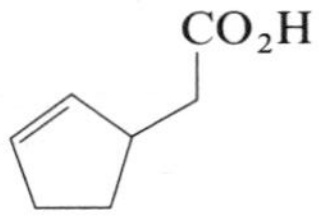 CO_2H	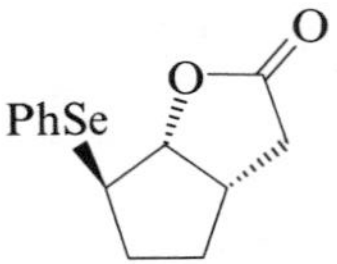 (73–100) PhSe	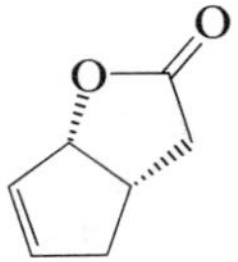 (92)	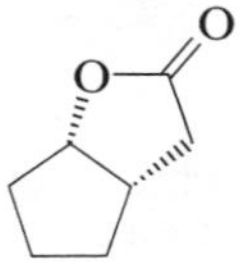 (97)	29, 41, 42

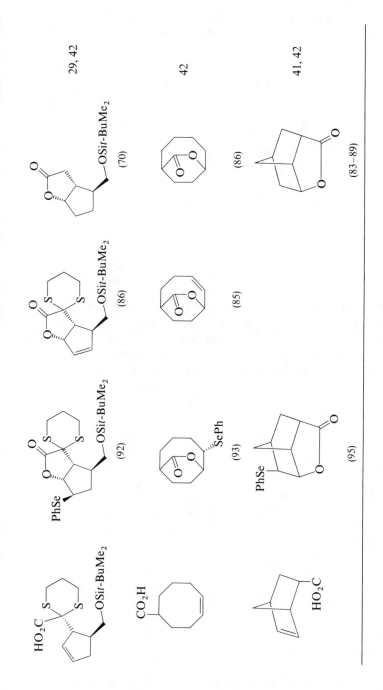

29, 42

42

41, 42

(70)

(86)

(83–89)

(86)

(85)

(92)

(93)

(95)

Rollinson et al.[46] applied this methodology to the synthesis of the naturally occurring lactone, 32, Eq. (26). In this case the lactone with *cis* OH and Me was selectively formed.

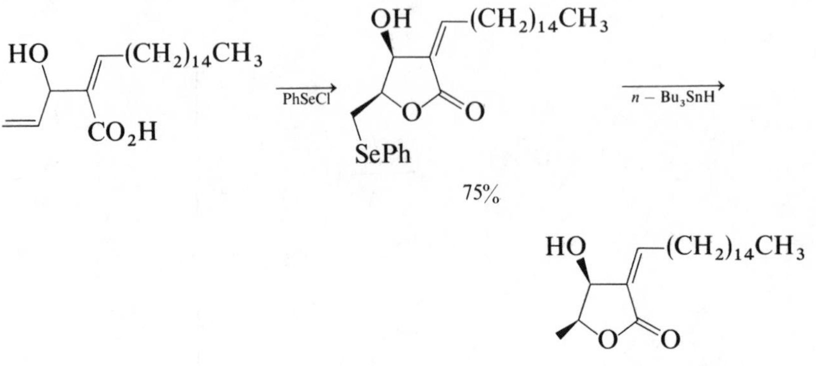

$$\tag{26}$$

Although PhSeCl or other organoselenium reagents cannot induce the formation of large ring lactones (macrolides) *N*-phenylselenophthalimide (*N*-PSP) and *N*-phenylselenosuccinimide (*N*-PSS) were found by Nicolaou et al.[29] to be effective for such transformations in the presence of catalytic amounts of camphor–sulfonic acid (CSA). An example is given in Eq. (27).

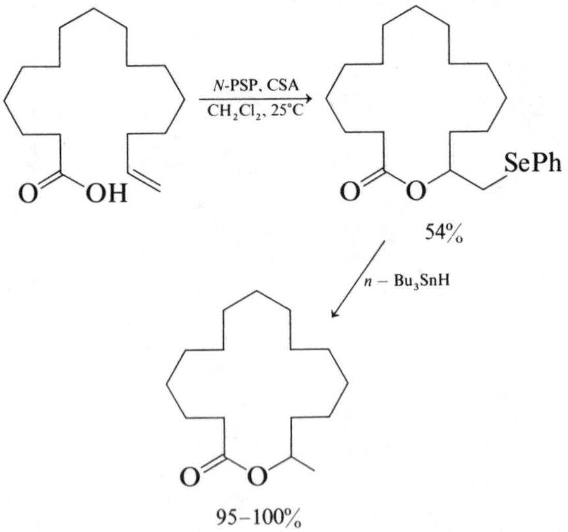

$$\tag{27}$$

A synthesis of lactones based on carbon–carbon bond formation, initiated from a PhSe group, was recently reported by Clive and Beaulieu.[47] This method involves the conversion of olefins to β-phenylseleno crotonates, with PhSeCl and silver crotonate, followed by treatment of these selenides with Ph_3SnH in the presence of a radical initiator (AIBN). The cyclization here presumably takes place via radical intermediates. An example is shown in Eq. (28).[47]

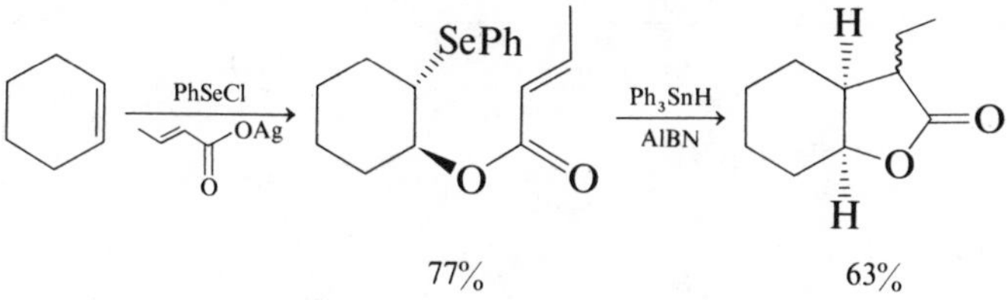

$$\text{(28)}$$

77%   63%

4. REVERSAL OF THE PHENYLSELENO-ETHERIFICATION AND LACTONIZATION REACTIONS

The previously described cyclizations of olefinic alcohols, Section 2, or olefinic carboxylic acids, Section 3, have the potential to be used for the selective and simultaneous blocking of an olefin and an OH or a CO_2H group, respectively. In order for these processes to be applied as a means of *protection* of these functionalities, a procedure for their reversion is required. Indeed, Nicolaou et al.[48] found that sodium in liquid ammonia can efficiently reverse both the phenylseleno-etherification[16] and the phenylseleno-lactonization[42] reactions. An alternative method, consisting of the use of chlorotrimethylsilane (Me_3SiCl) and sodium iodide (NaI) in acetonitrile (MeCN) was discovered by Clive and Kale.[49] Table 5 shows some examples of these reversals.

This reaction has found application in an overall conversion of geranyl acetate $\underline{33}$, to linalool, $\underline{36}$, in high yield (45%) as shown in Eq. (29). The conversion procedes via a phenylseleno etherification with PhSeOH to form $\underline{34}$,

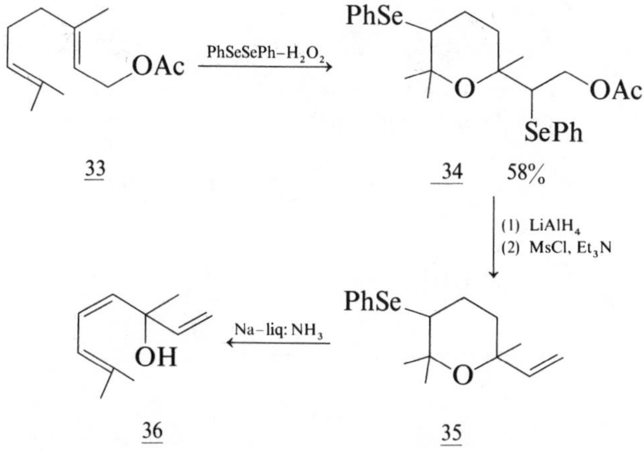

$$\text{(29)}$$

TABLE 5

REVERSAL OF THE PHENYLSELENO-ETHERIFICATION AND THE PHENYLSELENO-LACTONIZATION REACTIONS

Substrate	Conditions	"Decyclized" Product Yield (%)	References
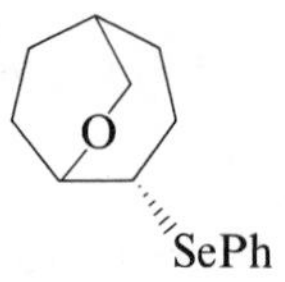	Na–liq NH_3	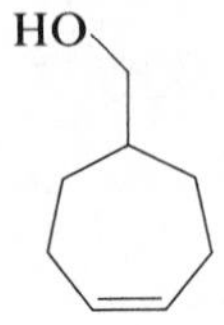 HO (78)	48
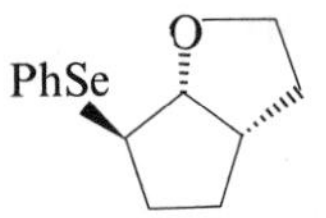	(1) Na–liq NH_3 (2) Me_3SiCl–NaI, MeCN	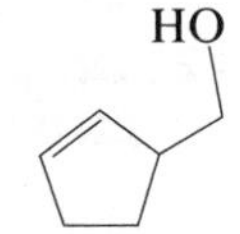 HO (82), (75)	48 49
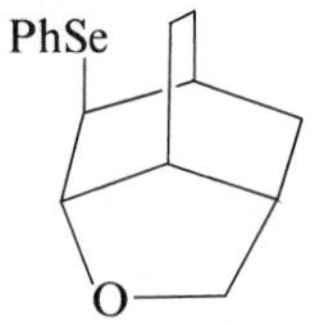	Me_3SiCl–NaI, MeCN	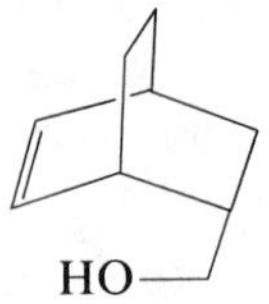 HO (78)	49
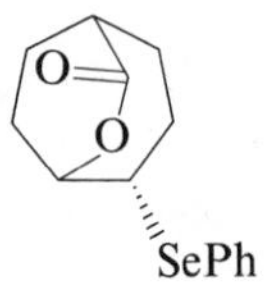	Na–liq NH_3	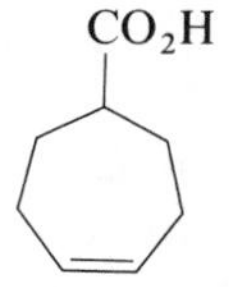 CO_2H (82)	48
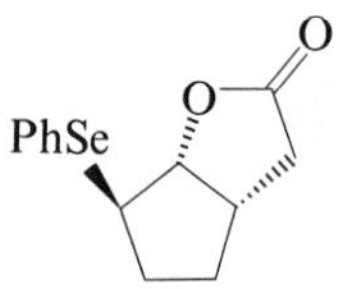	(1) Na–liq NH_3 (2) Me_3SiCl–NaI, MeCN	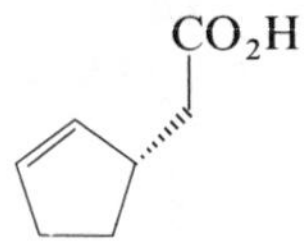 CO_2H (80), (84)	48 49
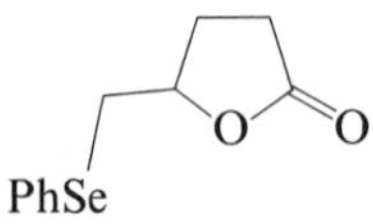	Na–liq NH_3	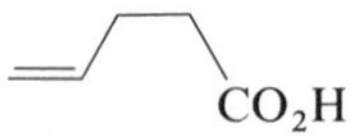 CO_2H (75)	49

elimination of one PhSe group to form the vinyl group of 35, followed by reversal of cyclization utilizing the other PhSe group.[34]

The combination of phenylseleno etherification with its reversal can also be used as a means of differentiation between similar hydroxyl groups. An example is given in Eq. (30), which shows the symmetrical diol, 37, being functionalized selectively on one site,[50] by temporarily engaging one OH while chemistry was performed on the other.

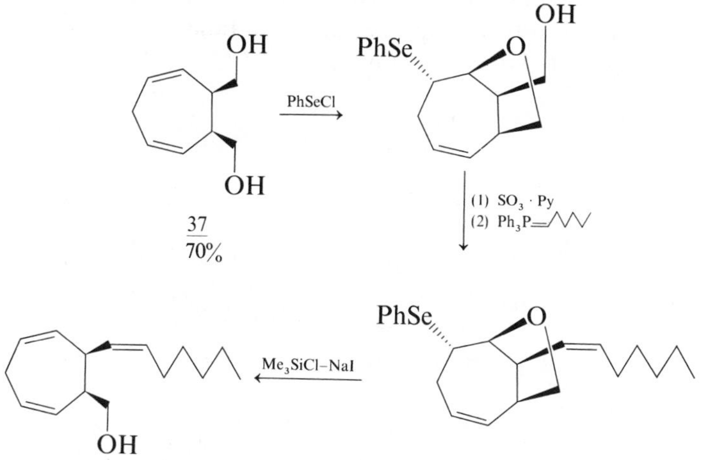

(30)

5. PHENYLSELENO-MEDIATED SYNTHESIS OF S-HETEROCYCLES

Organoselenium induced ring closures were also found by Nicolaou and co-workers[16,51] to be applicable for the synthesis of various types of S-hetero-cycles. Thiols, e.g., 38, and thioacetates, 39, can participate in such reactions giving cyclic phenylselenothioethers, 40, as exemplified in Eq. (31)[16,15] and Eq. (32).[16,51]

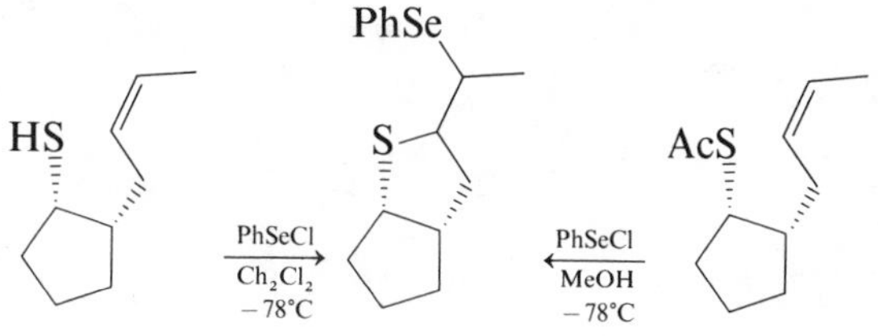

(31)

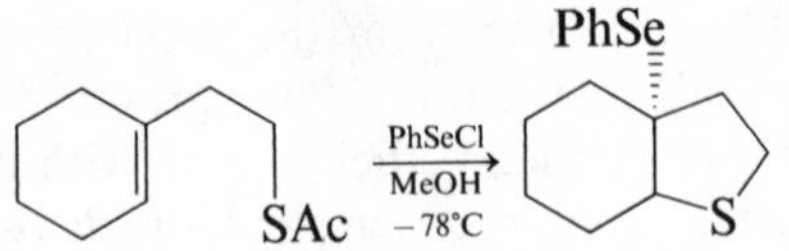

(32)

The oxidative elimination of the PhSe group from the phenylseleno thioethers is complicated by the presence of the thioether functionality. Depending on the oxidation method used, the product can be a sulfoxide or a sulfone, while the selenoxide elimination can proceed toward the sulfur or away from it. All of these possibilities with the appropriate conditions and the corresponding products as applied to $\underline{40}$ are given in Eq. (33).[16,51]

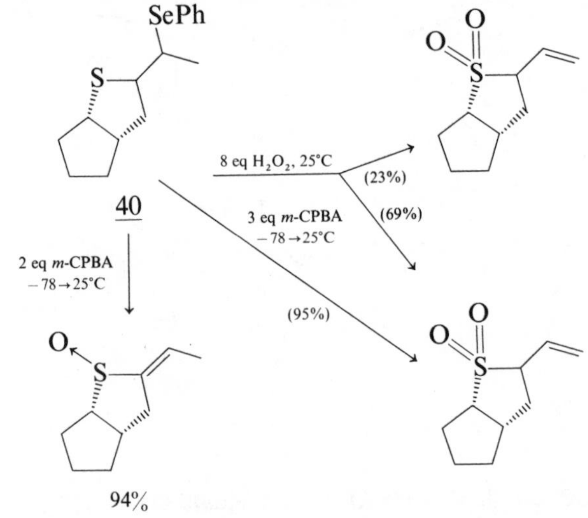

(33)

The geometry of the final olefinic bond depends on the geometry of the initial unsaturated substrate. For example, the trans isomer of $\underline{39}$ (i.e., $\underline{41}$), gives the (Z)-sulfoxide, $\underline{42}$, Eq. (34).[16,51]

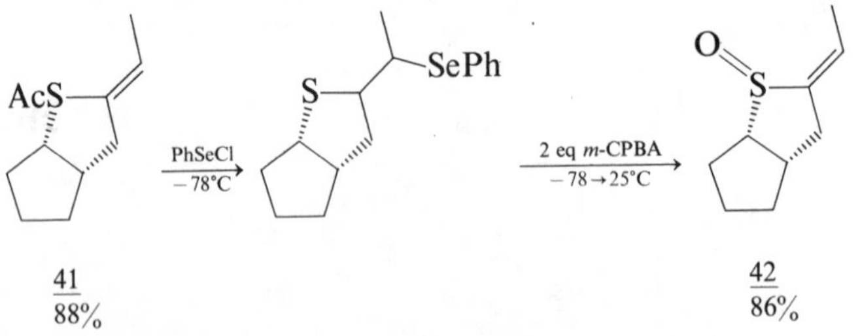

(34)

Nicolaou et al.[51,52] applied this methodology for the synthesis of the novel prostacyclin analogs, $\underline{43}$ (α-SO and β-SO isomers) $\underline{44}$ and $\underline{45}$ (α-SO and β-SO isomers), $\underline{46}$ and $\underline{47}$ (6α and 6β isomers).

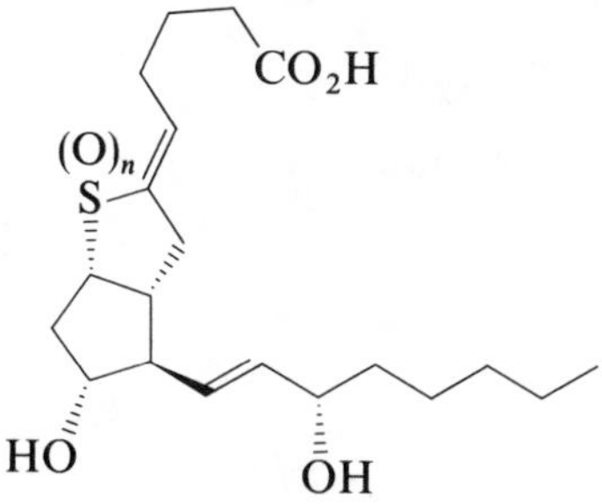

43, n = 1
44, n = 2

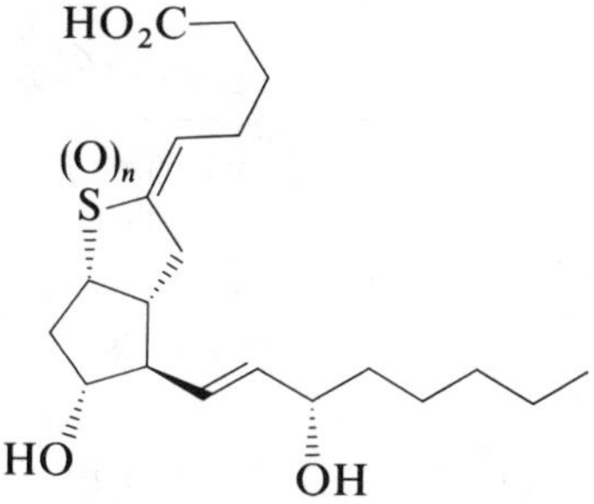

45, n = 1
46, n = 2

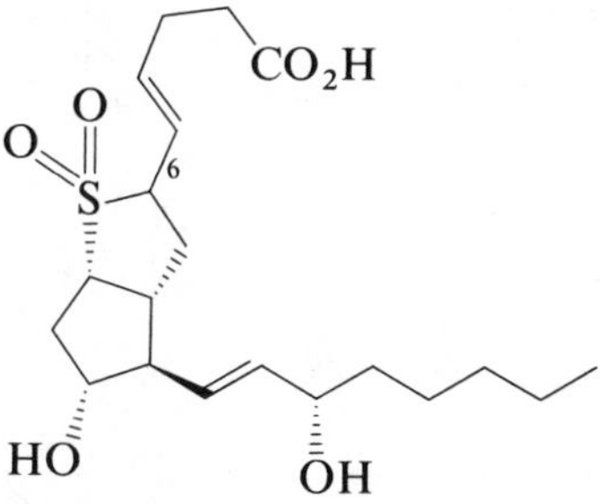

47

Reductive removal of the PhSe group from cyclic phenylselenothioethers can be accomplished with the tin hydride method as shown in Eq. (35).[16]

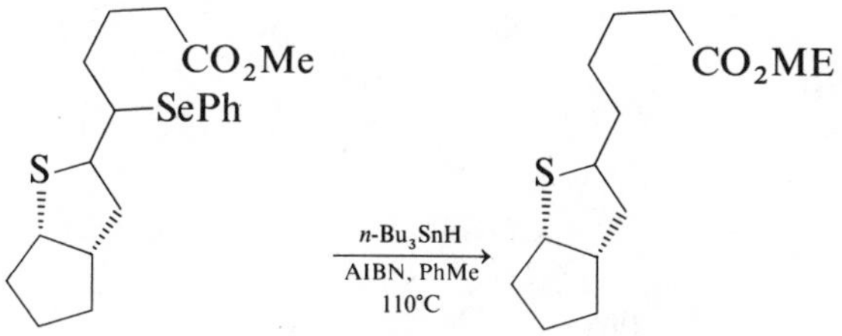

$$\xrightarrow[\substack{\text{AIBN, PhMe} \\ 110°C}]{n\text{-Bu}_3\text{SnH}}$$

76%

(35)

6. PHENYLSELENO-MEDIATED SYNTHESIS OF *N*-HETEROCYCLES

Clive and co-workers[53,54] have extended the organoselenium methodology to the synthesis of *N*-heterocycles, a process with potential applications to alkaloid synthesis. It was found that olefinic primary amines do not cyclize readily, while their urethane derivatives do.[53] Also, the presence of silica gel in the reaction mixture was shown to increase the yields of the cyclic products.[54]

TABLE 6

SYNTHESIS OF *N*-HETEROCYCLES

Substrate	Cyclized Product Yield (%)	Reduction Product Yield (%)	References
			53, 54, 22
	(77–93)	(88)	
			53, 54, 22
	(73–85)	(80)	
			53, 54, 22
	(59–82)	(72)	
			53, 54, 22
	(52–87)	(64)	
			54, 22
	(94)	(97)	
			54
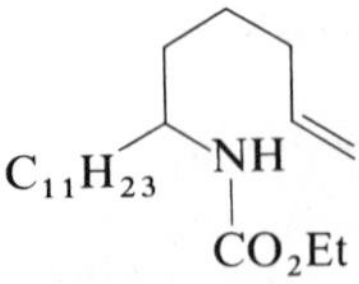	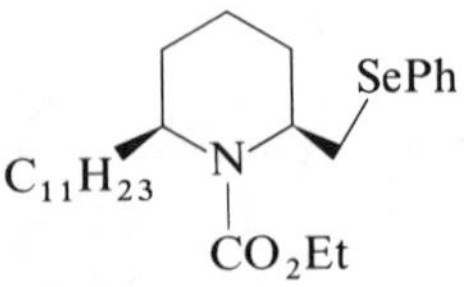 (84)	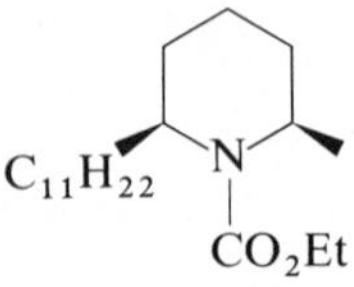 (96)	

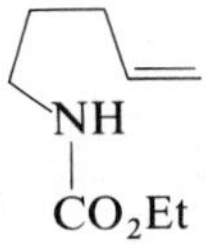

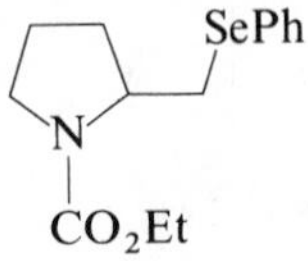

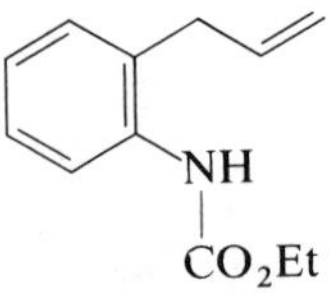

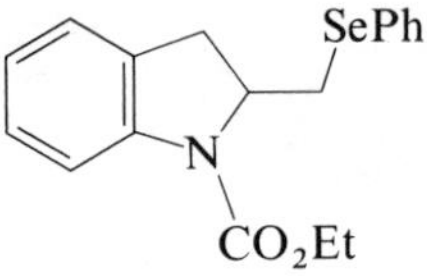

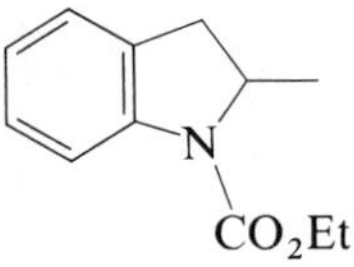

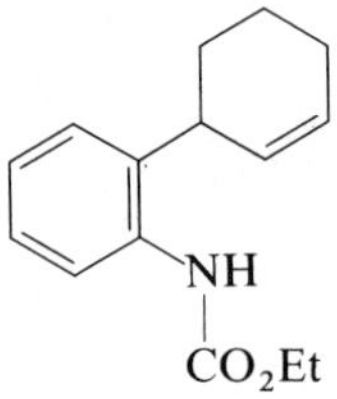

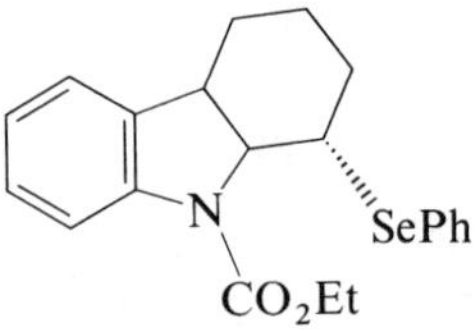

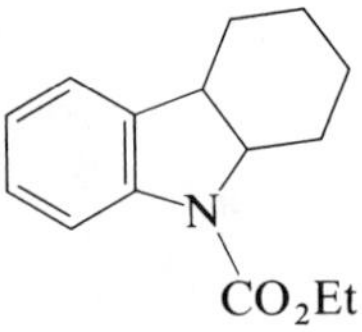

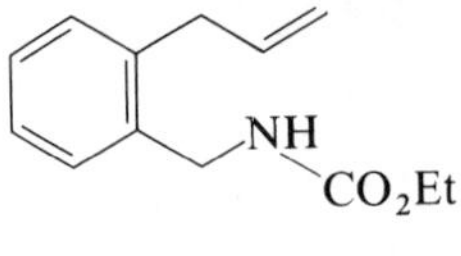

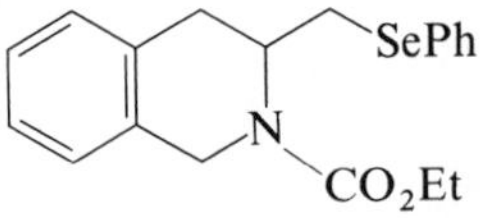

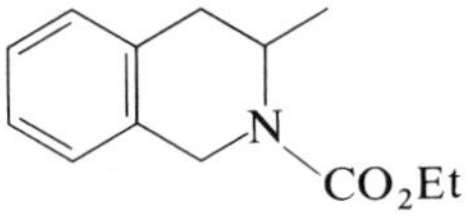

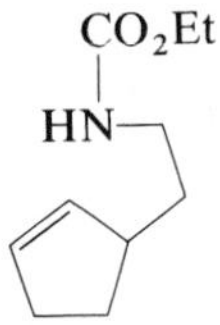

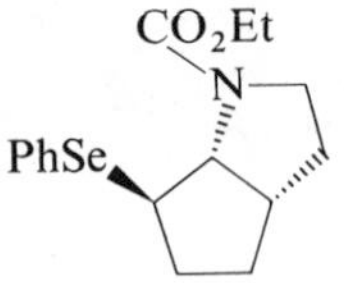

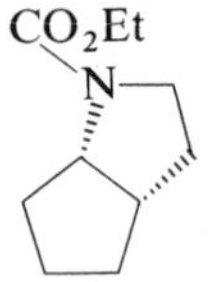

TABLE 6 (*Continued*)

SYNTHESIS OF *N*-HETEROCYCLES

Substrate	Cyclized Product Yield (%)	Reduction Product Yield (%)	References
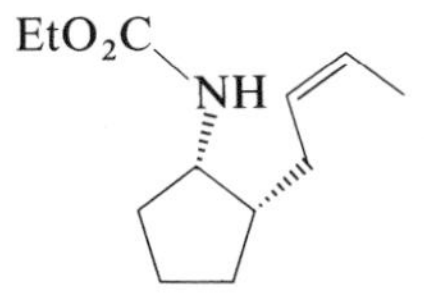	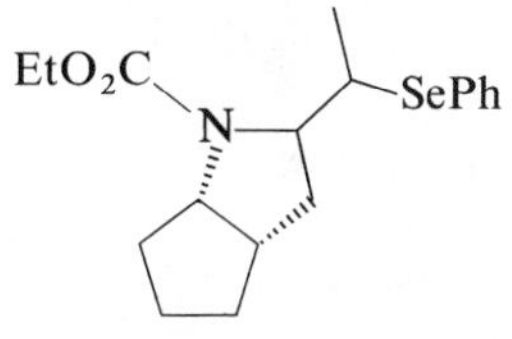 (90)		29
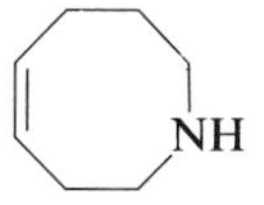	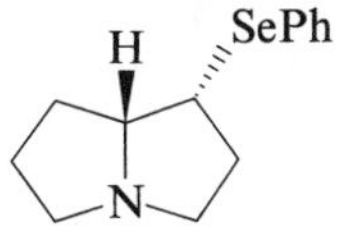 (63)		55

Removal of the PhSe group from these cyclized products could be done reductively with Ph_3SnH.[53,54,23]. Table 6 shows some examples of these transformations.

Using *N*-phenylselenophthalimide (*N*-PSP)[29] as the cyclizing reagent and tri-*n*-butylallylstannane (*n*-Bu$_3$SnCH$_2$CH=CH$_2$) in the presence of AIBN as the reducing agent, Webb and Danishefsky[56] have recently converted olefinic urethanes such as 48 to allyl substituted *N*-heterocycles, 49, Eq. (36).[56] The overall process was termed intramolecular *ureidoallylation* of olefinic bonds.

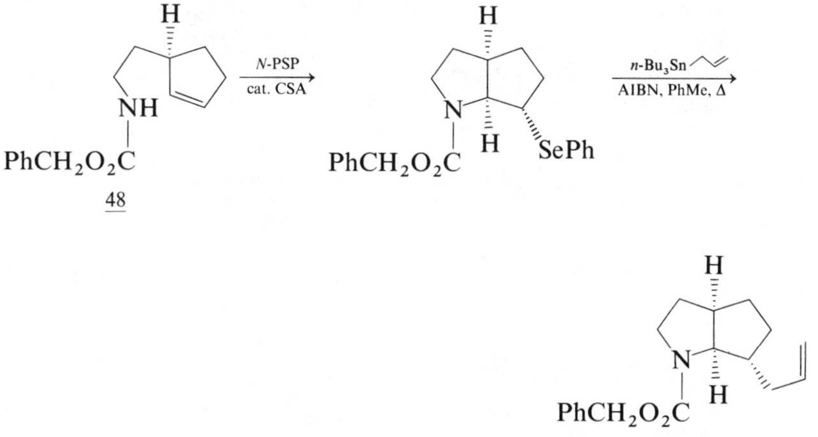

(36)

7. PHENYLSELENO-CARBOCYCLIZATION REACTIONS

The participation of an *olefinic bond as a nucleophile* in an intramolecular attack to an episelenonium ion leads to the formation of a new carbocyclic ring, and the generation of a carbonium ion, which can be captured by another nucleophile or the solvent. This type of cyclization turned out to be quite facile and synthetically useful.

If the episelenonium ion involved is formed, as usual, by addition of an electrophilic organoselenium reagent to an olefin, the overall process is a conversion of an acyclic diene to a functionalized carbocycle. Clive et al.[57] first observed such a transformation, with diene <u>50</u>, Eq. (37).

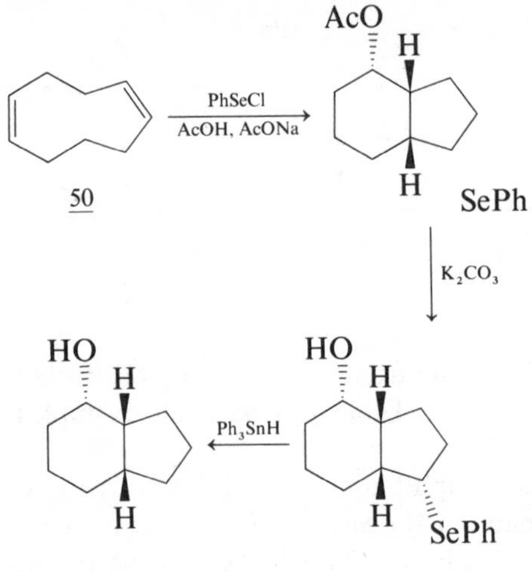

$$(37)$$

Phenylselenenyl iodide (PhSeI), prepared *in situ* by addition of iodine (I_2) to diphenyldiselenide (PhSeSePh) was found recently by Toshimitsu et al.[58] to be quite effective for this type of carbocyclizations. Since the reactions were carried out in acetonitrile (MeCN), the products had a MeCONH group attached to them, resulting from solvolysis of the intermediate carbonium ion. Two characteristic examples are given in Eqs. (38) and (39).

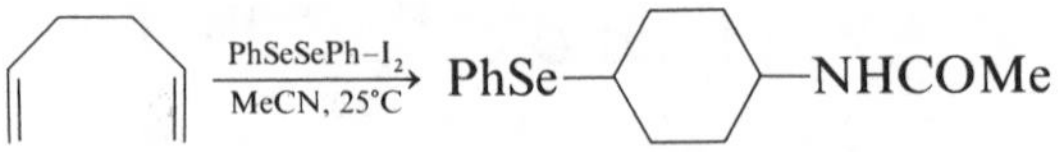

$$(38)$$

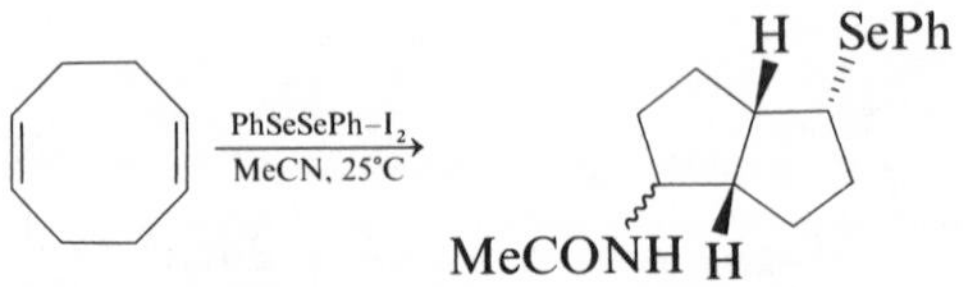

$$(39)$$

Kametani and co-workers[59–61] have developed an alternative route to carbocycles based on the generation of episelenonium ions from β-hydroxyselenides and trifluoroacetic acid (CF_3CO_2H). The required β-hydroxyselenides could be prepared from the corresponding olefin by epoxidation and nucleophilic opening of the epoxides with sodium phenylselenolate (PhSeNa), formed *in situ* by sodium borohydride reduction of PhSeSePh. This methodology was nicely applied to geranyl acetate, 51, and linanyl acetate, 52, with the results shown in Eq. 40[59] and Eq. 41.[60]

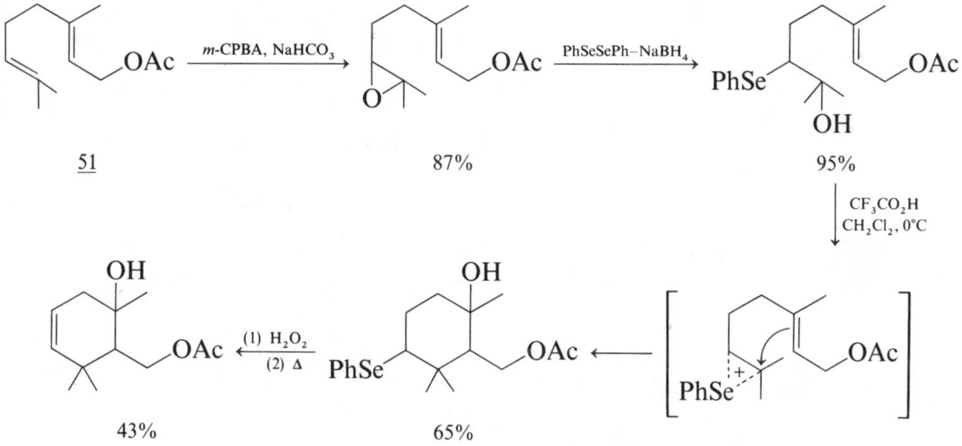

(40)

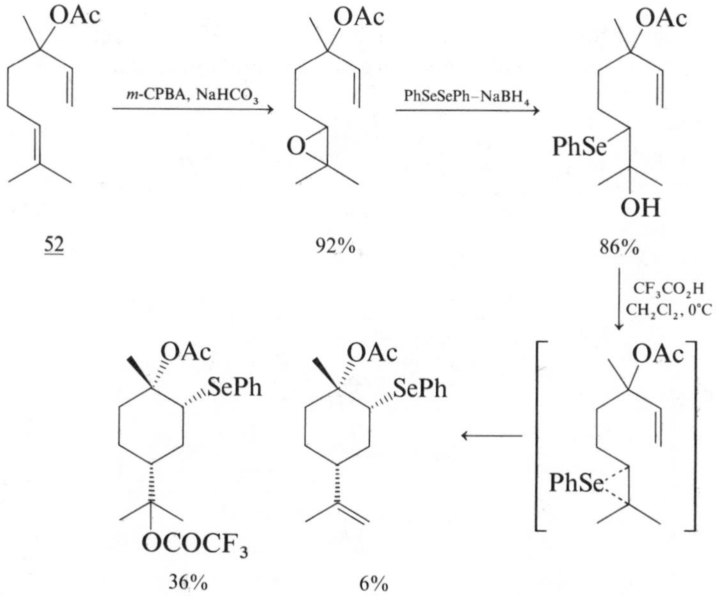

(41)

Intramolecular capture of the carbonium ion, formed during carbocyclizations, by an internal nucleophile such as OH, CO_2H, and so on, leads to the simultaneous formation of another ring. This was the case with nerolidol, 53, which was converted to caparrapi oxide, 54, Eq. (42), in a biogenetic manner.[61]

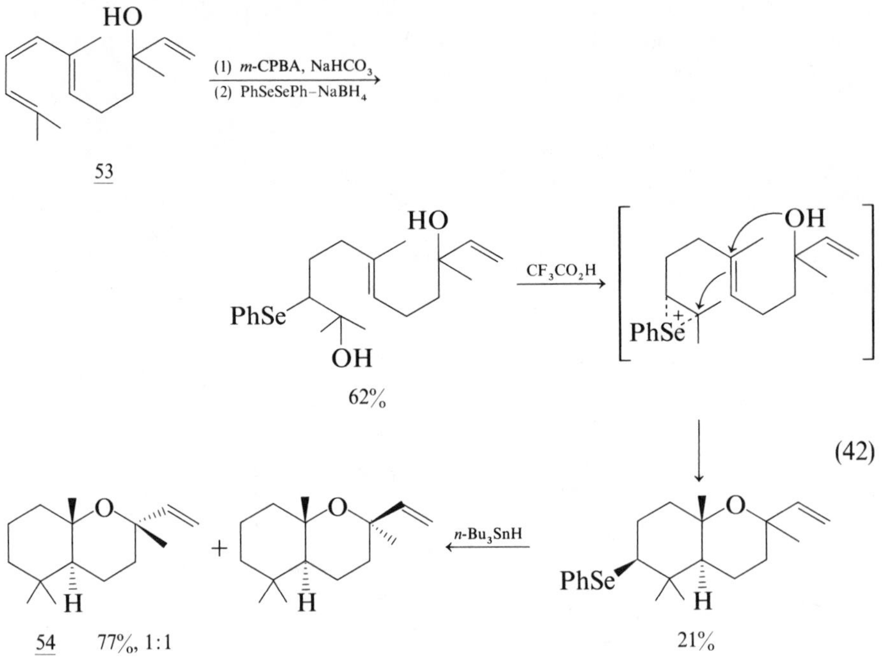

(42)

Direct conversion of unsaturated carboxylic acid, 55, to bicyclic phenylseleno lactone, 56, however, via a combined organoselenium induced carbocyclization—

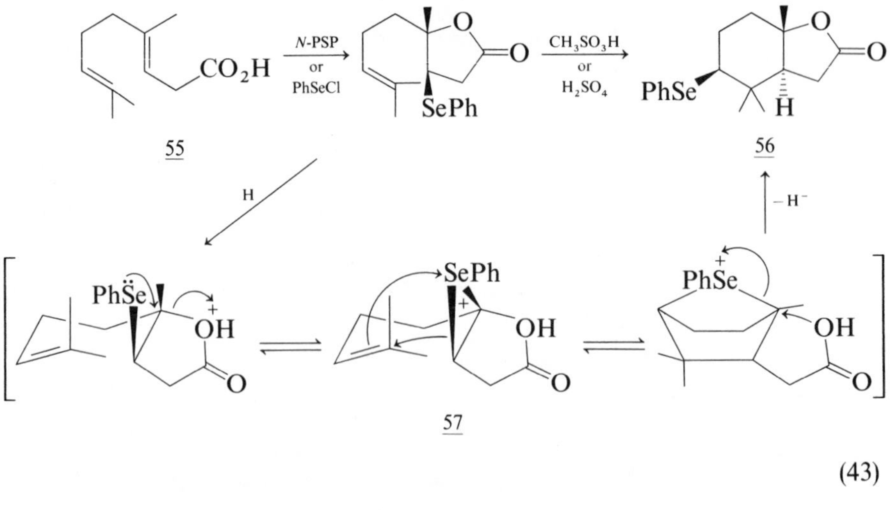

(43)

lactonization, could not be realized under a variety of conditions.[62] This could only be achieved stepwise by acid catalyzed regeneration of the episelenonium ion, 57 Eq. (43).[62]

Ley and co-workers[30–33] have successfully subjected the enolic type of olefinic bonds of β-dicarbonyl systems to organoselenium induced carbocyclizations. Several conditions were found for this operation including addition of N-PSP, under the catalysis of zinc iodide (ZnI_2)[(30)] or tin tetrachloride $(SnCl_4)$,[33] addition of $PhSeSbF_6$ at low temperature,[31] or acidic treatment of the α-phenylseleno-β-dicarbonyl compounds.[32,63] The kinetic products of these reactions are usually the corresponding O-cyclized ones, Section 2, which are converted to the C-cyclized products upon prolonged reaction times and strong acidic catalysis.[31,32] Table 7 presents some typical examples of this novel ring closure.

Recently, this methodology was applied by Ley and Murray[64] in an elegant synthesis of hirsutene, 58, outlined in Eq. (44).

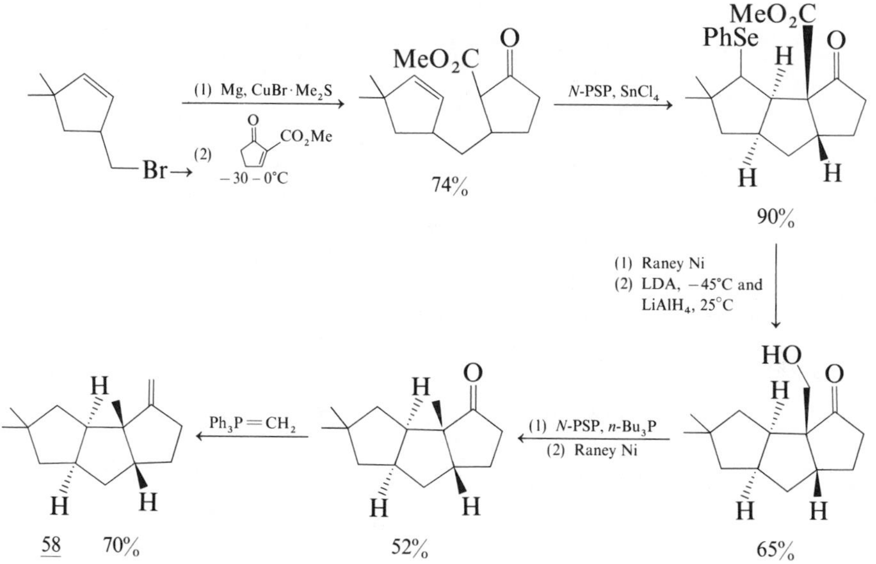

(44)

The synthesis of cyclopropanes, mediated by N-PSP, was observed by Nicolaou et al.[29] and is shown in Eq. (45).

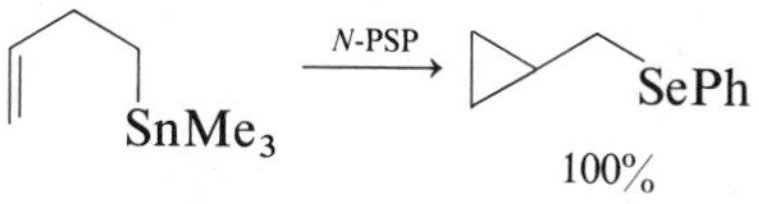

(45)

A conceptually different carbocyclization sequence, which utilizes a PhSe containing synthon, 60, was recently reported by Gravel et al.[65] Thus, the

TABLE 7
SYNTHESIS OF CYCLIC β-DICARBONYL COMPOUNDS

Substrate	Conditions	Cyclized Product Yield (%)	References
	N-PSP, ZnI$_2$ CH$_2$Cl$_2$, 25°C	(66)	30
	N-PSP, ZnI$_2$ CH$_2$Cl$_2$, 25°C	(67)	30
	N-PSP, SnCl$_4$ CH$_2$Cl$_2$, 25°C	(83)	33
	PhSeCl, AlCl$_3$ CH$_2$Cl$_2$, 25°C	(84)	63
	PhSeSbF$_6$, CH$_2$Cl$_2$, −78°C	(75)	31
	cat. SnCl$_4$	(77)	32

dianion of ketoester, 59, was alkylated with 60 giving selenide, 61, which upon treatment with trifluoroacetic acid at 0°C cyclized to give 62, Eq. (46).[65] Oxidative removal of the PhSe group of 62 produced 63.

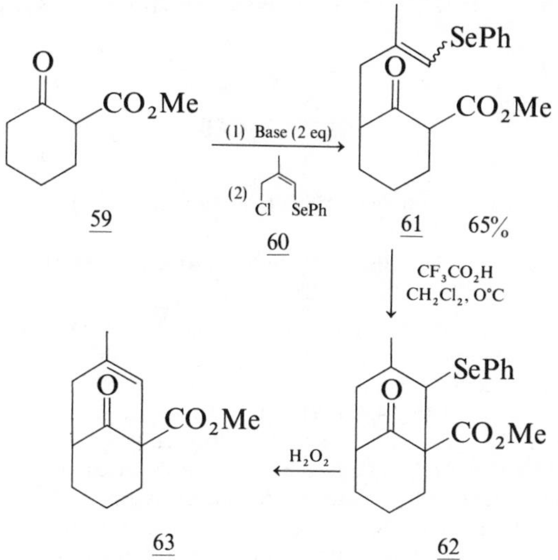

$$(46)$$

8. CONCLUSIONS

These ventures in organoselenium chemistry, in search of ways for ring synthesis, resulted in some very gratifying results. Not only were the initial goals set in our laboratories realized, but several useful applications were also found, stimulating further developments and new ideas in both our and other laboratories. Many of the cyclization reactions described previously are characterized by their mildness, efficiency, regio-, and stereoselectivity. The incorporation of a PhSe group to the newly generated ring, resulting from internal nucleophilic capture of an intermediate episelenonium ion, is a welcome feature. This functionality has evolved as a synthetically valuable group, readily convertible to olefins via oxidation–selenoxide *syn* elimination, or easily eliminated by reductive cleavage.

The wide variety of rings accessible by these methods demonstrate clearly the versatility and scope of this dimension of organoselenium chemistry. Further applications to the synthesis of natural products or other target molecules are anticipated.

ACKNOWLEDGMENTS

We wish to extend our appreciation to the former members of the group: W. E. Barnette, Z. Lysenko, R. L. Magolda, S. P. Seitz, W. J. Sipio, and to S. Webber for their contributions to the work described in this article. Our appreciation is also extended to Merck Sharp and Dohme, USA, the A. P. Sloan Foundation, the Camille and Henry Dreyfus Foundation, the Petroleum Research Fund, administered by the American Chemical Society, and the U. S. National Institutes of Health, for their generous financial support.

REFERENCES

1. For a review of earlier work on these types of cyclizations see: V. I. Staninets and E. A. Shilov, *Russ. Chem. Rev.*, **40**, 272 (1971).

2. (a) J. Klein, *J. Org. Chem.*, **23**, 1209 (1985); (b) M. F. Ansell and M. H. Palmer, *Q. Rev.*, **18**, 211 (1964); (c) R. D. Barry, *Chem. Rev.*, **64**, 229 (1964).

3. (a) For a review on the halo-lactonization reaction see: M. D. Dowle and D. I. Davies, *Chem. Soc. Rev.*, **8**, 171 (1979); (b) For some cyclizations induced by Br$^+$ see: A. Bresson, G. Dauphin, J. M. Geneste, A. Kergomard, and A. Lacourt, *Bull. Soc. Chim. Fr.*, 1080 (1971); E. Demole and P. Enggist, *Helv. Chim. Acta*, **54**, 456 (1971); I. Monkovic, Y. G. Perron, R. Martel, W. J. Simpson, and J. A. Gylys, *J. Med. Chem.* **16**, 403 (1973); H. Wong, J. Chapuis, and I. Monkovic, *J. Org. Chem.*, **39**, 1042 (1974); (c) For some cyclizations induced by I$^+$ see: D. L. H. Williams, *Tetrahedron Lett.*, 2001 (1967); J. N. Labos, Jr., and D. Swern, *J. Org. Chem.*, **37**, 3004 (1972); H. J. Gunther, V. Jager, and P. S. Skell, *Tetrahedron Lett.*, 2539 (1977); P. A. Bartlett and J. Myerson, *J. Am. Chem. Soc.*, **100**, 3950 (1978); (d) For an application of the iodo-etherification reaction as a means of protection of a hydroxyl group see: K. C. Nicolaou, N. A. Petasis, R. E. Zipkin, and J. Uenishi, *J. Am. Chem. Soc.*, **104**, 5555 (1982).

4. (a) R. M. Moriarty and K. Kapadia, *Tetrahedron Lett.*, 1165 (1964); (b) R, M. Moriarty, H. G. Walsh, and H. Gopal, *ibid.* 4363 (1966).

5. (a) H. Stetter and H. J. Meissner, *Tetrahedron Lett.*, 4599 (1966); (b) R. C. Larock, *Angew. Chem Int. Ed.*, **17**, 27 (1978) and references cited therein; (c) S. Danishefsky, E. Taniyama, and R. R. Webb, II, *Tetrahedron Lett.*, **24**, 11,15 (1983).

6. (a) V. Simonidesz, Z. Gombos-Visky, G. Kovacs, E. Baitz-Gacs, and L. Radics, *J. Am. Chem. Soc.*, **100**, 6756 (1978); (b) Y. Yamada, H. Sanjoh, and K. Iguchi, *Tetrahedron Lett.*, **423**, 1323 (1979).

7. (a) K. C. Nicolaou and Z. Lysenko, *J. Chem. Soc. Chem. Commun.*, 293 (1977); see also Ref. 40; (b) B. M. Trost, M. Ochiar, and P. G. McDougal, *J. Am. Chem. Soc.*, **100**, 7103 (1978).

8. K. B. Sharpless, K. M. Gordon, R. F. Lauer, D. W. Patrick, S. P. Singer, and M. W. Young, *Chem. Scr.*, **8A**, 9 (1975).

9. (a) H. J. Reich, *Organoselenium Oxidations*, in W. S. Trahanovsky, Ed., *Oxidation in Organic Chemistry, Part C*, Academic, New York, 1978, p. 1; (b) H. J. Reich, *Acc. Chem. Res.*, **12**, 22 (1979).

10. (a) D. L. J. Clive, *Tetrahedron*, **34**, 1049 (1978); (b) D. L. J. Clive, *Aldrichimica Acta*, **11**, 43 (1978).

11. K. B. Sharpless and R. F. Lauer, *J. Org. Chem.*, **39**, 429 (1974).

12. For a review see: K. C. Nicolaou, *Tetrahedron*, **37**, 4097 (1981).

13. D. L. J. Clive, G. Chittattu, N. J. Curtis, W. A. Kiel, and C. K. Wong, *J. Chem. Soc. Chem. Commun.*, 725 (1977).

14. (a) K. C. Nicolaou, W. E. Barnette, and R. L. Magolda, *J. Chem. Res.*, (S), 202 and (M), 2444 (1979), and references cited therein; (b) For a review on other methods for the synthesis of Prostacyclins see: K. C. Nicolaou, G. P. Gasic, and W. E. Barnette, *Angew. Chem. Int. Ed.*, **17**, 293 (1978).

15. K. C. Nicolaou and Z. Lysenko, *Tetrahedron Lett.*, 1257 (1977).

16. K. C. Nicolaou, R. L. Magolda, W. J. Sipio, W. E. Barnette, Z. Lysenko, and M. M. Joullie, *J. Am. Chem. Soc.*, **102**, 3784 (1980).

17. D. L. J. Clive, G. Chittattu, and C. K. Wong, *Can. J. Chem.*, **55**, 3894 (1977).

18. (a) K. C. Nicolaou, W. E. Barnette, and R. L. Magolda, *J. Am. Chem. Soc.*, **103**, 3480 (1981); (b) K. C. Nicolaou and W. E. Barnette, *J. Chem. Soc. Chem. Commun.*, 333 (1977).

19. (a) E. J. Corey, G. E. Keck, and I. Szekely, *J. Am. Chem. Soc.*, **99**, 2006 (1977); (b) E. J. Corey, H. L. Pearce, I. Szekely, and M. Ishigura, *Tetrahedron Lett.*, 1023 (1978).

20. Z. Lysenko, F. Ricciardi, J. E. Semple, P. C. Wang, and M. M. Joullie, *Tetrahedron Lett.*, 2679 (1978).

21. G. A. Kraus and M. J. Taschner, *J. Org. Chem*, **45**, 1175 (1980).

22. (a) D. L. J. Clive, G. J. Chittattu, V. Farina, W. A. Kiel, S. M. Menchen, C. G. Russell, A. Singh, C. K. Wong, and N. J. Curtis, *J. Am. Chem. Soc.*, **102**, 4438 (1980); (b) D. L. J. Clive, G. Chittattu, and C. K. Wong, *J. Chem. Soc. Chem. Commun.*, 41 (1978).

23. Y. Naruta, H. Uno, and K. Maruyama, *J. Chem. Soc. Chem. Commun.*, 1277 (1981).

24. K. C. Nicolaou, W. J. Sipio, R. L. Magolda, S. Seitz, and W. E. Barnette, *J. Chem Soc. Chem. Commun.*, 1067 (1978).

25. K. C. Nicolaou and N. A. Petasis, "Selenium in Natural Products Synthesis," CIS, Philadelphia, 1984.

26. G. A. Kraus and B. Roth, *J. Org. Chem.*, **45**, 4825 (1980).

27. S. Current and K. B. Sharpless, *Tetrahedron Lett.*, 5075 (1978).

28. S. V. Ley and B. Lygo, *Tetrahedron Lett.*, **23**, 4625 (1982).

29. (a) K. C. Nicolaou, D. A. Claremon, W. E. Barnette, and S. P. Seitz, *J. Am. Chem. Soc.*, **101**, 3704 (1979); (b) K. C. Nicolaou, N. A. Petasis, and D. A. Claremon, *Tetrahedron*, **41**, 4835 (1985).

30. W. P. Jackson, S. V. Ley, and J. A. Morton, *J. Chem. Soc. Chem. Commun.*, 1028 (1980).

31. W. P. Jackson, S. V. Ley, and A. J. Whittle, *J. Chem Soc. Chem. Commun.*, 1173 (1980).

32. W. P. Jackson, S. V. Ley, and J. A. Morton, *Tetrahedron Lett.*, **22**, 2601 (1981).

33. S. V. Ley, B. Lygo, H. Molines, and J. A. Morton, *J. Chem. Soc. Chem. Commun.*, 1251 (1982).

34. R. M. Scarborough, Jr., A. B. Smith, III, W. E. Barnette, and K. C. Nicolaou, *J. Org. Chem.*, **44**, 1742 (1979).

35. (a) A. Toshimitsu, S. Aoai, S. Uemura, and M. Okano, *J. Org. Chem.*, **46**, 3021 (1981); (b) S. Uemura, A. Toshimitsu, T. Aoai, and M. Okano, *J. Chem. Soc. Chem. Commun.*, 610 (1979); (c) S. Uemura, A. Toshimitsu, T. Aoai, and M. Okano, *Chem. Lett.*, 1359 (1979).

36. S. Uemura, A. Toshimitsu, T. Aoai, and M. Okano, *Tetrahedron Lett.*, **21**, 1533 (1980).

37. T. R. Hoye and A. J. Caruso, *J. Org. Chem.*, **46**, 1198 (1981).

38. P. D. Kane and J. Mann, *J. Chem. Soc. Chem. Commun.*, 224 (1983).

39. Reviews: (a) K. C. Nicolaou, *Tetrahedron*, **33**, 683 (1977); (b) S. Masamune, G. S. Bates, and J. W. Concoran, *Ang. Chem. Int. Ed.*, **16**, 587 (1977); (c) T. G. Back, *Tetrahedron*, **33**, 3041 (1977).

40. E. J. Corey, N. M. Weinshenker, T. K. Schaaf, and W. Huber, *J. Am. Chem. Soc.*, **91**, 5675 (1969).

41. K. C. Nicolaou and Z. Lysenko, *J. Am. Chem. Soc.*, **99**, 3185 (1977).

42. K. C. Nicolaou, S. P. Seitz, W. J. Sipio, and J. F. Blount, *J. Am. Chem. Soc.*, **101**, 3884 (1979).

43. D. L. J. Clive and G. Chittattu, *J. Chem. Soc. Chem Commun.*, 484 (1977).

44. D. L. J. Clive, C. G. Russell, G. Chittattu, and A. Singh, *Tetrahedron*, **36**, 1399 (1980).

45. D Goldsmith, D. Liotta, C. Lee, and G. Zima, *Tetrahedron Lett.*, 4801 (1979).

46. S. W. Rollinson, R. A. Amos, and J. A. Katzenellenbogen, *J. Am. Chem. Soc.*, **103**, 4114 (1981).

47. D. L. J. Clive and P. L. Beaulieu, *J. Chem. Soc. Chem. Commun.*, 307 (1983).

48. K. C. Nicolaou, W. J. Sipio, R. L. Magolda, and D. A. Claremon, *J. Chem. Soc. Chem. Commun.*, 83 (1979).

49. D. L. J. Clive and V. N. Kale, *J. Org. Chem.*, **46**, 231 (1981).

50. K. C. Nicolaou and S. Webber, unpublished results.

51. K. C. Nicolaou, W. E. Barnette, and R. L. Magolda, *J. Am. Chem. Soc.*, **100**, 2567 (1978).

52. K. C. Nicolaou, W. E. Barnette, and R. L. Magolda, *J. Am. Chem. Soc.*, **103**, 3486 (1981).

53. D. L. J. Clive, C. K. Wong, W. A. Kiel, and S. M. Menchen, *J. Chem. Soc. Chem. Commun.*, 379 (1978).

54. D. L. J. Clive, V. Farina, A. Singh, C. K. Wong, W. A. Kiel, and S. M. Menchen, *J. Org. Chem.*, **45**, 2120 (1980).

55. S. R. Wilson and R. A. Sawicki, *J. Org. Chem.*, **44**, 287 (1979).

56. R. R. Webb II and Danishefsky, *Tetrahedron Lett.*, **24**, 1357 (1983).

57. D. L. J. Clive, G. Chittattu, and C. K. Wong, *J. Chem. Soc. Chem. Commun.*, 82 (1982).

58. A. Toshimitsu, S. Uemura, and M. Okano, *J. Chem. Soc. Chem. Commun.*, 82 (1982).

59. (a) T. Kametani, K. Suzuki, H. Kurobe, and H. Nemoto, *Chem. Pharm. Bull.*, **29**, 105 (1981); (b) T. Kametani, K. Suzuki, H. Kurobe, and H. Nemoto, *J. Chem. Soc. Chem. Commun.*, 1128 (1979).

60. (a) T. Kametani, H. Kurobe, and H. Nemoto, *J. Chem. Soc. Perkin Trans. 1.* 756 (1981); (b) T. Kametani, H. Kurobe, and H. Nemoto, *J. Chem. Soc. Chem. Commun.*, 762 (1980).

61. T. Kametani, K. Fukumoto, H. Kurobe, and H. Nemoto, *Tetrahedron Lett.*, **22**, 3653 (1981).

62. (a) A. Rouessac, F. Rouessac, and H. Zamarlik, *Tetrahedron Lett.*, **22**, 2641 (1981); (b) F. Rouessac and H. Zamarlic, *Tetrahedron Lett.*, **22**, 2643 (1981).

63. See also M. Alderdice and L. Weiler, *Can. J. Chem.*, **59**, 2239 (1981).

64. S. V. Ley and P. J. Murray, *J. Chem. Soc. Chem. Commun.*, 1252 (1982).

65. D. Gravel, R. Deziel, and L. Bordeleau, *Tetrahedron Lett.*, **24**, 699 (1983).

3

Seleninic Anhydrides and Acids in Organic Synthesis

STEVEN V. LEY

Department of Chemistry, Imperial College, London, England

CONTENTS

1. INTRODUCTION

Prior to the early 1970s organic synthesis had been restricted to the use of selenium and selenium dioxide as the prime sources of selenium reagents. However, since the excellent compilation of selenium chemistry by Klayman and

Günther[1] and the need in modern synthesis to develop new reagents that show either improved selectivity, versatility, or novelty of reaction, interest in organoselenium reagents has increased dramatically. This chapter summarizes the preparation and properties of seleninic anhydrides and acids and discusses how they can be used to achieve a very wide range of synthetic transformations.[2]

Many of the reactions described are unique and have no counterpart in sulfur chemistry. The reactions often take place under mild conditions such that other sensitive functional groups remain intact. Although these reagents are very versatile, selectivity can often be achieved by appropriate choice of the solvent or temperature. Benzeneseleninic anhydride and benzeneseleninic acid are now commercially available and their cost in some reactions can be minimized by the use of catalytic cycles.

The drawback to using these reagents is their inherent toxicity and the volatility and unpleasant odors associated with *some* of the reaction by-products. In the future these problems may be overcome by the use of polymer bound systems.[3]

2. PREPARATION AND PROPERTIES OF SELENINIC ANHYDRIDES

Seleninic anhydrides can be prepared by a number of methods with varying levels of generality. Most anhydrides are hygroscopic, colorless solids that should be protected from moist atmospheres. By heating seleninic acids, preferably *in vacuo*, essentially quantitative yields of the anhydride are produced.[4-8] Alternatively, probably the most convenient practical method involves reaction of the diselenide with ozone at low temperature (-50 to $-10°C$).[5] During this reaction a number of partially oxidized species must be involved, none of which have been properly detected [Eq. (1)]. Benzeneseleninic anhydride produced by this method is an easily handled odorless solid, mp $164-165°C$, which may be stored in a desiccator for long periods of time without significant decomposition. Because of its hygroscopic nature, it is advisable to periodically reactivate benzeneseleninic anhydride by heating under vacuum at $130°C$ for a few hours if the reagent is in frequent use in the laboratory.

$$\text{PhSeSePh} \xrightarrow[-78°C]{\text{O}_3} \left[\underset{\text{PhSeSePh}}{\overset{\overset{\text{O}}{\|}}{}} \longrightarrow \text{PhSeOSePh} \longrightarrow \underset{\text{PhSeOSePh}}{\overset{\overset{\text{O}}{\|}}{}} \right]$$

$$\longrightarrow \underset{\text{PhSeOSePh}}{\overset{\text{O}\quad\text{O}}{\overset{\|\quad\|}{}}} \tag{1}$$

Oxidation of diselenides to anhydrides using anhydrous *t*-butyl hydroperoxide is also possible in some examples.[4,5] Kuwajima and co-workers have suggested that t-BuOOH reacts with diphenyl diselenide to produce the less oxidized benzeneselenenic anhydride.[9] More recent mechanistic studies refute these proposals and provide evidence for substantial formation of the seleninic anhydride.[10] Some care must be taken in performing these reactions as it has additionally been shown that benzeneseleninic anhydride reacts further with t-BuOOH to give the corresponding peroxyseleninate derivative.[11] This work also cautions against the use of hydrogen peroxide for the conversion of diphenyl diselenide to the seleninic acid owing to the production of what was thought to be $PhSeO_2H \cdot H_2O_2$, which explodes on warming to 53–55°C. No such problems arise when other oxidants are used.

Catalytic generation of benzeneseleninic anhydride by oxygen transfer from iodoxybenzenes to diphenyldiselenide does have many practical advantages that are especially noticeable during dehydrogenation reactions. Other less general routes that have been reported are the oxidation of trifluoromethaneselenenyl chloride with nitrogen dioxide[12,13] or disproportionation of the product derived from benzeneselenol with benzeneseleninyl chloride.[5] Oxidation of 2,2′-pyridyl-benzoselenophene with hydrogen peroxide in acetic acid is a special reaction for the preparation of 2,2′-dicarboxybenzeneseleninic anhydride and has not been applied to other systems.[14]

3. PREPARATION AND PROPERTIES OF SELENINIC ACIDS

In general seleninic acids are stable and are easily prepared. They tend to be weaker acids than the corresponding carboxylic acids with pK_a values around 4.5.[15] Although a large number of seleninic acids are now known,[16] benzene-seleninic acid is the only compound that has found wide application in synthesis. Probably the simplest and most reliable method for the preparation of seleninic acids involves the treatment of diselenides with nitric acid.[4,15] The product of this reaction is usually a crystalline hydronitrate salt from which the free acid is obtained after neutralization [Eq. (2)]. Similar oxidation of selenols is also possible although many of the selenols are socially unacceptable as starting materials[17] because of their unpleasant odors. Oxidation of diselenides to the seleninic acids using hydrogen peroxide has found very common usage,[18] although some problems with this method have been noted.[11] Oxidation using potassium permanganate,[19] bromine water,[20] or peracids[21] have been reported, but these methods do not seem to be generally accepted. Both alkyl and aryl selenocyanates under a variety of oxidation conditions also afford seleninic acids.[8,16k, 16p,18a,22] Hydrolysis of seleninoyl chlorides,[23] selenenyl halides,[16l]

$$(RSe)_2 \xrightarrow{\text{HNO}_3} \left[R\,Se \overset{OH}{\underset{OH}{<}} \right]^+ NO_3^- \xrightarrow{\text{Base}} RSeO_2H \qquad (2)$$

and selenium trihalides[20,24] gives the corresponding acid, these methods being limited by the lower accessibility of the starting materials. Reaction of alkyl Grignard reagents with SeO_2 produces the seleninic acid[25] after acidic work-up although the method is not applicable to arylseleninic acids. Benzeneseleninic acid may be obtained from phenylselenoglycolic acid by treatment with hydrogen peroxide.[26]

Most seleninic acids are unstable to heat and decompose either by loss of water to give the anhydride,[4,5] or further, to give selenoxides,[6] selenols,[16g,17] and selenium dioxide.

4. OXIDATION OF PHENOLS

A major driving force behind the discovery of benzeneseleninic anhydride as an organic oxidant was the need to develop a selective reagent for the oxidation of alkylphenols to o-hydroxy-dienones. This reaction was required for the introduction of the 12a-hydroxy substituent into a phenolic ring- A precursor during a projected synthesis of the antibiotic tetracycline, 1. While a variety of methods effected this transformation on simple model systems, none were compatible with more elaborate phenols, especially those additionally containing either ester or amide functional groups. Benzeneseleninic anhydride on the other hand proved to be the unique reagent for selective ortho oxidation of phenolate anions derived from 2 and 3 giving the hydroxy-dienones, 4 and 5, in excellent yield [Eq. (3)].[28] The mechanism of the reaction involves initial formation of a seleninic ester which after [2,3]-sigmatropic shift to the ortho position, affords the products. Benzeneseleninic anhydride also proved to be a general reagent

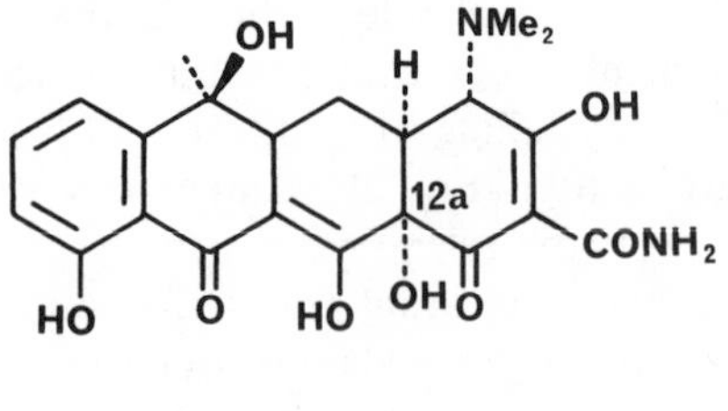

(1)

(3)

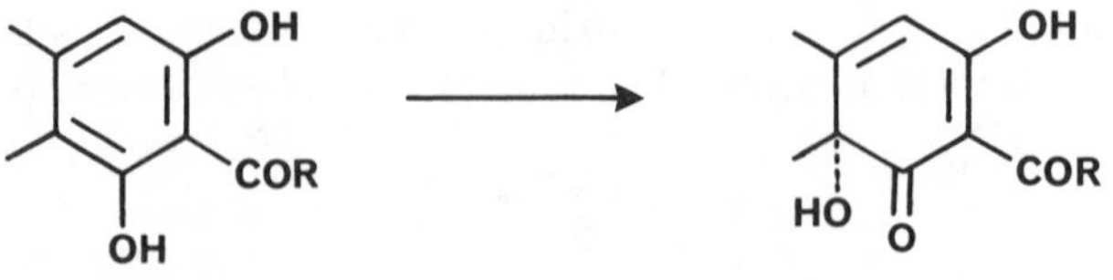

(2) R = OMe

(3) R = NH$_2$

(4) R = OMe

(5) R = NH$_2$

for the conversion of alkyl substituted phenols to o-hydroxy-dienones in four other examples. The initially formed o-hydroxy-dienone products from these simple phenols were isolated as their corresponding dimers [Eq. (4)]. An interesting application of the use of benzeneseleninic anhydride for the synthesis of a naturally occurring o-hydroxy-dienone, 7, has been reported.[29] Oxidation of the phenol, 6, with the anhydride in refluxing methylene chloride gave the naphthalenone, 7, in 85% yield. This compound was shown to have cyto-kinetic activity towards leukocytes and had previously been isolated from cotton [Eq. (5)].

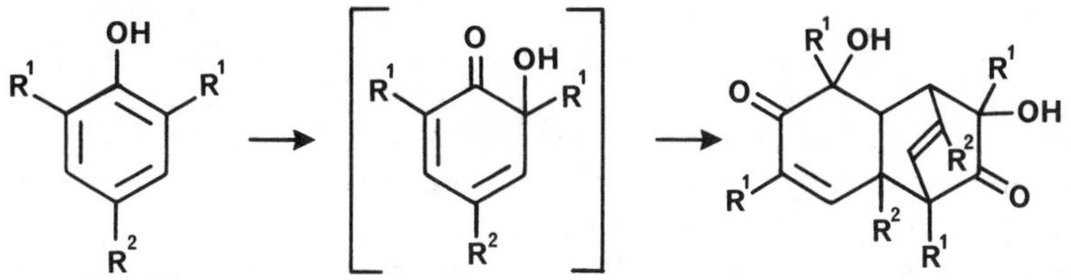

(4)

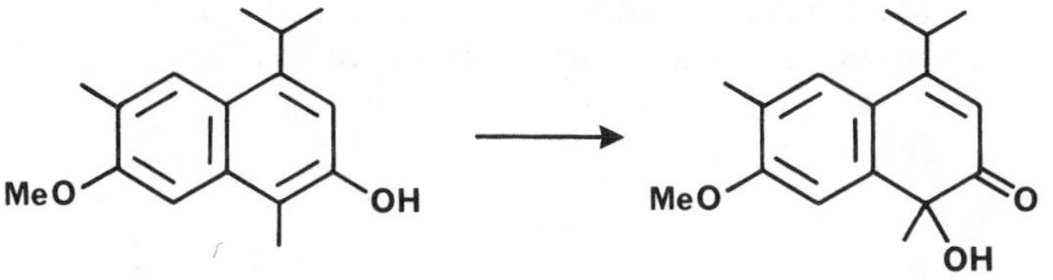

(5)

(6)   (7)

In view of the high ortho selectivity shown by the anhydride towards phenols it is not surprizing that phenols unsubstituted in the ortho position might be a source of o-quinones. Indeed further work[30] showed that a number of phenols, in tetrahydrofuran (THF) solutions, were oxidized at an elevated temperature (50°C) to give respectable yields of the corresponding o-quinones, even in cases where the para positon was unblocked. This last observation is important as there are very few literature methods that show this selectivity [Eq. (6)–(9)]. More recent work[31] has shown that in many examples by-products are also produced

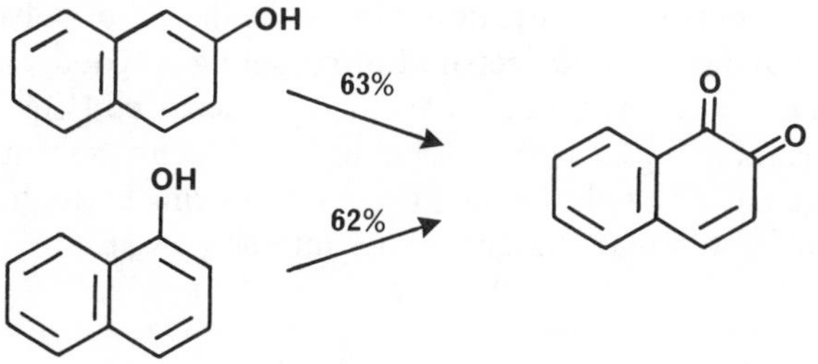

(6)
(7)

$$\text{Thymol} \qquad \xrightarrow{60\%} \qquad \qquad (8)$$

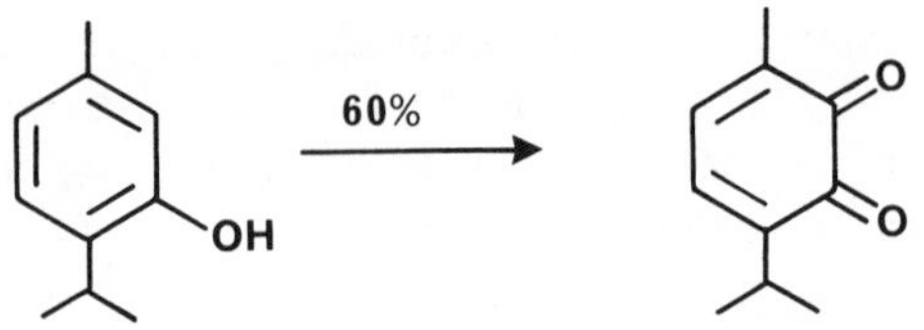

$$\text{Carvacrol} \qquad \xrightarrow{59\%} \qquad \qquad (9)$$

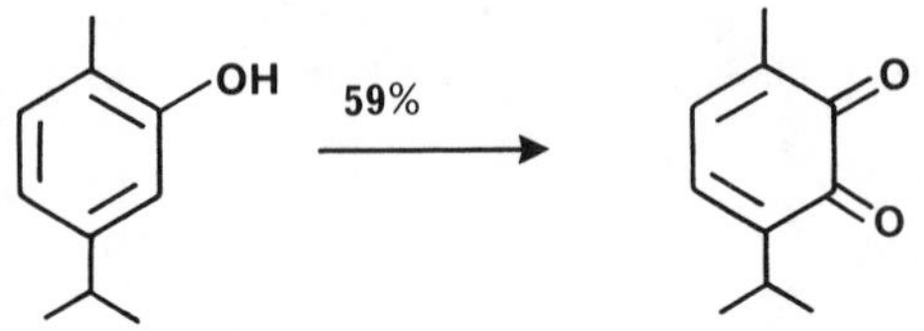

in these reactions. For example, with carvacrol and thymol up to 15% of the *p*-quinone was also formed. With other phenols such as **8**, the *p*-hydroxylated derivative was the major product together with the *o*-quinone and some selenated material as minor products [Eq. (10)]. Oxidation with the anhydride of simple hydroquinones and catechols to corresponding quinones proceeds in excellent yield and compares very favorably with other literature methods. In some cases the use of mesitylseleninic anhydride[32] gave improved yields of quinones, for example, oxidation of 2,4-di-tert-butylphenol gave the *o*-quinone in 84% yield. The reagent was, however, less convenient to prepare.

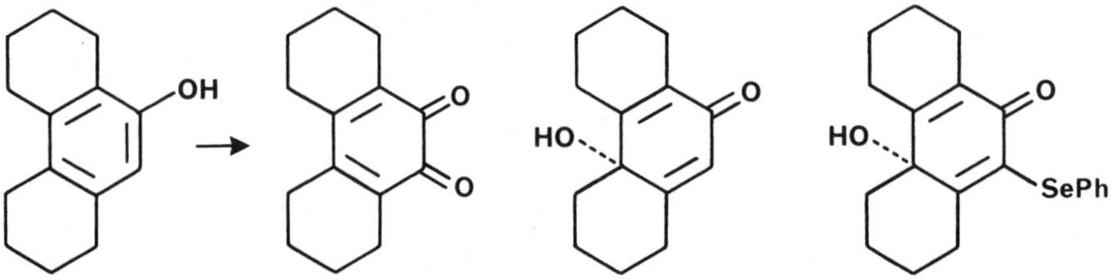

$$(\mathbf{8}) \qquad\qquad 37\% \qquad\qquad 44\% \qquad\qquad 19\% \qquad (10)$$

During studies directed at the synthesis of highly carcinogenic metabolites of 7,12-dimethylbenz[*a*]anthracene, Sukumaran and Harvey[33] further exploited the use of benzeneseleninic anhydride for the preparation of *o*-quinones. The key oxidation of the phenol, **9**, to the *o*-quinone was achieved in 80% yield [Eq. (11)]. In a more detailed investigation[34] of the oxidation of polycyclic phenols, a number of *o*-quinones were prepared in moderate to good yields. However, with certain substrates anomalous side reactions occurred [Eqs. (12) and (13)]. Oxidation of [2.2] and [3.3]paracyclophanols with benzeneseleninic anhydride has also been studied[35] with the intention of producing cyclophan-*o*-quinones, which show intramolecular charge transfer interactions [Eq. 14]. Interestingly,

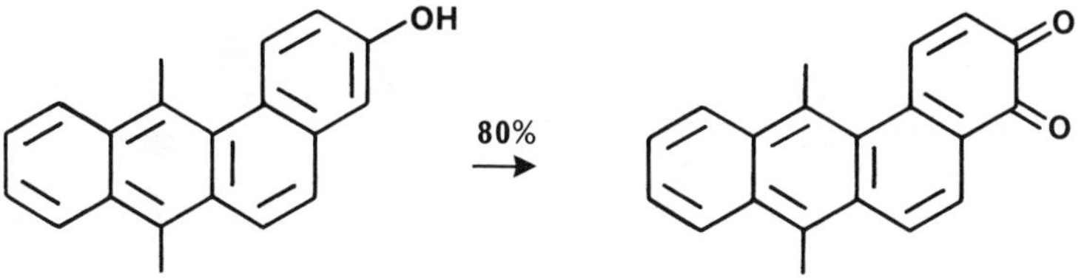

(11)

(12)

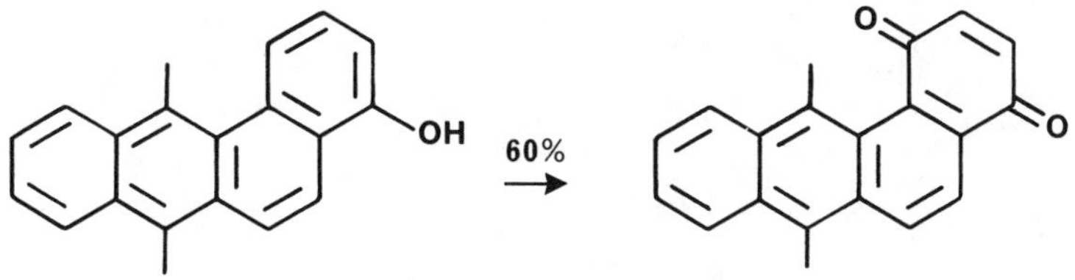

(13)

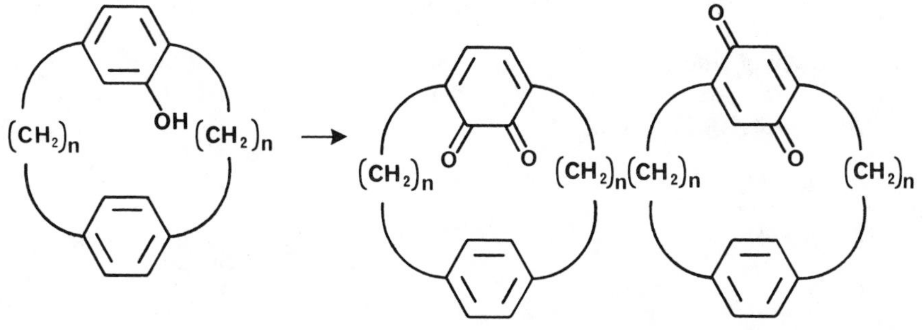

(14)

the more strained [2.2]paracyclophanol reacted slowly with the anhydride and required reflux in benzene for a period of 5 h for complete conversion to occur.

A novel and potentially useful reaction was noticed during the reaction of phenols with the anhydride in the presence of hexamethyldisilazane,[36] in which the major product was a phenylselenoimine rather than the expected hydroxy-dienones or quinones. Once again these reactions were highly orthoselective showing a *syn* configuration[37] of the phenylseleno group with respect to the neighboring oxygen atom. Typical examples of this process demonstrate the

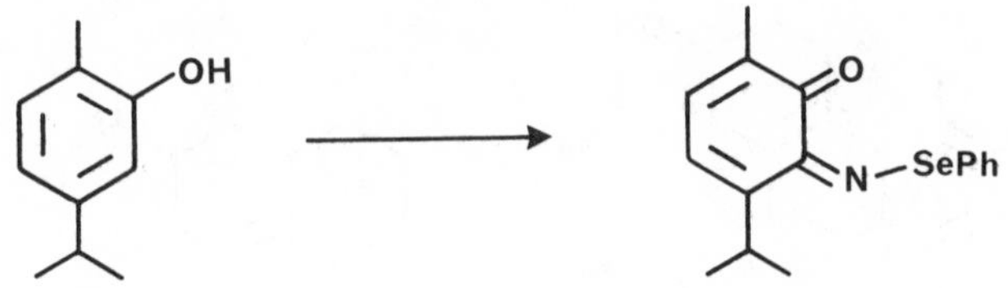

(15)

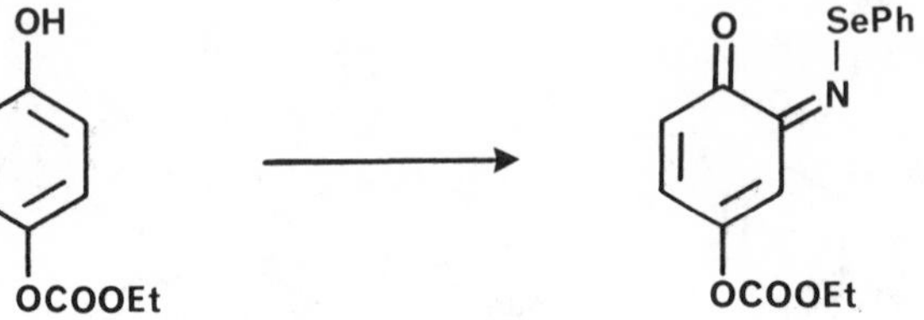

(16)

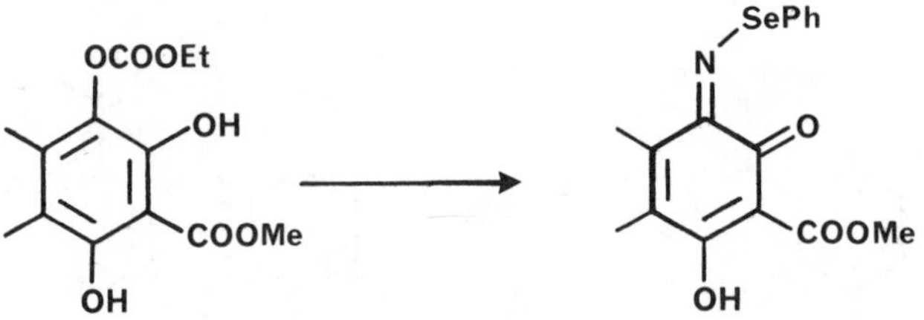

(17)

versatility and mildness of the reaction [Eqs. (15)–(17)]. From a synthetic point of view these reactions are important as the products may either be reductively acetylated using zinc and acetic anhydride or converted directly to the amino-phenol by treatment with benzenethiol. The process therefore constitutes a new amination procedure for phenols (Scheme 1).

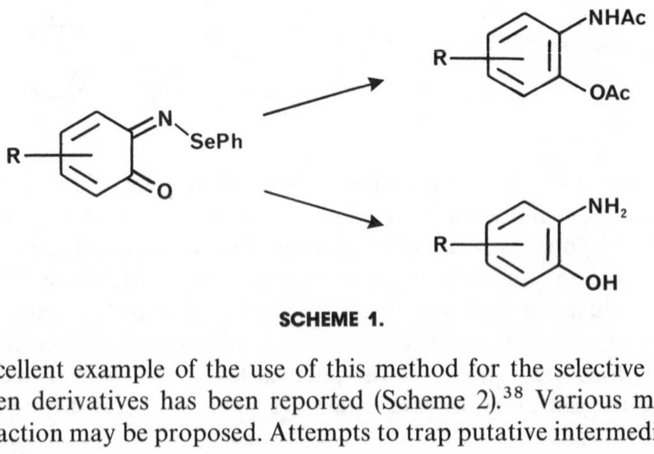

SCHEME 1.

An excellent example of the use of this method for the selective amination of estrogen derivatives has been reported (Scheme 2).[38] Various mechanisms for the reaction may be proposed. Attempts to trap putative intermediates using

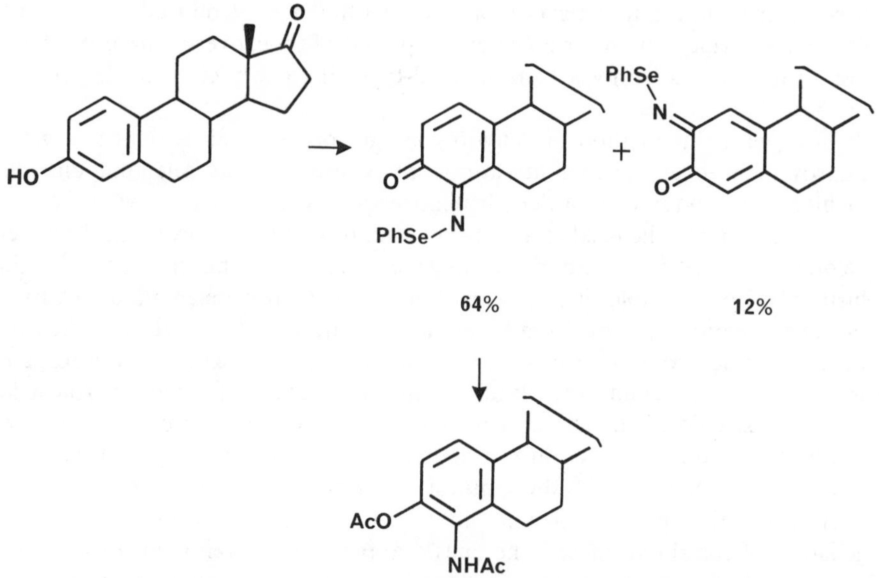

64%   12%

SCHEME 2.

either phenylazide or nitrile oxides were unsuccessful and led instead to the formation of either selenoimines [Eq. (18)] or selenoamides [Eq. (19)].[39]

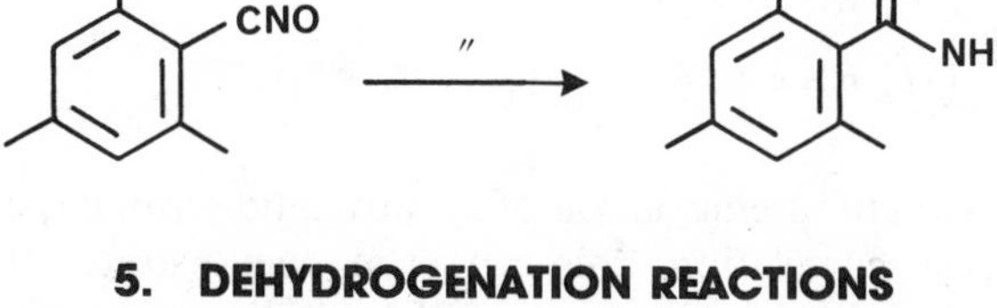

$$Ph-N_3 \xrightarrow[HN(SiMe_3)_2]{(PhSeO)_2O}$$ (18)

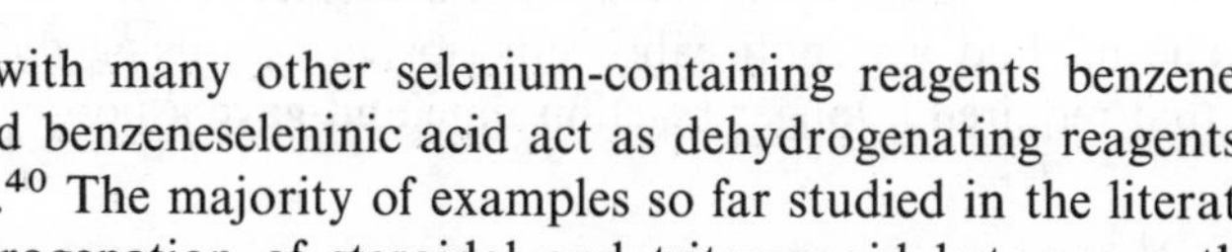

    (19)

5. DEHYDROGENATION REACTIONS

In common with many other selenium-containing reagents benzeneseleninic anhydride and benzeneseleninic acid act as dehydrogenating reagents for carbonyl species.[40] The majority of examples so far studied in the literature refer to the dehydrogenation of steroidal and triterpenoid ketones as these lead directly to biologically active compounds.

In typical reactions the ketone was treated with anhydride, in chlorobenzene as the solvent, at temperatures between 95 and 132°C. The quantity of reagent used determines the number of double bonds that are introduced. If an excess of the anhydride is used for an extended period of time, overoxidized products are formed that arise by a benzylic acid-type ring contraction of ring A, i.e., are A-nor-2,3-diones.

In general this method of dehydrogenation of steroidal ketones is a great improvement over existing literature procedures such as selenium dioxide, dichlorodicyanoquinone, or dehydrohalogenation (Table 1). In view of the cost of the reagent and the need to avoid side reactions a catalytic cycle has been developed that avoids these problems, therefore extending the method still further.[12] For example, it has been shown that benzeneseleninic anyhydride may be generated *in situ* from catalytic quantities of diphenyl diselenide by oxygen transfer from iodoxybenzene or *m*-iodoxybenzoic acid; the latter reagent is preferred as it facilitates work-up and allows recovery of *m*-iodobenzoic acid and the diselenide. The dehydrogenation of steroidal ketones using these new catalytic methods proceeds in excellent yields. This catalytic system has also found useful application in the synthesis of 4a-methyl-4aH-fluorenes.[41]

A report that benzeneseleninic anhydride was ineffective in the dehydrogenation of certain spirocyclic ketones[42] appears, upon reinvestigation, to be erroneous.[12] Consequently the dienones 10 or 11 are converted to their corresponding unsaturated derivatives using either the stoichiometric or the catalytic benzeneseleninic anhydride methods in excellent yield [Eq. (20)].

$$(20)$$

(10) n = 1

(11) n = 2

During the structural elucidation of an aromatic sesquiterpene involving a chemical synthesis correlation, Polonsky et al. demonstrated the effectiveness of the anhydride in the dehydrogenation of a decalenone to the 1,4-dione [Eq. (21)].[43] The method was noticeably superior to one using dichlorodicyanoquinone that required a longer reaction time and gave a poorer yield.

$$(21)$$

TABLE 1
REACTION OF KETONES WITH BENZENESELENINIC ANHYDRIDE (BSA)

Ketone	Reaction Conditions			Products (%)		
	Temperature (°C)	BSA(eq)	Time	Enone	A-Nor-2,3-dione	¹2-Phenylseleno Ketone
Lanostanone	95	1	45 min	1-enone (67)	(13)	(4)
	100	2	18 h	(38)	(41)	
	100	1ᵃ	150 min	(64)	(10)	
Cholest-4-en-3-one	132	1	40 min	1,4-dienone (92)		
4,4-Dimethylcholest-5-en-3-one	95–100	1	25 min	1,5-dienone (58)	(29)	
	95–100	2	19 h	(23)	(33)	
α-Amyrone	95–100	1	25 min	1-enone (74)		
	95–100	2	17 h	(32)	(42)	
β-Amyrone	95–100	1	15 h	1-enone (54)	(8)	
	95–100	2	18 h	(27)	(46)	
Hecogenin acetate	132	2	50 min	9,(11)-enone (91)		
	100	2	160 min	(81)		
Lupeone	95–100	1	15 min	1-enone (58)		
Cholestan-3-one	132	2	3 h	1,4-dienone (83)		
Cholest-1-en-3-one	95	1	45 min	1,4-dienone (76)		
Cholesta-4, 6-dien-3-one	95	1	1 h	1,4,6-trienone (>50)		

ᵃ Benzeneseleninic acid.

Dehydrogenation of 4-azacholestan-3-one occurs on treatment with the anhydride in diglyme although similar reactions with acyclic amides fail.[44] However, in a more detailed study, further examples of azasteroid δ-lactams were investigated and shown to conform to the normal dehydrogenation pattern.[45] Reaction of seven-membered ring lactams in a similar fashion led to the formation of imides as the major products [Eq. (22)].

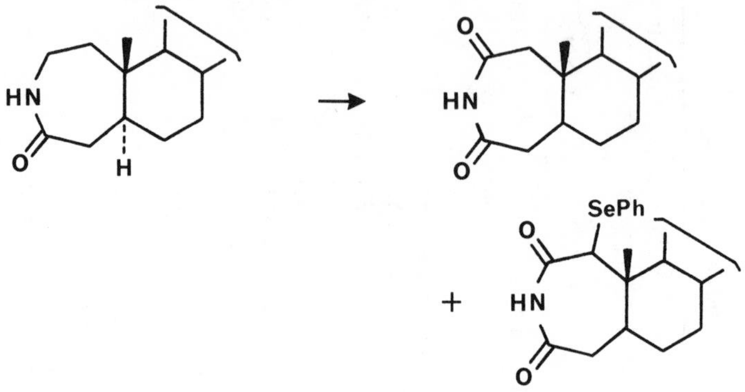

$$(22)$$

Related dehydrogenation reactions of steroidal δ-lactones are also possible but attempts to dehydrogenate the smaller ring γ-lactones or acyclic esters were unsuccessful.[46] During this study it was noticed that reaction with the anhydride over extended periods of time caused further oxidation to take place to give angular hydroxylated products [Eq. (23)]. Other examples of hydroxylation of ketones using benzeneseleninic anhydride have been observed and are discussed later.

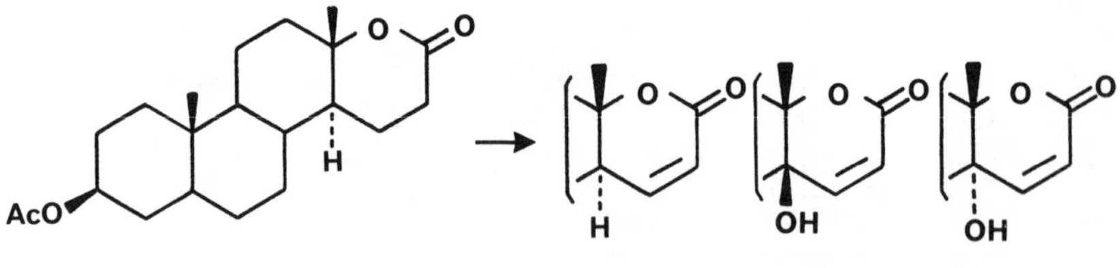

16%   23%   19%

$$(23)$$

The anhydride has also proved to be an effective dehydrogenation reagent for indolines and tetrahydroisoquinolines.[47] When the three-position of the indoline was unblocked the resulting indole was sufficiently reactive to undergo further reaction with a selenating species present in the mixture[48] to give 3-phenylseleno derivatives as the major products. The phenylseleno substituent could be readily removed by subsequent treatment with nickel boride to give the unsubstituted indole [Eq. (24)].

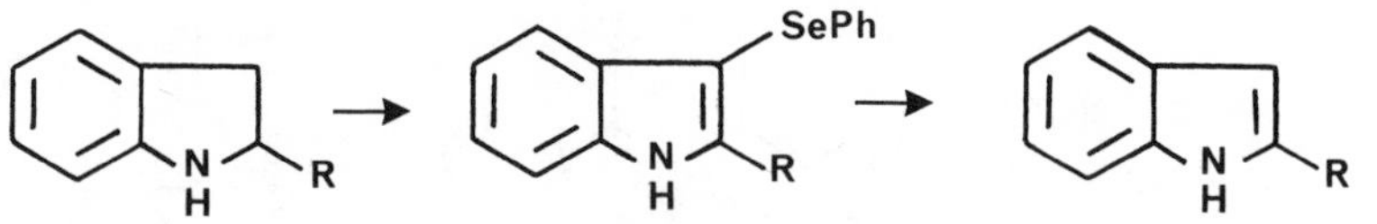

$$(24)$$

6. ANGULAR HYDROXYLATION REACTIONS

In a series of papers Yamakawa has described the remarkably efficient introduction of angular hydroxyl groups into polycyclic ketones using benzeneseleninic anhydride in refluxing toluene or chlorobenzene as solvent.[49-52] For example, with the tricyclic ketone, 12, the hydroxylated product was obtained in 70% yield without formation of any of the corresponding enone [Eq. (25)].[49] A number of other simple model compounds were studied in which

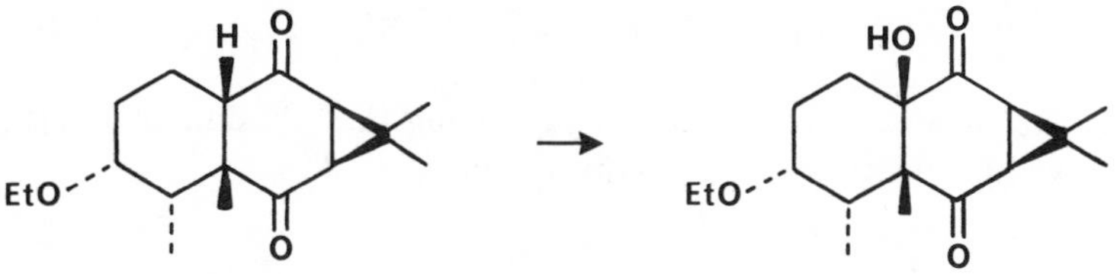

$$(25)$$

it was shown that prior formation of the enolate anion by treatment with NaH, followed by reaction with the anhydride favored the formation of the α-hydroxylated product, whereas in the presence of AlCl$_3$, dehydrogenation was the preferred reaction pathway.[50] The use of this angular hydroxyation process during the synthesis of furanoeremophilane natural products further illustrates the selectivity and mildness of the reaction conditions since other chemically sensitive functional groups survived the treatment.[51,52] Chemical modification of quassinoid compounds is of interest especially for the preparation of potentially useful antitumor agents. Khôi and Polonsky have shown that chaparrinone triacetate reacts with benzeneseleninic anhydride to afford products derived from dehydrogenation and angular hydroxyation (Scheme 3).[53] Deacetylation of these products gave novel compounds for biological evaluation.

Hydroxylation of enamidic Δ^5-4-azasteroids is also possible using benzeneseleninic anhydride (or benzeneseleninic acid) as the oxidant. However, other

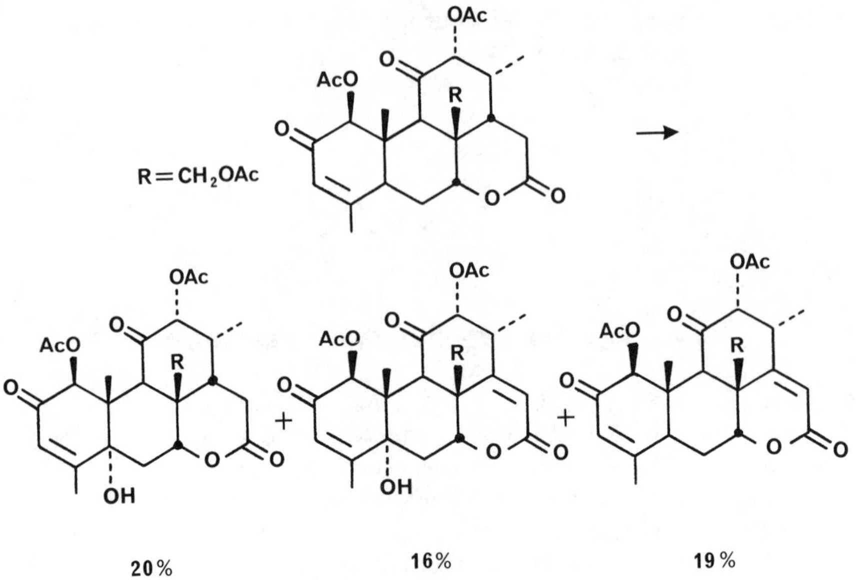

R=CH₂OAc

20%   16%   19%

SCHEME 3.

products are formed in the reaction suggesting that Pummerer intermediates are involved in the mechanistic pathway [Eq. (26)].[54]

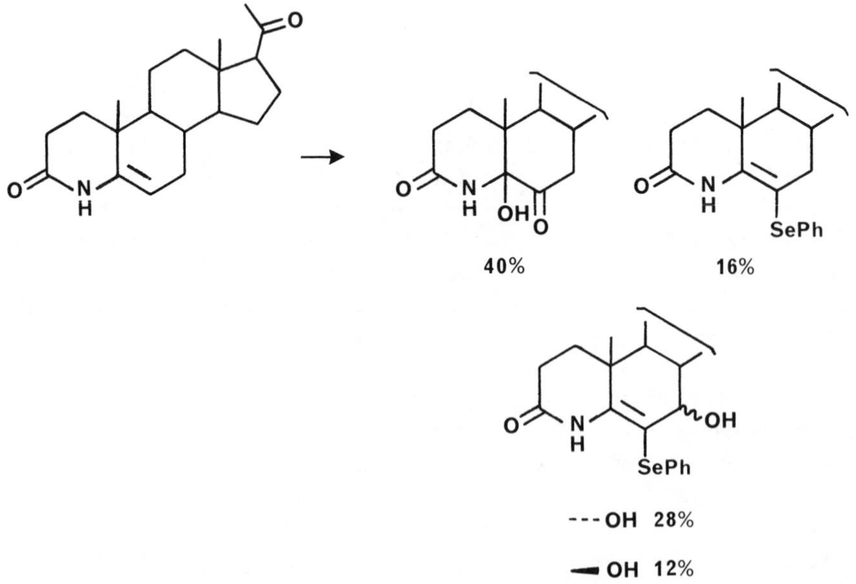

40%   16%

---OH 28%

—OH 12%

$$(26)$$

7. OXIDATION OF ALCOHOLS

Methods for effecting the oxidation of alcohols to carbonyl compounds that claim advantages over existing procedures continue to be introduced. This apparently simple but very important transformation has stimulated the discovery of a plethora of new reagents. The complexity of their nature, however, is such that there remains a need to find alternative and more refined oxidants.

Seleninic anhydrides react at room temperature with alcohols to form stable and isolable seleninic esters.[5,13] 1-Anthraquinone seleninic acid has been reported to be an oxidant for ethanol and hydroquinone,[55] but it has not been developed as a general reagent. On the contrary, benzeneseleninic anhydride smoothly converts alcohols to the corresponding carbonyl derivatives at higher temperatures, ranging from 65–130°C [Eqs. (27) and (28)].[56] As an added bonus in some of these reactions the known propensity for the anhydride to also effect dehydrogenation of carbonyl compounds may be utilized to provide unsaturated products thus achieving in one step an operation that is otherwise a multistep process [Eqs. (29) and (30)].

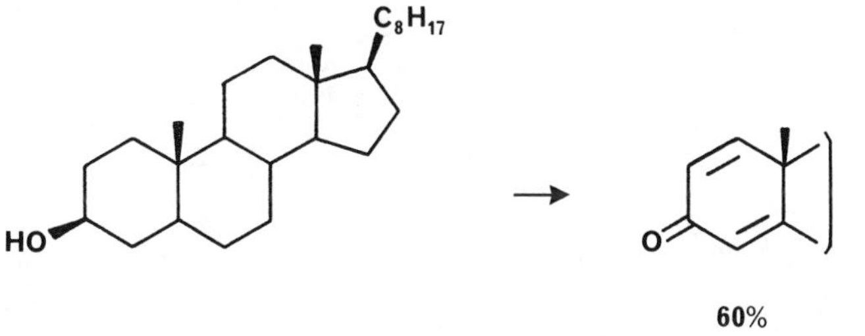

(27)

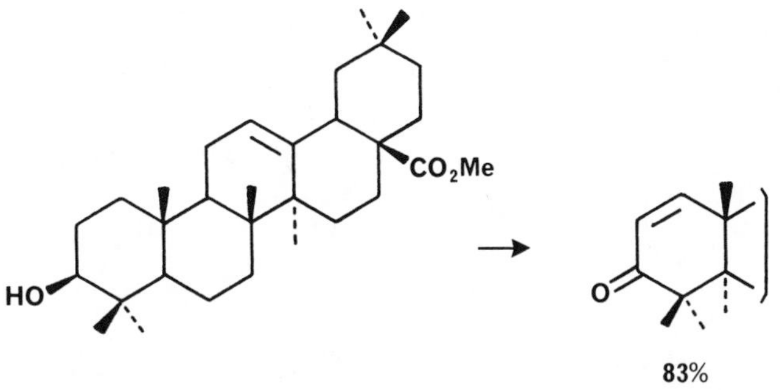

60%

(28)

83%

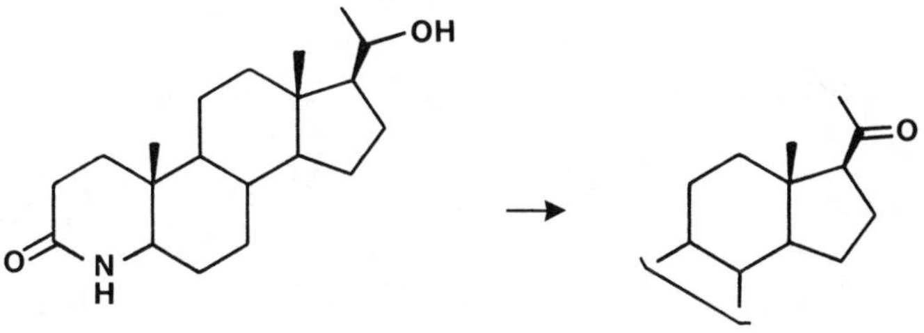

(29)

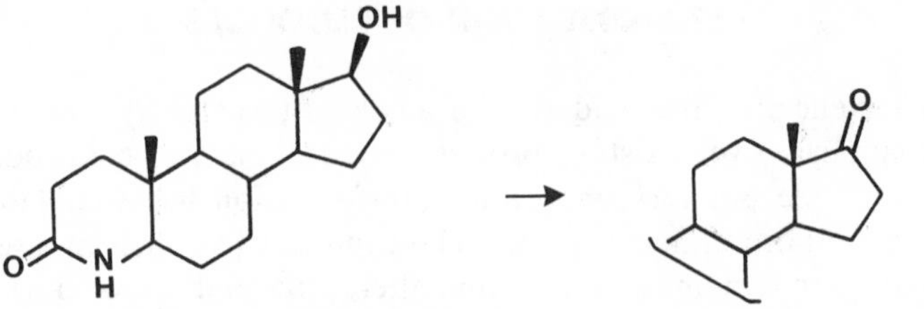

(30)

Oxidation of benzylic alcohols by the anhydride proceeds efficiently to give the carbonyl compound but, importantly, primary alcohols are not overoxidized to the carboxylic acids (Table 2). In competitive studies *p*-nitrobenzyl alcohol was shown to be oxidized approximately twice as fast as the *p*-methoxy compound.[57]

The mechanism for the reaction undoubtedly involves an intermediate seleninyl ester that fragments at higher temperatures by a process analogous to

TABLE 2
OXIDATION OF BENZYLIC ALCOHOLS WITH BENZENESELENINIC ANHYDRIDE

Alcohol	Product	Molecular Equivalent Anhydride	Solvent[a]	Time	(%) Yield
$PhCH_2OH$	PhCHO	2	C_6H_6	20 min	(99.5)[d]
$p\text{-}NO_2C_6H_4CH_2OH$	$p\text{-}NO_2C_6H_4CHO$	0.5	C_6H_6	8 min	(97)
$p\text{-}ClC_6H_4CH_2OH$	$p\text{-}ClC_6H_4CHO$	0.5	$C_6H_6^{b}$	1 day	(57)
$p\text{-}MeOC_6H_4CH_2OH$	$p\text{-}MeOC_6H_4CHO$	0.5	C_6H_6	15 min	(99)
$p\text{-}MeC_6H_4CH_2OH$	$p\text{-}MeC_6H_4CHO$	1	$C_6H_6^{b}$	2 days	(85)
PhCHOHPh	PhCOPh	1	THF^{c}	3 h	(85)
$PhCH{=}CH{\cdot}CH_2OH$	$PhCH{=}CHCHO$	0.33	$PhCl^{b}$	10 h	(53)[d]
PhCHOHCOPh	PhCOCOPh	1	THF	3 h	(92)
PhCHOHCHOHPh	PhCOCOPh	2	THF	3 h	(77)
$PhCHOHCO_2Me$	$PhCOCO_2Me$	0.5	C_6H_6	10 min	(97)
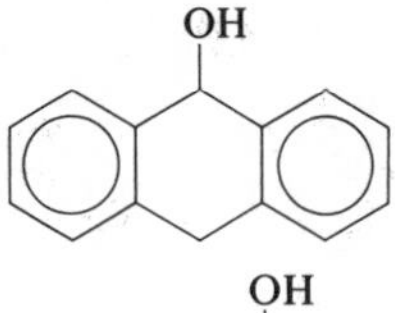	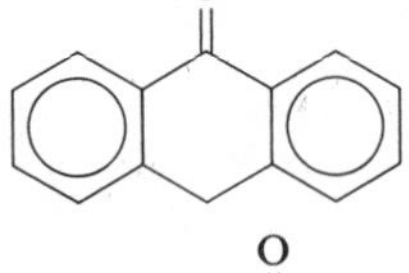	0.5	C_6H_6	5.5 h	(83)
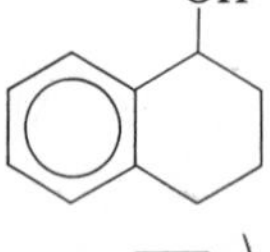	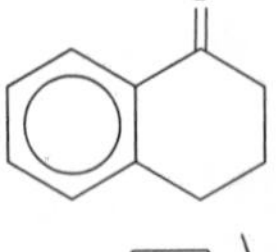	0.33	$C_6H_6^{b}$	18 h	(48)
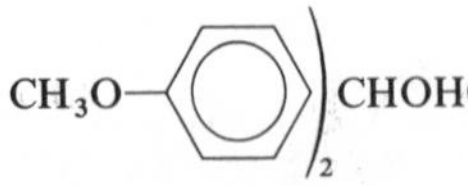	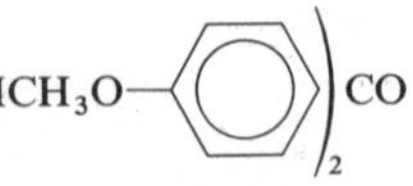	1	THF	8 min	(99)

[a] At reflux under nitrogen unless otherwise stated.
[b] At room temperature.
[c] THF = tetrahydrofuran.
[d] Isolated as the 2,4-dinitrophenyl hydrazine derivative.

the *syn* elimination of selenoxides (Scheme 4). Diphenyl diselenide is a readily isolated by-product of the reaction, and may be recycled via oxidation back to the anhydride. In principle these oxidation reactions only require 1/3 equivalents of the anhydride owing to disproportionation reactions of the presumed inter-mediate seleninic and selenenic acids [Eq. (31)]. These conclusions are confirmed in practice although the reaction only reaches completion after a long period of time. As a compromise consistent with economy, speed, and yield, it was found that the use of 0.5 equivalents of anhydride gave optimum results. In these reactions 1/3 of an equivalent of benzeneseleninic acid should also be produced as a side product [Eq. (32)]; this too is experimentally observed.[57]

$$3\,ArCH_2OH \ + \ \left(PhSeO\right)_2O \ \longrightarrow \ 3\,ArCHO \ + \ \left(PhSe\right)_2 \ + \ 3\,H_2O \quad (31)$$

$$2\,ArCH_2OH \ + \ \left(PhSeO\right)_2O \ \longrightarrow \ 2\,ArCHO \ + \ \tfrac{2}{3}\left(PhSe\right)_2 \ + \ \tfrac{2}{3}\underset{O}{\overset{O}{PhSeOH}}$$

$$+ \ \tfrac{5}{3}\,H_2O \qquad\qquad (32)$$

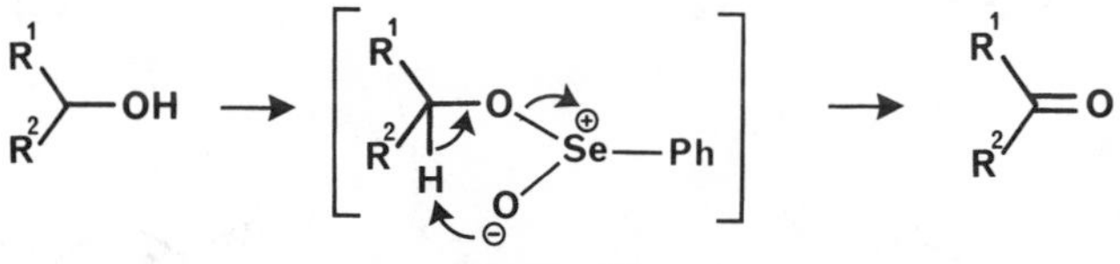

SCHEME 4.

In related studies a highly efficient method for the oxidation of hydroxyl groups has been developed that employs the use of *t*-butylhydroperoxide and catalytic quantities of diaryl diselenides.[58] For most examples the use of bis-(2,4,6-trimethylphenyl) diselenide has been recommended as the reagent of choice giving the best yields of aldehydes from allylic or aliphatic alcohols. Particularly impressive in this work from a synthetic point of view was the ability to oxidize hydroxyl groups in the presence of phenylseleno (and phenylthio) substituents as in compound 13 [Eq. (33)]. Classical oxidation of 13 using the Collins reagent gave a very poor yield of product. Several other examples were studied that underline the generality of the oxidation method [Eqs. (34)–(38)].[59]

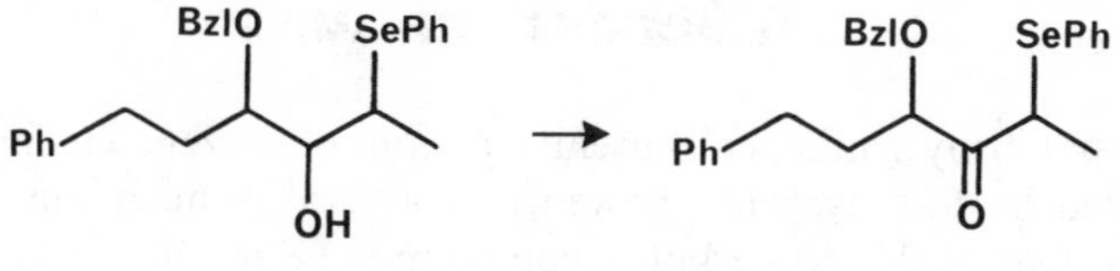

(13)   (33)

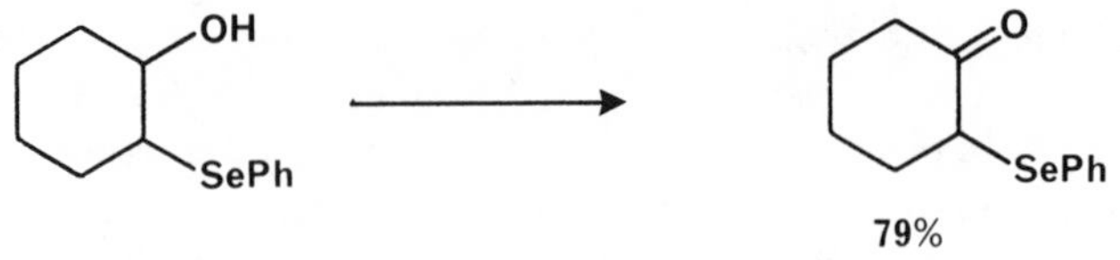

(34)

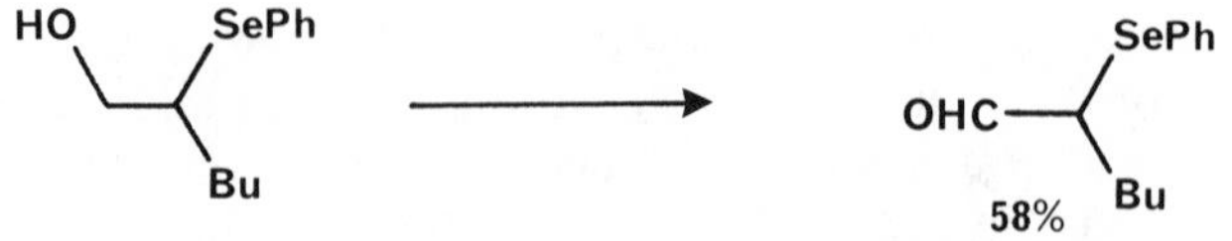

79%

(35)

58%

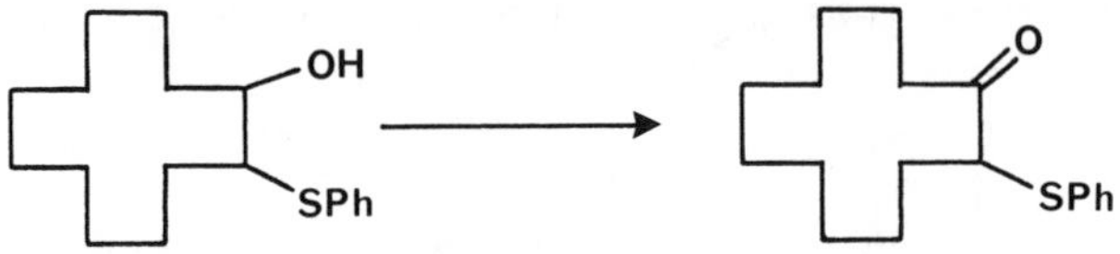

(36)

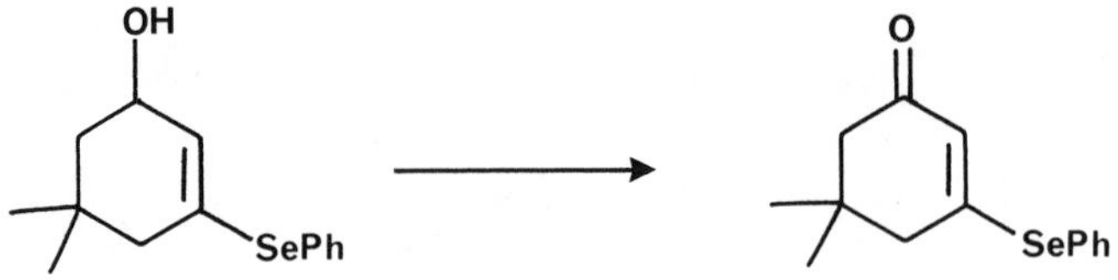

(37)

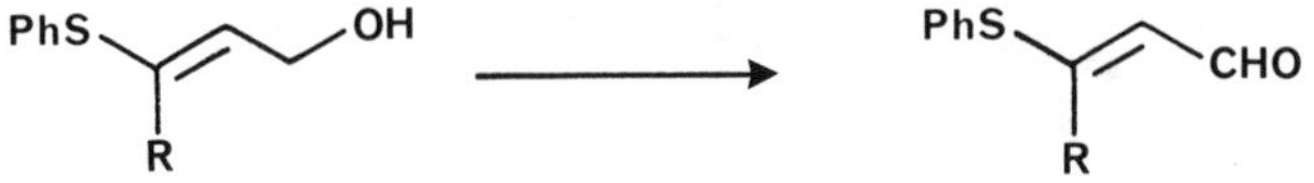

(38)

The oxidation of the substituted allylic alcohols [Eqs. (37) and (38)] is particularly noteworthy, as other literature procedures, the best of which being a large excess of manganese dioxide, gave only a moderate yield of product. A more detailed report of these results has been presented including comments as to the possible reaction intermediates.[60]

Finally, Taylor and Flood have shown that oxidation of alcohols may be achieved in excellent yields using polymer bound benzeneseleninic acid as catalyst in the presence of t-butylhydroperoxide.[3a] The catalyst is stable to the reaction conditions and may be recycled with no apparent loss of activity.

8. REACTIONS WITH AMINES

One of the first truly synthetically useful reactions of benzeneseleninic anhydride was reported by Czarny, who showed that several primary amines could be oxidized in high yields to carbonyl compounds [Eqs. (39)–(41)].[61] However, the reaction was limited to the preparation of nonenolizable ketones. The mech-

Ph$_2$CHNH$_2$ $\longrightarrow$ (39)

(40)

(41)

anism of the reaction parallels the oxidation of alcohols discussed earlier. Interestingly, when benzylamine was used as a substrate, with an excess of the anhydride, a 96% yield of benzonitrile was obtained. With only one equivalent of anhydride some benzaldehyde was also generated. Similar reactions utilizing benzeneseleninyl chloride as the oxidant, with the same amines, gave comparable results.[62] Related dehydrogenation reactions of cyclic amines[47,48] with benzeneseleninic anhydride are discussed in the early part of this review.

While phenols react with the anhydride to give a useful array of products, the same is not true for anilines where in general very complex mixtures of products are obtained. Two interesting exceptions are the reactions of o-phenylenediamine and 2,4,6-trimethylaniline, which give cleaner mixtures allowing isolation of major products. o-Phenylenediamine reacts rapidly at −17°C to give the novel bis-phenylselenodiimine [Eq. (42)] whereas 2,4,6-trimethylaniline reacts at room temperature, by an alternative route, to give the azo derivative in moderate yield [Eq. (43)].[63]

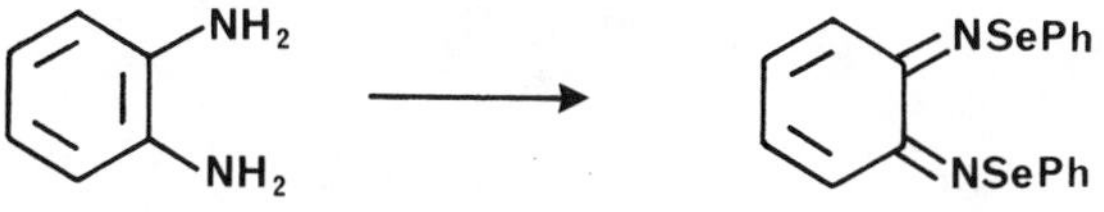

(42)

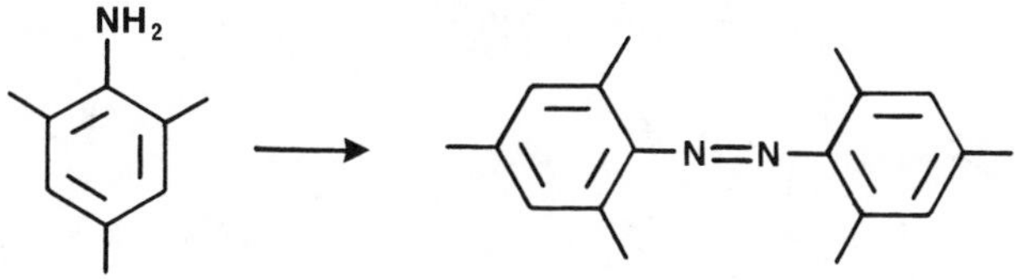

(43)

An important application of benzeneseleninic anhydride is its capacity to regenerate ketones from ketone hydrazone, oxime, and semicarbazone derivatives.[64,65] In a comparative study some indication of the relative ease with which

TABLE 3
CONVERSION OF BENZOPHENONE DERIVATIVES
TO BENZOPHENONE WITH BENZENESELENINIC
ANHYDRIDE

Derivative	(%) Yield	Time
Phenylhydrazone	(90)	3 h
p-Nitrophenylhydrazone	(56)	3 days
2,4-Dinitrophenylhydrazone	NR[a]	3 days
Tosylhydrazone	(95)	20 min
Oxime	(89)	3 h
Semicarbazone	(89)	2 h
N,N-Dimethylhydrazone	NR[a]	23 h
O-Methyloxime	NR[a]	24 h

[a] NR = no reaction.

benzophenone derivatives are unmasked to the parent ketone may be seen in Table 3.

Attention should be drawn, however, to deprotection reactions that resisted the existing literature methodology. For example, it was shown that p-nitro-phenylhydrazones of cholesta-1,4-dienone or the 1,4,6-trienone (prepared by

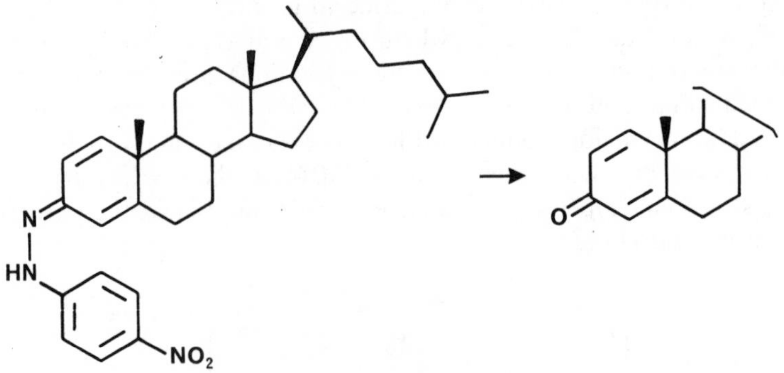

(44)

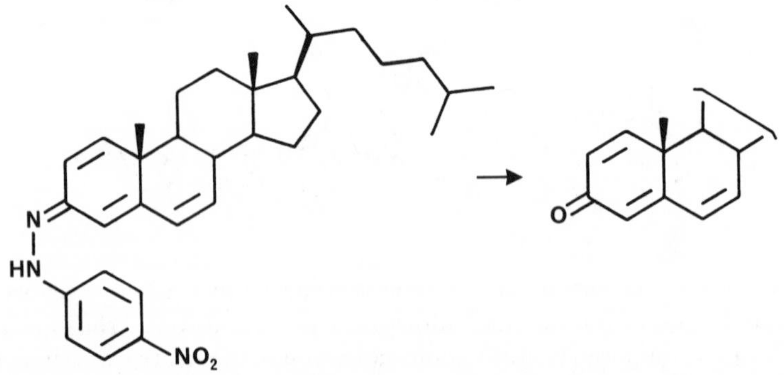

(45)

electron transfer dehydrogenation) reacted with the anhydride at room temperature over a few hours to give the ketonic products in excellent yields [Eqs. (44) and (45)].

These significant results prompted a more substantial study with aldehyde derivatives and other nitrogen containing species [65,66] where noticeable differences arose when compared with similar ketone derivatives. To illustrate these differences it is pertinent to consider 2-furanaldehyde derivatives. Treatment of either the phenyl or *p*-nitrophenyl hydrazone with $(PhSeO)_2O$ gave acylazo derivatives rather than regenerating the parent aldehyde [Eq. (46)]. If the aldehyde

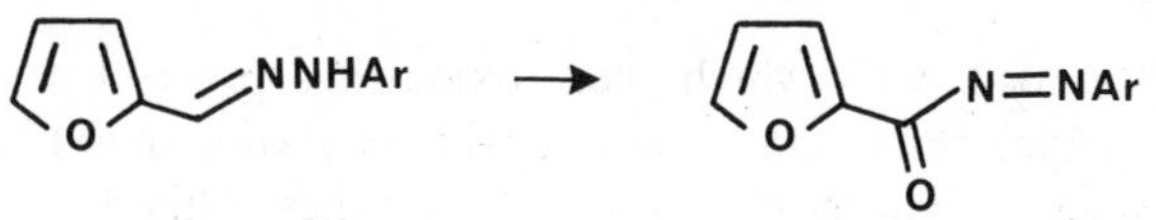

$$\text{(46)}$$

$$Ar = Ph \text{ or } \underline{p}NO_2Ph$$

is required, then tosyl hydrazones or oximes must be used as precursors. Mechanistically the acylazo compounds are obtained by [2,3]sigmatropic shift of an intermediate azoseleninic species followed by loss of benzeneselenol (Scheme 5). Several other aldehydes were investigated giving comparable results. An alternative preparation of the acylazo compounds by oxidation of acylhydrazides using *N*-bromosuccinimide has been reported[67] previously with yields ranging from 40–80%. By comparison benzeneseleninic anhydride was much more efficient and oxidized the same acylhydrazide in yields varying between 70–100%.[64,65]

These oxidation reactions may be compared with an early observation of Rheinbold and Giesbrecht [16j] where it was reported that hydrazine hydrate was oxidized by benzeneseleninic acid, presumably to form diimide. These studies have been confirmed by the trapping of the *in situ* generated diimide with cinnamic acid.[68]

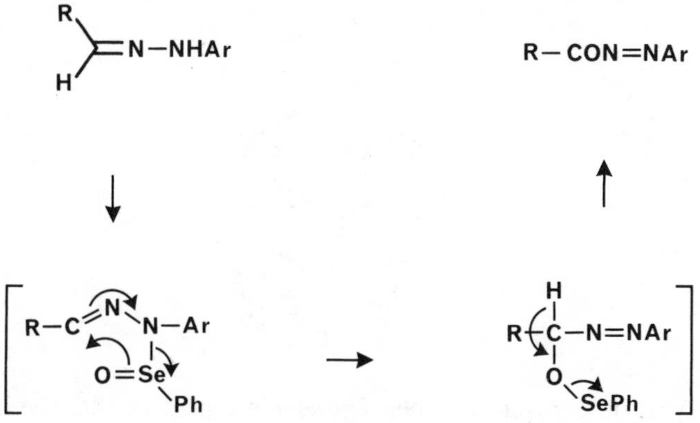

SCHEME 5.

Many other nitrogen-containing compounds are oxidized by the anhydride. Hydroxylamines afford nitroso derivatives [Eq. (47)] while hydrazo compounds are converted to their azo analogs [Eq. (48)].[65,66]

$$R-\underset{\underset{H}{|}}{N}-OH \longrightarrow R-N=O \qquad (47)$$

$$R-NH-NH-R \longrightarrow R-N=N-R \qquad (48)$$

In view of the ease with which these oxidations proceed it was attractive to use the anhydride as an *in situ* oxidant for the generation of N-phenyltriazoline dione. This reagent is commonly employed as a dienophile for reactive or unstable dienes and sometimes as a protecting group for the diene moiety. Traditionally it is prepared by oxidation of the corresponding hydrazide with either N_2O_4, t-BuOCl, DMSO/Tol NCO, $Pb(OAc)_4$, NBS, or Br_2/H_2O with varying degrees of success. Reaction of the N-phenylurazole with benzeneseleninic anhydride in THF solution produced a deep-red solution of the N-phenyltriazoline dione after 3–4 min. Dienes, such as ergosterol derivatives, were added to give high yields of the Diels–Alder adducts [Eq. (49)].[65,69]

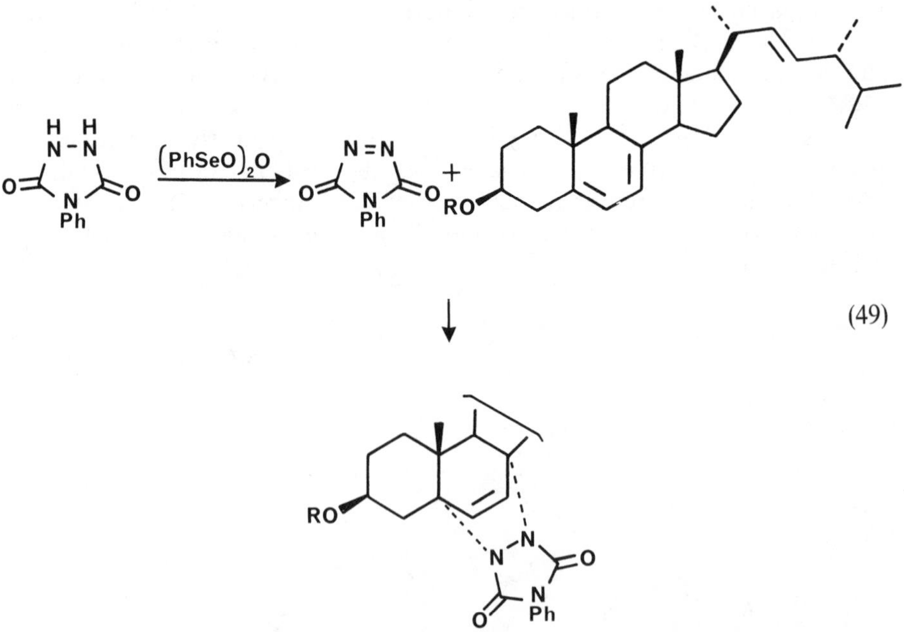

$$(49)$$

N-Acylhydrazines react with benzeneseleninic acid in the presence of triphenylphosphine to give selenolesters in high yield [Eq. (50)].[68,70] The products

$$R-\overset{\overset{\displaystyle O}{\|}}{C}NHNH_2 \quad \xrightarrow[Ph_3P]{PhSeO_2H} \quad R-\overset{\overset{\displaystyle O}{\|}}{C}SePh \qquad (50)$$

of the reaction are of special interest as acyl transfer agents; the presence of the phosphine in the reaction encourages higher levels of selenating reagent in the reaction mixtures. From the eleven examples studied it is clear that the procedure is tolerant of both aryl, heteroaryl, and alkyl substituents containing hindered, alkenyl, or potentially labile functional groups such as t-butyldimethylsilyl ethers. The method also has advantages over existing procedures for the preparation of selenolesters that employ air sensitive or malodorous reagents.[71,72] While the reactions proceed well in methylene chloride solution a change to more polar solvents tended to lower the yield of selenolester; nevertheless useful precursors for macrolide synthesis could be prepared [Eq. (51)].[68]

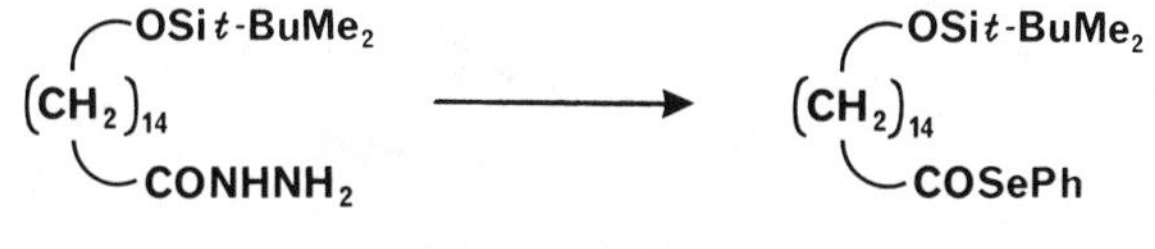

$$\begin{array}{c}\ulcorner OSi\,t\text{-BuMe}_2 \\ (CH_2)_{14} \\ \llcorner CONHNH_2\end{array} \quad \longrightarrow \quad \begin{array}{c}\ulcorner OSi\,t\text{-BuMe}_2 \\ (CH_2)_{14} \\ \llcorner COSePh\end{array} \qquad (51)$$

Similar oxidation of sulfonylhydrazides with the anhydride provided an excellent and very convenient preparation of selenosulfonates.[73]

The conversion of 1,1-disubstituted hydrazines to tetrazenes is fairly readily achieved by many oxidants, but good results may also be obtained using benzeneseleninic acid in methanol.[74] Of several mechanistic pathways that may be proposed for this transformation the authors favor Pummerer-like reaction of intermediate seleninamide (Scheme 6). Selenium dioxide was also shown to be a suitable oxidant but the yields were less reliable. Anomalous results were observed in cases where the hydrazine contained an aryl substituent, or a leaving group such as sulfonyl, or where it was particularly hindered.

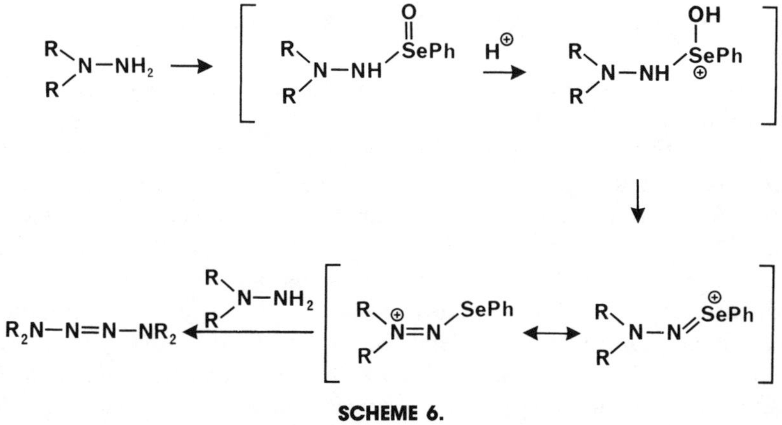

SCHEME 6.

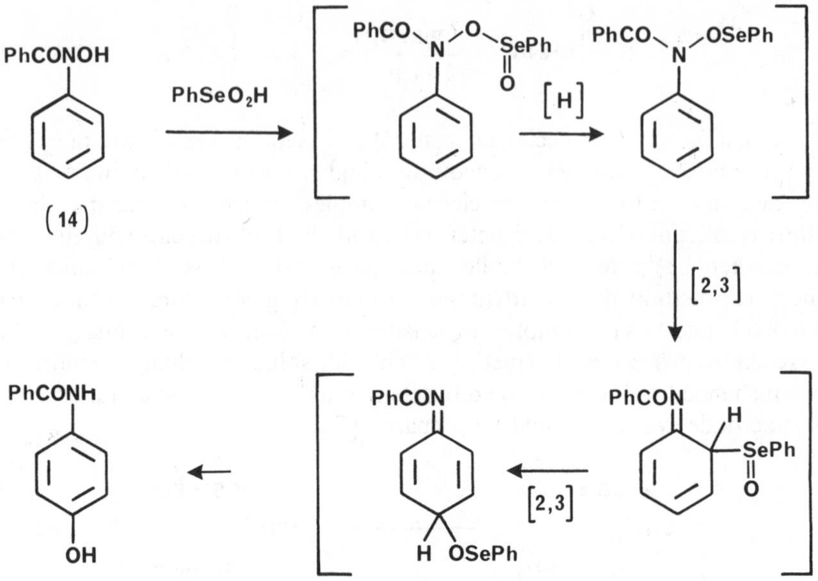

SCHEME 7.

Treatment of *p*-nitrophenylhydrazine with benzeneseleninic anhydride at 0°C gave some nitrobenzene (27%), however, the major product was shown to be *p*-nitrophenylselenobenzene.[65]

Rearrangement reactions of *N*-arylhydroxamic acids catalyzed by seleninic acids (and phenylselenyl chloride) have also been noted. In a typical example the hydroxamic acid, 14, with 0.05 eq of phenylseleninic acid gave the benzanilide product. Arguments in favor of the mechanism shown in Scheme 7 are presented.[75] The products of the reaction vary depending upon the substituents in the aryl ring undergoing the rearrangement.

9. REACTIONS WITH THIOACETALS

The 1,3-dithiolane unit is commonly employed in organic synthesis to protect the carbonyl group but of equal importance is the ability to selectively remove it at the end of the synthetic sequence. Although many methods to effect this deprotection are now known, finding new procedures remains a priority. The electrophilic character and the high oxidizing power of benzeneseleninic anhydride make it an attractive candidate for achieving these deprotection reactions.

Treatment of several 1,3-dithiolanes with the anhydride at between room temperature and 40°C smoothly regenerates the parent carbonyl compound.[76] A prominent feature of these deprotection reactions is that very sterically hindered 1,3-dithiolanes, such as those derived from fenchone and 2,2,6-trimethylcyclohexanone, are readily deprotected [Eqs. (52) and (53)]. More

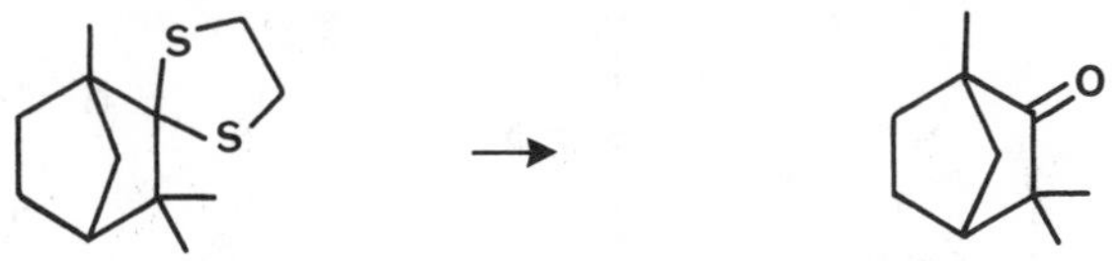

$$(52)$$

$$(53)$$

striking, however, is that certain compounds, for example, 15, underwent de-protection with the anhydride in good yield, whereas with many standard reagents only very poor results were obtained [Eq. (54)].

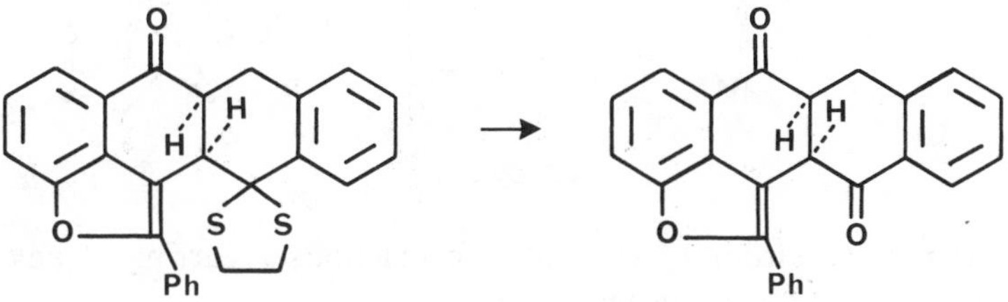

$$(54)$$

(15)

If the reaction was monitored by ^{1}HNMR or IR spectroscopy rapid formation of the carbonyl products was observed even under scrupulously dry conditions and hydrolytic work-up was therefore unnecessary. This fact coulld be of some use in synthesis where hydrolytically sensitive groups are also present.

The mechanism of the deprotection reaction does not involve intermediate formation of sulfoxides or radicals, rather it is thought that initial selenination of sulfur occurs followed by dithiolane ring opening and trapping by phenyl-seleninate anion to eventually afford the carbonyl compound (Scheme 8).

In another study the use of benzeneseleninic anhydride to deprotect seleno-acetals has been compared with more conventional methods such as mercury(II) chloride/calcium carbonate in wet acetonitrile, wet copper(II) chloride in acetone, or hydrogen peroxide in THF. In four of the six examples investigated benzeneseleninic anhydride was the superior reagent.[77]

Other work has also demonstrated the clear advantages of utilizing the anhydride deprotection method.[78,79] During a synthetic approach towards anthracyclinone derivatives conversion of the 1,3-dithiolane compound 16, to the ketone proceeded in 24% yield [Eq. (55)]. Mercury based reagents failed due to interaction with the furan ring while others either did not react or gave substantial decomposition and uncharacterizable products.[78] Similar observa-tions were made during efforts directed towards tetracycline where the dithiolane

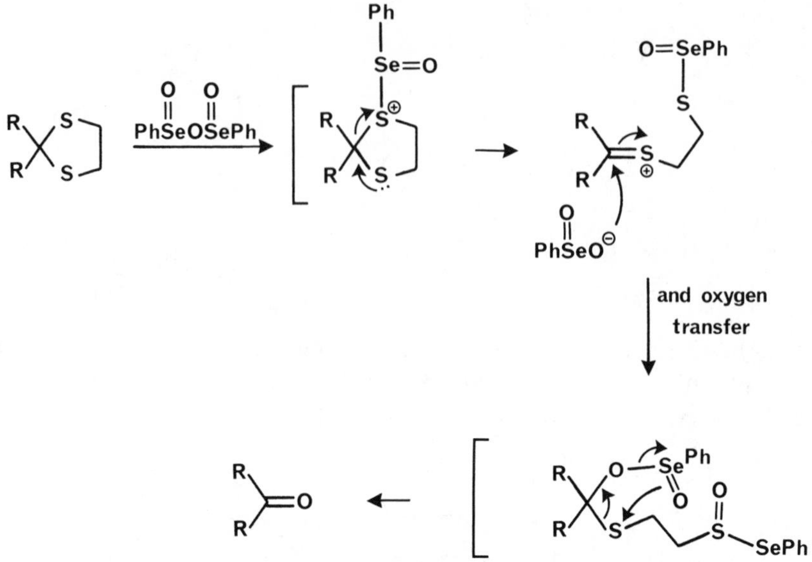

SCHEME 8.

group was removed from <u>17</u> without side reactions occurring at any of the other potentially reactive sites [Eq. (56)].

A further interesting example of the use of the anhydride to deprotect thioacetals appeared during the preparation of 11α-hydroxy-13-oxa-prostanoic acid.[80] Here deprotection of <u>18</u> took place in the presence of a double bond,

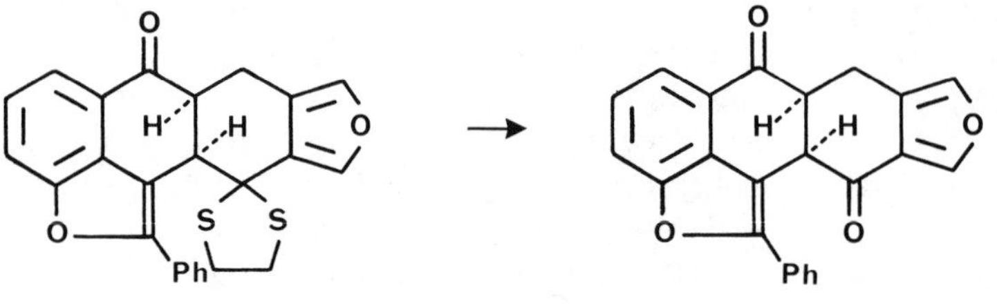

$$(55)$$

$$(16)$$

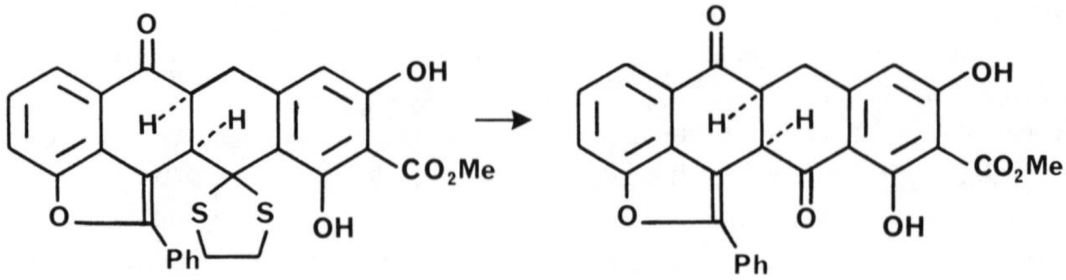

$$(17)$$

$$(56)$$

an oxyacetal, and an allylic ether, all of these being groups that might have caused problems [Eq. (57)]. Additionally this reaction was buffered by the co-addition of propylene oxide to eliminate traces of acids. Buffering with pyridine has also been used for highly acid sensitive substrates.[79]

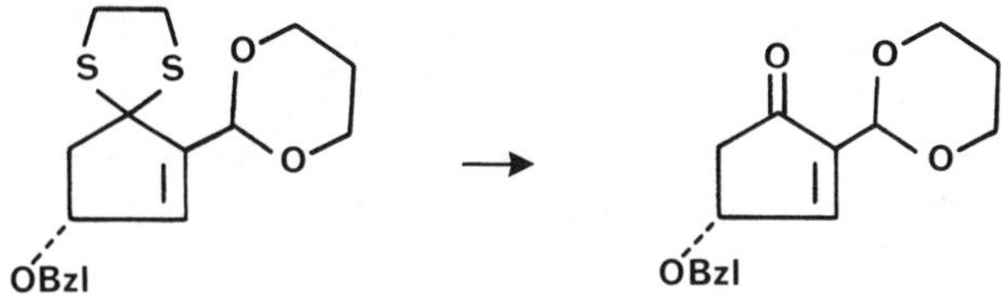

(57)

(18)

Similar deprotection of 1,3-dithianes is possible, but these reactions are often slower and few examples have been investigated. Reaction with 1,3-oxathiolanes on the other hand is rapid yet contaminating phenylselenated by-products cause problems.[76]

In summary, benzeneseleninic anhydride appears to be a good reagent for the deprotection of hindered ketone, 1,3-dithiolanes and is a reagent worth considering in synthetic strategies often ahead of some of the more commonly used methods.

10. REACTIONS WITH THIOCARBONYLS

In view of the results obtained with thioacetals and benzeneseleninic anhydride it would be expected that thiocarbonyl derivatives could also be converted to the parent oxo derivatives in a similar fashion.

In order to properly assess the scope of the reagent a range of xanthates, thioesters, thioureas, thiocarbonates, and thiones were studied. Generally good yields of the oxo products are obtained after a few hours at room temperature [Eqs. (58)–(61)].[81] The limitation of this method is that readily enolized thiones

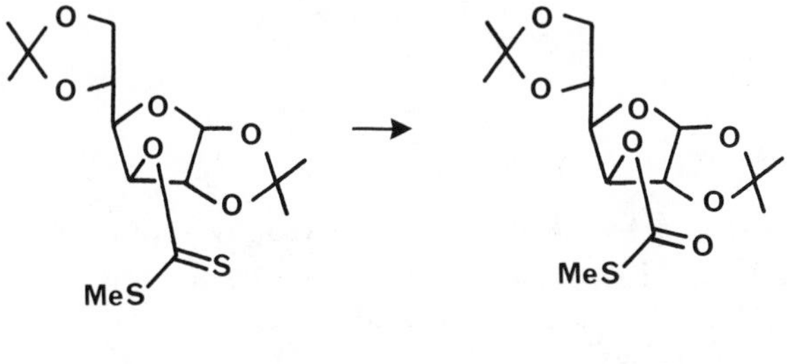

(58)

(59)

$$(60)$$

$$\underset{\text{S}}{\text{RNH}\overset{\|}{\text{C}}\text{NHR}} \longrightarrow \underset{\text{O}}{\text{RNH}\overset{\|}{\text{C}}\text{NHR}} \qquad (61)$$

such as thiocamphor react to give a mixture of products, in this case the products are camphor (9%), 3-*endo*-phenylseleninylcamphor (36%), and camphor quinone (54%) [Eq. (62)].

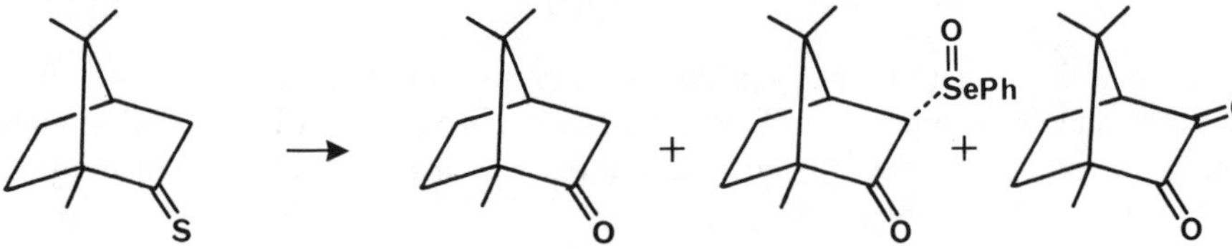

$$(62)$$

Examples of conversion of selenoester[81] and telluroester[82] to their oxo compounds with the anhydride are also known. For the telluroester, 19, this was used as structural evidence in that the product was a known compound [Eq. (63)].[82]

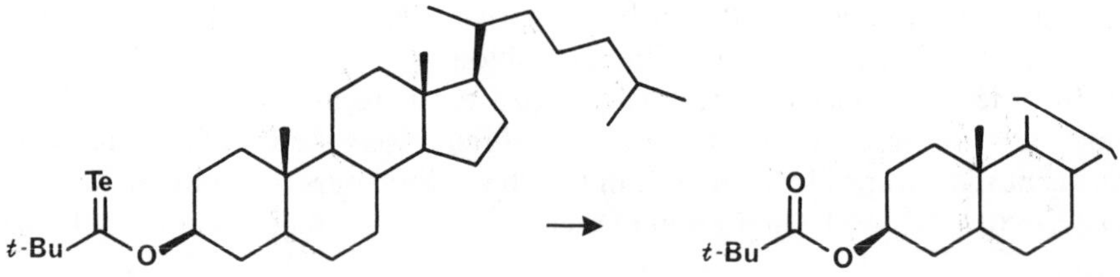

$$\left(\mathbf{19}\right)$$

$$(63)$$

Isoquinoline thiones, which were prepared by photochemical cyclization of thioenamides, could be readily transformed by the anhydride to their oxo-analogs; attempted direct photocyclization of oxo enamides merely resulted in acyl transfer (Scheme 9).[83]

Benzeneseleninic anhydride compares well with other literature methods for the thiocarbonyl–carbonyl conversion, but it is not universally applicable. Nevertheless, its ready availability, stability, and ease of use offer distinct advantages over other procedures.

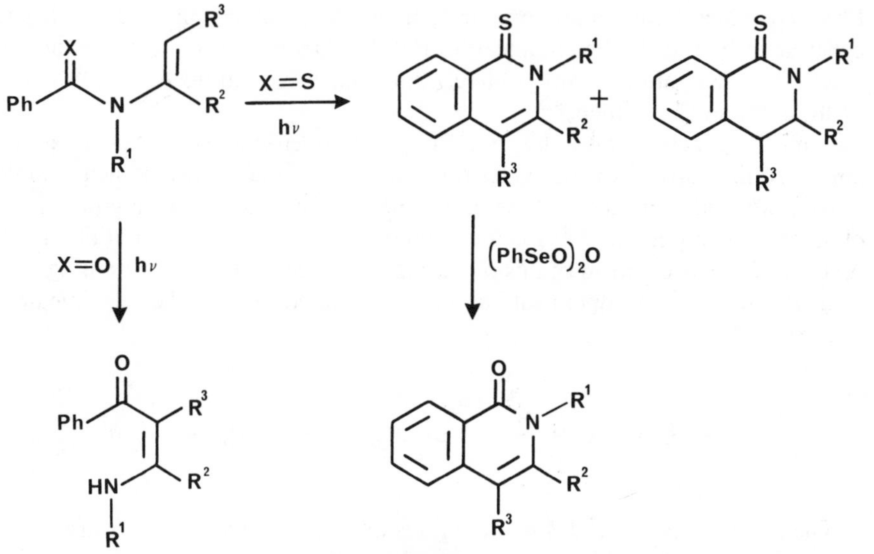

SCHEME 9.

11. REACTIONS WITH THIOLS AND SULFIDES

In 1955 Rheinboldt and Giesbrecht first reported the reaction of seleninic acids with thiols.[84] These authors suggested that the reaction products varied depending upon the stoichiometry [Eqs. (64)–(66)]. During a very careful recent study of the mechanism of this reaction it was shown that benzeneseleninic acid reacts with alkanethiols (3 eq.) to afford phenylseleno sulfides, together with the disulfide [Eq. (67)].[85] The phenylseleno sulfides were characterized by isolation and comparison with authentic samples. The same stoichiometry was observed at all pHs at which the reaction was studied. Upon standing in solution the phenylseleno sulfides disproportionate very slowly to diphenyldiselenide and the disulfide.[86] Evidence was presented for two distinct stages occurring during the reaction of thiols with seleninic acids both of which exhibit first-order kinetics with respect to thiol concentration. In the first stage the thiol and

$$\text{ArSeO}_2\text{H} + 2\text{RSH} \longrightarrow \text{ArSeOH} + (\text{RS})_2 + \text{H}_2\text{O} \qquad (64)$$

$$2\text{ArSeO}_2\text{H} + 6\text{RSH} \longrightarrow (\text{ArSe})_2 + 3(\text{RS})_2 + 4\text{H}_2\text{O} \qquad (65)$$

$$\text{ArSeO}_2\text{H} + 4\text{RSH} \longrightarrow \text{ArSeH} + 2(\text{RS})_2 + 2\text{H}_2\text{O} \qquad (66)$$

$$\text{PhSeO}_2\text{H} + 3\text{RSH} \longrightarrow \text{PhSeSR} + (\text{RS})_2 + 2\text{H}_2\text{O} \qquad (67)$$

$PhSeO_2H$ react to form an intermediate having a λ_{max} of 265 nm, and which was believed to be the thiolseleninate PhSe(O)SR. This intermediate subsequently reacts with the thiol to initiate a reaction sequence leading to PhSeSR and $(RS)_2$ as the final products.[85]

While benzeneseleninic anhydride[76] or p-chlorobenzeneseleninic acid[87] alone fail to oxidize sulfides to sulfoxides rapid reaction occurs between the seleninic acids and sulfides in the presence of p-toluenesulfonic acid to give good yields of the sulfoxides [Eq. (68)].[87] Further oxidation to the sulfone was not observed. A study of the mechanism of this oxidation indicated that the rate-limiting step is nucleophilic attack by the sulfide on a protonated form of the seleninic acid $[ArSe(OH)_2^+]$.

$$\underset{R'}{\overset{R}{\diagdown}}S \;+\; \tfrac{2}{3}ArSeO_2H \;\overset{H^{\oplus}}{\longrightarrow}\; \underset{R'}{\overset{R}{\diagdown}}S{=}O \;+\; \tfrac{1}{3}\left(ArSe\right)_2 \;+\; \tfrac{1}{3}H_2O \qquad (68)$$

The conversion of sulfides to sulfones is an important synthetic transformation that can be effected by a range of oxidizing reagents. However, the use of *in situ* generated peroxyseleninic acid to achieve this is especially useful as it is cheap, easy to use, and may be employed when the substrate also contains other easily oxidized functional groups. Both Reich et al.[88] and Nicolaou et al.[89] have independently developed systems for the oxidation of sulfides by peroxyseleninic acids. Other uses of peroxyseleninic acids in synthesis are discussed in the following section. Reaction of excess hydrogen peroxide with seleninic acids[88,90] or diselenides[89] is thought to establish an equilibrium in which the formation of reactive peroxyseleninic acids occurs.

Several interesting applications of the method are known. For example, during the construction of novel sulfur-containing analogs of prostacyclin (PGI), it was shown that the sulfide, 20, was selectively oxidized to the sulfone in 93% yield [Eq. (69)].[89] Key steps during the synthesis of the ionophore antibiotic X-14547A also relied on this oxidation strategy.[91,92] In one of these syntheses oxidation of the sulfide, 21, to the sulfone was smoothly achieved using the *in situ* generated peroxyseleninic acid, whereas other reagents, such as *m*-chloro-

$$(69)$$

$$(20)$$

peroxybenzoic acid or oxone, had failed owing to the reactivity of the pyrrole ring and the olefinic bond [Eq. (70)].[92]

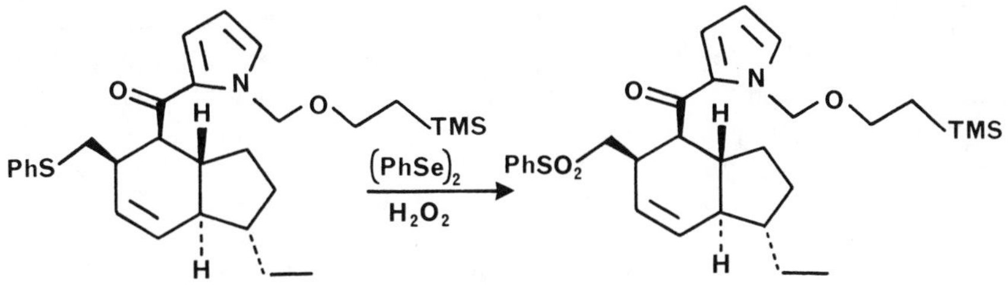

(21)

$$(70)$$

12. REACTIONS WITH SULFINIC ACIDS

Gancarz and Kice have shown that arenesulfinic acids are rapidly oxidized at 0°C in acetonitrile by benzeneseleninic acid to give, in addition to the arenesulfonate salt, Se-phenyl areneselenosulfonate [Eq. (71)].[93]

$$2\,PhSeO_2H + 2\,ArSO_2H \longrightarrow PhSe\overset{\overset{O}{\|}}{\underset{\underset{O}{\|}}{S}}Ar + ArSO_3^{\ominus}\,PhSeO_2H_2^{\oplus} + H_2O \quad (71)$$

The same products may be obtained by reaction of 1 mole of $ArSO_2H$ with 0.5 mole of benzeneseleninic anhydride. The initial step in these reactions is thought to be a redox process that affords phenylseleninic acid and the arenesulfonic acid. The more acidic sulfonic acid protonates a second molecule of seleninic acid to give the salt. The benzeneselenic acid is then consumed by reaction with the sulfinic acid to give the Se-phenyl areneselenosulfonate.

Since these areneselenosulfonates are potentially useful in synthesis,[94,95] ways of improving its formation were sought,[95] the best of which proved to be a change of solvent to ethanol. This meant that in contrast with acetonitrile as solvent the sulfinic acid salt was no longer precipitated so that all the seleninic acid was consumed in the reaction with the sulfinic acid. In this way yields of up to 90% of the areneselenosulfonate could be realized.

13. REACTIONS WITH PHOSPHINES AND PHOSPHITES

Reactions of seleninic anhydrides or seleninic acids with phosphorus-containing compounds are few. Both hypophosphorous acid and phosphorous acid reduce seleninic acids to either selenenic acids or further to diaryl diselenides depending on the reaction conditions.[96] This reduction reaction may be usefully employed in the preparation of β-hydroxyselenides, which are versatile intermediates for further reactions. The unstable selenenic acids obtained by treatment of either methane- or benzeneseleninic acid with aqueous hypophosphorous acid may be efficiently trapped with alkenes in yields ranging from 63–91% [Eq. (72)].[97] This work also reports similar preparation of the selenenic acid using diethyl phosphite as the reductant.

$$RSeO_2H + H_3PO_2 \longrightarrow \left[RSeOH \right] \xrightarrow{\qquad} \text{(72)}$$

Triphenylphosphine reacts with arylseleninic acids to give triphenylphosphine oxide with the same stoichiometry as noted previously in the oxidation of sulfides to sulfoxides. The main kinetic difference is the relatively rapid rate of reaction of the phosphines even in the absence of added acids.[87]

14. OXIDATION OF BENZYLIC HYDROCARBONS

Oxidation of aromatic side chains is often difficult to control owing to the vigorous nature of the reagents generally used. This is especially true of selective oxidation of aromatic methyl substituents to aldehydes. Probably the best single reagent is ceric ammonium nitrate under strongly acidic conditions although this too has serious limitations.[98]

The high reactivity and polarized nature of the bonds in benzeneseleninic anhydride pointed towards its possible used as an oxidant for benzylic activated hydrocarbons.[99,100] While this is indeed possible it does have limitations such as overoxidation or the simultaneous formation of selenated by-products.

Initial results[99] confirmed that xylenes are oxidized by the anhydride to give monoaldehydes only by heating the neat mixtures under reflux for periods of around 10 h. The yields in these reactions are moderate, ranging from 42–66%. If excess anhydride was used substantial formation of the carboxylic acid occurred even if oxygen was rigorously excluded from the reaction vessel. Similar

51%   14%

(73)

oxidation of 1,8-dimethylanthracene has been studied. Here, reaction takes place to give both the monooxidized product and a small amount of the dialdehyde [Eq. (73)]. On the other hand, treatment of 1-methylnaphthalene with benzene-seleninic anhydride affords a mixture of products that varied depending on the quantity of anhydride used (Scheme 10).[57,100]

When the studies were extended to incorporate aromatic rings bearing either electron-donating or electron-withdrawing substituents other effects were noted. Firstly, substitution with nitro groups completely inhibited the oxidation reaction. Conversely, reaction of p-methoxytoluene occurred rapidly to

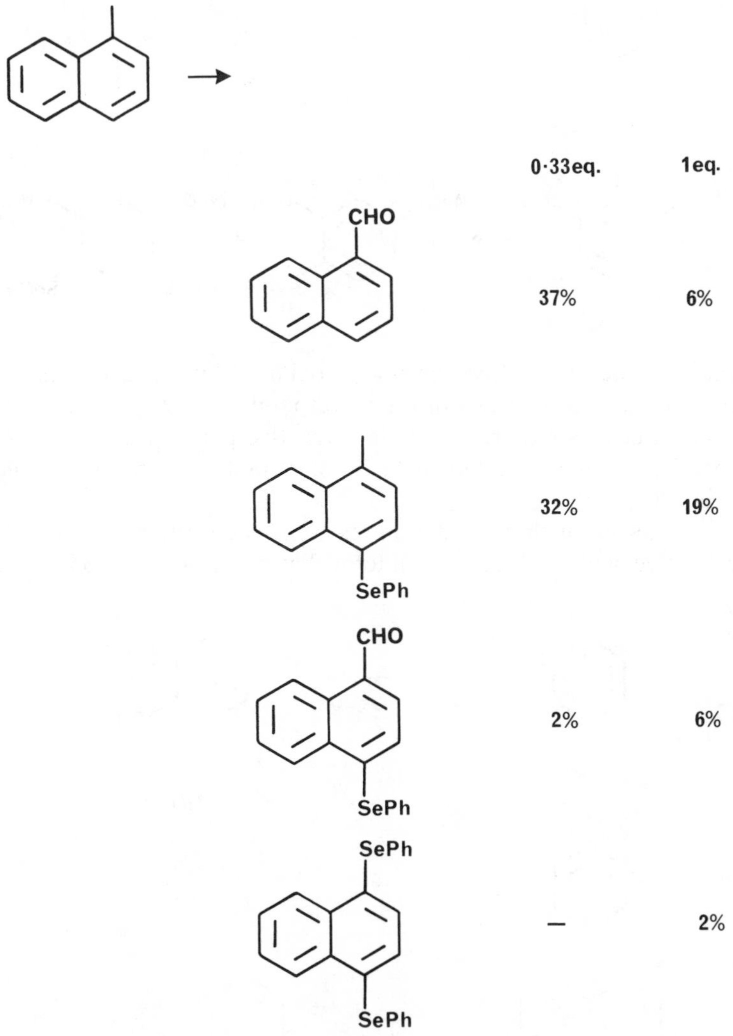

SCHEME 10.

give some monoaldehyde, yet the major product was always 1-methoxy-2-ben-zeneselenyl-4-methylbenzene [Eq. (74)]. Similar results were obtained using 2,6-dimethoxytoluene as the substrate [Eq. (75)]. The high levels of selenation in these reactions were thought to be due to initial seleninylation to afford selen-oxides that subsequently undergo loss of oxygen to give the final products. Some precedent for this hypothesis existed in that other selenoxides had been shown to decompose in a similar fashion on heating.[101,102]

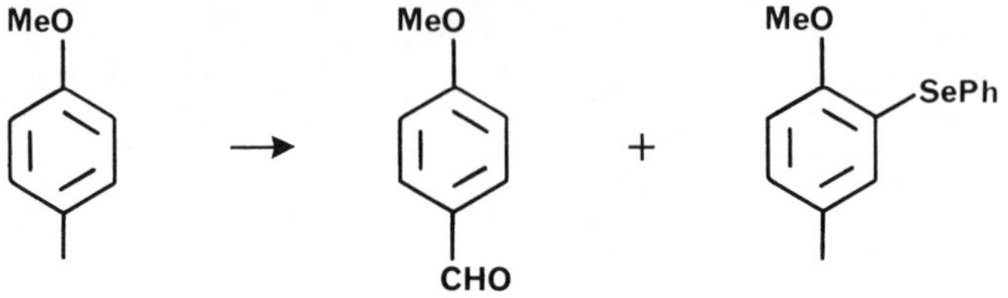

(74)

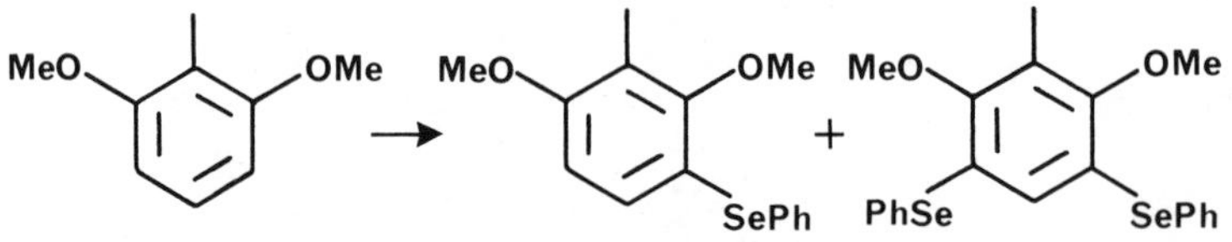

(75)

Several heterocyclic methylarenes were oxidized by benzeneseleninic anhy-dride to give reasonable yields of the corresponding aldehydes. Generally the methyl substituent is required to be in a reactive position of the heterocyclic ring; 3-methylpyridine was shown to be inert under the reaction conditions [Eqs. (76)–(80)].[99,100]

The mechanism for these oxidation reactions is uncertain although the pos-tulated involvement of Pummerer intermediates is not unreasonable. In line

(76)

(77)

(78)

$$(79)$$

$$(80)$$

with this proposal Johnstone and co-workers have shown that benzylphenyl selenoxides decompose on warming ($>110°C$) to aldehydes (Scheme 11).[103] Comparison of the oxidation method using benzeneseleninic anhydride showed it to be noticeably superior to equivalent reactions with selenium dioxide.

Oxidations of other arene activated hydrocarbons were studied but these were frequently complicated by secondary oxidation reactions.[99,100] Diphenylmethane reacts slowly at 90°C over 5 days to give benzophenone in 90% yield. Related oxidation of ethylbenzene gave acetophenone as the major product although this was accompanied by phenylglyoxal (10%) and a small amount of a selenated product [Eq. (81)]. In accord with the formation of phenylglyoxal, camphor is oxidized to camphorquinone in 74% yield using the anhydride [Eq. (82)]. A more complicated reaction occurs during the oxidation of tetralin, where many products are formed, none of which are obtained in high yield. These include α-tetralone, 1,2-naphthoquinone, 1,4-naphthoquinone together with some selenated material.

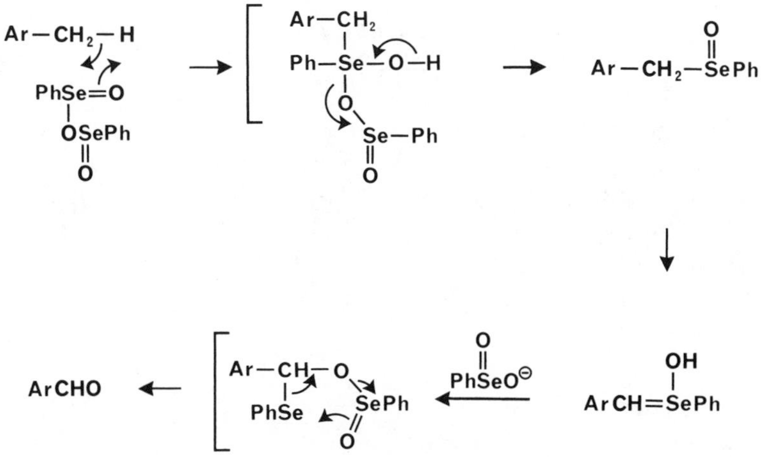

SCHEME 11.

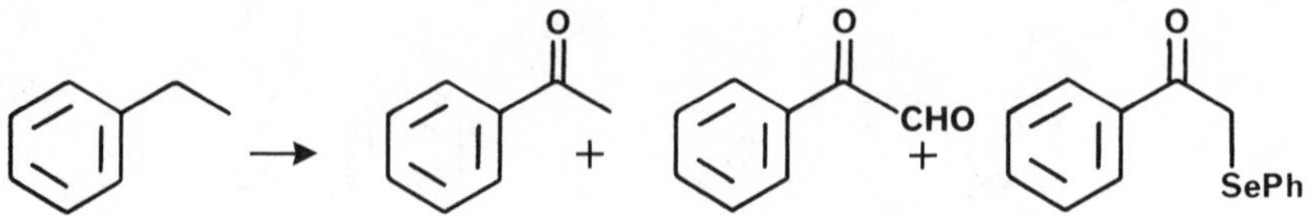

(81)

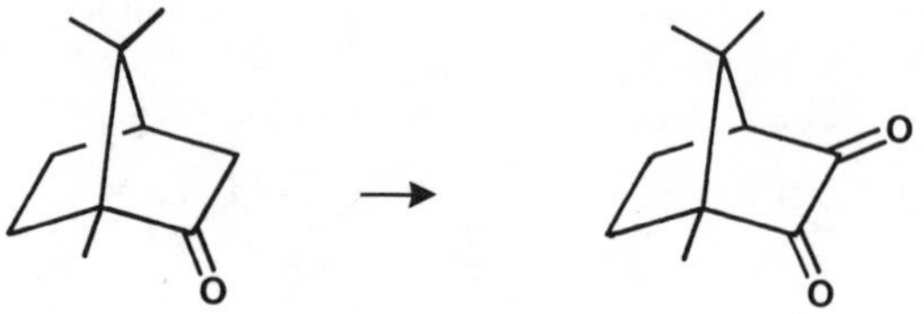

(82)

15. OXIDATION OF ALKENES

Two pioneering studies described the *in situ* generation of selenenic acids by comproportionation (reverse of disproportionation) of diselenides and seleninic acids.[104,105] The selenenic acids formed are trapped by addition to alkenes to give β-hydroxy selenides that are useful for further transformations.

More specific work directed at the oxidation of alkenes shows how combinations of benzeneseleninic anhydride and diphenyl diselenide form intermediates that react to give α-phenylseleno ketones [Eq. (83)];[9] this new procedure complements the established selenation of ketone enolates for the preparation of these compounds. When the substrate was a terminal alkene formation of 1-phenylseleno ketones rather than 2-phenylseleno aldehydes was favored, especially when dimethyl sulfoxide was used as the solvent.

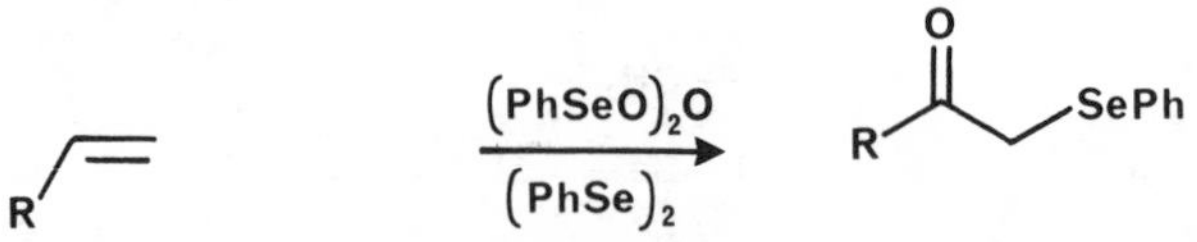

(83)

For allylic alcohol derivatives this regioselectivity could be reversed so that 2-phenylseleno aldehydes were the predominant products in all the examples studied with consistently high yields [Eq. (84)].[106] Moreover, the directing ability of the ether substituent is not restricted to terminal olefins and indeed governs the regiochemistry in the oxyselenation of all proximal alkene bonds [Eqs. (85)–(87)].

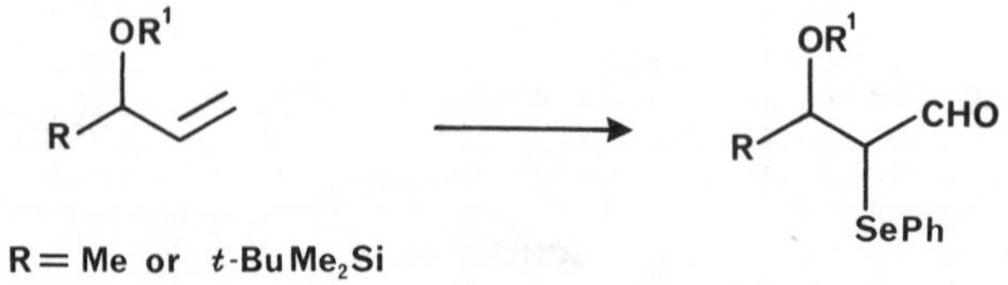

(84)

R = Me or t-BuMe$_2$Si

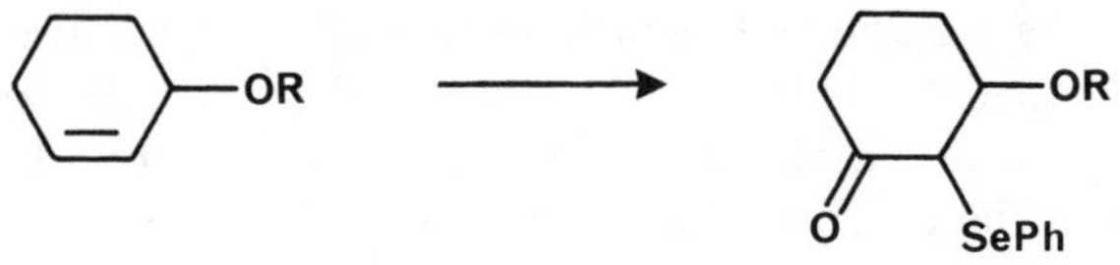

$$(85)$$

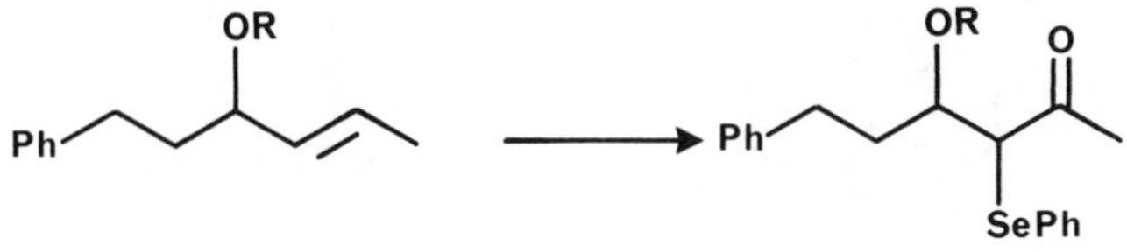

$$(86)$$

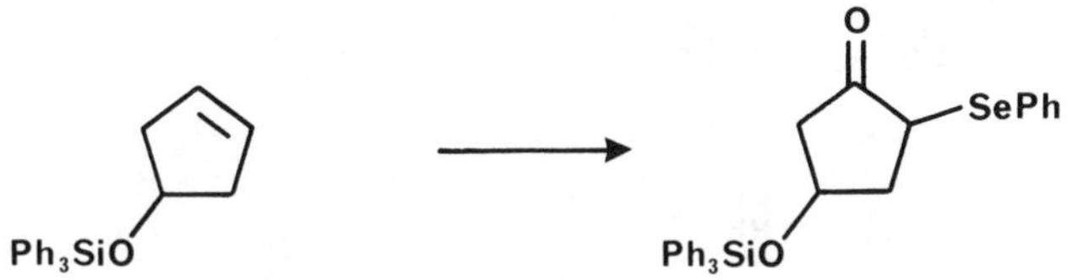

$$(87)$$

α-Selenocarbonyl derivatives may also be obtained from vinyl selenides simply by reaction with seleninic acids or anhydrides in chloroform at temperatures between 20 and 50°C [Eq. (88)].[107] This paper discusses the various

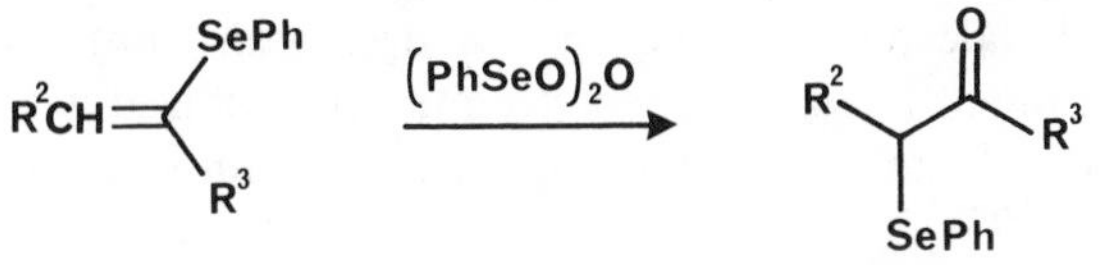

$$(88)$$

synthetic applications of the derived α-seleno-carbonyl compounds. An isolated report utilizes benzeneseleninic anhydride in the presence of borontrifluoride etherate to effect oxidation of an allylsilane in which [2,3]sigmatropic rearrangement of an intermediate allylic selenoxide affords the product (Scheme 12).[108]

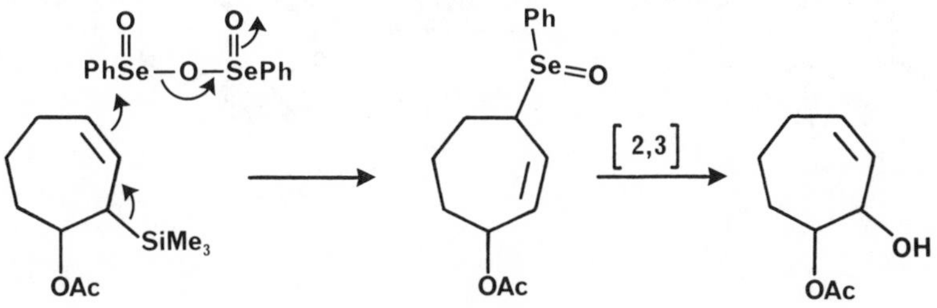

SCHEME 12.

Few examples are known in which alkenes are oxidized directly by seleninic anhydrides or acids. Treatment of acenaphthalene in THF for 18 h provides an excellent yield of the related dione, and anthracene affords anthraquinone, while *trans*-stilbene gives benzil [Eqs. (89)–(91)].[57]

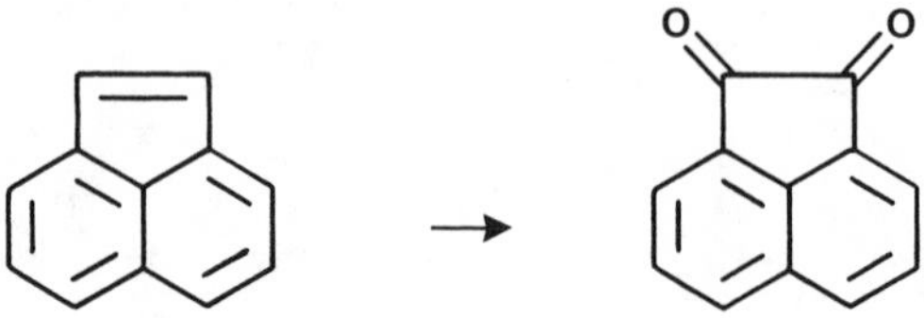

(89)

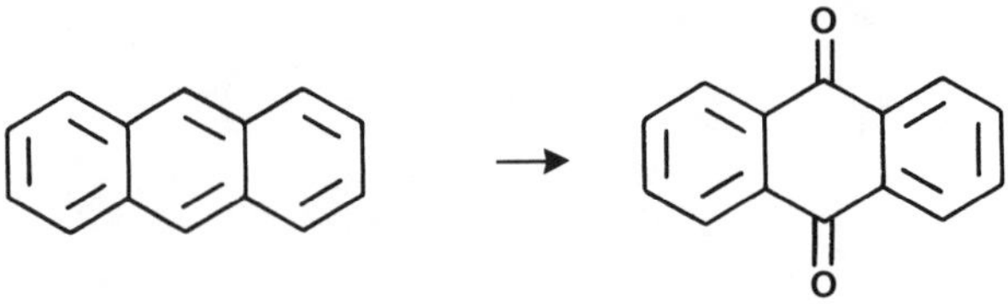

(90)

(91)

During studies on the dehydrogenation of lupeol with the anhydride it was observed that the product formed in 60% yield also contained an oxidized side chain [Eq. (92)]. The ready oxidation of the isoprene unit was confirmed in a separate experiment in which lupeol acetate was oxidized with the anhydride to give the enal as the major product together with minor amounts of selenated compounds [Eq. (93)].[57]

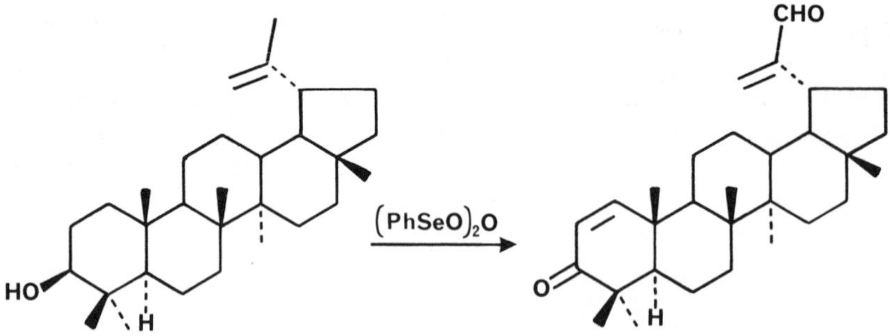

(92)

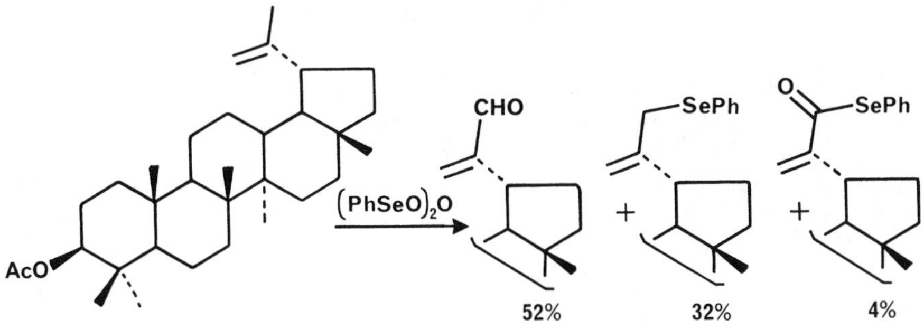

$$52\% \qquad 32\% \qquad 4\%$$

$$(93)$$

16. EPOXIDATION REACTIONS

Epoxides have been detected as by-products during *syn*-elimination reactions of selenoxides when an excess of hydrogen peroxide had been used as an oxidant in their preparation.[90,104] A systematic search for the cause of these epoxidation reactions resulted in the development of a synthetically useful oxidizing reagent formulated as a peroxyseleninic acid, RSe(O)OOH. The reagent, which has been discussed earlier, is formed by reaction of hydrogen peroxide with a seleninic acid. This species is an excellent epoxidizing reagent for alkenes [Eqs. (94) and (95)].[88,90,104] The regioselectivity observed in these epoxidation reactions is such that the procedure complements transition metal catalyzed epoxidation with *t*-butylhydroperoxide, since the olefinic bond that is furthest removed from the hydroxyl function [Eq. (96)] is preferentially oxidized. Normal selectivity for the most substituted double bond is also observed. The best results are obtained when the mixtures are buffered to pH 7 using phosphate to avoid formation of diols.

$$(94)$$

$$(95)$$

$$(96)$$

17. BAEYER-VILLIGER REACTIONS

Baeyer–Villiger reactions have been observed unexpectedly during dehydro-genation reactions of steroidal ketones using a selenation/*syn*-elimination se-quence [Eq. (97)].[109] The cause of this was thought to be the combination of hydrogen peroxide and benzeneseleninic acid. Very similar observations were recorded by Williams and Leber during dehydrogenation studies with estrone methyl ether.[110]

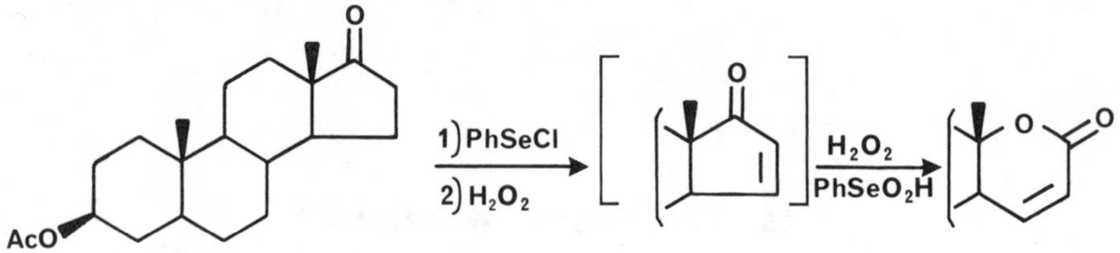

$$(97)$$

In a more rational approach Grieco et al. used peroxybenzeneseleninic acid to achieve the Baeyer–Villiger reaction with a number of five- and six-ring ketones [Eqs. (98) and (99)].[111] It has since been used as a key reagent in the total synthesis of the pseudoguaianolide psilostachyin C [Eq. (100)].[112]

$$(98)$$

$$(99)$$

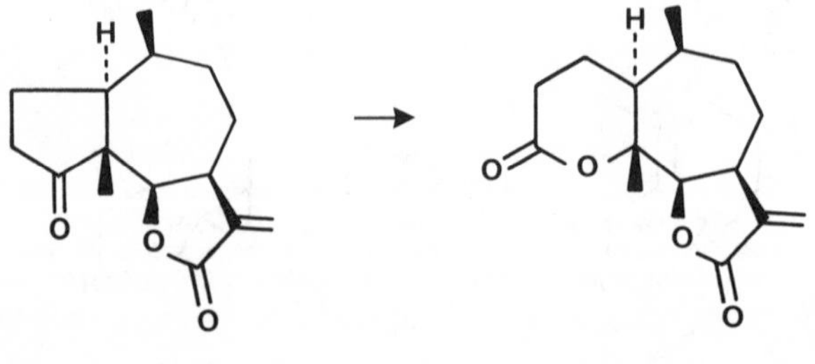

$$(100)$$

The future of this reagent system, however, may well lie in the development of polymer bound seleninic acid as it has also been shown to effect the Baeyer–Villiger reaction in the presence of hydrogen peroxide.[39]

REFERENCES

1. *Organic Selenium Compounds: Their Chemistry and Biology*, D. L. Klayman and W. H. H. Günther, Eds., Wiley, Interscience, London, 1973.

2. (a) For relevant reviews see T. W. Campbell, H. G. Walker, and G. M. Coppinger, *Chem. Rev.*, **50**, 279 (1952); (b) H. J. Reich, in *Oxidation in Organic Chemistry*, W. S. Trahanovsky, Ed., Academic, New York, 1978, p. 1; (c) D. L. J. Clive, *Aldrichimica Acta*, **11**, 43 (1978) and *Tetrahedron*, 1049 (1978); (d) D. H. R. Barton and S. V. Ley, *Further Perspectives in Organic Chemistry*, Ciba Foundation Symposium Elsevier, 1978, p. 53; (e) H. J. Reich, *Acc. Chem. Res.*, **12**, 22, (1979); (f) P. D. Magnus, in *Comprehensive Organic Chemistry*, Vol. 3, D. H. R. Barton, and W. D. Ollis, Eds. Pergamon, Oxford 1979, p. 491; (g) S. V. Ley, *Ann. Rep.*, **77**, 233, (1980); (h) R. Okazaki, and K.-T. Kang, *Yuki Kagaku Kyobaishi*, **38**, 1223, (1980); (i) *Selenium in Natural Products Synthesis*, K. C. Nicolaou and N. A. Petasis, CIS, Philadelphia, 1984; (j) D. Liotta, *Acc. Chem. Res.*, **17**, 28, (1984).

3. (a) R. T. Taylor, and L. A. Flood, *J. Org. Chem.*, **48**, 5160, (1983); (b) R. Michels, M. Kato, and W. Heitz, *Makromol. Chem.*, **177**, 2311, (1976); (c) M. Kato, R. Michels, and W. Heitz, *J. Polym. Sci. Polym. Lett. Ed.*, **14**, 413, (1976).

4. H. W. Doughty, *Am. Chem. J.*, **41**, 326, (1909).

5. G. Ayrey, D. Barnard, D. T. Woodbridge, *J. Chem. Soc.*, 2089 (1962).

6. V. V. Kozlov and V. M. Pronyakova, *Zh. Org. Khim.*, **1**, 493, (1965).

7. R. Lesser and R. Weiss, *Ber.*, **46**, 2640, (1913).

8. R. Paetzold and D. Lienig, *Z. Anorg. Allg. Chem.*, **335**, 289, (1965).

9. M. Shimizu, R. Takeda, and I. Kuwajima, *Tetrahedron Lett.*, 419 (1979).

10. R. A. Gancarz and J. L. Kice, *Tetrahedron Lett.*, 1661 (1981).

11. A. J. Bloodworth and D. J. Lampham, *J. Chem. Soc. Perkin Trans. 1*, 471, (1983).

12. D. H. R. Barton, J. W. Morzycki, W. B. Motherwell, and S. V. Ley, *J, Chem. Soc. Chem. Commun.*, 1044 1981); D. H. R. Barton, C. R. A. Godfrey, J. W. Morzycki, W. B. Motherwell, and S. V. Ley, *J. Chem. Soc. Perkin Trans. 1*, 1947 (1982).

13. R. Paetzold, S. Borek, and E. Wolfram, *Z. Anorg. Allg. Chem.*, **353**, 53, (1967).

14. C. A. Buehler, J. O. Harris, and W. F. Arendale, *J. Am. Chem. Soc.*, **72**, 4953, (1950).

15. D. De Filippo and F. Momicehioli, *Tetrahedron*, 5733 (1969).

16. (a) B. Rathke, *Ann. Chem. Pharm.*, **152**, 181, (1969); (b) C. L. Jackson, *Ber.*, **7**, 1277, (1874); (c) C. L. Jackson, *Ann.*, **179**, 1, (1875); (d) F. Krafft and R. E. Lyons, *Ber.*, **29**, 424 (1896); (e) M. Stoecker and F. Krafft, *Ber*, **39**, 2197, (1906); (f) F. L. Pyman, *J. Chem. Soc.*, **115**, 166, (1919); (g) G. T. Morgan and F. H. Burstall, *J. Chem. Soc.*, 1096 (1929), 173 (1931); (h) R. Poggi and G. Speroni, *Gazz Chim.* **64**, 501, (1934); (i) H. J. Backer and W. van Dam, *Rec. Trav. Chim. Pays-Bas*, **49**, 482 (1930), **54**, 531, (1935); (j) H. Rheinboldt and E. Giesbrecht, *Chem. Ber.*, **88**, 666, (1955); (k) J. W. Dale, H. J. Emeléus, and R. N. Hazeldine, *J. Chem. Soc.*, 2939 (1958); (l) M. Giua and R. Bianco, *Gazz. Chim.*, **89**, 693, (1959); (m) R. Paetzold and E. Rönsch, *Z. Anorg. Allg. Chem.*, **338**, 195, (1965); (n) L.-B. Agenäs, *Acta. Chem. Scand.*, **19**, 764 (1965); (o) E. Rebane, *Ark. Kemi*, **26**, 345, (1966); (p) W. H. H. Günther and M. N. Salzman, *Ann. N.Y. Acad. Sci.*, **192**, 25, (1972).

17. (a) F. Wöhler, and J. Dean, *Ann. Chem. Pharm.*, **97**, 1, (1856); (b) F. L. Pyman, *J. Chem. Soc.*, **115**, 166, (1919); (c) E. H. Shaw, Jr., and E. E. Reid, *J. Am. Chem. Soc.*, **48**, 520, (1926); (d) D. G. Foster, *J. Am. Chem. Soc.*, **61**, 2972, (1939).

18. (a) J. D. McCullough and E. S. Gould, *J. Am. Chem. Soc.*, **71**, 674, (1949); 1109, **73**, (1951); (b) D. L. Klayman, *J. Org. Chem.*, **30**, 2454, (1965); (c) D. L. Klayman and J. W. Lown, *J. Org. Chem.*, **31**, 3396, (1966).

19. G. T. Morgan and J. C. Elliot, *Proc. Chem. Soc.*, **30**, 248, (1914).

20. L. Pichat, M. Herbert, and M. Thiers, *Tetrahedron*, **12**, 1, (1961).

21. J. D. McCullough, T. W. Campbell, and E. S. Gould, *J. Am. Chem. Soc.* **72**, 5753, (1950).

22. (a) J. Loevenich, H. Fremdling, and M. Föhr, *Ber.*, **62**, 2856, (1929). (b) M. C. Thompson, and E. E. Turner, *J. Chem. Soc.*, 29 (1938); (c) V. V. Kozlov and S. E. Suvorova, *Zh. Obshch. Khim.*, **31**, 3034, (1961).

23. R. Paetzold and E. Wolfram, *Z. Anorg. Allg, Chem.*, **353**, 167, (1967).

24. (a) O. Behaghel and H. Seibert, *Ber.*, **65**, 812, (1932); (b) D. G. Foster, *J. Am. Chem. Soc.*, **55**, 822, (1933); (c) O. Behaghel and H. Seibert, *Ber.*, **66**, 708, (1933). (d) H. J. Backer and H. J. Winter, *Rec. Trav. Chim. Pays-Bas*, **56**, 492, (1937); (e) O. Behaghel and K. Hofmann, *Ber.*, **72**, 582, (1939); (f) H. Rheinboldt and E. Giesbrecht, *Chem. Ber.*, **89**, 631, (1956).

25. D. D. Karve, *J. Ind. Chem. Soc.*, **2**, 141, (1925).

26. W. H. Porritt, *J. Chem. Soc.*, 27 (1927).

27. M. L. Bird, and F. Challenger, *J. Chem. Soc.*, 570 (1942).

28. D. H. R. Barton, P. D. Magnus, and M. N. Rosenfeld, *J. Chem. Soc. Chem. Commun.*, 301 (1975); D. H. R. Barton, S. V. Ley, P. D. Magnus, and M. N. Rosenfeld, *J. Chem. Soc. Perkin Trans. 1*, 567 (1977).

29. P. W. Jeffs and G. Lynn, *Tetrahedron Lett.*, 1617 (1978).

30. D. H. R. Barton, A. G. Brewster, S. V. Ley, and M. N. Rosenfeld, *J. Chem. Soc. Chem. Commun.*, 985 (1976).

31. D. H. R. Barton, A. G. Brewster, S. V. Ley, C. M. Read, and M. N. Rosenfeld, *J. Chem. Soc. Perkin Trans. 1*, 1473 (1981).

32. A. Emke, Diploma of Imperial College thesis, Imperical College, London 1979.

33. K. B. Sukumaran and R. G. Harvey, *J. Am. Chem. Soc.*, **101**, 1353, (1979).

34. K. B. Sukumaran and R. G. Harvey, *J. Org. Chem.*, **45**, 4407, (1980).

35. Y. Miyahara, T. Inazu, and T. Yoshino, *Tetrahedron Lett.*, **23**, 2189, (1982).

36. D. H. R. Barton, A. G. Brewster, S. V. Ley, and M. N. Rosenfeld, *J. Chem. Soc. Chem. Commun.*, 147 (1977).

37. A. Atkinson, A. G. Brewster, S. V. Ley, R. S. Osborn, D. Rogers, D. J. Williams, and K. A. Woode, *J. Chem. Soc. Chem. Commun.*, 325 (1977).

38. J. S. Holker, E. O'Brien, and B. K. Park, *J. Chem. Soc. Perkin Trans. 1*, 1915 (1982).

39. P. J. Murray, Ph.D. Thesis, London 1983.

40. D. H. R. Barton, D. J. Lester, and S. V. Ley, *J. Chem. Soc. Chem. Commun.*, 130 (1978); D. H. R. Barton, D. J. Lester, and S. V. Ley, *J. Chem. Soc. Perkin Trans. 1*, 2209 (1980).

41. D. Neuhaus and C. W. Rees, *J. Chem. Soc. Chem. Commun.*, 318 (1983).

42. R. E. Zipkin, N. R. Natale, I. M. Taffer, and R. O. Hutchins, *Synthesis*, 1035 (1980).

43. J. Polonsky, Z. Varon, H. Jacquemin, D. M. X. Donnelly, and M. J. Meegan, *J. Chem. Soc. Perkin Trans. 1*, 2065 (1980).

44. T. G. Back, *J. Chem. Soc. Chem. Commun.*, 278 (1978).

45. T. G. Back, *J. Org. Chem.* **46**, 1442, (1981).

46. D. H R. Barton, R. A. H. F. Hui, S. V. Ley, and D. J. Williams, *J. Chem. Soc. Perkin Trans. 1*, 1919 (1982).

47. D. H. R. Barton, X. Lusinchi, and P. Milliet, *Tetrahedron Lett.*, 4949 (1982).

48. For a full discussion see D. H. R. Barton, X. Lusinchi, and P. Milliet, *Tetrahedron Lett.*, **41**, 4949, (1982).

49. K. Yamakawa, T. Satoh, N. Ohba, and R. Sakaguchi, *Chem. Lett.*, 763 (1979).

50. K. Yamakawa, T. Satoh, N. Ohba, R. Sakaguchi, S. Takita, and N. Tamura, *Tetrahedron*, **37**, 473, (1981).

51. K. Yamakawa, and T. Satoh, *Heterocycles*, **15**, 337, (1981).

52. K. Yamakawa, T. Satoh, and S. Takita, *Heterocycles*, **17**, 259, (1982).

53. N. Khôi and J. Polonsky, *Helv. Chim. Acta*, **64**, 1540, (1981).

54. T. G. Back and N. Ibrahim, *Tetrahedron Lett.*, 4931 (1979); T. G. Back, N. Ibrahim, and D. J. McPhee, *J. Org. Chem.*, **47**, 3283, (1982).

55. O. Behaghel and, W. Müller, *Ber.*, **68**, 1540 (1935).

56. D. H. R. Barton, A. G. Brewster, R. A. H. F. Hui, D. J. Lester, S. V. Ley, and T. G. Back, *J. Chem. Soc. Chem. Commun.*, 952 (1978).

57. R. A. H. F. Hui, doctoral thesis, Imperial College, London, 1981.

58. M. Shimizu and I. Kuwajima, *Tetrahedron Lett.*, **19**, 2801, (1979).

59. M. Shimizu, H. Urabe, and I. Kuwajima, *Tetrahedron Lett.*, **22**, 2183, (1981).

60. I. Kuwajima, M. Shimizu, and H. Urabe, *J. Org. Chem.*, **47**, 837 (1982).

61. M. R. Czarny, *J. Chem. Soc. Chem. Commun.*, 81 (1976).

62. M. R. Czarny, *Syn. Commun.*, **6**, 285, (1976).

63. D. J. Lester, doctoral thesis, Imperial College, London, 1978.

64. D. H. R. Barton, D. J. Lester, and S. V. Ley, *J. Chem. Soc. Chem. Commun.*, 445 (1977).

65. D. H. R. Barton, D. J. Lester, and S. V. Ley, *J. Chem. Soc. Perkin Trans. 1*, 1212 (1980).

66. D. H. R. Barton, D. J. Lester, and S. V. Ley, *J. Chem. Soc. Chem. Commun.*, 276 (1978).

67. H. Bock, *Chem. Ber.*, **99**, 3337, (1966).

68. T. G. Back, S. Collins, and R. G. Kerr, *J. Org. Chem.*, **46**, 1564, (1981).

69. D. H. R. Barton, X. Lusinchi, and J. S. Ramirez, *Tetrahedron Lett.*, **24**, 2995, (1983).

70. T. G. Back and S. Collins, *Tetrahedron Lett.*, **20**, 2661, (1979).

71. S. Masamune, Y. Hayase, W. Schilling, W. K. Chan, and S. G. Bates, *J. Am. Chem. Soc.*, **99**, 6756, (1977).

72. A. P. Kozikowski and A. Ames, *J. Org. Chem.*, **43**, 2735, (1978).

73. T. G. Back and S. Collins, *Tetrahedron Lett.*, **21**, 2213, (1980).

74. T. G. Back and R. G. Kerr, *Can. J. Chem.*, **60**, 2711, (1982).

75. T. Frejd and K. B. Sharpless, *Tetrahedron Lett.*, **19**, 2239, (1978).

76. D. H. R. Barton, N. J. Cussans, and S. V. Ley, *J. Chem. Soc. Chem. Commun.*, 751 (1977); *J. Chem. Soc. Perkin Trans. 1*, 1654 (1980).

77. A. Burton, L. Hevesi, W. Dumont, A. Cravador, and A. Krief, *Synthesis*, 877 (1979).

78. D. H. R. Barton, C. C. Dawes, G. Franceschi, M. Foglio, S. V. Ley, P. D. Magnus, W. L. Mitchell, and A. Temperelli, *J. Chem. Soc. Perkin Trans. 1*, 643 (1980).

79. D. H. R. Barton, M. T. Bielska, J. M. Cardoso, N. J. Cussans, and S. V. Ley, *J. Chem. Soc. Perkin Trans. 1*, 1840 (1981).

80. J.-C. Barrière, A. Chiaroni, J. Cléophax, S. D. Géro, C. Riche, and M. Vuilhorgne, *Helv. Chim. Acta*, **64**, 1140, (1981).

81. D. H. R. Barton, N. J. Cussans, and S. V. Ley, *J. Chem. Soc. Chem. Commun.*, 393 (1978); *J. Chem. Soc. Perkin Trans. 1*, 1650 (1980).

82. A. G. M. Barrett, D. H. R. Barton, and R. W. Read, *J. Chem. Soc. Perkin Trans. 1*, 2191 (1980).

83. A. Couture, R. Dubiez, and A. Lablache-Combier, *J. Chem. Soc. Chem. Commun.*, 842 (1982).

84. H. Rheinboldt and E. Giesbrecht, *Chem. Ber.*, **88**, 1037, (1955).

85. J. L. Kice and T. W. S. Lee, *J. Am. Chem. Soc.*, **100**, 5094, (1978).

86. G. Bergson and G. Nordstrom, *Ark. Kemi*, **17**, 569, (1961).

87. L. G. Faehl and J. L. Kice, *J. Org. Chem.*, **44**, 2357, (1979).

88. H. J. Reich, F. Chow, and S. L. Peake, *Synthesis*, 299 (1978).

89. K. C. Nicolaou, R. L. Magolda, and W. E. Barnette, *J. Chem. Soc. Chem. Commun.*, 375 (1978);
K. C. Nicolaou, W. E. Barnette, and R. L. Magolda, *J. Am. Chem. Soc.* **103**, 3486, (1981);
K. C. Nicolaou, R. L. Magolda, W. J. Sipio, W. E. Barnette, Z. Lysenko, and M. M. Joullie,
J. Am. Chem. Soc., **102**, 3784, (1980).

90. P. A. Grieco, Y. Yokoyama, S. Gilman, and M. Nishizawa, *J. Org. Chem.*, **42**, 2034, (1977).

91. K. C. Nicolaou, D. A. Claremon, D. P. Papahatjis, and R. L. Magolda, *J. Am. Chem. Soc.*,
103, 6969, (1981).

92. M. P. Edwards, S. V. Ley, S. G. Lister, and B. D. Palmer, *J. Chem. Soc. Chem. Commun.*, 630
(1983).

93. R. A. Gancarz and J. L. Kice, *Tetrahedron Lett.*, **21**, 1697 (1980).

94. T. G. Back and S. Collins, *Tetrahedron Lett.*, **21**, 2215, (1980); *J. Org. Chem.*, **46**, 3249, (1981).

95. R. A. Gancarz and J. L. Kice, *J. Org. Chem.*, **46**, 4899, (1981).

96. H. Rheinboldt and E. Giesbrecht, *Chem. Ber.*, **88**, 1974, (1955).

97. D. Labar, A. Krief, and L. Hevesi, *Tetrahedron Lett.*, 3967 (1978).

98. L. Syper, *Tetrahedron Lett.*, 4493 (1966).

99. D. H. R. Barton, R. A. H. F. Hui, D. J. Lester, and S. V. Ley, *Tetrahedron Lett.*, 3331 (1979).

100. D. H. R. Barton, R. A. H. F. Hui, and S. V. Ley, *J. Chem. Soc. Perkin Trans. 1*, 2179 (1982).

101. W. R. Gaythwaite, J. Kenyon, and H. Phillips, *J. Chem. Soc.*, 2287 (1928).

102. M. Sevrin, W. Dumont, and A. Krief, *Tetrahedron Lett.*, 3835 (1977).

103. I. D. Entwistle, R. A. W. Johnstone, and J. H. Varley, *J. Chem. Soc. Chem. Commun.*, 61 (1976).

104. T. Hori and K. B. Sharpless, *J. Org. Chem.*, **43**, 1689, (1978).

105. H. J. Reich, S. Wollowitz, J. E. Trend, F. Chow, and D. F. Wendelborn, *J. Org. Chem.*, **43**,
1697, (1978).

106. M. Shimizu, R. Takeda, and I. Kuwajima, *Tetrahedron Lett.*, 3461 (1979); *Bull. Chem. Soc.
Jpn.*, **54**, 3510 (1981).

107. A. Cravador and A. Krief, *J. Chem. Soc. Chem. Commun.*, 951 (1980).

108. P. Magnus, F. Cooke, and T. Sarkar, *Organomet.* **1**, 562, (1982).

109. K. B. Sharpless, K. M. Gordon, R. F. Lauer, D. W. Patrick, S. P. Singer, and M. W. Young,
Chem. Scr. 9, **8A** (1975).

110. J. R. Williams and J. D. Leber, *Synthesis*, 427 (1977).

111. P. A. Grieco, Y. Yokoyama, S. Gilman, and Y. Ohfune, *J. Chem. Soc. Chem. Commun.*, 870
(1977).

112. P. A. Grieco, Y. Ohfune, and G. Majetich, *J. Am. Chem. Soc.*, **99**, 7393, (1977).

4

Nucleophilic Selenium

ROBERT MONAHAN, DAVID BROWN, LILADHAR WAYKOLE, AND DENNIS LIOTTA

Department of Chemistry, Emory University, Atlanta, Georgia

CONTENTS

1. PROPERTIES OF ORGANOSELENIUM COMPOUNDS

Often chemists refer to the properties of organosulfur compounds in order to rationalize the properties of organoselenium compounds. This is a reasonable approach, considering the chemistry of these elements is qualitatively similar. However, some features of selenium compounds make them more suitable for synthetic transformations.[1]

For example, carbon–sulfur bonds are much stronger than carbon–selenium bonds. Therefore, processes involving cleavage of the latter type of bond occur at a much greater rate than the former (e.g., [2,3]sigmatropic rearrangements, *syn* eliminations).

In addition, selenols and selenide anions are more potent nucleophiles than the corresponding sulfur compounds; this makes the formation of carbon–selenium bonds often a simple process. Selenide anions are also less basic than thiolate ions and, as would be expected, better leaving groups.[2]

According to hard–soft acid–base theory, neutral and anionic organoselenium compounds are predicted to be potent (soft) nucleophiles. Only a few studies have been reported, however, that quantify the actual nucleophilicity of these species. For perspective, phenyl selenide anion (sodium phenyl selenolate) was shown to be over 10^4 times more reactive than methoxide, 10^3 times more reactive than iodide, and seven times more reactive than thiophenoxide.[3] Similar information can be inferred from photoelectron spectroscopy data.[4]

2. GENERATION OF SELENIUM NUCLEOPHILES

Sodium phenyl selenide has been the most widely used nucleophilic selenium reagent to date.[5] It is available by a variety of procedures. The phenyl selenide anion's potency as a nucleophile, however is highly dependent on the method of generation.[6] Typically, diphenyl diselenide is treated with sodium borohydride in ethanol. The resulting clear, colorless, homogeneous solution contains a sodium phenyl selenide–borane complex that is satisfactory for epoxide opening but is not sufficiently nucleophilic for ester cleavage at the carbinol carbon [Eqs. (1)–(3)]. Apparently the nucleophilicity of phenyl selenide anion, generated in this manner, is compromised due to complexation with borane. The more reactive uncomplexed reagent can be easily prepared by other reductive procedures (*vide infra*).

$$\text{NaBH}_4 + \text{PhSeSePh} \xrightarrow{\text{EtOH}} 2\,\text{PhSeNa·BH}_3 + \text{H}_2 \qquad (1)$$

$$\text{PhSeNa·BH}_3 \xrightarrow[\text{2) H}_3\text{O}^{(+)}]{\text{1) }\triangle\!\!\!\diagup R} \text{PhSe}\diagdown\!\!\diagup\!\!\diagdown\text{OH}_R \qquad (2)$$

$$\text{PhSeNa·BH}_3 \xrightarrow{\text{THF}} \text{N.R.} \qquad (3)$$

Sodium borohydride reduction of diphenyl diselenide is highly exothermic and liberates larger amounts of hydrogen making it an unattractive method for large scale preparation of the phenyl selenide anion. A more convenient source of large quantities of ethanolic sodium phenyl selenide is the reduction of diphenyl diselenide with sodium hydroxymethyl sulfoxylate (Rongalite, $\text{NaOSOCH}_2\text{OH}$).[7] Benzeneselenol and sodium phenyl selenide are available from diphenyl diselenide by reaction with hypophosphorous acid.[8]

o-Nitro- and 2,4-dinitrophenyl selenolates are prepared from the corresponding selenocyanates by reduction with sodium borohydride in ethanol.[9] These nucleophiles have been widely used in synthesis since their alkylation products undergo oxidative elimination under very mild conditions. Among the numerous applications of this method is the synthesis of the vernolepin analog, $\underline{1}$ [Eq. (4)].[10]

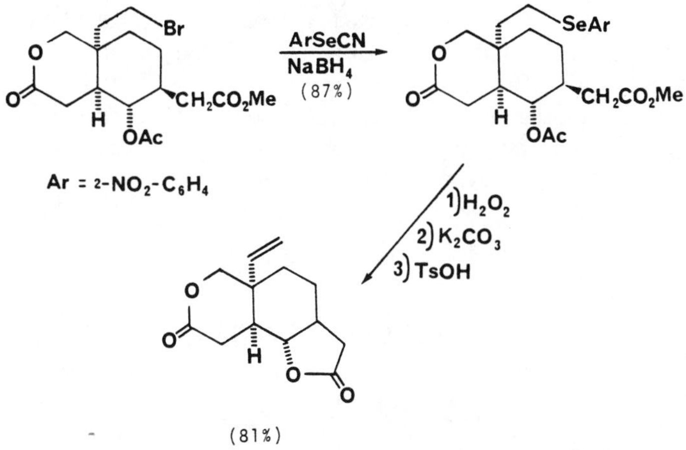

$$\text{Ar} = \text{2-NO}_2\text{-C}_6\text{H}_4$$

(4)

(81%)

Alkyl halides (or sulfonates), when reacted with selenium nucleophiles, produce aryl alkyl selenides. These electrophiles are usually prepared from the alcohol. An efficient alternative to this two-step process is the direct alcohol–selenide conversion developed by Grieco et al. [Eq. (5)], and utilized in the synthesis of estradiol.[11] Many other uses of this reaction in natural products synthesis can be found in the recent literature.[12]

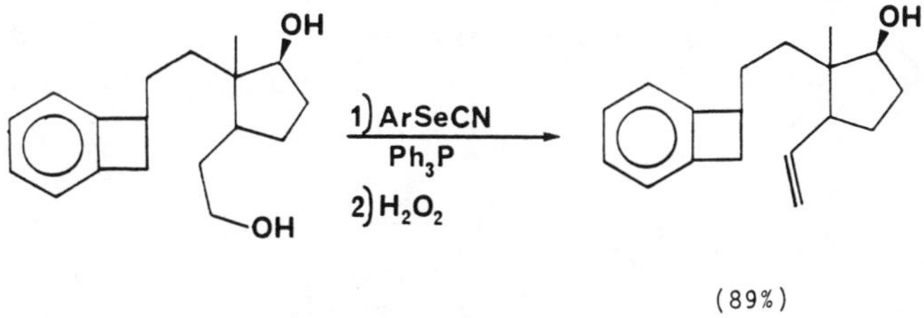

(5)

(89%)

Arylselenols may be generated electrochemically.[13] In a typical procedure diphenyl diselenide was electrolyzed between platinum electrodes in the presence of chlorotrimethylsilane, tetraethylammonium tosylate, and mesityl oxide in methanol [Eq. (6)]. Electrochemically generated phenyl selenide anion initially

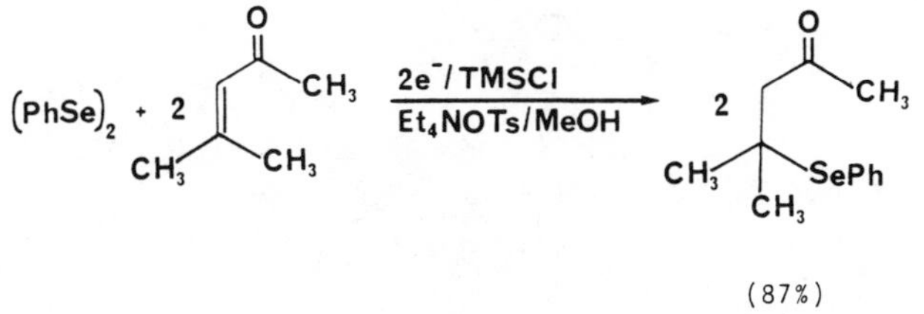

(6)

(87%)

forms phenylselenotrimethylsilane which, after methanolysis, reacts with the electrophile.

Although metal phenyl selenides are potent nucleophiles in protic solvents, their reactivity is enhanced in polar aprotic solvents such as N,N-dimethyl-formamide (DMF) and hexamethylphosphoric triamide (HMPA). Two general procedures for generating uncomplexed aprotic solutions of phenyl selenide anion are the reduction of diphenyl diselenide with sodium or potassium metal in tetrahydrofuran (THF) [Eq. (7)][14] and the reduction of diphenyl diselenide with sodium or potassium hydride in THF [Eq. (8)].[15] After a suitable reflux period a precipitate forms that is typically solubilized with HMPA and treated with an electrophile. The reactive species from the two methods seem to be identical. If the diselenide is substituted with reducible functionality, the latter method may be preferable. Alternatively, the selenols may be deprotonated with a metal hydride [Eq. (9)].[16]

$$(PhSe)_2 \quad \xrightarrow{\text{2 Na/THF}} \quad 2\ NaSePh \qquad\qquad (7)$$

$$PhSeH \quad \xrightarrow{\text{NaH/THF}} \quad NaSePh \quad + \quad H_2 \qquad (8)$$

$$(PhSe)_2 \quad \xrightarrow{\text{2 NaH/THF}} \quad 2\ NaSePh \quad + \quad H_2 \qquad (9)$$

Potassium phenyl selenide may be generated in the presence of an electrophile from phenylselenotrimethylsilane (PhSeTMS) and potassium fluoride (or iodide)/18-crown-6 in THF [Eq. (10)].[17] This method is potentially useful for effecting selective anionic attack on substrates with two or more electrophilic centers since the nucleophile is slowly produced and maintains a small steady state concentration throughout most of the reaction period. The crown ether facilitates both nucleophile generation and substitution by coordinating with the potassium ion.

$$ (10) $$

(73%)

Phenylselenotrimethylsilane may be prepared either from lithium phenyl selenide (benzeneselenol and MeLi) by quenching with chlorotrimethysilane

$$ (11) $$

$$ (12) $$

[Eq. (11)] or from the selenol by reaction with chlorotrimethylsilane in the presence of a catalytic amount of chloro-[tris-(triphenylphosphine)] rhodium [Eq. (12)].[18]

Lithium alkyl selenides are especially attractive reagents. They are easily prepared from selenols by reaction with alkyl lithium and are highly soluble in diethyl ether and THF.[19] The reaction of elemental selenium with lithium enolates produces β-keto lithium selenides that can be alkylated[20] and oxidatively eliminated to yield α,β-unsaturated carbonyl compounds [Eq. (13)].

$$(13)$$

A few general conclusions about the reactivity of metal phenyl selenides can be made: (1) The more ionic the species the greater its reactivity; experimentally this is supported by the generally higher yields obtained with sodium phenyl selenide as compared to the lithium reagent (especially in ester cleavage).[21] (2) Selenium nucleophiles are stronger in aprotic solvents than in protic solvents. (3) Species that selectively complex with the counterion increase the reactivity of the anion. The following reactivity gradient has been proposed for the phenyl selenide anion: NaSePh/18-crown-6/THF > NaSePh/HMPA/THF > LiSePh/HMPA/THF > LiSePh/THF $\approx$ LiSePh/Et$_2$O > PhSeSePh/NaBH$_4$/EtOH/THF.[16]

3. REACTIONS OF SELENIUM NUCLEOPHILES— SYNTHETIC APPLICATIONS

3.1. Epoxides

Ethanolic sodium phenyl selenide reacts with epoxides to produce the β-hydroxyalkyl selenide in moderate to high yield by normal S_N2 opening. Long reaction times and elevated temperatures may be required for hindered epoxides but the reaction is generally successful. Table 1 contains some typical examples.

Alkyl and phenyl selenols react with epoxides in basic or neutral media regioselectively to yield products derived from attack at the less hindered carbon atom. Catalytic amounts of boron trifluoride present in the reaction mixture give rise to a mixture of products [Eq. (14)].[27]

$$(14)$$

TABLE 1[a]

Entry	Substrate	Products	(%) Yield	References
1			(50)	22
2			(95)	22
3[b]			(40)	22
4[b]			(51)	22
5			(NA)[c]	23
6			(86)	24, 25
7			(62)	26
8			(85)	22

[a] All reactions utilize the phenyl selenide anion except Entry 5.
[b] Cholestane skeleton.
[c] NA = not available.

Mildly acidic conditions, produced by stirring reactants with aluminum oxide (alumina), have been successfully used with a number of nucleophiles, including benzeneselenol, to effect epoxide opening.[28] The reaction is regioselective, chemoselective, mild, and suitable for large scale work [Eq. (15)].

$$\text{(15)}$$

Triphenylselenoborane reacts with equimolar amounts of epoxide in chloroform solution at room temperature or below to furnish β-hydroxy selenides.[29] The products are dependent on the nature and number of substituents and, in some cases, the stereochemistry of the epoxide. For disubstituted epoxides the cis isomer is much more reactive.[30] This allows for possible chemical separation of epoxide mixtures. The rate of reaction of $(PhSe)_3B$ with terminal epoxides is particularly fast (<1 h at $20°C$). As with the phenyl selenide anion, attack occurs at the less substituted carbon atom. However, with styrene oxide, substitution occurs at the benzyl carbon; ethanolic sodium phenyl selenide gives a mixture of regioisomers [Eqs. (16) and (17)].[31]

$$\text{(16)}$$

$$\text{(17)}$$

Stirring the reaction mixture for long periods leads to significant amounts of olefin through decomposition of the β-selenoborate. Trisubstituted epoxides yield only allylic alcohols with saturated alcohols as by-products.

The opening of epoxides with phenyl selenide anion had been widely employed in natural products synthesis. Still et al. used this reaction in the preparation of **2**, a key intermediate in his eucannabinolide synthesis, from (+)-carvone [Eq. (18)].[32] The epoxide opening reaction has also been utilized in the synthesis of chorismic acid,[33] lycorine,[34] lycoricidine,[35] senepoxide,[36] kanamycin B analogs,[37] aphicholin,[38] and in a biomimetic approach to the amphilectane diterpenes.[39] The last application, recently published by Stevens and Albizati, is shown in Eqs. (19)–(21). The diastereomeric mixture of epoxides, **3**, was reacted with sodium phenyl selenide; two regioisomers were obtained. When **3** was protected as the t-butyldimethylsilyl ether, reaction occurred predominately at the tertiary carbon to yield the desired regioisomer, **4** [Eq.

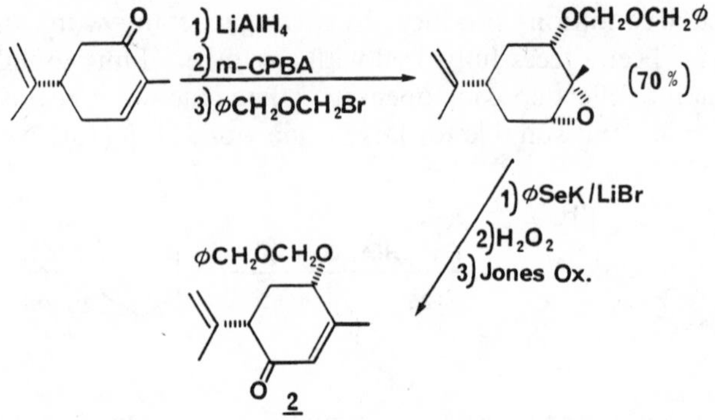

(18)

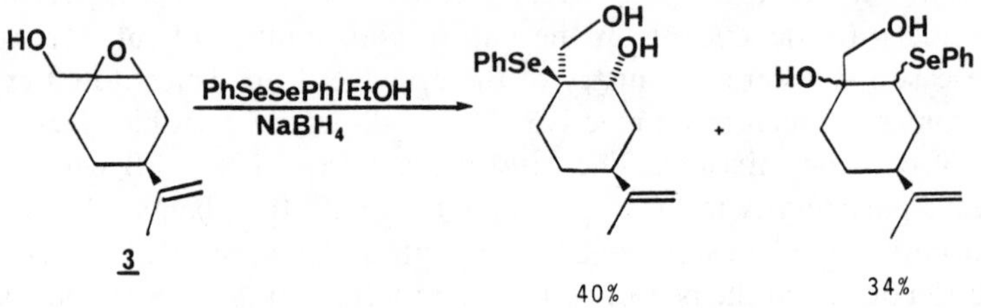

(19)

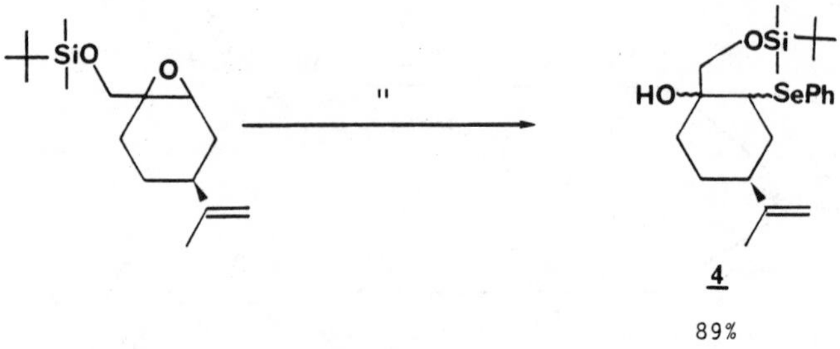

(20)

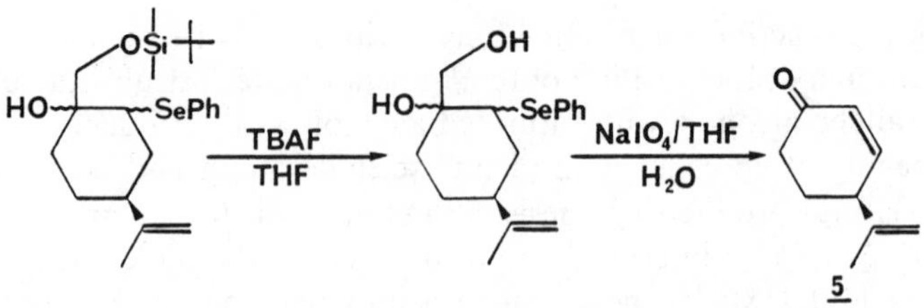

(21)

(20)]. After cleavage of the ether with fluoride, the resulting diol was treated with excess sodium periodate to produce the (S)-(−)-enone, 5 [Eq. (21)].

β-hydroxy selenides can be converted to allylic alcohols, olefins, bromohydrins, vinyl selenides, and epoxides. Allylic alcohols are available from the β-hydroxy selenides by oxidative elimination. This interconversion, first dem-

TABLE 2[a]

Entry	Epoxide	Allylic Alcohol	Yield (%)
1	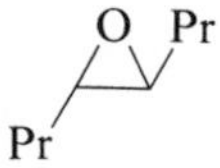	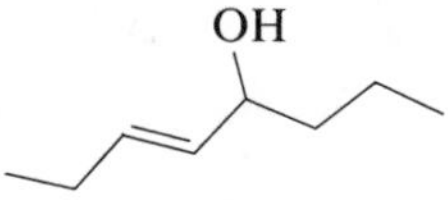	(98)
2	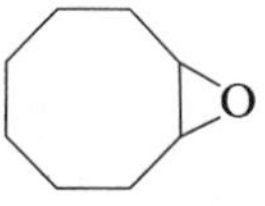	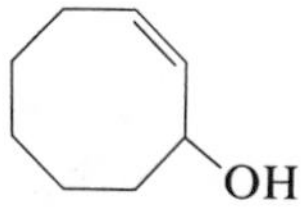	(75)
3[b]	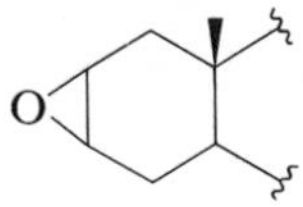	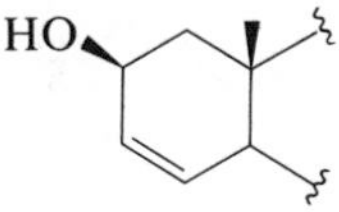	(51)
4[b]	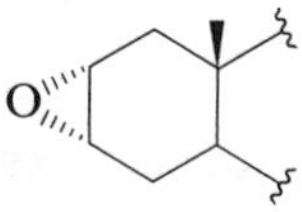	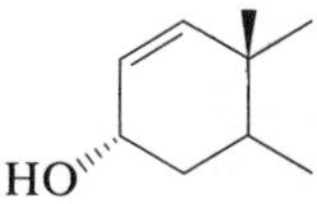	(40)
5	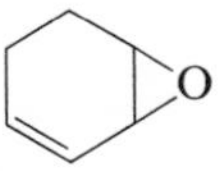	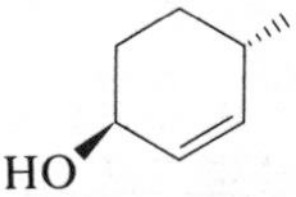	(50)

[a] All reactions utilize the phenyl selenide anion for step one and hydrogen peroxide for step two.
[b] Cholestane skeleton.

onstrated by Sharpless and Lauer, produces good to excellent yields of allylic alcohols generally free of by-products. Table 2 shows representative cases.[40]

Olefins can be obtained from β-hydroxy selenides by methods that convert the hydroxyl into a better leaving group: $MeSO_2Cl$/triethylamine (TEA),[41] TsOH,[42] $HClO_4$,[43] $SOCL_2$/TEA,[44] trimethylsilylchloride (TMSCl)/NaI,[43] PI_3/TEA,[44] and t-butyldimethylsilyl chloride (TBDMSCl)/NANH₃.[45] These reagents effect stereospecific trans elimination and the choice of which to employ should be made on the basis of the nature of additional functionality in the β-hydroxy selenide. This process is useful for the stereospecific deoxygenation of epoxides [Eq. (22)].[46]

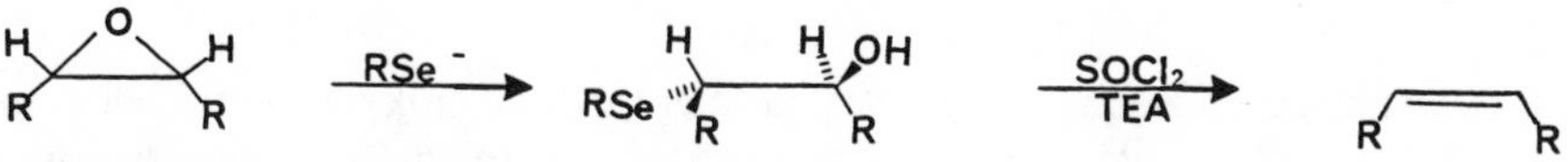

$$(22)$$

Alternatively, epoxides may be deoxygenated directly with the nucleophilic selenium reagents shown in Table 3.

TABLE 3

Selenium Reagent	References
KSeCN	46
$\phi_3P{=}Se$	47
$Bu_3P{=}Se$	48
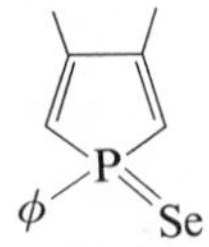	49
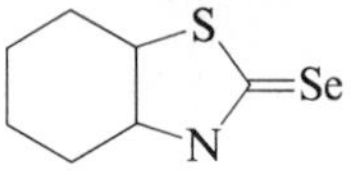	50

Van Ende and Krief have developed a method for the isomerization of di-substituted olefins [Eq. (23)].[51]

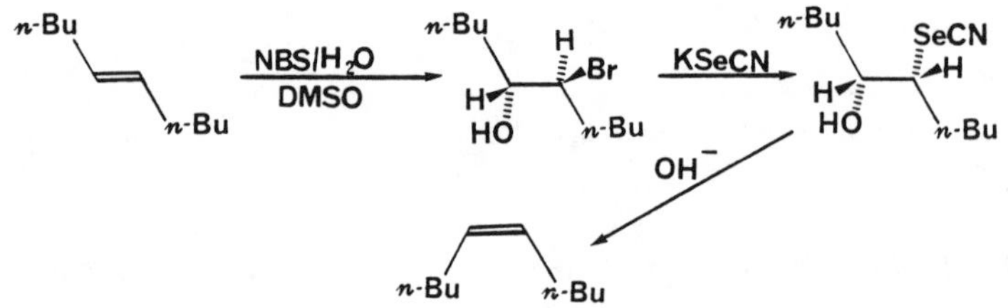

(23)

Bromohydrins may be prepared from β-hydroxy selenides by treatment with either one molar equivalent of Br_2 or two molar equivalents of N-bromosuccinimide in aqueous ethanol [Eq. (24)].[52]

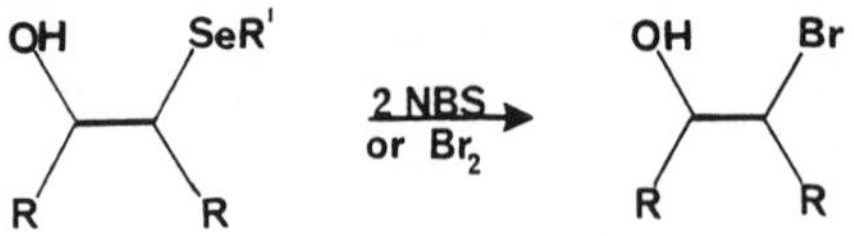

(24)

β-Hydroxy selenides may be converted to epoxides by alkylation on selenium with either dimethyl sulfate or methyl iodide (phenyl selenides require the pres-

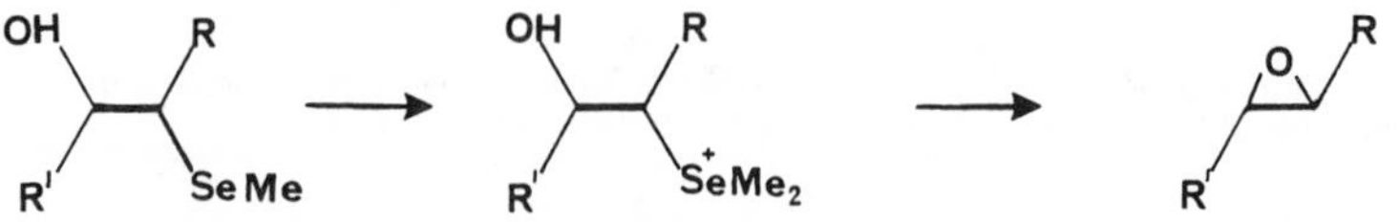

(25)

ence of silver ion) and deprotonation of the resulting species with sodium ethoxide in dimethylsulfoxide (DMSO) [Eq. (25)].[53]

3.2. Esters

Properly generated sodium phenyl selenide will cleave certain esters and lactones. The reaction usually involves displacement of carboxylate by attack of the nucleophile at the carbinol carbon. The preference for attack at the carbinol carbon is best understood in terms of hard–soft acid–base theory.[16] Soft nucleophiles, such as phenyl selenide anion, are predicted to prefer substitution at the carbinol carbon (soft–soft interaction) rather than addition to the carbonyl carbon (soft–hard interaction).

Sodium phenyl selenide, generated under the standard conditions of diphenyl diselenide reduction with sodium borohydride in ethanol, is highly nucleophilic. However, in refluxing THF this reagent does not cleave esters.[6] N,N-Dimethylformamide solutions of sodium phenyl selenide heated to 110–120°C do result in ester cleavage but, it is possible to effect this conversion under less vigorous conditions.[5d] If the anion is uncomplexed [i.e., generated as in Eqs. (7–9)], the reaction occurs in gently refluxing THF [Eq. (28)].[14]

$$\xrightarrow[\text{EtOH}/\,78\,°c]{(\phi Se)_2\,/\,NaBH_4} \quad \text{N.R.} \tag{26}$$

$$\xrightarrow[\text{DMF}/110-120\,°c]{(\phi Se)_2\,/\,NaBH_4} \quad \phi Se\diagdown\diagup\diagdown CO_2H \tag{27}$$

$$\xrightarrow[\text{THF}/\text{HMPA}/\sim 65\,°c]{(\phi Se)_2\,/\,Na} \quad '' \tag{28}$$

The degree of solvation of the phenyl selenide anion significantly effects the success of the ester cleavage reaction. If butyrolactone is allowed to react with sodium phenyl selenide in THF/HMPA, after 3 h of reflux, one obtains an 85% yield of acid. If HMPA is replaced by a 0.05 molar equivalent of 18-crown-6 the same yield of acid results from stirring 3 h at room temperature [Eq. (29)].[14]

$$\xrightarrow[\text{3h}/\text{rt}]{\phi SeNa/18-C-6} \quad \phi Se\diagdown\diagup\diagdown\diagup CO_2H \tag{29}$$

The nature of the counterion also effects the reactivity of the phenyl selenide anion. For example, valerolactone refluxed with uncomplexed sodium phenyl selenide in THF/HMPA for 3 h produces an 85% yield of γ-phenylselenenylbutanoic acid [Eq. (30)]. When lithium is substituted for sodium and the remainder

TABLE 4

Substrate	Product	Time (h)	(%) Isolated Yield
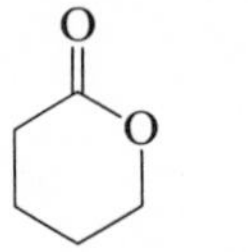	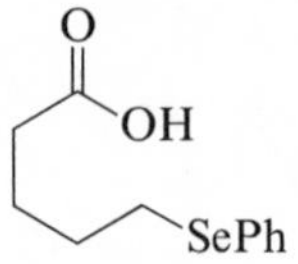	3	(85)
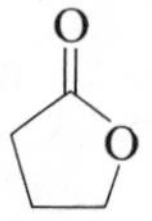	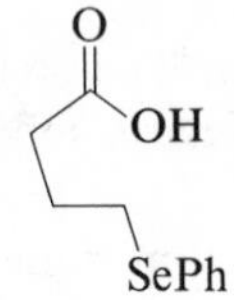	3	(85)
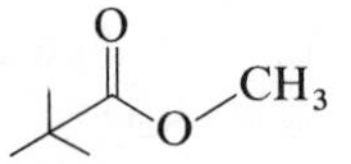	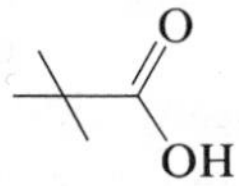	6	(98)
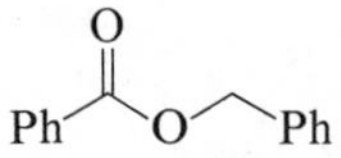	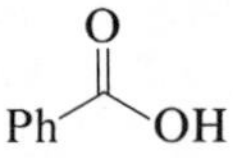	8	(96)
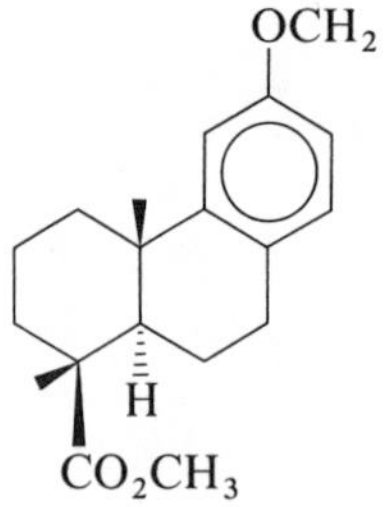	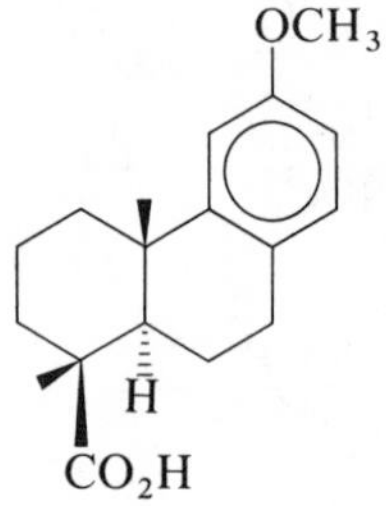	10	(100)
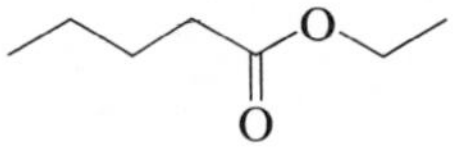	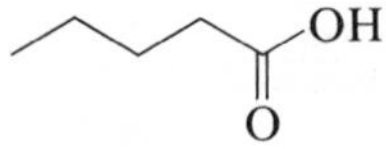	10	(99)
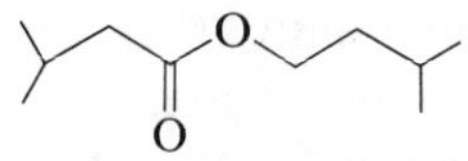	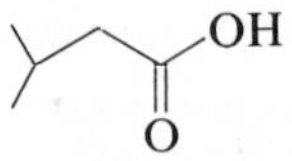	12	(92)
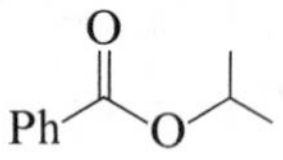	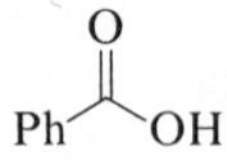	96	(92)
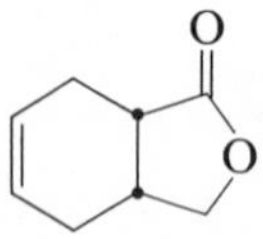	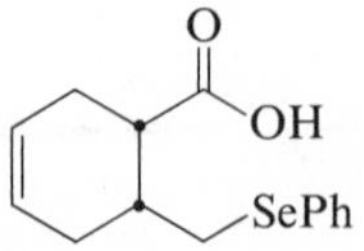	4	(75)
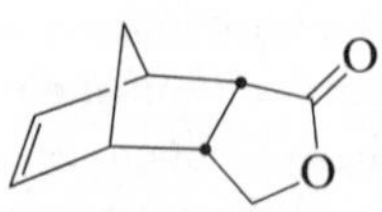	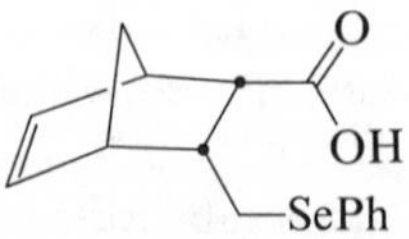	4	(80)

TABLE 4 (*Continued*)

Substrate	Product	Time (h)	(%) Isolated Yield
ε-caprolactone (7-membered lactone)	HOOC(CH₂)₄CH₂—SePh	14	(20)
5-methyl-γ-butyrolactone	HOOC(CH₂)₂CH(CH₃)—SePh	55	(96)
5-pentyl-γ-butyrolactone	HOOC—CH₂—CH(SePh)—(CH₂)₄CH₃	72	(20)
Ph—C(=O)—O—CH₂CH₃	Ph—C(=O)—OH	18	(99)
Ph—C(=O)—N(CH₃)H (N-methylbenzamide)		24	(0)
Ph—N(CH₃)—C(=O)—O—CH₃	Ph—NH—CH₃	10	(93)
Ph—C(=O)—N[CH(CH₃)—C(=O)—OCH₃]	Ph—C(=O)—N[CH(CH₃)—C(=O)—OH]	10	(94)
3-acetyl-γ-butyrolactone	CH₃—C(=O)—CH₂—CH₂—CH₂—SePh	4	(99)
3-acetyl-3-methyl-γ-butyrolactone	CH₃—C(=O)—CH(CH₃)—CH₂—CH₂—SePh	48	(94)
Ph—C(=O)—OCH₃	Ph—C(=O)—OH	3	(98)

of the reaction conditions held constant, only a 33% yield of acid is obtained [Eq. (31)]. The more highly ionic sodium reagent appears to be a more potent

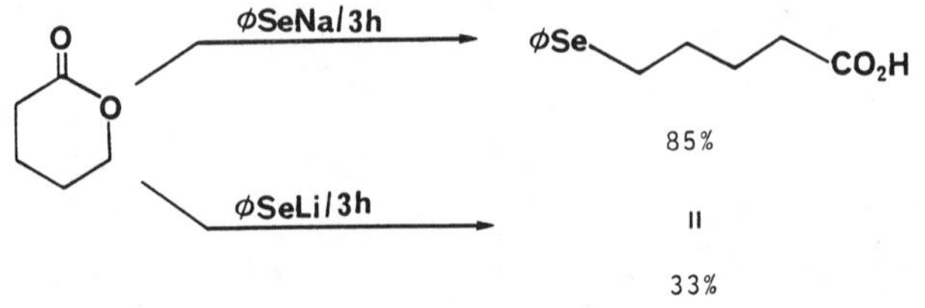

$$\phi SeNa/3h \qquad (30)$$

85%

$$\phi SeLi/3h \qquad (31)$$

33%

nucleophile.[14] Typical results of the ester cleavage reactions with phenyl selenide anion are given in Table 4.[16]

This ester–acid conversion has the advantage of being successful with sterically hindered esters that react sluggishly under normal saponification conditions. Structure 6 bears a methoxycarbonyl moiety adjacent to a quaternary center and is resistant to alkaline hydrolysis, presumably due to the severe steric interactions between the angular methyl group and the tetrahedral intermediate [Eq. (32)]. By contrast, no such steric crowding occurs during displacement at carbinol carbon [Eq. (33)].[16]

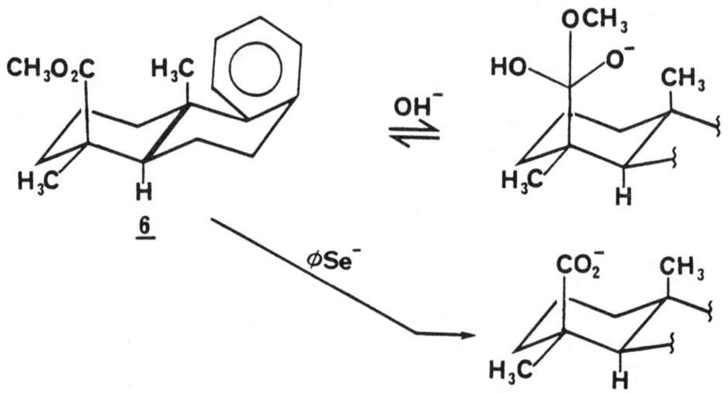

$$(32)$$

$$(33)$$

S_N2-type ester cleavage reactions typically employ nucleophiles such as amines, halides, cyanide, alkoxides, or thiocyanate and often work well only with methyl esters.[54] Phenyl selenide anion reacts with more heavily substituted esters and lactones (e.g., benzyl, isopropyl, and isoamyl) to produce the acid in high yield. However, secondary and tertiary lactones, with the exception of γ-valerolactone, are unaffected.[5d]

The phenyl selenide anion-induced cleavage of esters has been used in the synthesis of the following natural products: semburin,[55] isosemburin,[55] neoplanocin A,[56] and nanomycin [Eq. (34)].[57]

It is necessary to esterify selenides such as 7 (usually with diazomethane) since elimination will not occur in the presence of the free carboxyl group [Eq. (35)].[58] This sequence allows for the synthesis of acids (or esters) with unsaturation remote from the carbonyl; see Table 5.[59]

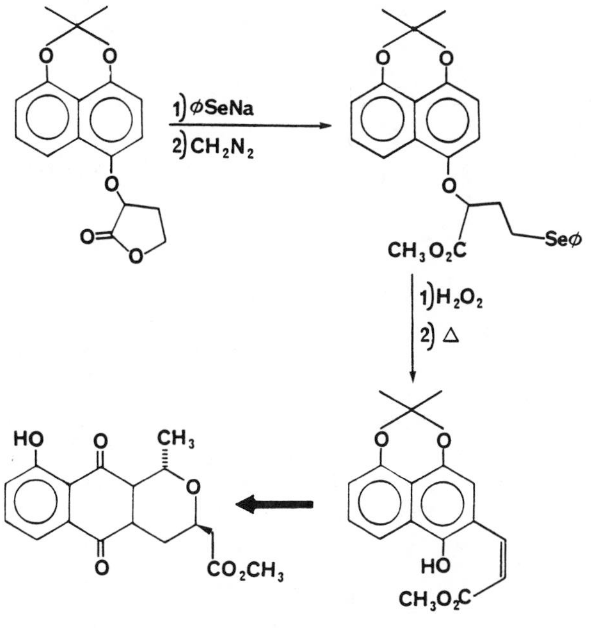

$$(34)$$

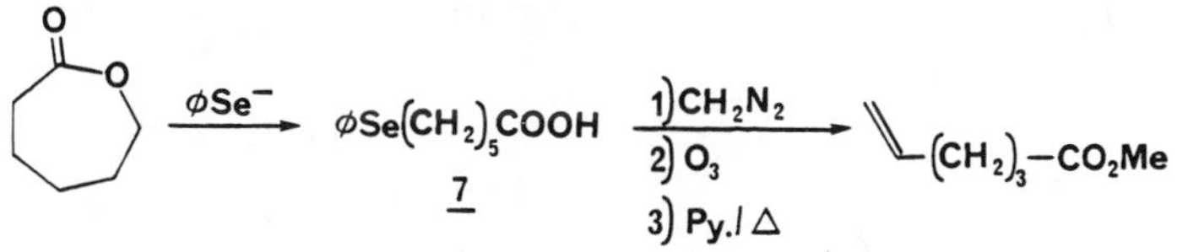

$$(35)$$

The ester cleavage reaction could conceivably occur by a variety of pathways including: (1) elimination, (2) acyl–oxygen cleavage, and (3) direct nucleophilic displacement at the carbinol carbon. The first possibility may be ruled out since the reaction of sodium phenyl selenide with isopropyl benzoate produces no propene. With this ester, although benzoic acid is obtained in 96% yield, no selenide is isolated. All other cleavages listed in Table 4 produced the selenide in a yield comparable to that of the acid. With isopropyl benzoate (and other secondary or otherwise hindered esters), it is believed that ester cleavage can occur by acyl–oxygen bond scission.[16]

To test this hypothesis, the ester, 8, was reacted with phenyl selenide anion in refluxing THF for 24 h. Structure 9 was isolated and neither the selenide nor olefin could be detected [Eqs. (36) and (37)].[16]

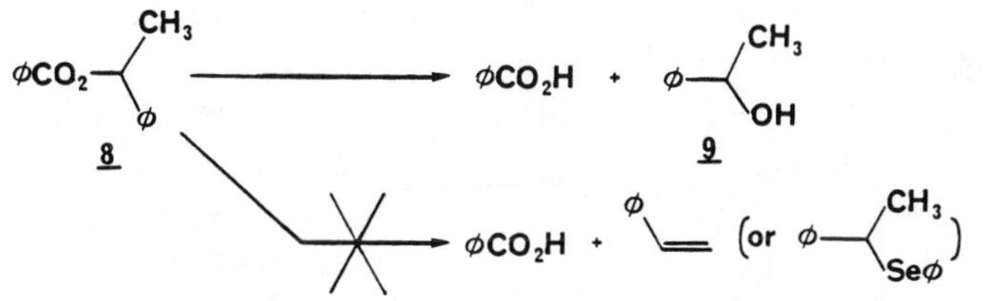

$$(36)$$

$$(37)$$

TABLE 5

Lactone	Phenyl Selenide	(%) Yield	Olefinic Ester	(%) Yield
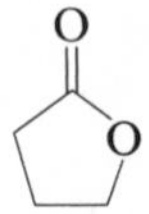	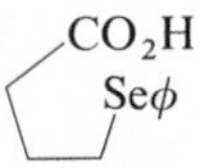 CO_2H, $Se\phi$	(98)	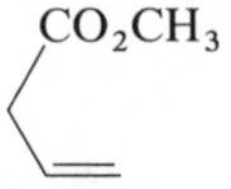 CO_2CH_3	(70)
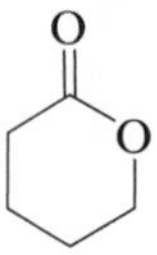	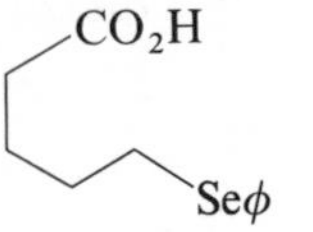 CO_2H, $Se\phi$	(95)	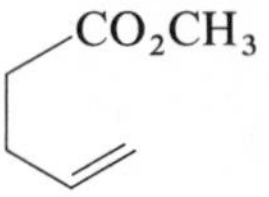 CO_2CH_3	(65)
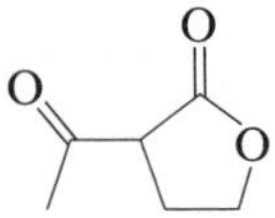	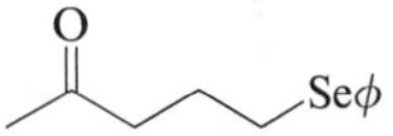 $Se\phi$	(92)	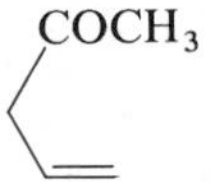 $COCH_3$	(60)
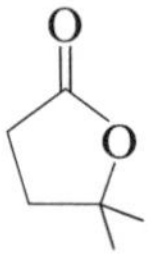	NR^a		...	
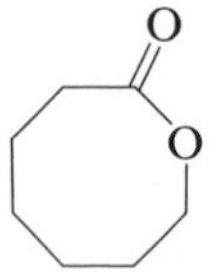	$HO_2C(CH_2)_6Se\phi$	(90)	$H_3CO_2C(CH_2)$ 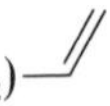	(71)

^{a} NR = no reaction.

3.3. Alkyl Halides, Sulfonates, and so on

Metal phenyl selenides react smoothly with primary and secondary alkyl halides to form alkyl phenyl selenides. In contrast, the reaction of carbon nucleophiles with the latter class of electrophiles tends to be troublesome. In connection with the synthesis of selenium containing peptides, the sodium salt of N-(Z)-selenocysteine diphenylmethyl ester, $\underline{10}$, was reacted with diphenyl-methyl bromide to yield $\underline{11}$ in 79% yield [Eq. (38)].^{60} The starting material

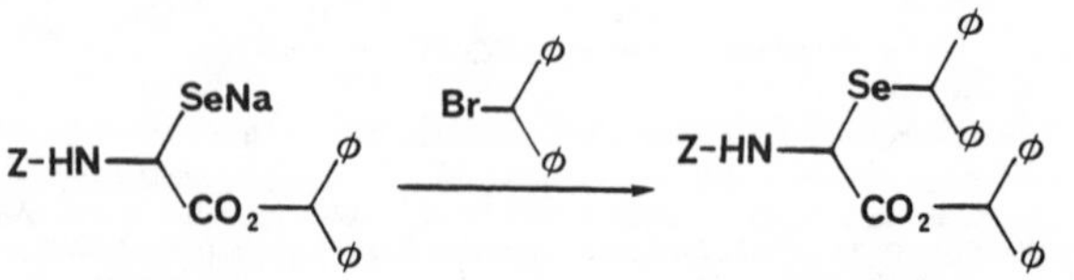

$$Z-HN-\underset{\underline{10}}{\overset{SeNa}{\underset{CO_2-\phi}{|}}}\phi \quad \xrightarrow{Br-\overset{\phi}{\underset{\phi}{|}}} \quad Z-HN-\underset{\underline{11}}{\overset{Se-\overset{\phi}{\underset{\phi}{|}}}{\underset{CO_2-\phi}{|}}}\phi \qquad (38)$$

was prepared from the corresponding tosylate by reaction with sodium hydrogen selenide.

Allylic halides, upon treatment with phenyl selenide anion, form phenyl allyl selenides by either S_N2 or S_N2' reaction (or both) depending on the nature of the halide. After oxidation, these selenoxides have two decomposition pathways: (1) the usual *syn* elimination to yield dienes[61] and (2) [2,3]sigmatropic rearrangement giving selenate esters that can be hydrolyzed to allylic alcohols.[62] The cholesteryl bromide, 12, upon reaction with sodium phenyl selenide, gave 13 in 88% yield and required 15 min for completion [Eq. (39)].[63] Oxidation resulted in approximately equal amounts of diene and allylic alcohol [Eq. (40)].

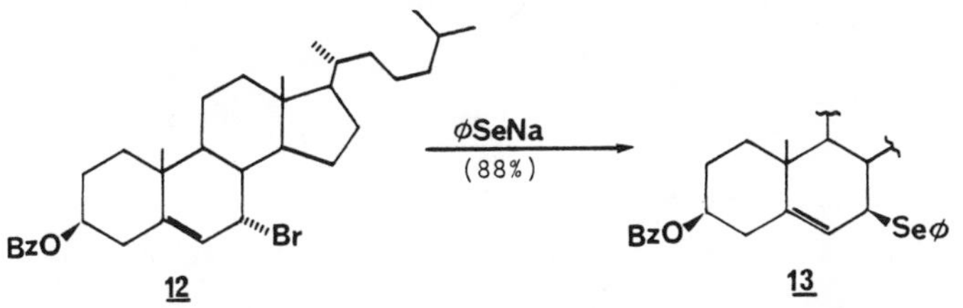

$$\text{(39)}$$

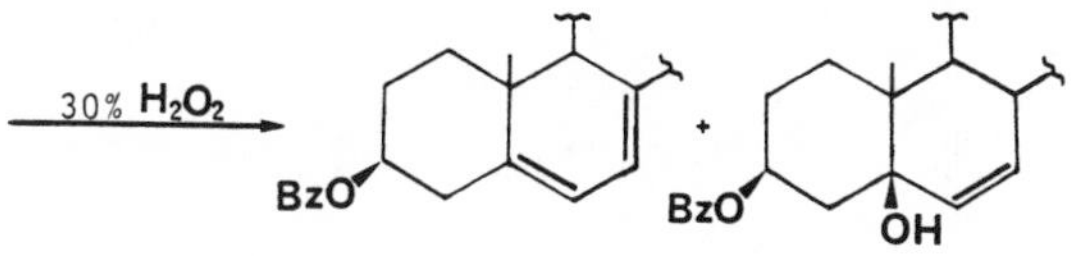

$$\text{(40)}$$

For many substrates this process is an alternative to the epoxide opening, oxidative elimination sequence demonstrated by Sharpless and Lauer for the synthesis of allylic alcohols.[40]

An especially useful synthesis of unsaturated carbonyl compounds involves the reaction of metal enolates with elemental selenium to produce β-keto metal selenides.[64] These nucleophiles may be alkylated with iodomethane and subjected to oxidative elimination to form α,β-unsaturated carbonyl compounds.[65] This sequence compliments the quenching of enolates with phenylselenenyl halides. Shown is an application of this method [Eq. (41)].[66]

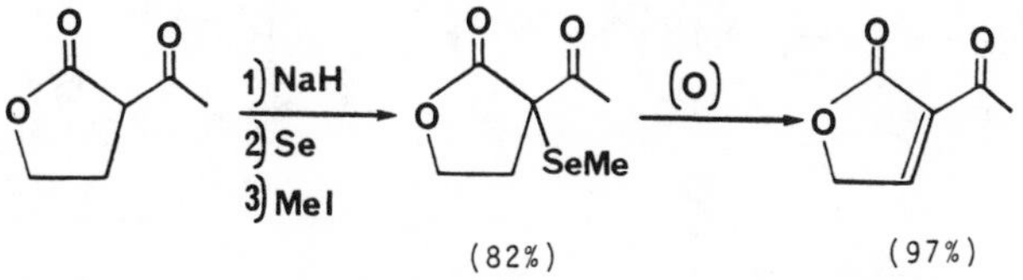

$$\text{(41)}$$

Critical to the success of the β-keto selenide anion alkylation is the presence of HMPA. Three to six molar equivalents were found to be optimal—enough to solubilize the enolate but not enough to encourage polyselenide formation. It is also important that the methylation time be kept brief so that further alkylation on selenium is minimized. Although alkyl selenide anions are known

TABLE 6

Substrate	Product	(%) Yield
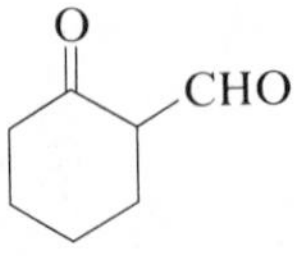 2-formylcyclohexanone	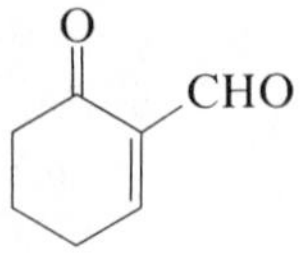 2-formylcyclohex-2-enone	(85)
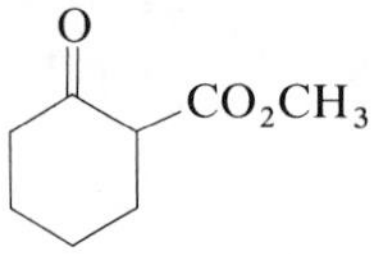 methyl 2-oxocyclohexanecarboxylate	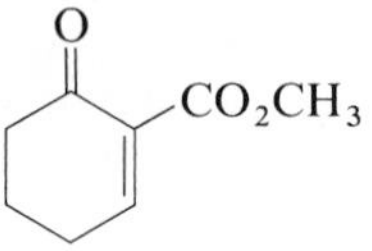 methyl 2-oxocyclohex-2-enecarboxylate	(98)
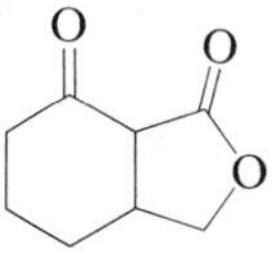 bicyclic keto-lactone	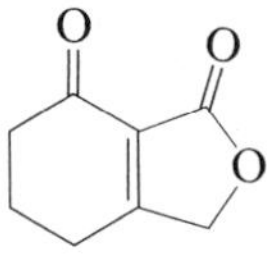 unsaturated bicyclic keto-lactone	(85)
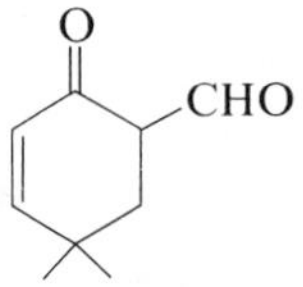 4,4-dimethyl-6-oxocyclohex-2-ene-carbaldehyde	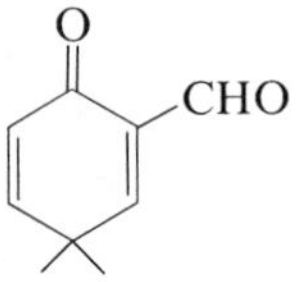 dienone CHO	(95)
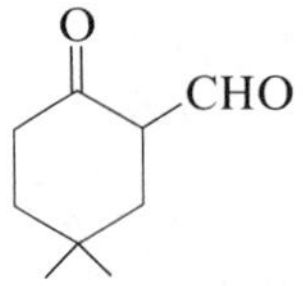 4,4-dimethyl-2-formylcyclohexanone	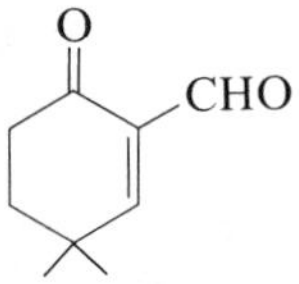 4,4-dimethyl-2-formylcyclohex-2-enone	(85)
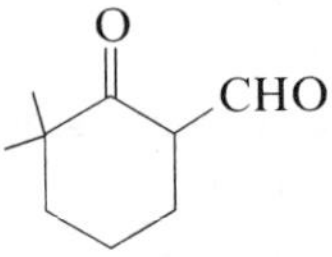 3,3-dimethyl-2-formylcyclohexanone	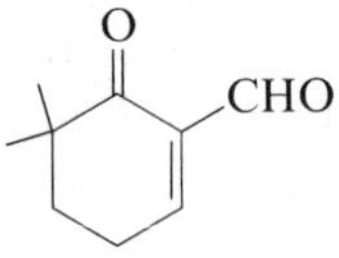 3,3-dimethyl-2-formylcyclohex-2-enone	(84)
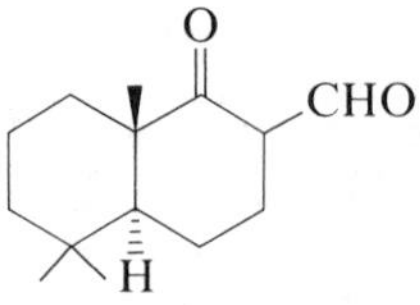 decalin keto-aldehyde	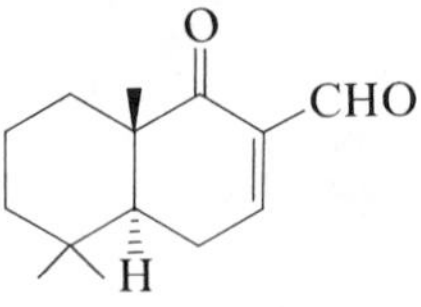 unsaturated decalin keto-aldehyde	(100)

to cleave lactones, this reaction does not occur under these conditions. Table 6 contains representative examples.[66]

A similar process has been utilized in the conversion of alkyl phosphonates to vinyl phosphonates [Eq. (42)].[67] Yields for the methylselenation step normally exceed 80%. It is interesting to note that, with an equimolar amount of oxidant, sulfoxide formation is not observed. The following scheme indicates a synthetic application of vinyl phosphonate, 14 [Eq. (43)].[67]

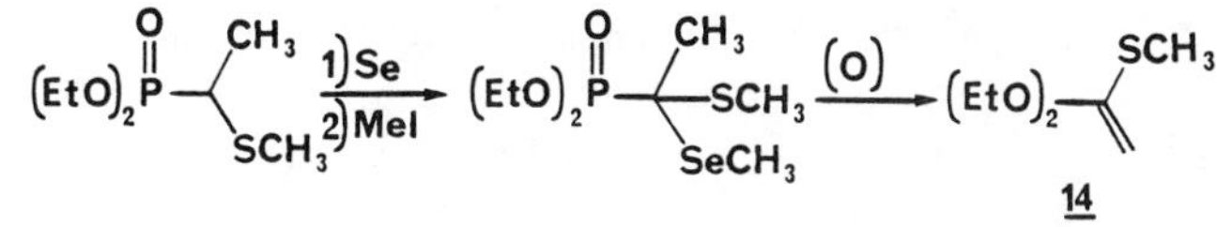

$$(EtO)_2\overset{O}{\underset{}{P}}-\overset{CH_3}{\underset{SCH_3}{C}} \xrightarrow[2)MeI]{1)Se} (EtO)_2\overset{O}{\underset{}{P}}-\overset{CH_3}{\underset{SeCH_3}{C}}-SCH_3 \xrightarrow{(O)} (EtO)_2\overset{SCH_3}{\underset{}{P}} \qquad (42)$$

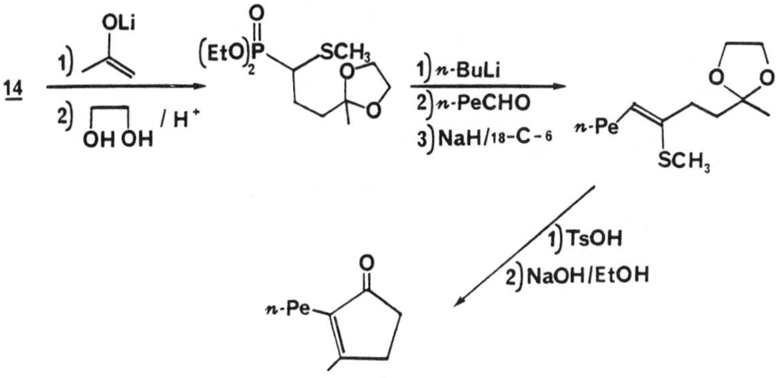

$$\underline{14}$$

$$\underline{14} \xrightarrow[2)\,\overset{}{\underset{OH\ OH}{\square}}\,/H^+]{1)\ \overset{OLi}{\underset{}{\diagup\!\!\diagup}}} (EtO)\overset{O}{\underset{}{P}}_2-\overset{SCH_3}{\underset{}{C}} \xrightarrow[3)NaH/18-C-6]{1)\,n\text{-BuLi}} \underset{n\text{-Pe}}{\overset{SCH_3}{\longrightarrow}} \qquad (43)$$

1) TsOH
2) NaOH/EtOH

$$n\text{-Pe}$$

Displacements of halides with the phenyl selenide anion have been widely used in the synthesis of natural products to introduce olefins under mild conditions. Examples include the synthesis of vernolepin,[68] vernopelin analogs,[69] clavulanic acid analogs,[70] the antibiotic bicyclomycin,[71] pederine,[72] paliclavine,[73] 5-epipaliclavine [Eq. (44)],[73] moenocinol,[74] and diplodialide A [Eq. (45)].[75]

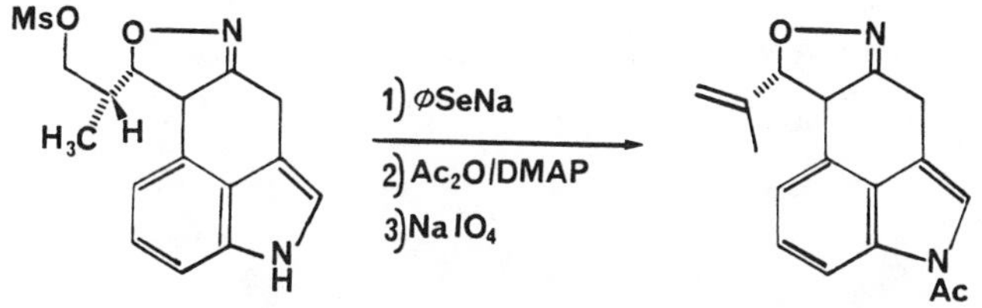

$$\xrightarrow[3)NaIO_4]{\substack{1)\,\phi SeNa \\ 2)\,Ac_2O/DMAP}} \qquad (44)$$

$$\xrightarrow[(67\%)]{\phi SeNa} \qquad (45)$$

1) H$_2$O$_2$
2) CF$_3$CO$_2$H

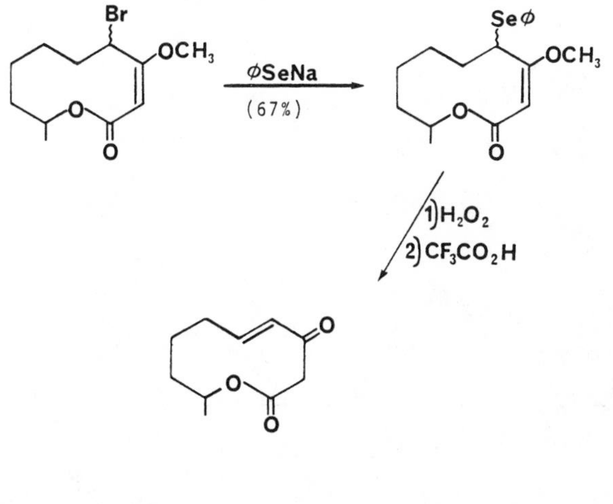

Acetates may be converted to the corresponding phenylselenides by reaction with PhSeTMS and a Lewis acid catalyst [Eqs. (46) and (47)].[76] In the case of allylic acetates, the product structure seems to be determined by double bond stability.[77]

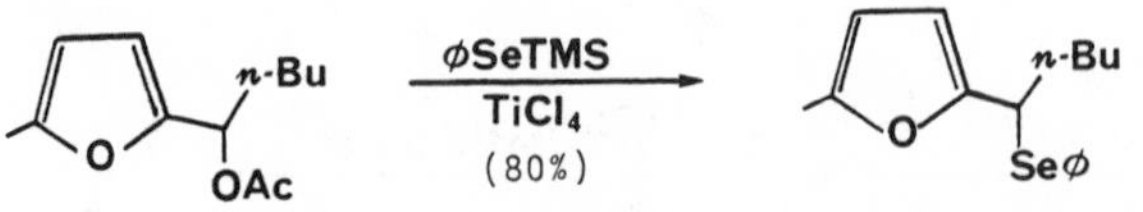

$$
\text{(46)}
$$

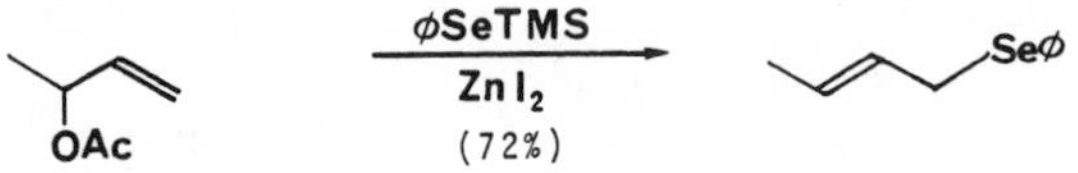

$$
\text{(47)}
$$

3.4. Unsaturated Carbonyl Compounds

β-Arylselenocarbonyl compounds may be obtained from unsaturated carbonyl compounds by the conjugate addition of arylselenols [Eqs. (48) and (49)].[78] The reaction occurs readily; many α,β-unsaturated carbonyl compounds undergo 1,4 addition with benzeneselenol without catalysis (acetylenic carbonyl

TABLE 7

Enone	Diselenide	(%) Yield of Adduct
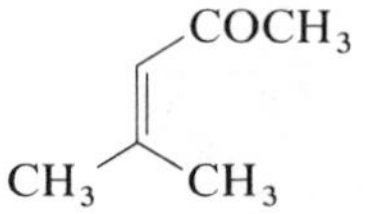	$(\phi Se)_2$	(87)
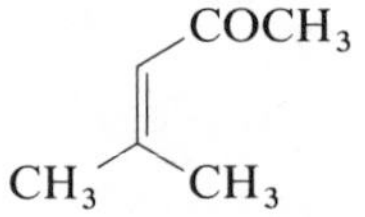	$(p\text{-}CH_3OC_6H_4Se)_2$	(92)
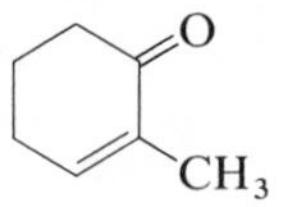	$(\phi Se)_2$	(87)
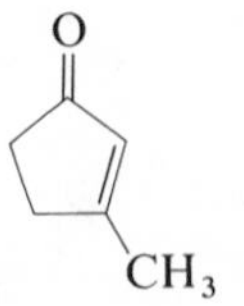	$(\phi Se)_2$	(47)
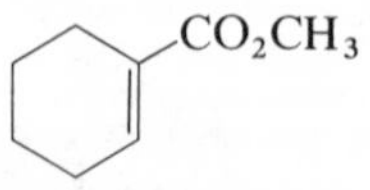	$(p\text{-}CH_3OC_6H_4Se)_2$	(62)

compounds do require base catalysis).[79] Stronger reagent systems include metal phenyl selenides and PhSeTMS with either Lewis acid or fluoride ion catalysis.

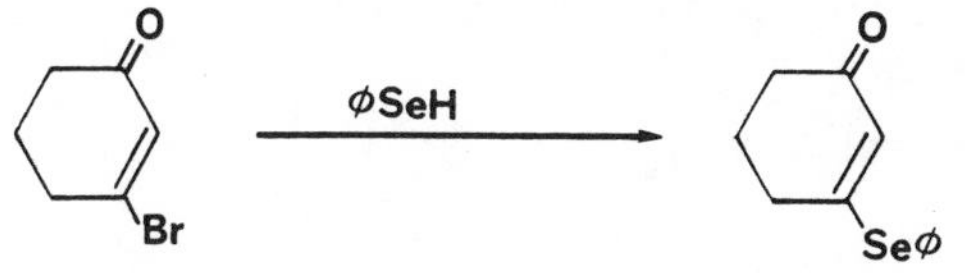

$$(48)$$

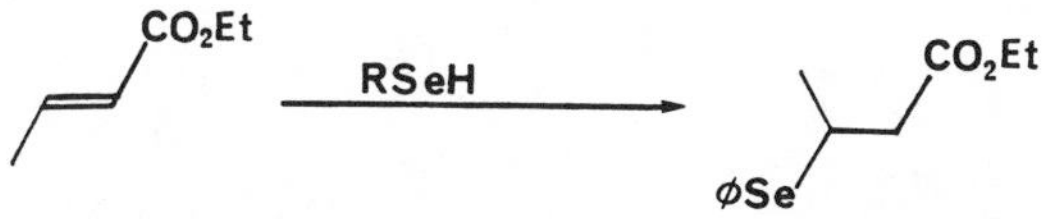

$$(49)$$

In the synthesis of the pheromone ipsenol, 16, the phenylselenenyl group was introduced by reaction of complexed phenyl selenide anion (PhSeSePh/ $NaBH_4$) with α-methylene lactone, 15. Phenyl selenide anion was later eliminated [Eq. (50)].[80] This method could potentially be used as a method of protecting the sensitive α-methylene functionality.[81]

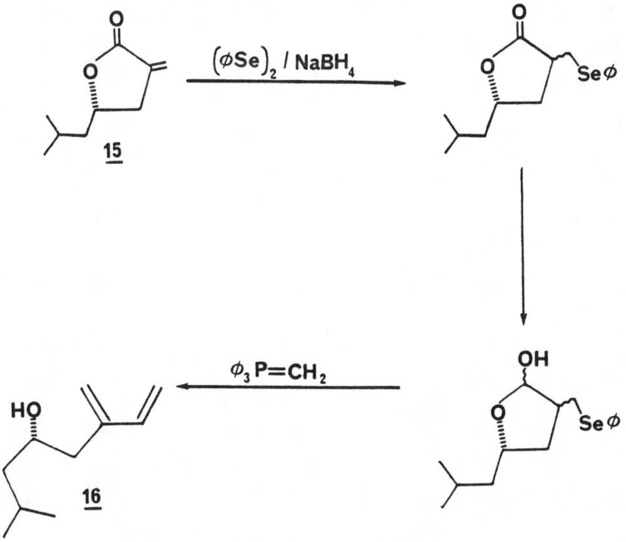

$$(50)$$

Table 7 contains examples of electrochemically generated arylselenol addition to Michael acceptors.[82]

α,β-Unsaturated ketones react with PhSeTMS in the presence of an appropriate catalyst, to produce positionally defined silyl enol ethers of β-phenylseleno ketones [Eq. (51)].[83] Aldehydes react similarly. Typical catalysts are BF_3, $ZnCl_2$, Ph_3P, and TMSOTf. Nuclear magnetic resonance studies of the reaction of PhSeTMS with unsaturated aldehydes have shown that, after a

TABLE 8

Substrate	Catalyst	Adduct	$(E)/(Z)$	(%) Yield
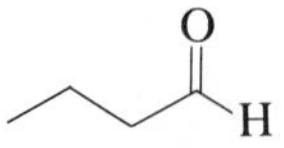	BF_3	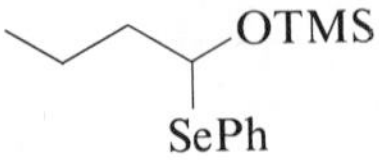	$\cdots$	(100)
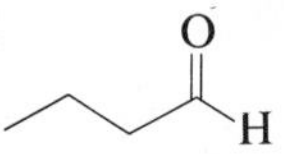	PPh_3	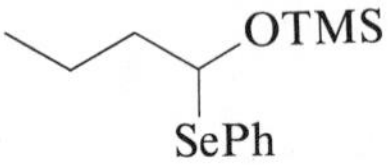	$\cdots$	(100)
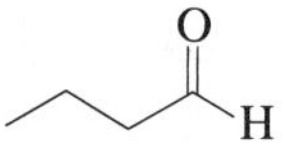	$ZnCl_2$	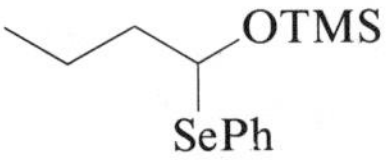	$\cdots$	(100)
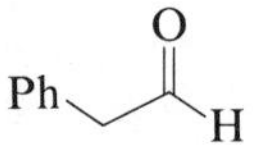	$ZnCl_2$	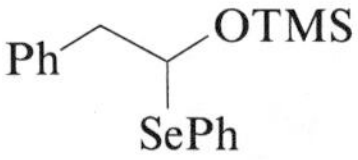	$\cdots$	(100)
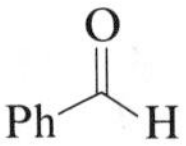	$\cdots$	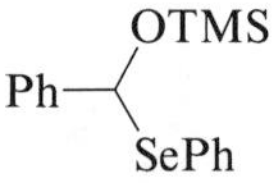	$\cdots$	(100)
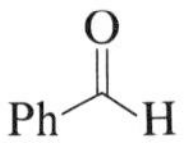	$ZnCl_2$	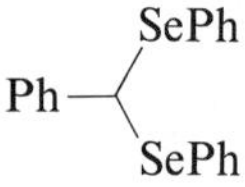	$\cdots$	(99)
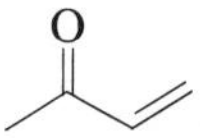	PPh_3	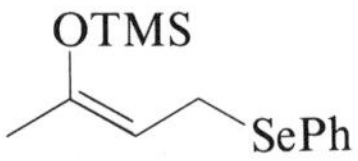	3/2	(90)
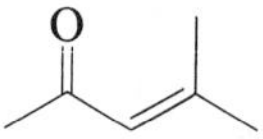	PPh_3	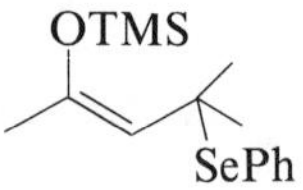	$\cdots$	(70)

short reaction time, both 1,2 and 1,4 adducts are present. However, with time the 1,4 adduct predominates, indicating a thermodynamic preference for this product.[84] Table 8 contains examples of this reaction.[84]

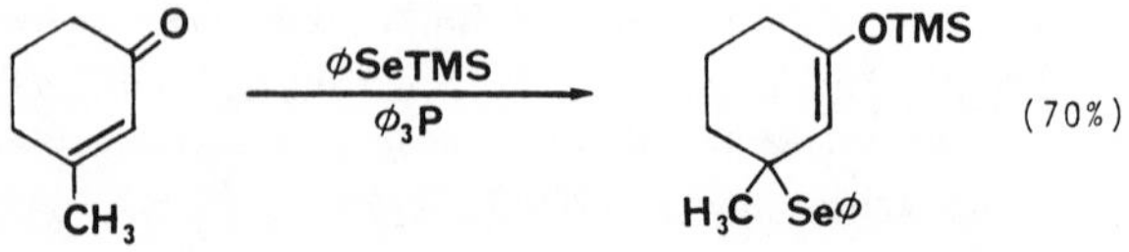

$$(51)$$

TABLE 8 (*Continued*)

Substrate	Catalyst	Adduct	(*E*)/(*Z*)	(%) Yield
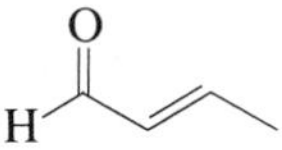	PPh$_3$	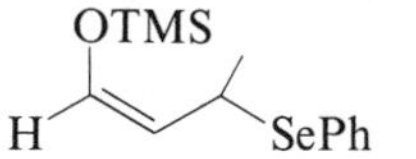	2/1	(100)
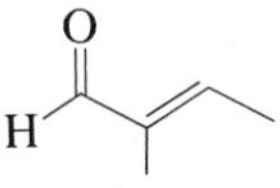	PPh$_3$	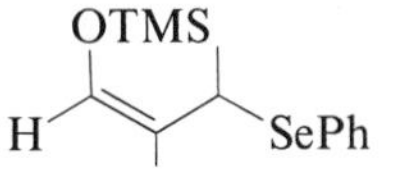	6/1	(95)
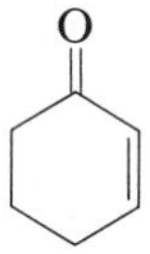	PPh$_3$	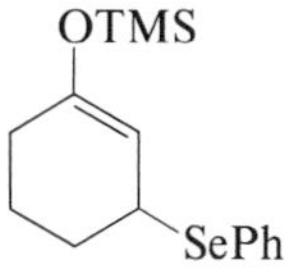	$\cdots$	(100)
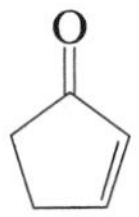	PPh$_3$	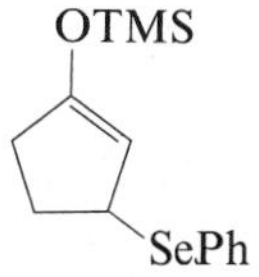	$\cdots$	(100)
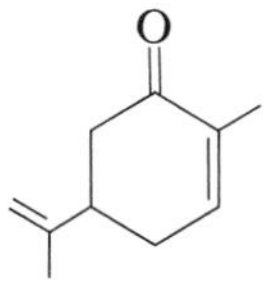	PPh$_3$	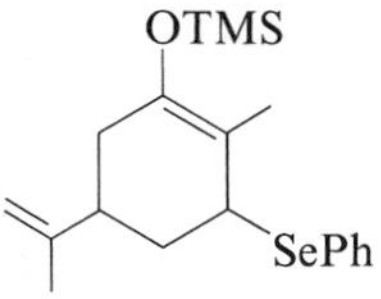	$\cdots$	(100)
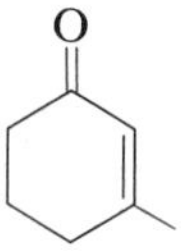	PPh	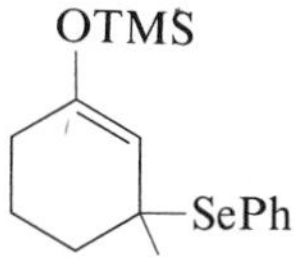	$\cdots$	(70)

Potassium phenyl selenide, generated from PhSeTMS/KF (or KI)/18-crown-6, adds in conjugate fashion to unsaturated carbonyl compounds, typically in high yield [Eq. (52)].[17]

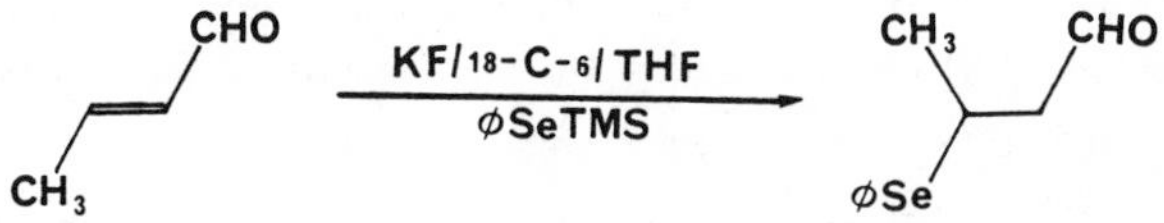

$$ \text{Eq. (52)} \qquad (73\%) $$

Noyori et al. has developed a one-pot procedure to accomplish the conversion of an enone to an α-(dialkoxy alkyl)- or α-(alkoxy alkyl)enone.[85] The first step in the process is the conjugate addition of PhSeTMS to the enone

TABLE 9

Enone	Trapping Acetal	Product	(%) Yield
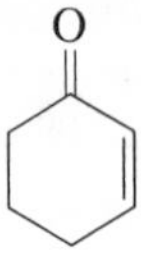	$C_6H_5CH(OCH_3)_2$	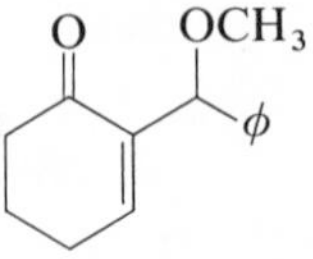	(75)
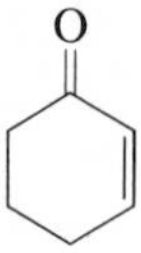	$C_6H_5C(CH_3)(OCH_3)_2$	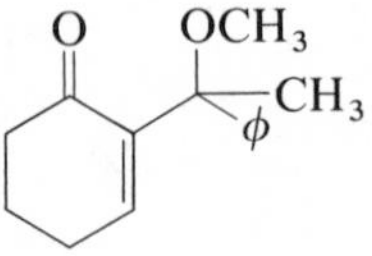	(30)
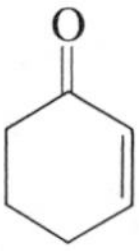	$C_6H_5CH{=}CH(OCH_3)_2$	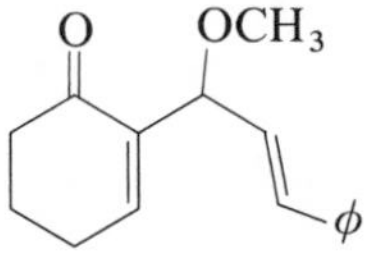	(83)
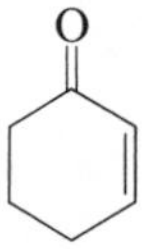	$CH(OCH_3)_3$	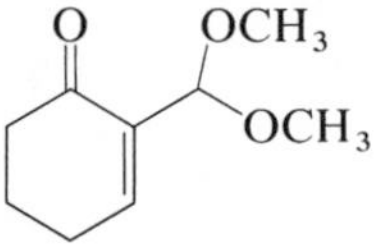	(76)
	$CH(OCH_3)_3$	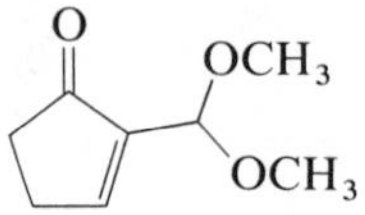	(58)
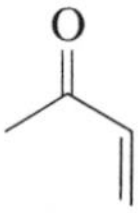	$C_6H_5CH(OCH_3)_2$	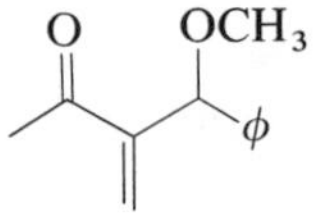	(57)

in the presence of a catalytic amount of TMSOTf. This reagent activates the enone by O-silylation and, after conjugate addition of the selenide, desilylates the resulting selenium cation to regenerate the catalyst. The β-(phenylseleno) silyl enol ether is then treated with an acetal or ortho ester (depending upon the desired product). Oxidative elimination produces the new enone [Eq. (53)]. The reaction succeeds with a variety of enones including methyl vinyl ketone, cylohexenone, and cyclopentenone (Table 9).[85]

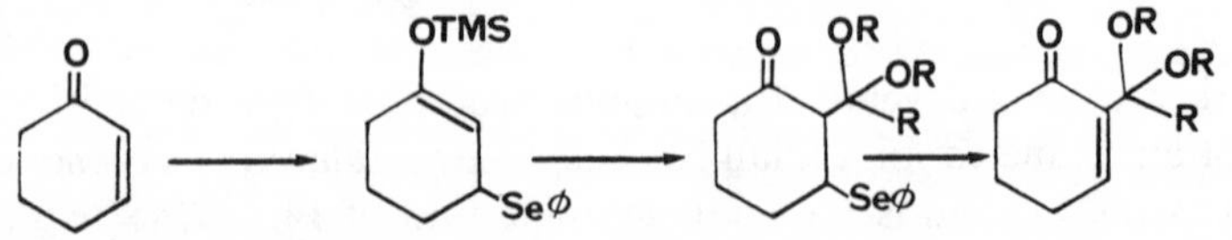

(53)

TABLE 10

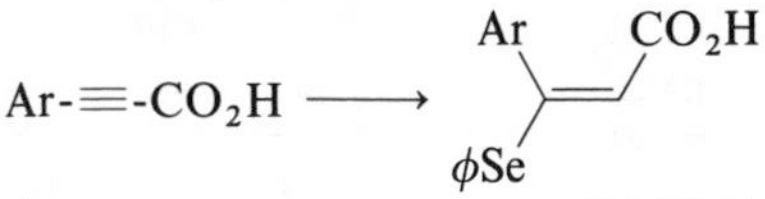

$$Ar-\equiv-CO_2H \longrightarrow \underset{\phi Se}{\overset{Ar \quad CO_2H}{\diagdown}}$$

Ar	(%) Yield
C_6H_5	(81)
$4\text{-}CH_3OC_6H_4$	(96)
$2,5\text{-}(CH_3)_2C_6H_3$	(40)
$2,5\text{-}(CH_3O)_2C_6H_3$	(68)
$2\text{-}CH_3OC_{10}H_6$	(55)

Arylselenols add in conjugate fashion to arylpropiolates under basic conditions to give α-substituted cinnamates of predominately (Z) stereochemistry [Eq. (54)].[86] Mixing a neat solution of arylselenol and propiolate results in free-radical addition producing α-(arylseleno) cinnamates [Eq. (55)]. Both reactions are general for a variety of acetylenic esters (Table 10).[86]

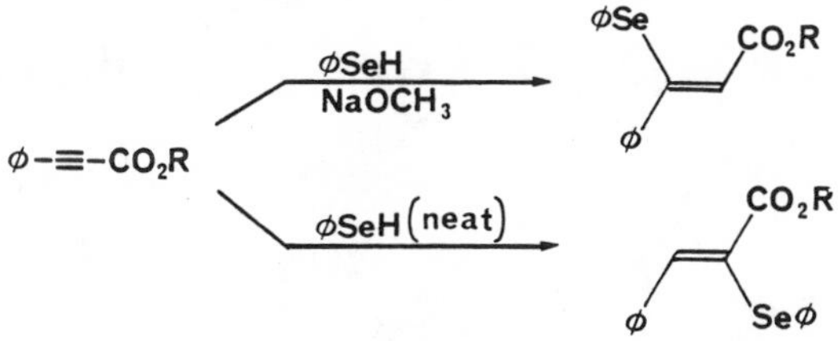

$$\phi-\equiv-CO_2R \quad \xrightarrow[\text{NaOCH}_3]{\phi SeH} \quad \underset{\phi}{\overset{\phi Se \quad CO_2R}{\diagdown}} \tag{54}$$

$$\xrightarrow[]{\phi SeH\,(\text{neat})} \quad \underset{\phi}{\overset{CO_2R}{\diagdown}} Se\phi \tag{55}$$

Solutions of sodium phenyl selenide (DMF, 120°C) will cleave monoactivated cyclopropanes in low to moderate yield.[87] In some cases the corresponding lithium reagent results in higher yields (Table 11). This reaction was employed by Masamune et al. to construct the useful synthetic intermediate, 17 [Eq. (56)].[88] In the case of 18, no cyclopropyl cleavage was observed; the acid was obtained in 81% yield [Eq. (57)].[87]

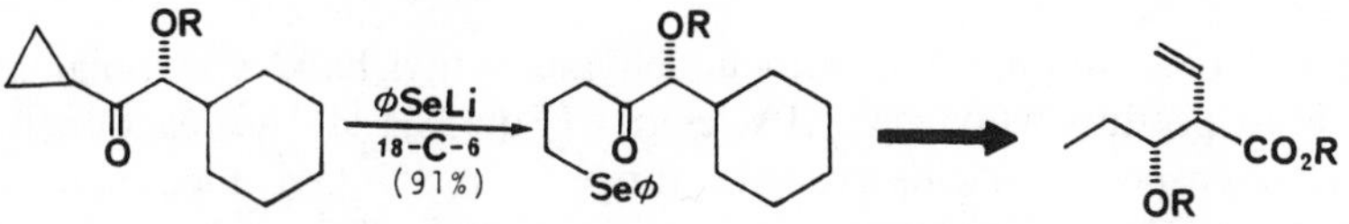

$$\tag{56}$$

R= TBDMS

17

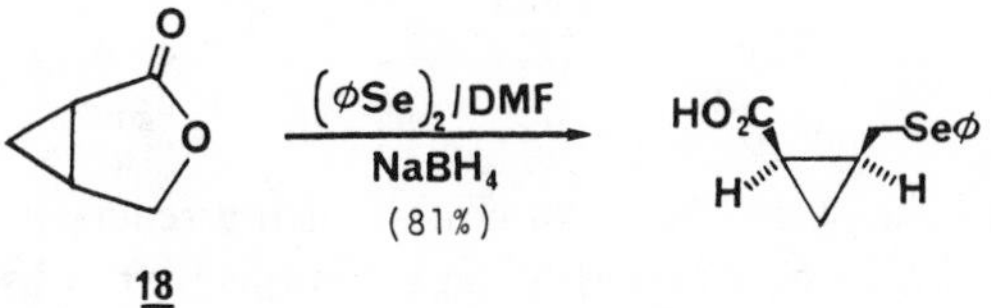

$$\tag{57}$$

18

TABLE 11

Substrate	NaBH$_4$/DMF/120°C (%) Yield	PhSeLi/PhH (%) Yield	Selenide
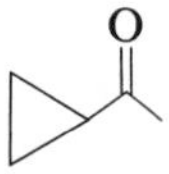 (cyclopropyl methyl ketone)	(13)	(90)	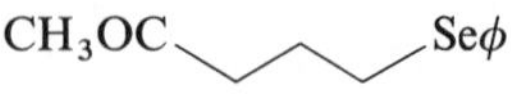 CH$_3$OC⌒⌒Seϕ
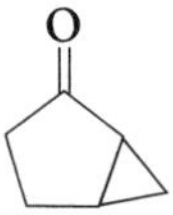 (bicyclo[3.1.0] ketone)	(10)	(47)	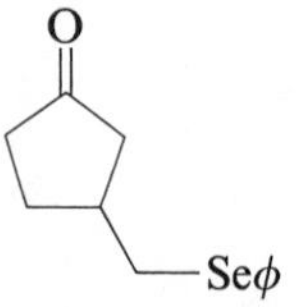 (cyclopentanone with CH$_2$—Seϕ)
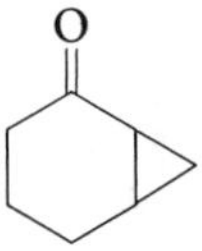 (bicyclo[4.1.0] ketone)	(12)	(64)	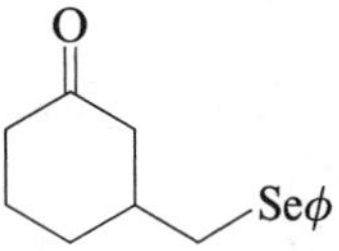 (cyclohexanone with ⌒Seϕ)
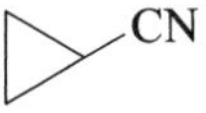 ▷—CN	(47)	(0)	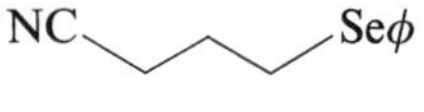 NC⌒⌒Seϕ
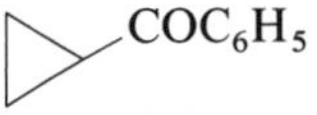 ▷—COC$_6$H$_5$	(6)	(6)	 H$_5$C$_6$O$_2$C⌒⌒Seϕ
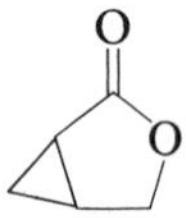 (bicyclic lactone)	(81)	(0)	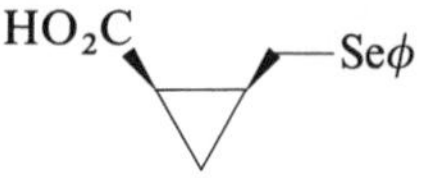 HO$_2$C▷—Seϕ
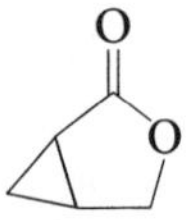 (bicyclic lactone)	(0)	(0)	⋯

3.5. Vinyl and Aryl Halides

Phenyl selenide anion reacts with unactivated vinyl halides in polar aprotic solvents to produce vinyl phenyl selenides [Eq. (58)].[89] The reaction is completely stereospecific. Yields typically range from 65–85%. Ideal solvents are HMPA and DMF. Experimental results are given in Table 12.[89]

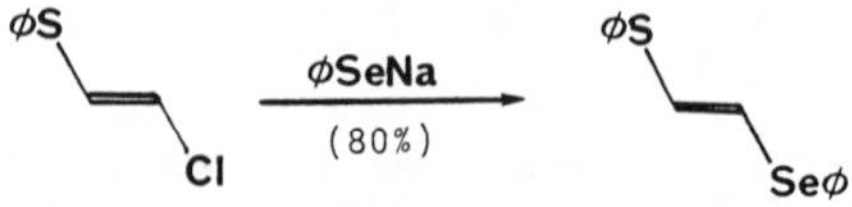

$$\phi S\diagup\diagdown_{Cl} \quad \xrightarrow[\text{(80\%)}]{\phi SeNa} \quad \phi S\diagup\diagdown_{Se\phi} \tag{58}$$

On the basis of the stereochemical course of the reaction, Tiecco et al. has suggested that it can be considered to be a bimolecular substitution in which

TABLE 12

Halide	Se Anion	Vinyl Selenide	Yield (%)
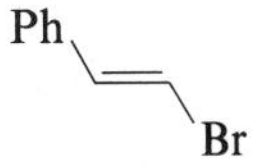 Ph—CH=CH—Br (E)	MeSe	(E)-PhCh=CHSeMe	(82)
	PhSe	(E)-PhCH=CHSePh	(82)
	1-$C_{10}H_7$Se	(E)-PhCH=CHSe$C_{10}H_7$	(75)
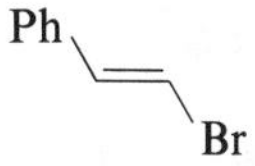 Ph—CH=CH—Br (Z)	MeSe	(Z)-PhSH=CHSeMe	(67)
	PhSe	(Z)-PhCH=CHSePh	(60)

the nucleophile attacks the vinylic carbon bearing the halogen. This is analogous to related reactions of alkanethiolate anions.[90]

An unusual carbocyclization, based on vinyl selenides, is shown in Eq. (59).[91]

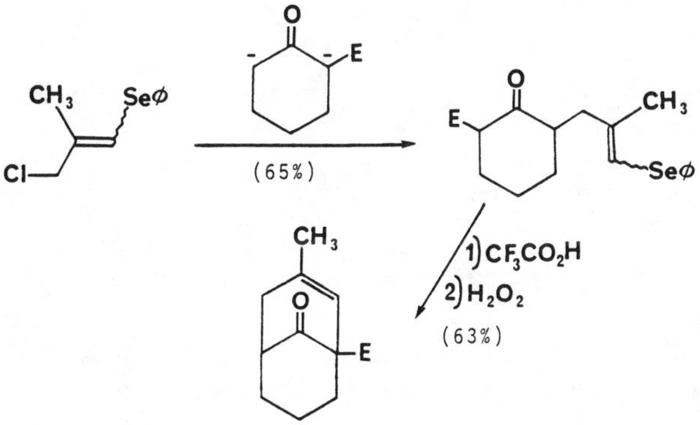

$$(59)$$

Methyl aryl selenides may be prepared by reacting lithium methyl selenide with unactivated aryl halides. Typically long reaction times and high temperatures are required. Yields are generally moderate (45–85%). A general method for the preparation of aryl alkyl selenides was recently reported by Tiecco et al.[89b] The phenyl halide is treated with three equivalents of MeSeLi. Phenyl methyl selenide is immediately formed and rapidly dealkylated by excess MeSeLi. Addition of alkyl halide yields the alkyl phenyl selenide [Eq. (60)].

$$CH_3Li + Se \longrightarrow CH_3SeLi$$

$$\phi X + CH_3SeLi \longrightarrow \phi SeCH_3 + LiX$$

$$\phi SeCH_3 + CH_3SeLi \longrightarrow \phi SeLi + (CH_3)_2Se$$

$$\phi SeLi + CH_3I \longrightarrow \phi SeCH_3 + LiI$$

$$(60)$$

Aryl alkyl selenides may also be prepared from aromatic amines by diazotization followed by decomposition of the intermediate diazonium salt in the presence of aqueous selenocyanate salts.[92] The selenocyanate is treated with a Grignard reagent to obtain the aryl alkyl selenide.[93] This method has been applied in the synthesis of selenated amino acids for use as pancreatic imaging agents [Eq. (61)].[93]

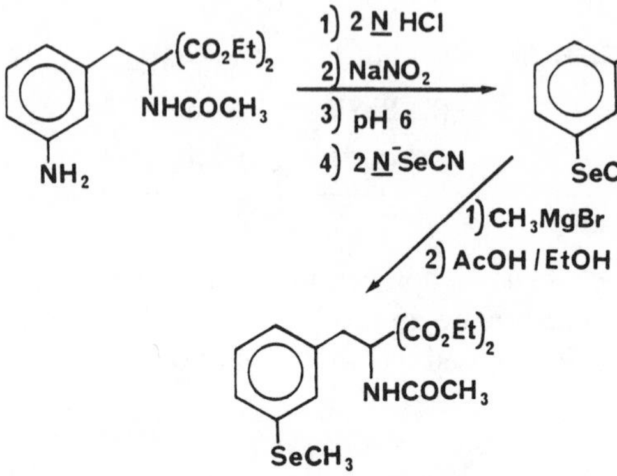

(61)

Standard methods for the synthesis of diaryl selenides involve nucleophilic substitution on an aryl halide by an aryl selenide anion. As stated previously the necessary reaction conditions are vigorous. The use of copper (I) iodide seems to accelerate the reaction.[94] Although the mechanism of this reaction is unclear, Suzuki et al. suggests that *in situ* formed copper (I) aryl selenolate may be the nucleophilic species. Yields of various diaryl selenides are given in Table 13.[94]

Diaryl selenides have been prepared from the haloarene and phenyl selenide anion in liquid ammonia under irradiation, probably by the $S_{RN}1$ mechanism.[95] Under the same experimental conditions there is no dark reaction. The standard photostimulated $S_{RN}1$ reaction mechanism adapted to the present case is shown

TABLE 13

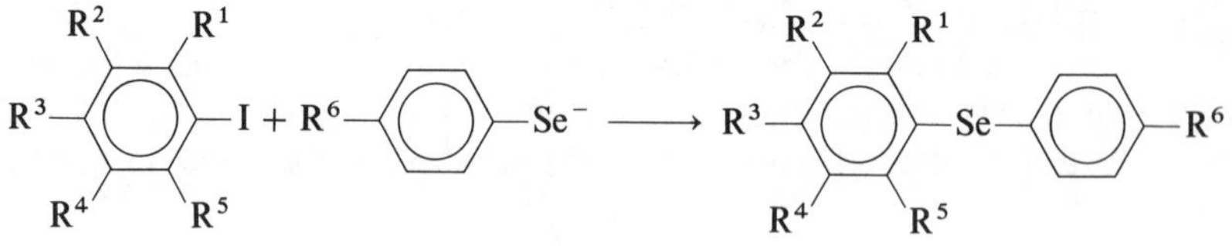

R^1	R^2	R^3	R^4	R^5	R^6	(%) Yield
Me	Me	Me	Me	Me	H	(60)
Me	Me	Me	Me	Me	Cl	(50)
Me	Me	H	Me	Me	H	(56)
Me	Me	Me	Cl	Me	H	(43)
Me	Me	Cl	Me	Me	H	(52)
Me	Me	Me	Me	Br	H	(54)
NO	H	H	H	H	H	(76)

[Eq. (62)]. Photons probably induce electron transfer from the phenyl selenide anion to the haloarene.

$$ArX + PhSe^- \longrightarrow (ArX)^{\overline{\cdot}} + PhSe\cdot$$

$$(ArX)^{\overline{\cdot}} \longrightarrow Ar\cdot + X^-$$

$$Ar\cdot + PhSe^- \longrightarrow (ArSePh)^{\overline{\cdot}}$$

$$(ArSePh)^{\overline{\cdot}} + ArX \longrightarrow ArSePh + PhX$$

(62)

3.6. Miscellaneous

3.6.1. AMINES

Tertiary amines may be converted to selenides by treatment with phenylselenolates in the presence of a ruthenium catalyst [Eq. (63)].[96] Primary and secondary amines do not undergo the transformation since these amines produce a Schiff base complex (instead of an iminium ion complex) upon treatment with metal catalyst. The scope of the reaction can be extended to include primary and secondary amines provided the nitrogen–hydrogen bonds are protected with the trimethylsilyl group [Eq. (64)].

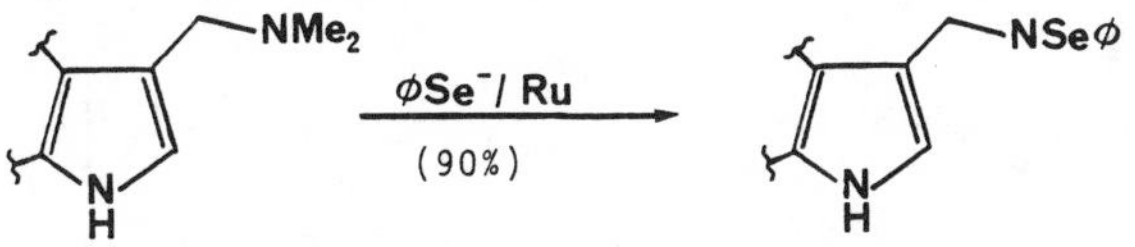

(63)

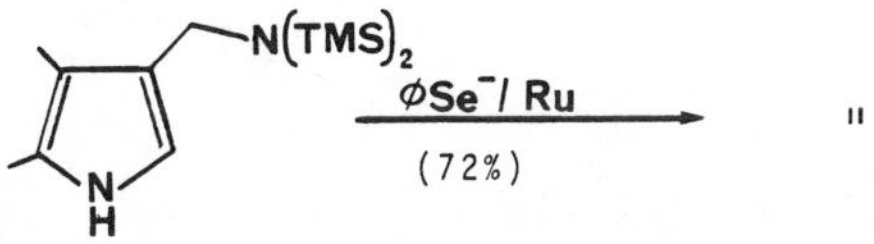

(64)

β-Phenylselenation of ketones (formally equivalent to the Michael addition of benezeneselenolate to an enone) can be accomplished by treatment of the ketals of Mannich bases with benzeneselenolate and ruthenium catalyst [Eq. (65)].[96]

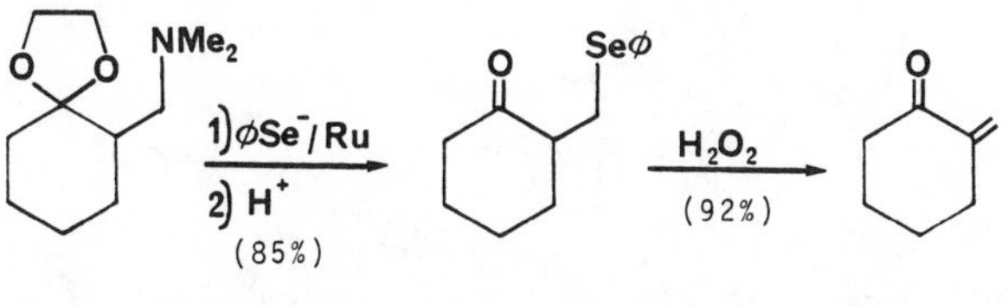

(65)

Cyclic nitrogen compounds undergo cleavage to produce aminoselenides that are often versatile synthetic intermediates. For example, canadine, when reacted with benzeneselenolate/ruthenium catalyst, gave the aminoselenide, 19, in 67% yield [Eq. (66)].[96] No cleavage at the other benzylic positions was observed.

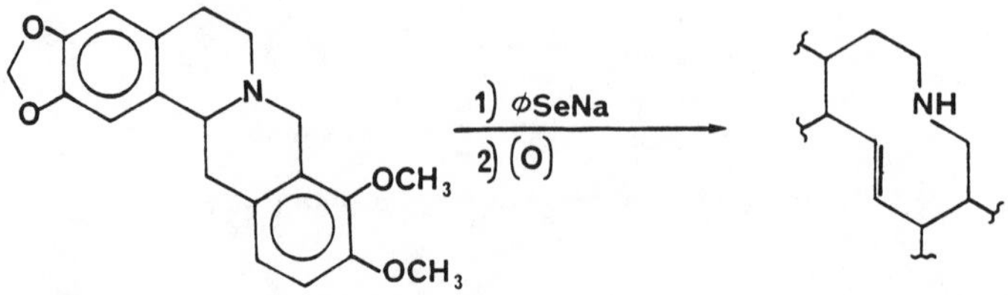

$$(66)$$

Mechanistically, this reaction probably involves the nucleophilic attack of selenolate anion on the iminium ion complex (formed from the insertion of ruthenium into a carbon–hydrogen bond adjacent to the nitrogen). Subsequent hydride transfer from the metal to the carbon adjacent to selenium and concurrent displacement of the dialkylamide, produces the aryl alkyl selenide [Eq. (67)].[96]

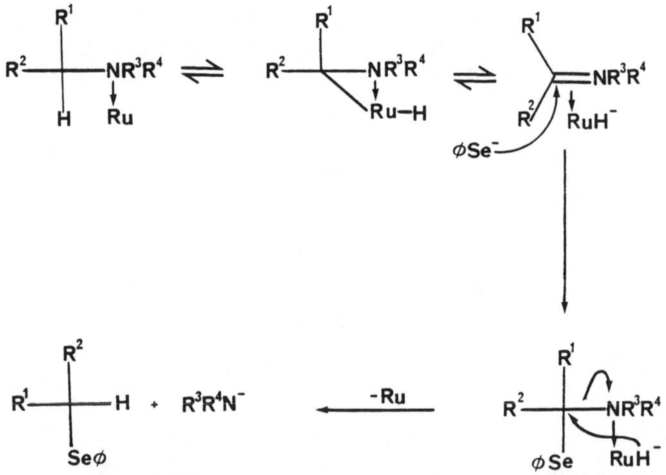

$$(67)$$

Sodium phenyl selenide has been used for the *N*-demethylation of alkaloid methyltrialkylammonium chlorides [Eq. (68)].[97] The reaction is successful for

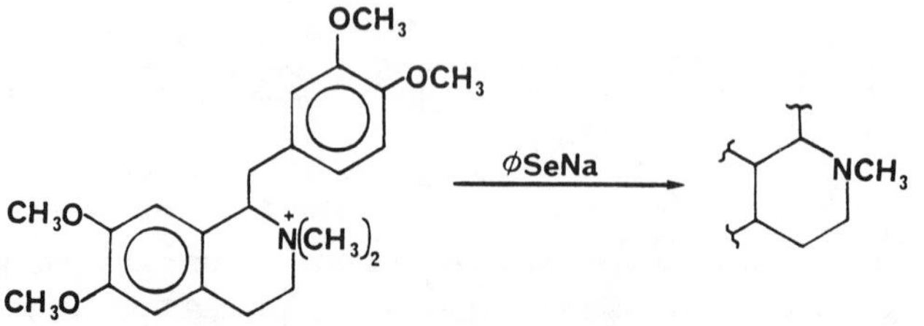

$$(68)$$

substrates in which the heteroatom is contained in one ring. If the heteroatom is part of a bicyclic system or conjugated with an aromatic ring, the reaction fails. This demethylation is quicker and cleaner than the analogous reaction with thiophenoxide.

Selenides can also be obtained from ditosylamides by reaction with phenyl-selenolate anion [Eq. (69)].[98]

(69)

3.6.2. ALCOHOLS

Clarenbeau and Krief recently reported a simple preparation of tertiary alkyl, secondary and tertiary benzyl, and allyl selenides from the corresponding alcohols. This method, which is unsuccessful for primary and secondary alcohols, yields selenides without the complication of significant olefin formation. It consists simply of treating a solution of alcohol and selenol with an acid catalyst (e.g., H_2SO_4 and $ZnCl_2$) [Eq. (70)].[99]

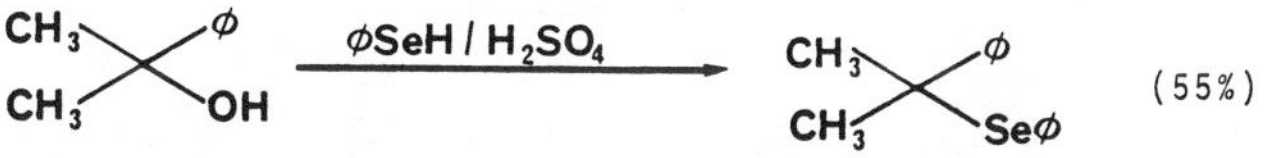

(70)

This compliments the selenide preparation of Grieco et al. [Eq. (71)], which is the preferred method of conversion for primary and unhindered alcohols.[100]

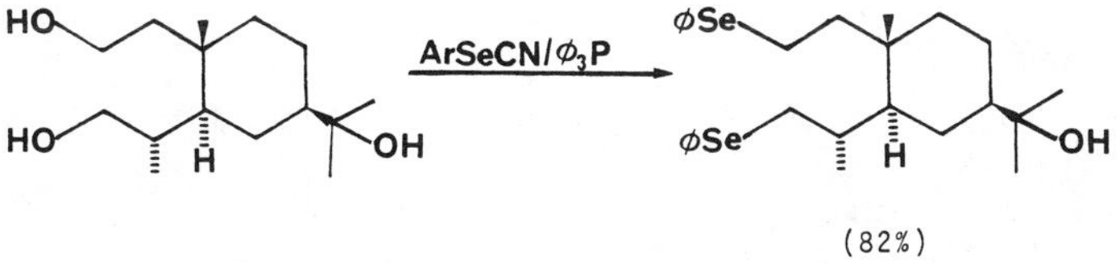

(71)

3.6.3. CARBONYL COMPOUNDS

In certain instances it is possible to directly convert ketones to the corresponding selenides under conditions expected to form the selenoketal [Eq. (72)].[101] The course of the reaction (reduction or ketalization) is substrate dependent. There is some evidence to suggest that the ability of the substrate to stabilize

(72)

TABLE 14

$$R^1 \underset{R^2}{\diagup}=O \xrightarrow{CH_3SeH} R^1 \underset{R^2}{\diagup}-(SeCH_3)_2 + R^1 \underset{R^2}{\diagup}-SeCH_3$$

R^1	R^2	Overall Yield (%)		Acetal	Selenide
C_9H_{19}	Me	91		0	100
$C_{10}H_{21}$	H	98		100	0
i-Pr	i-Pr	47		17	83
Ph	H	49		0	100
p-NO$_2$C$_6$H$_4$	Me	89		85	15
Ph	Me	98		0	100
p-MeOC$_6$H$_4$	Me	78	(26 h)	0	100
p-MeOC$_6$H$_4$	Me	98	(0.3 h)	70	30

electron deficient (cationic or free-radical) centers favors reduction. Table 14 contains some representative data.[101]

REFERENCES

1. H. J. Reich in *Oxidation of Organic Compounds, Part C*, W. Trahanovsky, Ed., Academic, New York, 1978.

2. C. J. M. Stirling, *Acc. Chem. Res.*, **12**, 198 (1979).

3. R. G. Pearson, H. Sobel, and J. Songsted, *J. Am. Chem. Soc.*, **90**, 319 (1968).

4. A. D. Baker, G. H. Armen, Y. Guang-di, D. Liotta, N. Flannagan, C. Barnum, M. Saindane, G. C. Zima, and J. Johnston, *J. Org. Chem.*, **46**, 4127 (198).

5. (a) K. B. Sharpless and R. F. Lauer, *J. Am. Chem. Soc.*, **95**, 2697 (1973); (b) K. B. Sharpless, M. W. Young, and R. F. Lauer, *Tetrahedron Lett.*, 1979 (1973); (c) P. A. Grieco and M. Miyashita, *Tetrahedron Lett.*, 1869 (1974); (d) R. M. Scarborough and A. B. Smith, III, *Tetrahedron Lett.*, 4361 (1977).

6. (a) D. Liotta, W. Markiewicz, and H. Santiesteban, *Tetrahedron Lett.*, 4365 (1977); (b) D. Liotta and H. Santiesteban, *Tetrahedron Lett.*, 4369 (1977).

7. H. J. Reich, F. Chow, and S. K. Shah, *J. Am. Chem. Soc.*, **101**, 6638 (1979).

8. W. G. Salmond, M. A. Barta, A. M. Cain, and M. C. Sabala, *Tetrahedron Lett.*, 1683 (1977).

9. (a) K. B. Sharpless and M. W. Young, *J. Org. Chem.*, **40**, 947 (1975); (b) P. A. Grieco, J. A. Noguez, and Y. Masaki, *J. Org. Chem.*, **42**, 495 (1977).

10. P. A. Grieco, J. A. Noguez, and Y. Masaki, *Tetrahedron Lett.*, 4213 (1975).

11. P. A. Grieco, Y. Masaki, and D. Boyler, *J. Am. Chem. Soc.*, **97**, 1597 (1975).

12. (a) W. C. Still, *J. Amer. Chem. Soc.*, **101**, 2493 (1979); (b) G. Stork and D. J. Morgans, Jr., *J. Amer. Chem. Soc.*, **101**, 7110 (1979); (c) P. A. Grieco, E. Williams, H. Tanaka, and S. Gilmer, *J. Org. Chem.*, **45**, 3537 (1980); (d) D. J. Morgans, Jr., *Tetrahedron Lett.*, 3721 (1981); (e) R. L. Snowdon, *Tetrahedron Lett.*, 101 (1981); (f) S. Takano, K. Morimoto, Masuda, and K. Ogasawara, *Chem. Pharm. Bull.*, **30**, 4238 (1982).

13. S. Torii, T. Inokuchi, and N. Hasegawa, *Chem Lett.*, 639 (1980).

14. D. Liotta, *Acc. Chem. Res.*, **17**, 28 (1984).

15. P. Dowd and P. Kennedy, *Synth. Commun.*, **11**, 935 (1981).

16. D. Liotta, U. Sunay, H. Santiesteban, and W. Markiewicz, *J. Org. Chem.*, **13**, 2605 (1981).

17. (a) M. R. Detty, *Tetrahedron Lett.*, 5087 (1978); (b) M. R. Detty, *Tetrahedron Lett.*, 4189 (1979).

18. (a) D. Liotta, P. B. Paty, J. Johnston, and G. Zima, *Tetrahedron Lett.*, 5091 (1978); (b) N. Miyoshi, H. Ishii, K. Kondo, S.. Murai, and N. Sonada, *Synthesis*, 300, (1979); (c) M. R. Detty and G. P. Wood, *J. Org. Chem.*, **45**, 80 (1980); (d) M. R. Detty and M. D. Seidler, *J. Org. Chem.*, **46**, 1283 (1981).

19. A. B. Smith, III, and R. M. Scarborough, Jr., *Tetrahedron Lett.*, 1649 (1978).

20. D. Liotta, M. Saindane, C. Barnum, H. Ensley, and P. Balakrishnan, *Tetrahedron Lett.*, 3043 (1981).

21. D. Liotta, *Acc. Chem. Res.*, **17**, 28 (1984).

22. K. B. Sharpless and R. F. Lauer, *J. Am. Chem. Soc.*, **95**, 2697 (1973).

23. K. B. Sharpless, K. M. Gordon, R. F. Lauer, D. W. Patrick, S. P. Singer, and M. W. Young, *Chem. Scr.*, **8A**, 9 (1975).

24. D. N. Jones, D. Mundy, and R. D. Whitehouse, *J. Chem. Soc. Chem. Commun.*, 36 (1970).

25. R. Walter and J. Roy, *J. Org. Chem.*, **36**, 2561 (1971).

26. J. Remion, W. Dumont, and A. Krief, *Tetrahedron Lett.*, 1385 (1976).

27. L. A. Khazemova and V. M. Al'bitskaya, *Zhur. Org. Khim.*, **6**, 935 (1970).

28. G. H. Posner and D. Z. Rogers, *J. Am. Chem. Soc.*, **99**, 8208 (1979).

29. A. Cravador and A. Krief, *Tetrahedron Lett.*, 2491 (1981).

30. (a) M. J. Towsend and K. B. Sharpless, *Tetrahedron Lett.*, 3313 (1972); (b) D. L. J. Clive and S. M. Menchen, *J. Org. Chem.*, **45**, 2347 (1980).

31. (a) K. B. Sharpless and R. F. Lauer, *J. Am. Chem. Soc.*, **95**, 2697 (1973); (b) N. Miyoshi, K. Kondo, S. Murai, and N. Sonada, *Chem. Lett.*, 909 (1979).

32. W. C. Still, S. Murata, G. Revial, and K. Yoshihara, *J. Am. Chem. Soc.*, **105**, 1153 (1983).

33. (a) D. A. McDowan and G. A. Berchtold, *J. Am. Chem. Soc.*, **104**, 1153 (1982); (b) J. H. Hoove, P. P. Policastro, and G. A. Berchtold, *J. Amer. Chem. Soc.*, **105**, 6264 (1983).

34. Y. Tsudu, T. Sano, J. Taka, K. Isobe, J. Toda, H. Irie, H. Tanaka, S. Takagi, M. Yamaki, and M. Murata, *J. Chem. Soc. Chem. Commun.*, **933** (1975).

35. S. Ohta and S. Kimoto, *Tetrahedron Lett.*, 2279 (1975).

36. R. H. Schlessinger and A. Lopes, *J. Org. Chem.*, **46**, 5252 (1981).

37. S. Toda, S. Nakagawa and T. Naito, *J. Antibiot.*, **30**, 1002 (1977).

38. B. M. Trost, Y. Nishimura, and K. Yamamoto, *J. Am. Chem. Soc.*, **101**, 1328 (1979).

39. R. V. Stevens and K. F. Albizati, *J. Org. Chem.*, **50**, 632 (1985).

40. K. B. Sharpless and R. F. Lauer, *J. Am. Chem. Soc.*, **95**, 2697 (1973).

41. H. J. Reich and F. Chow, *J. Chem. Soc. Chem. Commun.*, **1135** (1975).

42. (a) J. Remion, W. Dumont, and A. Krief, *Tetrahedron Lett.*, 3743 (1976); (b) J. Remion and A. Krief, *Tetrahedron Lett.*, 3743 (1976).

43. D. L. J. Clive and V. N. Kale, *J. Org. Chem.*, **46**, 231 (1981).

44. S. Halazy and A. Krief, *J. Chem. Soc. Chem. Commun.*, 1136 (1979).

45. K. C. Nicolaou, W. J. Sipio, R. L. Mogolda, and D. A. Clarenon, *J. Chem. Soc., Chem. Commun.*, 83 (1979).

46. (a) J. M. Behan, R. A. W. Johnstone, and M. J. Wright, *J. Chem. Soc. Perkin* 1, 1216 (1975); (b) A. M. Leonard-Coppens and A. Krief, *Tetrahedron Lett.*, 1385 (1976).

47. D. L. J. Clive and C. V. Denyer, *J. Chem. Soc. Chem. Commun.*, **253** (1973).

48. T. H. Chan and J. R. Tinkelbine, *Tetrahedron Lett.*, 2091 (1974).

49. L. F. Matley and G. Muller, *Compt. Rend.*, **281**, 881 (1975).

50. M. V. Calo, L. Lopez, A. Mincuzzi, and G. Pesce, *Synthesis*, 200 (1976).

51. D. Van Ende and A. Krief, *Tetrahedron Lett.*, 2709 (1975).

52. (a) M. Sevrin, W. Dumount, L. Henesi, and A. Krief, *Tetrahedron Lett.*, 2647 (1976); (b) G. Holzle and W. Jenny, *Helv. Chim. Acta*, **41**, 712 (1958).

53. (a) D. Van Ende, W. Dumont, and A. Krief, *Angew. Chem. Int. Ed. Eng*, **14**, 700 (1975); (b) D. Van Ende and A. Krief, *Tetrahedron Lett.*, 457 (1976); (c) W. Dumont and A. Krief, *Agnew. Chem. Int. Ed.*, **14**, 350 (1975).

54. (a) J. McMurray, *Org. React.*, **24**, 187, (1977); (b) P. A. Bartlett and W. S. Johnson, *Tetrahedron Lett.*, 4459 (1970); (c) T. R. Kelley, H. M. Dali, and W. Tsang, *Tetrahedron Lett.*, 3589 (1977).

55. S. Takano, N. Tamuro, K. Ogasawora, Y. Nakagawa, and T. Soka, *Chem. Lett.*, **933** (1982).

56. M. Anita, K. Adachi, Y. Ito, H. Sawii, and M. Ohno, *J. Am. Chem. Soc.*, **105**, 4049 (1983).

57. T. Kametani, Y. Takenchi, and E. Yoshi, *J. Org. Chem.*, **48**, 2360 (1983).

58. (a) D. Liotta, U. Sunay, H. Santiesteban, and W. Markiewicz, *J. Org. Chem.*, **46**, 2605 (1981); (b) D. Liotta, W. Markiewicz, and H. Santiesteban, *Tetrahedron Lett.*, 4365 (1977). (c) Z. Ahmed and M. P. Cava, *J. Am. Chem. Soc.*, **105**, 682 (1983).

59. (a) A. B. Smith and R. M. Scarborough, *Tetrahedron Lett.*, 4361 (1977); (b) D. Liotta and H. Santiesteban, *Tetrahedron Lett.*, 4369 (1977).

60. J. Roy, W. Gordon, I. L. Schwartz, and R. Walker, *J. Org. Chem.*, **35**, 511 (1982).

61. (a) D. N. Jones, D. Mundy, and R. D. Whitehouse, *J. Chem. Soc. Chem. Commun.*, **86**, (1970); (b) K. B. Sharpless, W. M. Young, and R. L. Lauer, *Tetrahedron Lett.*, 1979 (1973).

62. K. B. Sharpless and R. F. Lauer, *J. Am. Chem. Soc.*, **94**, 7154 (1972).

63. D. W. G. Salmond, M. A. Cain, and M. C. Sobala, *Tetrahedron Lett.*, 1683 (1977).

64. D. Liotta, G. Zima, B. Barnum, and M. Saindane, *Tetrahedron Lett.*, 3643 (1980).

65. D. Liotta, C. Barnum, R. Puleo, G. Zima, C. Bayer, and H. Kazer, III, *J. Org. Chem.*, **46**, 2920 (1981).

66. D. Liotta, M. Saindore, C. Barnum, H. Ensley, and P. Balakrishva, *Tetrahedron Lett.*, 3043 (1981).

67. A. M. Mikolajczyk, S. Grzejszezak, and K. Kovbacz, *Tetrahedron Lett.*, 3097 (1981).

68. (a) M. Lsobe, H. Lio, T. Kavai and T. Goto, *J. Am. Chem. Soc.*, **100**, 1940 (1978); (b) M. Isobe, H. Iio, T. Kawai, and T. Goto, *Tetrahedron Lett.*, 703 (1977).

69. P. A. Grieco, J. A. Nogeuer, and Y. Mosaki, *Tetrahedron Lett.*, 4213 (1975).

70. P. H. Bentley and E. Hunt, *J. Chem. Soc. Chem. Commun.*, 518 (1978).

71. S. Nakatsuka, K. Yamada, K. Yoshida, O. Asono, Y. Mirakoni, and T. Goto, *Tetrahedron Lett.*, 5627 (1983).

72. F. Matsuda, N. Tomiyoshi, M. Yanagiwa, and T. Matsumoto, *Tetrahedron Lett.*, 1277 (1983).

73. A. P. Kozikowski and Y. Y. Chen, *J. Org. Chem.*, **46**, 5248 (1981).

74. P. A. Grieco, Y. Masai, and D. Boyler, *J. Am. Chem. Soc.*, **97**, 1597 (1975).

75. T. Wakamatsu, K. Akassaka, and Y. Ban, *J. Org. Chem.*, **44**, 2008 (1979).

76. L. Kuwajima, S. Hoshita, J. Tanaka, and M. Shimizer, *Tetrahedron Lett.*, 3209 (1980).

77. N. Mijoshi, H. Lshii, S. Murai, and N. Sonoda, *Chem. Lett.*, **87**, 3 (1979).

78. (a) A. A. Anciaux, A. Eman, N. Dumont, and A. Krief *Tetrahedron Lett.*, 1617 (1975); (b) *Ibid.*, 1613 (1975).

79. (a) H. Gilman and L. F. Cason, *J. Am. Chem. Soc.*, **73**, 1074 (1951); (b) G. Zydansky, *Arkiv. Kemi.*, **21**, 211 (1963).

80. K. Moni, *Tetrahedron*, **32**, 1101 (1976).

81. P. A. Grieco and M. Mijashita, *Tetrahedron Lett.*, 1869 (1974).

82. S. Torii, T. Inokuchi, and N. Hasegawa, *Chem. Lett.*, 639 (1980).

83. (a) R. D. Miller and D. R. McKeon, *Tetrahedron Lett.*, 2305 (1979); (b) M. R. Detty, *Tetrahedron Lett.*, 4189 (1979).

84. D. Liotta, P. B. Patty, J. Johnston, and G. Zima, *Tetrahedron Lett.*, 5091 (1978).

85. B. M. Suzuki, T. Kawagishi, and R. Nayoni, *Tetrahedron Lett.*, 1809 (1981).

86. D. H. Wadsworth and M. R. Detty, *J. Org. Chem.*, **45**, 4611 (1980).

87. A. B. Smith, III, and R. M. Scarborough, *Tetrahedron Lett.*, 1649 (1978).

88. S. Masamune, T. Kaiho, and D. S. Garvey, *J. Am. Chem. Soc.*, **104**, 5521 (1982).

89. (a) M. Tiecco, L. Testaferri, M. Lingoli D. Chianelli, and M. Montanucci, *Tetrahedron Lett.*, 4975 (1984); (b) M. Tiecco, M. Testaferai, D. Tingali, D. Chianelli, and M. Montanucci, *J. Org. Chem.*, **48**, 4289 (1983).

90. M. Tiecco, L. Testaferri, M. Lingoli; D. Chianelli, and M. Montanucci, *J. Org. Chem.*, **48**, 4795 (1983).

91. D. Gravel, R. Dezal, and L. Bordeleau, *Tetrahedron Lett.*, 699 (1983).

92. (a) C. A. Loeschon, C. J. Kelley, R. N. Hanson, and M. A. Davis, *Tetrahedron Lett.*, 3387 (1984); (b) J. Gosselek and E. Walters, *Chem. Ber.*, **95**, 1237 (1960); (c) H. Bauer, *Chem. Ber.*, **46**, 92 (1913).

93. B. Greenberg, E. S. Gould, and W. Dulod, *J. Am. Chem. Soc.*, **78**, 4028 (1956).

94. H. Suzuki, H. Abe, and A. Osuka, *Chem. Lett.*, 151 (1981).

95. A. B. Pierini and R. A. Rossi, *J. Organomet. Chem.*, 144 (1978).

96. S. I. Murahaswi and T. Yano, *J. Am. Chem. Soc.*, **102**, 2456 (1980).

97. V. Simanek and A. Klasek, *Tetrahedron Lett.*, 3039 (1969).

98. P. Muller and M. P. N. Thi, *Helv. Chim. Acta*, **63**, 2168 (1980).

99. M. Clarenbeau and A. Krief, *Tetrahedron Lett.*, 3625 (1984).

100. P. A. Grieco, T. Takigawa, and W. J. Schillinger *J. Org. Chem.*, **45**, 2247 (1980).

101. C. A. Cravador, A. Krief, and L. Hevesi, *J. Chem. Soc. Chem. Commun.*, 451 (1980).

5

Selenium-Stabilized Carbanions

HANS J. REICH

Department of Chemistry, University of Wisconsin,
Madison, Wisconsin

CONTENTS

1. INTRODUCTION

As early as 1939 Gilman[1] followed up earlier successes in the preparation of several sulfur substituted lithium reagents by attempting to prepare analogous selenium substituted ones, but found that only cleavage products resulting from

243

lithium/selenium exchange could be detected [Eq. (1)].[2] The earliest indications that α-carbanionic selenium compounds could be generated were provided indirectly by base catalyzed isotopic exchange reactions $(CH_3)_3Se^+I^-$,[3] $C_6H_5SeCH_3$,[4a] selenophene[4b]), allylic isomerization of double bonds (allyl to vinyl selenides[5]), and aldol condensations of α-selenocarbonyl compounds.[6] It was not, however, until the publication of a key paper by Seebach and Peleties[7] in 1969 that it became clear that stable solutions of α-lithio selenides could be prepared, and that these anions were quite comparable in reactivity, basicity, and stability to their sulfur analogs. In fact, these authors introduced two now widely used techniques [Eq. (2)][7,8] for circumventing the inability to carry out classical alkyllithium metalation reactions on selenides: i.e., use of lithium dialkylamides as bases for metalation, and the cleavage of selenoacetals and ketals with alkyllithium reagents (*vide infra*). Subsequent work in several laboratories generalized these and introduced other methods for the preparation of selenium substituted carbanions, and identified numerous applications for the preparation of nonselenium-containing compounds.

$$C_6H_5-Se-CH_3 \xrightarrow[\text{(2) } CO_2]{\text{(1) } n\text{-BuLi/Et}_2O} C_6H_5-CO_2H + C_6H_5-Se-n\text{-}C_4H_9 \qquad (1)$$
$$28\% \qquad\qquad 26\%$$

$$\underset{C_6H_5Se}{\overset{\overset{\text{Li}}{|}}{\diagup}}\!\!\diagdown_{SeC_6H_5} \xleftarrow{\text{LiNR}_2} C_6H_5Se\diagup\!\!\diagdown SeC_6H_5 \xrightarrow{n\text{-BuLi}} C_6H_5Se\diagup\!\!\diagdown Li$$
$$(2)\ (\text{Refs. 7, 8})$$

2. COMPARISON OF SULFUR AND SELENIUM ACIDIFYING EFFECTS

From some of the work mentioned previously, as well as more recent studies, a number of direct comparisons between the kinetic and thermodynamic acidity of analogous sulfur and selenium compounds can be made (Table 1).[9,10a,10b,11] The only true pK_a measurements, those of Bordwell et al. on phenacyl phenyl sulfide and selenide,[9a] show that 1.5 pK_a units separate the two compound types. Other less direct measurements such as isotopic exchange rates, metalation rates, or equilibration of lithium reagents show similar results: In most cases the sulfur analog is slightly more acidic. The principal exceptions are sp^2 hybridized anions such as those prepared from vinyl selenides/sulfides and from selenophene/thiophene where the selenium analog is more acidic.

Needless to say, the S/Se rate and equilibrium differences are almost insignificant compared to the massive total stabilization of negative charge (10–15 pK_a units)[9a] provided by a sulfide or selenide substituent. Nevertheless, the experimental results seem to be at variance with a theoretical study that predicts a substantially higher stability (~4 kcal/mole) of selenium over sulfur substituted carbanions.[12]

TABLE 1

COMPARISON OF KINETIC AND THERMODYNAMIC ACIDITIES OF SELENIDES AND SULFIDES

Compound	Base/Solvent/T°C	Measured	Sulfur/ Selenium	Reference
$(CH_3)_3Y^+I^-$ (with Y in ring)	$NaOD/D_2O/62$	k_{isotop}	45	3
C_6H_5—Y—CD_3	KOt-$Bu/(CH_3)_2SO/25$	k_{isotop}	0.67	4(b)
	$KNH_2/NH_3/-33$	k_{istop}	10	4(a)
C_6H_5—Y(O_2)—CD_3	$R_2NH/CD_3OD/25$	k_{isotop}	~ 100	10(c)
m-$CF_3C_6H_4$—Y—CH_3	$LiTMP/THF/-78$	k_{deprot}	3.8	10(a)
C_6H_5—Y—$CH_2CH{=}C$	$LDA/THF/-78$	k_{deprot}	7.5	10(b)
C_6H_5—Y—$CH_2C{\equiv}CH$	$NaOEt/HOEt/25$	$k_{isom}[a]$	$12.6[b]$	11
C_6H_5—Y—$CH{=}C{=}CH_2$	$NaOEt/HOEt/25$	$k_{isom}[c]$ [2]	$7.0[b]$	11
$C_6H_5C(O)CH_2$—Y—C_6H_5	$NaCH_2S(O)CH_3/(CH_3)_2SO$	K_{eq}	32	9(a)
C_6H_5—Y—$CH{=}CH_2$	$LDA/THF/-78$	K_{eq}	0.21	10(b)
	$LiTMP/THF/-78$	k_{deprot}	0.37	10(b)
m-$CF_3C_6H_4$—Y—$CH{=}CH_2$	$LDA/THF/-78$	K_{eq}	0.37	10(b)
	$LiTMP/THF/-78$	k_{deprot}	0.42	10(b)

[a] Isomerization of propargyl to allenyl.
[b] These numbers were calculated from the $\Delta G^{\ddagger}$ values given.
[c] Isomerization of allenyl to 1-propynyl.
[d] pK_a (DMSO) = 17.1 (S); 18.6 (Se).

Spectroscopic studies of several α-lithio selenides and the analogous sulfides have also revealed details of bonding and ion pairing in these substances.[13]

3. PREPARATION OF SELENIUM SUBSTITUTED CARBANIONS

3.1. Metalation of Selenides

As already mentioned in the introduction, attempts to metalate most aryl and alkyl selenides with alkyllithium reagents result in a competitive or exclusive Li/Se exchange reaction rather than the desired Li/H exchange.[7,8,10a,14–18] Thus a technique that is widely applicable to the preparation of α-lithio sulfides can be extended to selenides in only a few cases (such as the metalation of selenophenes[19a] and benzoselenophenes,[19b] of phenyl trimethylsilyl selenide,[20] and of diethyl phenylselenomethylphosphonate[21]).

Lithium and other alkali metal dialkylamides behave as substantially less "selenophilic" bases than do alkyllithiums,[7] and thus often serve well for the deprotonation of selenium compounds. They do, however, have a lower pK_a cutoff so the range of selenides that can be metalated successfully is considerably smaller than that of sulfides.

Table 2[22–49] summarizes some of the selenides that have been successfully deprotonated together with the conditions used. Almost any C_6H_5—Se—

TABLE 2
EXAMPLES OF LITHIUM REAGENTS PREPARED BY DEPROTONATION OF SELENIDES[a]

Lithium Reagent	Base,[b] °C	Electrophiles, References
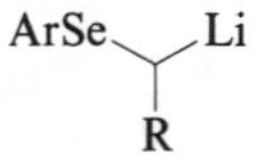 $\text{ArSe}-\overset{\displaystyle R}{\underset{\displaystyle }{\text{C}}}\text{H}-\text{Li}$ R = H[c]	n-BuLi, TMEDA, 0	K[7]
R = H[d]	LiTMP, −55	Si;[10a] RX[10a]
R = OCH₃[b]	LiTMP, −78	K, A[10a]
R = SC₆H₅[c]	LDA, −78	RX[22]
R = SeC₆H₅[c]	LDA, −78	A[7,10a,10b,23] K[7,10b] RX[7,24] Si;[35] Ep[26]
R = TeC₆H₅[d]	LDA, −78	H₂O[27]
R = Si(CH₃)₃	sec-BuLi, TMEDA, hexane	Si[28]
$\overset{\displaystyle O}{\|}$	LDA, TMEDA, −78	RX[29] Ep[30]
R = P(OC₂H₅)₂[c]	n-BuLi, −78	RX, A, K[21]
	NaH, HMPA	RX, A, K[21]
R = C₆H₅[c]	LDA, −78	A[10a] RX[10a,15] Ep[10a] Si[28]
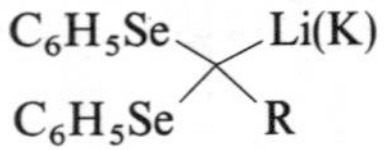 $\text{C}_6\text{H}_5\text{Se}-\overset{\displaystyle \text{Li(K)}}{\underset{\displaystyle R}{\text{C}}}-\text{Si(CH}_3)_3$ R = Si(CH₃)₃	LDA, −78–0	A[21,25] K[21]
R = CH₃	KDA, −78	RX, A, K, En, Ep[24]
	LiTMP, HMPA, −30	RX[18c]
$\text{C}_6\text{H}_5\text{Se}-\overset{\displaystyle \text{Li}}{\underset{\displaystyle \text{C}_6\text{H}_5}{\text{C}}}-\text{Si(CH}_3)_3$	LiN(C₂H₅)₂, 0	RX, Si[28]
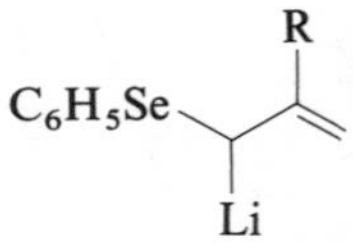 $\text{C}_6\text{H}_5\text{Se}-\overset{\displaystyle R}{\underset{\displaystyle \text{Li}}{\text{CH}}}-\text{C}(=\text{CH}_2)$ R = H	LDA, −78	En[31] Si;[14] Sn[32] RX[14] Ep[14] A[33a] BR₃[33b]
R = CH₃	LDA, −78	RX[10b] Sn[32]

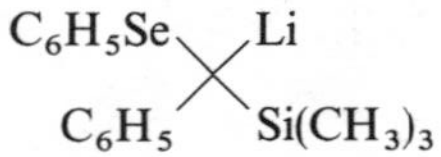

Substrate	R	Conditions	Reagents/Products
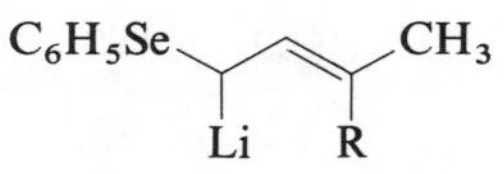 C_6H_5Se–C(Li)–CH=C(CH$_3$)–R (allylic)	R = Cl	LDA, −78	RX, Ep, Si[14]
	R = CH$_3$	LDA, 6, 20 min	Si[14]
	R = C$_6$H$_{11}$O	LDA, −55	RX[51]
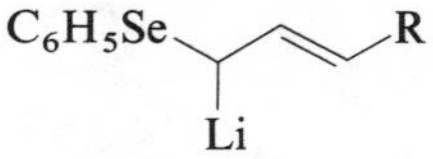 C_6H_5Se–C(Li)–CH=CH–R	R = CH$_3$	LDA, −78	RX[14]
	R = C$_6$H$_5$	LDA, −78	K[14]Si[34]
	R = SeC$_6$H$_5$	LDA, −78	A, K, Si, RX, Ep[35]
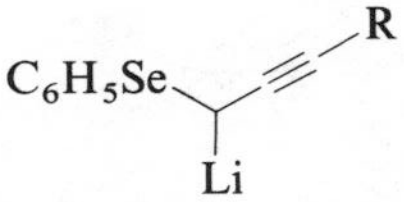 C_6H_5Se–C(Li)–C≡C–R	R = Li	LDA, −78	RX[36]
	R = CH$_3$	LDA, −78	RX[36]
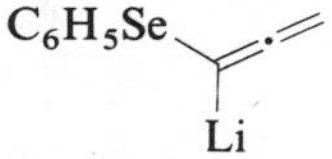 C_6H_5Se–C(Li)–CH=C=CH$_2$ (allene)		LDA, −78	RX[36]
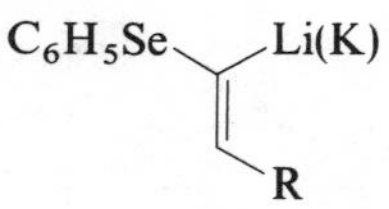 C_6H_5Se–C(Li(K))=CH–R	R = H	LDA, −78	S[10b]Si[10b]CO$_2$[10b]RX[10b,37]K[10b,37]A[38]
		KDA, −78	RX, A, En, Ep[24]
		LDA, HMPA, −78	RX[38]
	R = CH(CH$_3$)$_2$	KDA, −78	RX, A[24]
C_6H_5Se–CH(Li(K))–C(=O)–R	R = C$_6$H$_{13}$	t-C$_4$H$_9$OK, t-C$_4$H$_9$OH	RX[39]
		LDA, 25	RX[39]
	R = C$_6$H$_5$	LDA, −78–0	RX[23]

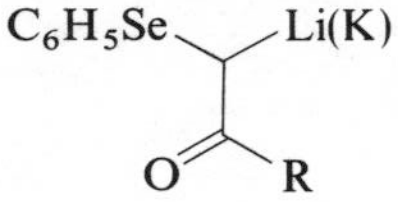

247

TABLE 2 (Continued)

Lithium Reagent	Base,[b] °C	Electrophiles, References
(structure: decalone with OCH$_3$, SeC$_6$H$_5$, Li, H, O, dioxolane)	LDA, HMPA, 0	RX[40]
(cyclopentanone with Li, SeC$_6$H$_5$)	LDA, HMPA	RX[41]
C$_6$H$_5$–CHLi–CO–OR R = CH$_3$ R = C$_6$H$_5$ R = Li	LDA, −78 LiICA, −78; DMSO LDA, −78	A[42], En[43] RX[44] A[10a]
C$_6$H$_5$Se–C(CH$_3$)(CHLi)–CO$_2$Li	2 LDA, 0–40	Ep, RX[45]

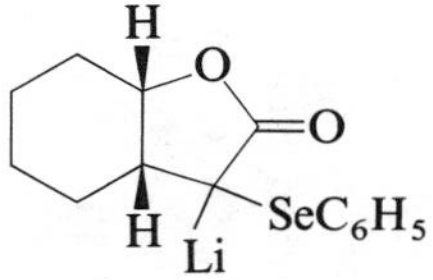

H, SeC$_6$H$_5$, H Li (bicyclic lactone structure) LDA, -78 RX[46]

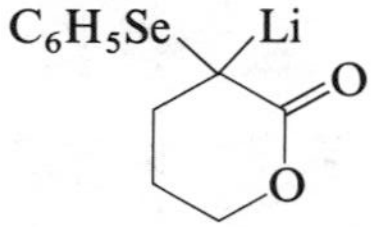

C$_6$H$_5$Se, Li (lactone structure) LDA A[42]

C$_6$H$_5$Se—Li(M), R—CN

R = H	NaOH, Bu$_4$NI	RX[47]
R = CH$_2$C$_6$H$_5$	NaOH, Bu$_4$NI	RX[47]
R = n-C$_6$H$_{13}$	LDA, -78	RX[48] En[48]

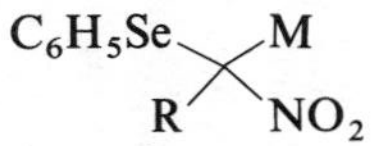

C$_6$H$_5$Se, M, R—NO$_2$ Ca(OH)$_2$ A[49]

$$RSe{\diagup}{\diagdown}H \xrightarrow{\text{base}} RSe{\diagup}{\diagdown}Li \xrightarrow{E} RSe{\diagup}{\diagdown}E$$

[a] Electrophiles: Aldehydes (A), Ketones (K), Acylating agents (Ac), Enone (En), Carbon dioxide (CO$_2$), Alkyl halide (RX), Epoxide (Ep), R$_3$SiCl (Si), R$_3$SnCl (Sn), Trialkylborane (BR$_3$).

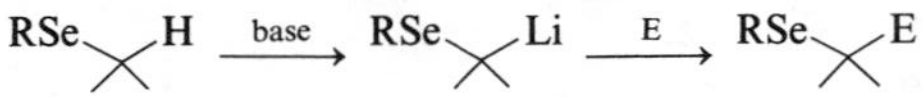

[b] Solvent is THF unless otherwise indicated.

[c] Ar = C$_6$H$_5$

[d] Ar = m-CF$_3$C$_6$H$_4$

CH_2—G function bearing a carbanion stabilizing group G can be deprotonated using lithium diisopropylamide (LDA) in tetrahydrofuran (THF) or related base system. This includes G equal to RS, RSe, RO, SiR_3, aryl, vinyl, alkynyl, carbonyl, carboxylate, cyano, and nitro. On the other hand, if a tertiary lithium (reagent is to be prepared (e.g., C_6H_5Se—$CLi(CH_3)G$), the choice of G groups sufficient to promote smooth metalation is more limited and side reactions become more important. For example, whereas selenides 1a, 2a, and 3a are deprotonated by LDA/THF, 1b, 2b, and 3b are not.[10a] Compound 3b can, however, be deprotonated with potassium diisopropylamide (KDA)/THF,[24] LDA/HMPA/THF,[18c] or with lithium 2,2,6,6-tetramethylpiperidide (LiTMP)/HMPA/THF[18c]. Both 4a and 4b are deprotonated by LDA/THF, but the amount of C—Se bond cleavage that occurs is greater for 4b (22%) than for 4a (5–8%).[23] Such results are not totally unexpected, since the substitution of alkyl for hydrogen invariably causes decreases in the rate of formation of carbanions and, in those cases where the carbanion is not extensively delocalized, it also substantially raises the pK_a of the carbon acid (e.g., 1,3-dithiane: $pK_a = 31.1$; 2-methyl-1,3-dithiane: $pK_a = 38.3^{50}$; phenyl methyl sulfone: $pK_a = 29.0$; phenyl ethyl sulfone: $pK_a = 31.0^{9b}$).

m-$CF_3C_6H_4Se$⁀⁀⁀R C_6H_5Se⁀⁀⁀

       R

            1   2

(a) R = H
(b) R = CH_3

C_6H_5Se⁀⁀⁀SeC_6H_5

       R

         3

C_6H_5—C(=O)—⁀SeC_6H_5

       R

         4

Attempts to deprotonate alkyl phenyl selenides have generally been unsuccessful (this includes phenyl cyclopropyl selenide),[10a] although methyl phenyl selenide can be deprotonated in low yield [BuLi, tetramethylethylenediamine (TMEDA), 0°C].[7] From the few cases where attempts to prepare methylseleno substituted carbanions by metalation have been reported it can be concluded that this group is significantly less acidifying than phenylseleno. For example, 1,1-bis (phenylseleno)alkanes are deprotonated under several conditions that cause extensive C—Se bond cleavage for the analogous 1,1-bis (methylseleno) alkanes.[18c]

Other cases where simple LDA/THF conditions are of marginal effectiveness are phenyl methoxymethyl, phenyl trimethylsilylmethyl, and phenyl propenyl selenides. Improved results can be achieved by replacing the phenylseleno group by the more strongly (10–20 times) acidifying m-trifluoromethylphenylseleno group;[10,28] by using lithium 2,2,6,6-tetramethylpiperidide instead of LDA;[10,18c] by introducing coordinating cosolvents such as tetramethylethylenediamine[29] or hexamethylphosphoric triamide (HMPA)[18c,52] or by utilizing potassium rather than lithium as counterion.[24] The latter two techniques do however suffer from the typical problems associated with strongly ionizing basic media: Reactions of carbanions with carbonyl compounds proceed less smoothly because of enolization, and 1,4 addition to enones often predominates over 1,2 addition.

3.2. The Lithium–Selenium Exchange Reaction

The bond between carbon and most of the main group third, fourth, and fifth row elements undergoes facile lithium/metalloid exchange to form a carbon–lithium bond. An approximate reactivity series for the ArLi/Ar'M exchange can be defined from various published[53] and unpublished data:[54] I > Te, Sb, Bi > Sn > Br > Se > Pb > Cl, S, P. Specifically, for the exchange of Tol—Li with Ph—M the relative rates of M = Pb, Se, Br, Sn are $\sim 10^{-2}$, 10^{-1}, 1, 10^3.[54] The rate of the process depends dramatically on the solvent, stability of the starting and product organolithium reagent, and to a lesser extent on pendant groups attached to M but not undergoing exchange [e.g., $(C_6H_5)_2$Se exchanges with Tol—Li five times as fast as does C_6H_5Se—C_4H_9]. Since these exchanges are equilibria, it follows that (1) the lithium reagent formed must be more stable than the one used to prepare it and (2) if there are several groups attached to M, then only the one that forms the most stable carbanion will be ejected. The position of selenium in the Li/M exchange series presented previously suggests that α-lithioselenides could be prepared from α-iodo, α-bromo, α-stannyl, α-telluro as well as α-seleno selenides. A few examples of Li/Br exchange for the preparation of α-lithio selenides have been reported (Table 3),[55] and one example of a Li/Te exchange,[27] but much more extensive studies have been carried

TABLE 3

PREPARATION OF α-LITHIO SELENIDES BY Li/Br EXCHANGE[a]

Lithium Reagent		Electrophiles, References
C_6H_5Se⟍⟋Li | R	R = H[b,c]	A,[55] K,[55] Si[20]
	R = CH_3[b]	A, K[55]
	R = n-C_6H_{13}[b]	A[55]
C_6H_5Se⟍⟋Li | CH_3 CH_3	[b]	A[55]

[a] Electrophiles: Aldehyde (A), Ketone (K), Trimethylsilyl chloride (Si).

$$C_6H_5Se\diagdown Br \xrightarrow[\text{THF, }-78°C]{\textit{n}\text{-BuLi}} C_6H_5Se\diagdown Li \xrightarrow{\text{E}} C_6H_5Se\diagdown E$$

[b] This reagent was also prepared by Li/Se exchange (Table 5).
[c] This reagent was also prepared by Li/H exchange (Table 2).

TABLE 4
PREPARATION OF SELENOACETALS AND RELATED COMPOUNDS

Starting Materials	Reagents	Product	Yield (%)	References
Selenoacetals				
CH_2I_2, CH_2Br_2, or CH_2Cl_2	C_6H_5SeNa	$(C_6H_5Se)_2CH_2$	(92)–(95)	7, 10[a]
CH_2N_2	$(C_6H_5Se)_2$, hv	$(C_6H_5Se)_2CH_2$	(100)	56
$C_2H_5O_2CCHN_2$	$(C_6H_5Se)_2$, Cu–Bronze	$(C_6H_5Se)_2CHCO_2C_2H_5$	(46)	57
C_6H_5Se⌐	C_6H_5SeH, $BF_3 \cdot Et_2O$	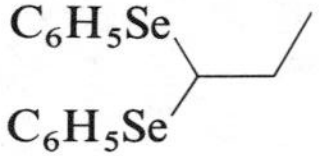 C_6H_5Se / C_6H_5Se	(55)	58
Selenoketals				
(4-tert-butylcyclohexanone)	C_6H_5SeH, $ZnCl_2$	(1-SeC₆H₅, 1-SeC₆H₅ cyclohexane)	(75)	59
	$B(SeR)_3$, CF_3CO_2H		(80)	60
$(C_6H_5Se)_2CHCH_3$	(1) KDA, THF, $-78°C$ (2) $C_6H_5CH_2Br$	$(C_6H_5Se)_2C(CH_3)CH_2C_6H_5$	(88)	24
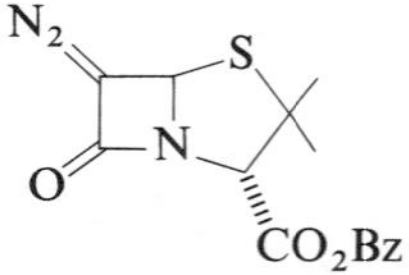 (penam diazo N_2, S, CO_2Bz)	$(C_6H_5Se)_2$, $BF_3 \cdot Et_2O$	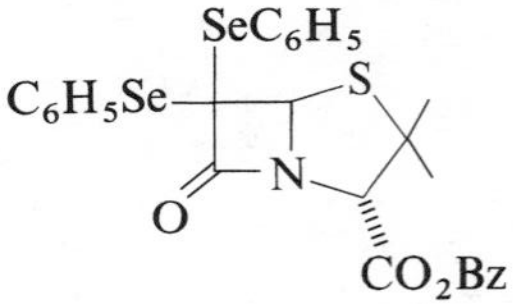 (penam bis-SeC₆H₅, CO_2Bz)	good	61

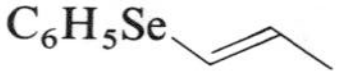

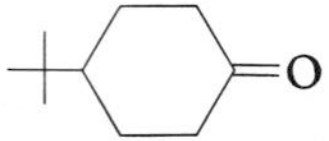

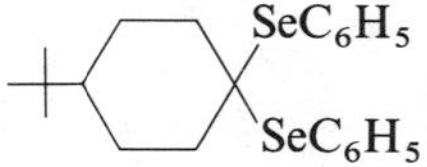

Substrate	Conditions	Product	Yield	Ref.
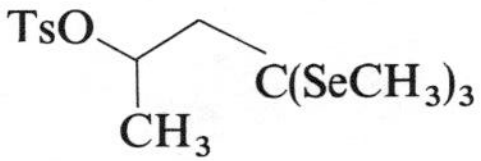 TsO–CH(CH$_3$)–C(SeCH$_3$)$_3$	n-BuLi, THF, -78–$20°$C	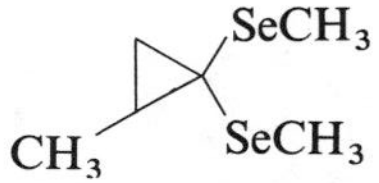 [cyclopropane with SeCH$_3$, SeCH$_3$, CH$_3$]	(60)	62
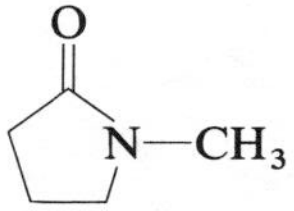 [pyrrolidinone, N–CH$_3$]	(1) LDA, THF (2) PhSeCl	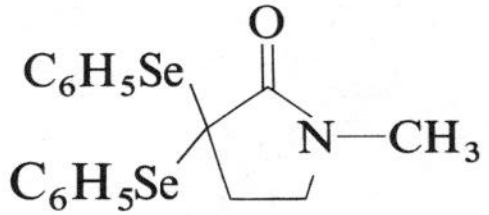 C$_6$H$_5$Se, C$_6$H$_5$Se, O, N–CH$_3$	(55)	63

Seleno-ortho esters

| (CH$_3$O)$_3$CH | C$_6$H$_5$SeH, BF$_3$·Et$_2$O | (C$_6$H$_5$Se)$_3$CH | (64) | 7 |

Selenoketene acetals

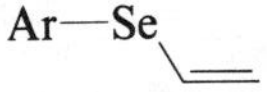

Ar–Se–CH=CH$_2$	(1) LDA, THF, $-78°$C (2) C$_6$H$_5$SeBr	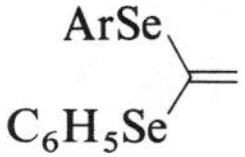 ArSe, C$_6$H$_5$Se	(89)[a]	10(b)
C$_6$H$_5$Se, C$_6$H$_5$Se–CH–Si(CH$_3$)$_3$	(1) LDA (2) C$_6$H$_5$CHO	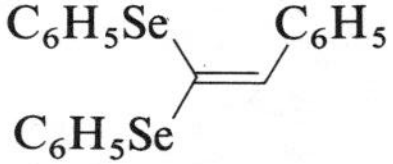 C$_6$H$_5$Se, C$_6$H$_5$Se, C$_6$H$_5$	(75)	25
(C$_6$H$_5$Se)$_3$C–n-C$_4$H$_9$	P$_2$I$_4$, NEt$_3$, 20°C	C$_6$H$_5$Se, C$_6$H$_5$Se, n-C$_3$H$_7$	(76)	64
[cyclohexanone, OTf, H]	KOt-C$_4$H$_9$, (C$_6$H$_5$Se)$_2$ DME, -50–$20°$C	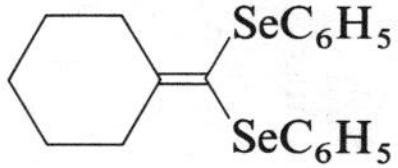 SeC$_6$H$_5$, SeC$_6$H$_5$ (cyclohexylidene)	(61)	63

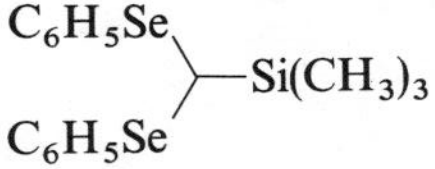

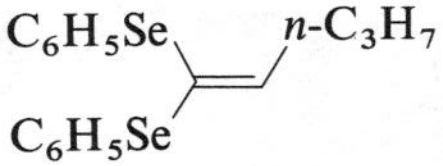

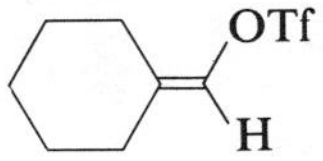

[a] Ar = m-CF$_3$C$_6$H$_4$

out on the Li/Se exchange. This is because the precursor selenoacetals and selenoketals are more easily prepared than α-bromo selenides, and generally undergo the exchange reaction in good yield. Table 4[56–64] summarizes typical preparations of selenoacetals, selenoketals, selenoortho esters, and selenoketene acetals.

Treatment of essentially all selenoacetals and selenoketals with n-butyllithium in THF at −78°C results in Li/Se exchange with the formation, in generally good yield, of the α-lithio selenide. Both phenylseleno and methylseleno substituted reagents have been prepared. Table 5[65–83] summarizes the major types, and indicates which electrophiles have been used with them.

The ease of nucleophilic attack at selenium (compared to, e.g., sulfur) has other chemical consequences as well. Thus the synthetically useful regio- and stereoisomerization of α-(phenylseleno)carbonyl compounds by deselenation–selenation can be easily carried out by treatment with appropriate nucleophiles.[41,52,84] See also Section 4.7.

The α-lithio selenides listed in Table 5 show advantageous characteristics for further transformations: They are excellent nucleophiles, tend to cause little

TABLE 5
EXAMPLES OF LITHIUM REAGENTS PREPARED BY Li/Se Exchange

Lithium Reagent R = CH_3 or C_6H_5	Electrophiles, References
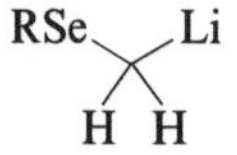 RSe—Li, H H	[b,c] A,[18b,65] K,[7,10a,65b,65c,66] En,[10a,67] RX,[68] Ep[69]
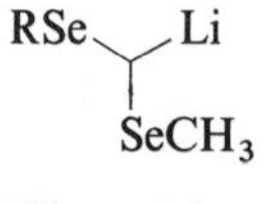 RSe—Li, $SeCH_3$	Si,[18c] A,[18c] En[70]
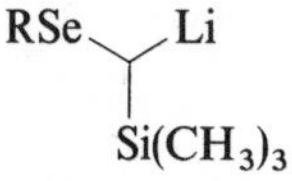 RSe—Li, $Si(CH_3)_3$	[c] A[71]
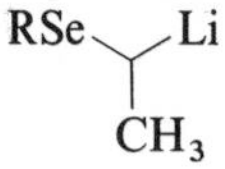 RSe—Li, CH_3	[b] Si,[72] S,[8,58] A,[18a,18b,65b] K,[8,18b,65a] Ep[69b]
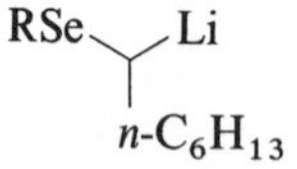 RSe—Li, n-C_6H_{13}	Si,[72] A,[65b,65c,73] K,[18b,65c] En,[67] Ac,[74] RX,[68,73a,75] Ep[69a,69b] Ox[69b]
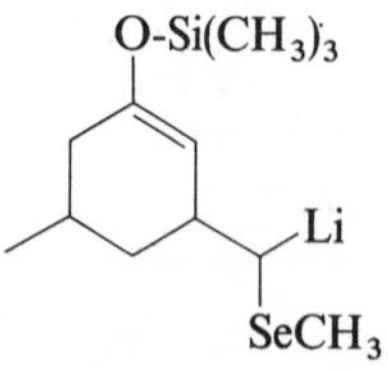 O-$Si(CH_3)_3$ ring, —Li, $SeCH_3$	RX[70]

TABLE 5 (*Continued*)

Lithium Reagent $R = CH_3$ or C_6H_5	Electrophiles, References
RSe Li / CH$_3$ SeR	RX,[7] En[70,76]
RSe Li / CH$_3$ CH$_3$	[b]Si,[8.5 8.72] A,[18a,18b,65c,77] K,[7,18a,18b,65c,66,77] En,[67] Ac[74]
RSe Li (bicyclic)	RX, A, D$_2$O[78]
RSe Li (cyclopropane)	Si,[79] A,[62,80] K,[66,69b,80] CO$_2$,[26] DMF,[73c] RX
RSe Li (cyclobutane)	A,[73c,82] K,[82] RX,[82b] Ep[82b]
RSe Li (cyclopentane)	K[77]
RSe Li (cyclohexane)	A,[8,18b] K,[18b] CO$_2$,[74] DMF,[74] Ac,[74] RX,[68,74] Ep[69a]
RSe Li / C$_6$H$_5$	Ac[16]
Li / n-C$_{10}$H$_{21}$	A, K, En, CO$_2$, DMF, Ac, RX[83]

[a] Electrophiles: Trialkylchlorosilane (Si); disulfide (S); aldehyde (A); ketone (K); enones (En); *N,N*-dimethylformamide (DMF); acid chloride (Ac); carbon dioxide (CO$_2$); alkyl halide (RX); epoxide (Ep); oxetane (Ox).

[b] These lithium reagents have also been prepared by Li/Br exchange (see Table 3).

[c] These lithium reagents have also been prepared by Li/H exchange (see Table 2).

enolization during reactions with enolizable ketones, and give acceptable yields of the expected products with aldehydes, ketones, enones, enals, carbon dioxide, dimethylformamide, acid halides, methylchloroformate, primary alkyl bromides, and iodides, epoxides, oxetanes, disulfides, and trialkylsilyl halides.

3.3. Metalation of Selenoxides

In contrast to the large body of research on the preparation and reactions of α-metallo selenides, relatively little work has been reported on the chemistry of metalated selenoxides.[20,26,85] The reasons for this are primarily the chemical properties of selenoxides. They are in general reactive compounds, difficult to handle and purify, and their α-lithio derivatives are less stable than analogous sulfur compounds, which have been very extensively studied.

α-Lithio selenoxides are best prepared using an *in situ* oxidation (*m*-chloroperbenzoic acid) deprotonation (LDA) sequence, since most alkyl aryl selenoxides are unstable at room temperature, and in addition are often strongly hygroscopic materials.[26] The procedure involves addition of approximately two equivalents of LDA to the THF solution of the selenoxide and *m*-chloroperbenzoic acid at low temperature. Primary, secondary, and tertiary α-lithio selenoxides can be prepared, and they can be alkylated, hydroxyalkylated, and acylated. The lithium *m*-chlorobenzoate present in solution does not seem to interfere in reactions of the α-lithio selenoxides.

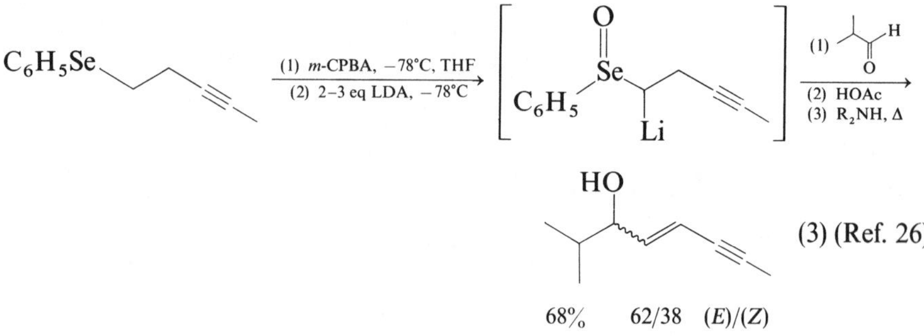

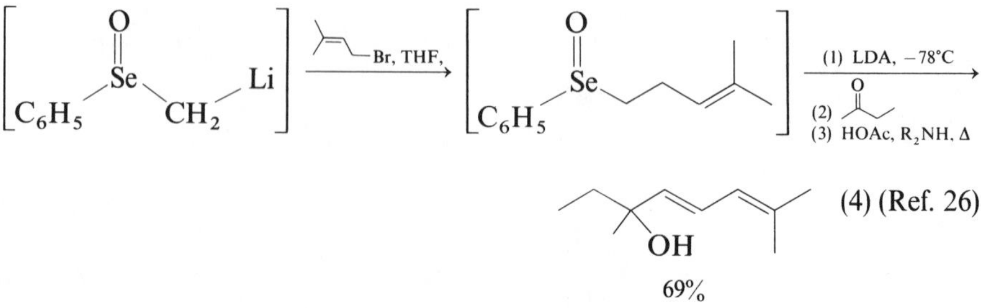

After derivation of the lithium reagent has been accomplished, the product selenoxide can be directly thermolyzed to give olefin, or the selenoxide can be reduced *in situ* to give a selenide as product. Some typical reactions are illustrated by Eqs. (3)–(5).[26]

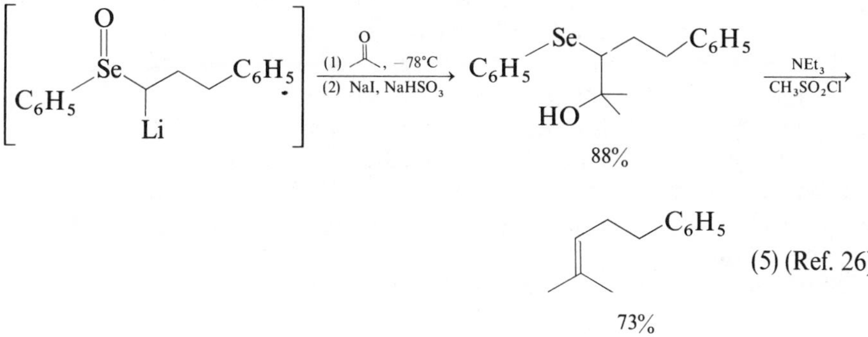

88%

$$C_6H_5$$

(5) (Ref. 26)

73%

A cyclobutane ring expansion procedure employing α-lithio selenoxides has been reported [Eq. (6)].

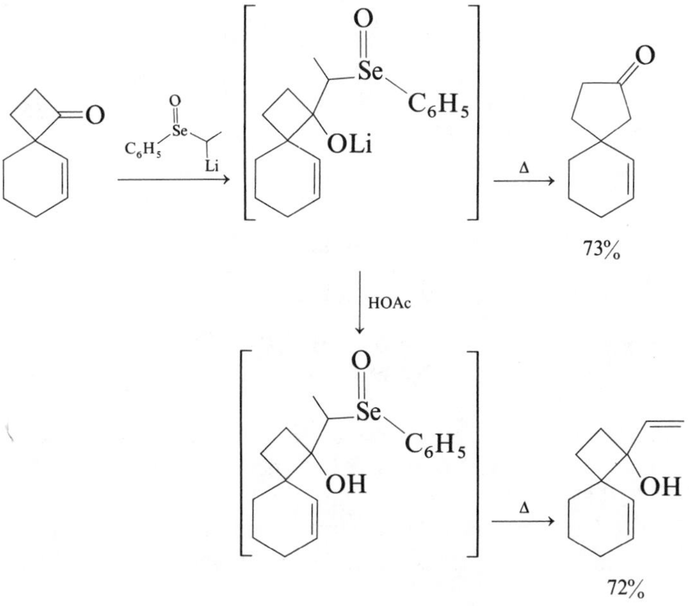

73%

HOAc

(6) (Ref. 85)

72%

This transformation is based on the observation that whereas thermolysis of the α-hydroxy selenoxide leads to allyl alcohol, the lithium alkoxide undergoes a pinacol like rearrangement. This reaction takes advantage of the leaving group ability of the selenoxide function (see also Section 4.5). Phenyl vinyl selenoxide can be deprotonated and methylated in only modest yield.[10b]

3.4. Metalation of Selenones

One of the more dramatic contrasts between sulfur and selenium chemistry is the comparison between sulfones and selenones. α-Lithio sulfones are perhaps the most widely used of all sulfur substituted carbanions because the sulfones are stable and easy to prepare, the metalations are convenient to carry out and the α-lithio sulfones show excellent nucleophilic properties. In contrast, selenones are difficult to prepare, requiring strong oxidants and careful control of reaction conditions. Although they are substantially more acidic than sulfones (see Table 1) their deprotonation and derivatization is difficult to carry out because of their ease of reduction and high reactivity as S_N2 alkylating agents ($C_6H_5SeO_2CH_3$ is more reactive than CH_3I). A modestly successful reaction is shown in Eq. (7).[10c,86] Note that the intermediate β-alkoxide selenone cyclizes under the reaction conditions to give an epoxide.

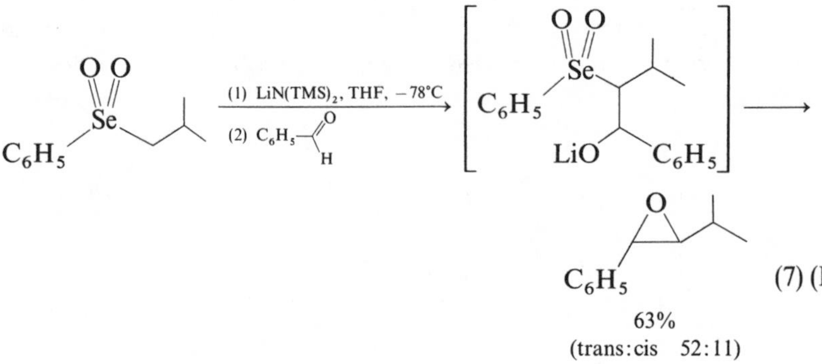

(7) (Ref. 10c)

63%
(trans:cis   52:11)

Metalated selenones are intermediates in several reactions of vinyl selenones with nucleophiles. Some examples are presented in Section 3.5.

3.5. Conjugate Addition to Vinyl Selenides, Selenoxides, and Selenones

Phenyl vinyl selenide reacts with strong bases by at least three distinct pathways, depending on base and solvent. With LDA metalation occurs predominantly.[10b,24,38,87] In THF at 0°C some metalation takes place, but the principal reaction is Li/Se exchange (giving n-butyl vinyl selenide and phenyllithium), whereas in THF at −78°C or in ether or dimethoxymethane at 0°C addition

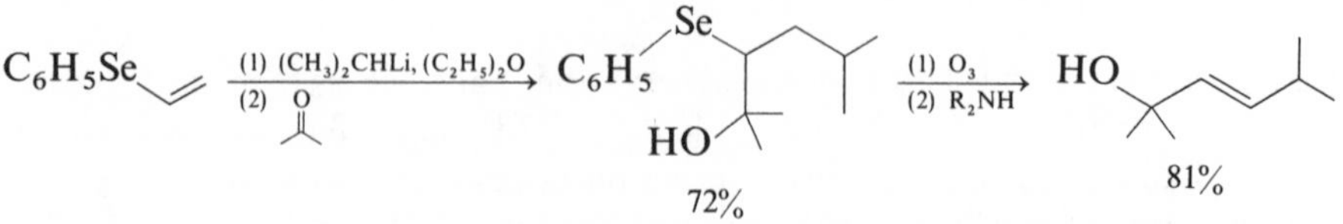

72%

81%

(8) (Ref. 87a)

to the double bond occurs to give an α-lithio selenide.[38,87] The anion formed in the latter reaction can be trapped with typical electrophiles [alkyl halides, halosilanes, aldehydes, ketones, nitriles, Eq. (8).]

A related reaction involving an addition–cyclization sequence leading to substituted pyrrolidines has been reported [Eq. (9)].[88]

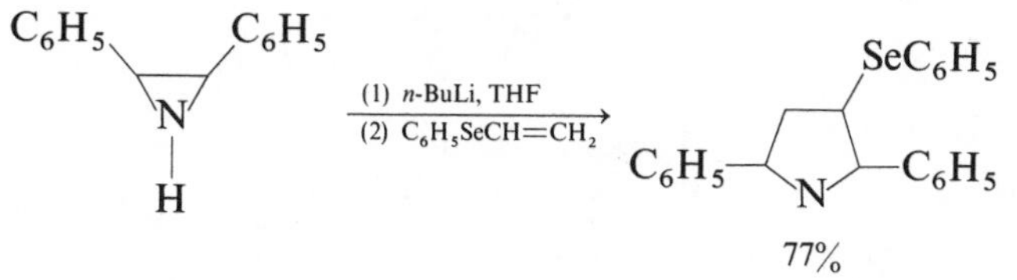

(9) (Ref. 88)

A particularly interesting application of the conjugate addition method has been reported by Liotta and co-workers [Eq. (10)].[41,52] Here proper control of the stereochemistry of the selenium group determines the regioselectivity of a selenoxide elimination.

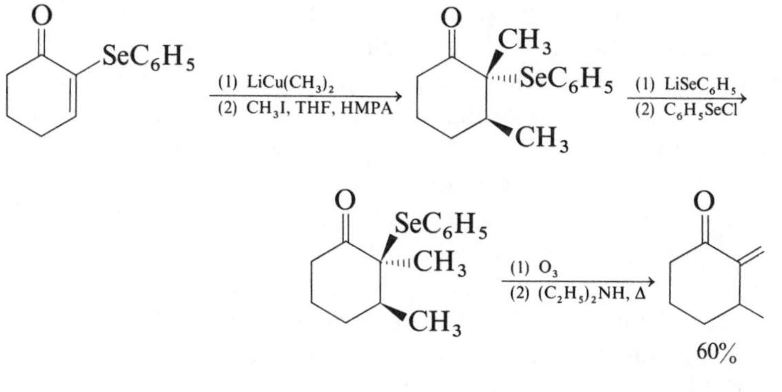

(10) (Ref. 52)

Vinyl selenoxides and selenones are very much more reactive in conjugate additions than are vinyl selenides, and several interesting applications have been reported. Treatment of ketone enolates with vinyl selenoxides[89] and selenones[89b,90] results in cyclopropanation reactions [Eq. (11)]. Several reactions of vinyl selenones that involve conjugate addition of alkoxide [Eq. (12)][89b,91] and amines [Eq. (13)] may also be useful for the preparation of oxetanes and aziridines.

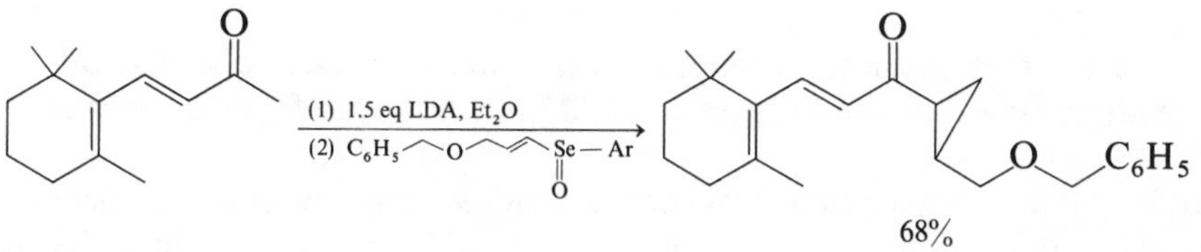

(11) (Ref. 89b)

(12) (Ref. 89b)

81%

(13) (Refs. 10c, 92)

87%

4. SYNTHETIC TRANSFORMATIONS BASED ON α-LITHIO SELENIDES

The goal of much of the effort in the area of selenium substituted carbanions is the utilization of the carbanionic center for the formation of C—C bonds. The role played by selenium is two fold. First, the seleno function stabilizes the adjacent carbanionic center, thus assisting the formation of a specific carbanion. Secondly, the selenium group can be used to carry out functional group modifications at the site of C—C bond formation.

The reactivity of α-litho selenides does not differ markedly from that of their sulfur analogs. In general, the anions show excellent nucleophilic properties and can almost always be alkylated with S_N2 reactive halides and epoxides. Good nucleophilicity is also demonstrated by the low degree of enolization observed when addition to enolizable carbonyl compounds is attempted.[18b,65a] The question of 1,2 versus 1,4 addition with α,β-unsaturated carbonyl compounds has been studied with several α-lithio selenides.[43,70,76,93] Typically 1,2 additions are favored in nonpolar media (ether, THF) whereas 1,4 additions are often observed in THF/HMPA or dimethoxyethane solution.

The tendency of certain α-lithio selenides (i.e., those prepared by deprotonation of α-phenylseleno carbonyl compounds) to alkylation in part on selenium (giving a transient selenonium ylid) rather than on carbon, should be noted.[30]

Some typical examples illustrating further transformations that can be carried out on the products are presented in Section 4.1

4.1. Oxidation to Selenoxides and Selenimides Followed by Selenoxide *Syn* Elimination or [2, 3] Sigmatropic Rearrangement

By far the most used reaction of organoselenium compounds is the selenoxide fragmentation. High yields of olefins are routinely obtained provided that the reaction is carried out under proper conditions.[73a,94] Sharpless et al.[44] and Grieco and Miyashita[46] reported on the alkylation–elimination of α-lithio esters [Eq. (14)] and lactones [Eq. (15)], and similar sequences have been carried out with metalated phenylseleno ketones[23,39,40] [Eq. (16)], phenylseleno carboxylic acids[10a,45] [Eq. (17)], and phenylseleno nitriles.[48] Along the same lines are the numerous reactions of metalated benzyl selenides that lead to substituted styrenes.[10a,15,28]

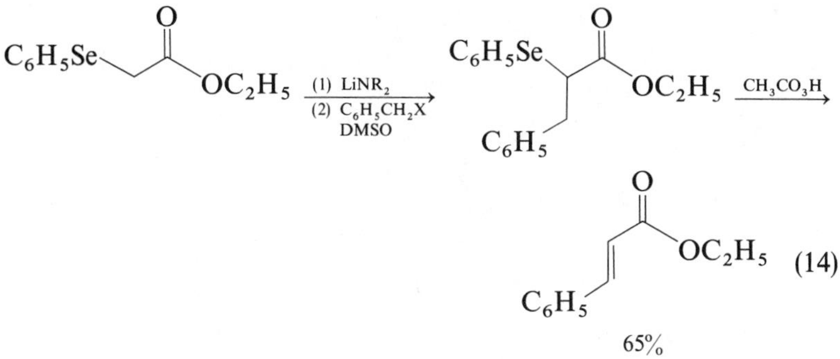

(14) (Ref. 44)

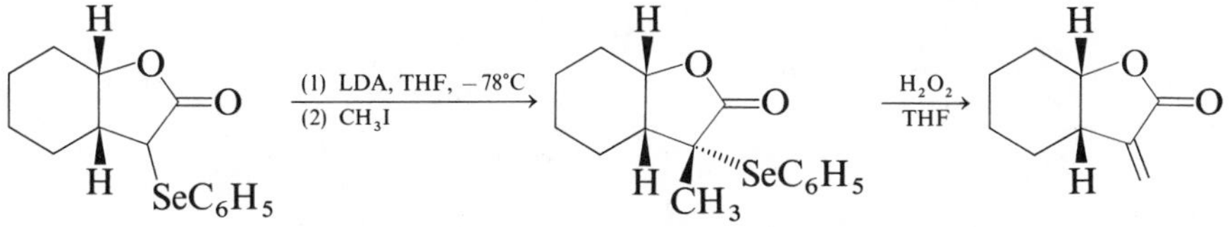

65%

(15) (Ref. 46)

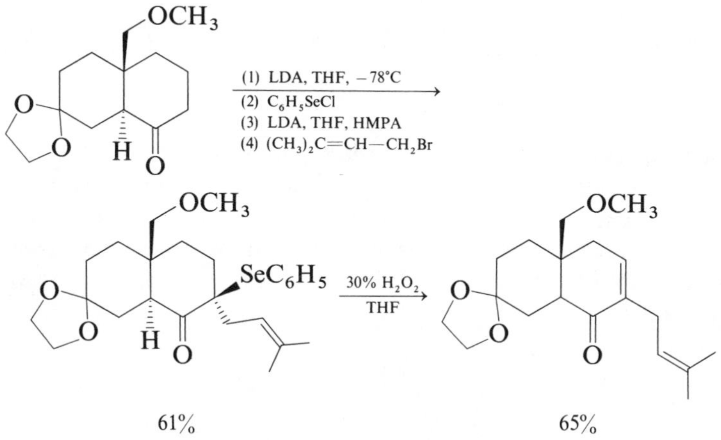

(16) (Ref. 40)

61% 65%

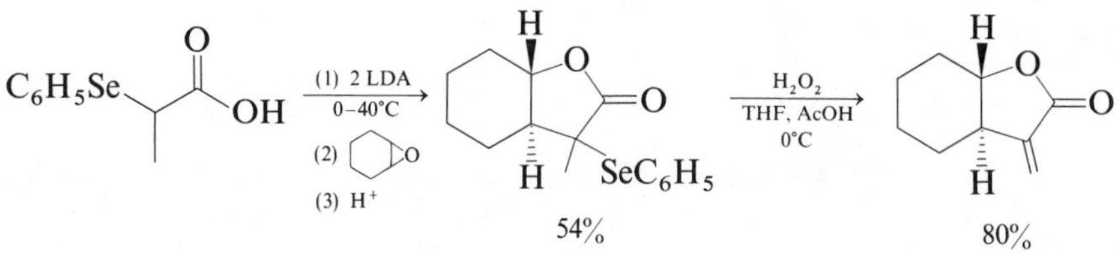

(17) (Ref. 45)

54% 80%

α-Lithio selenides can also be acylated and carboalkoxylated [Eq. (18)]. The reactions with aldehydes and ketones generally proceed in high yield, and give allyl alcohols as products after elimination [Eqs. (19) and (20)]. Little or no selenoxide elimination to give enols is observed. In these reactions the α-lithio selenides are functioning as vinyllithium equivalents.

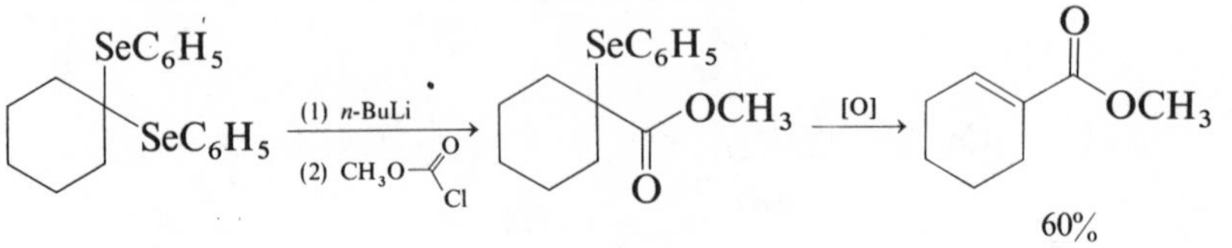

(18) (Ref. 74)

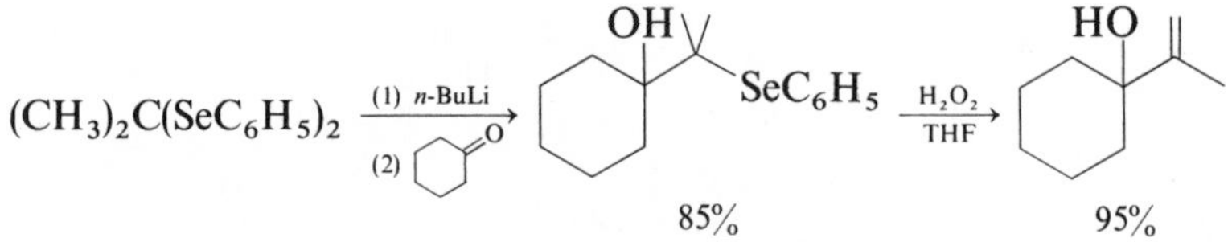

(19) (Ref. 8)

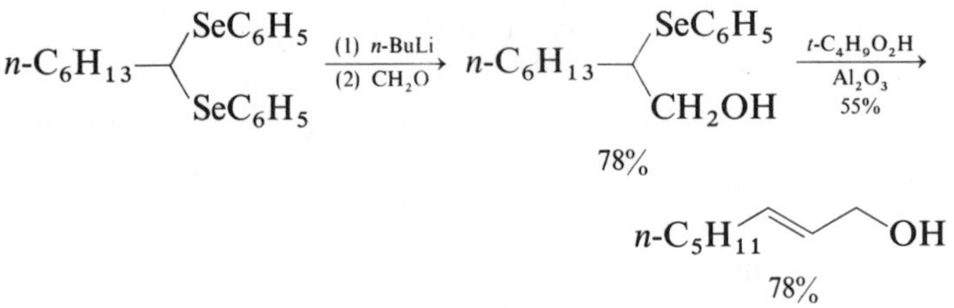

(20) (Ref. 73b)

An alternative procedure involves derivatizations of α-lithio selenoxides, which are then directly subjected to thermolysis. Some examples are presented in Section 3.3.

Thermolysis of vinyl selenoxides under suitable conditions leads to acetylenes and/or allenes [Eq. (21)]. A metalated vinyl selenide can thus be used as an acetylide equivalent.

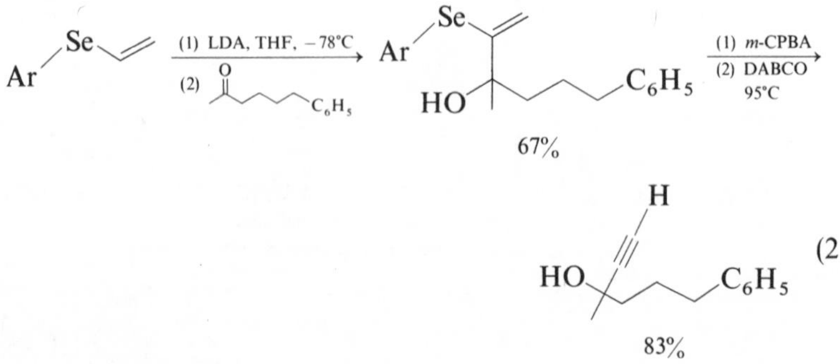

(21) (Ref. 37)

When the selenoxide is allylic, propargylic, or allenic, the preferred pathway is generally [2,3]sigmatropic rearrangement rather than selenoxide fragmentation. The products of such reactions can be allylic alcohols [Eq. (22)], enones [Eqs. (23) and (24)], or propargyl alcohols [Eq. (25)].

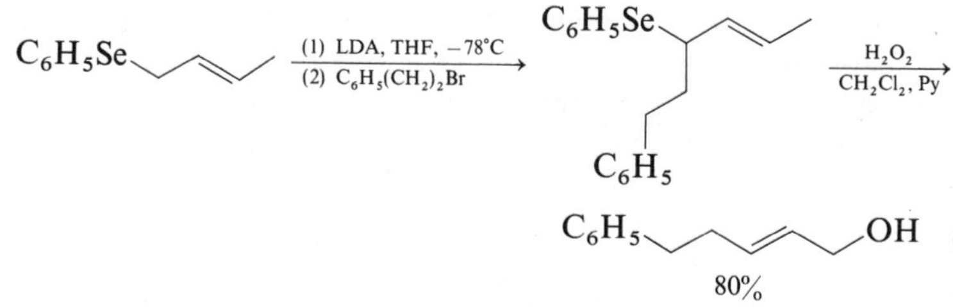

C_6H_5Se ⟶ $\xrightarrow[\text{(2) } C_6H_5(CH_2)_2Br]{\text{(1) LDA, THF, } -78°C}$ ⟶ C_6H_5Se ... C_6H_5 $\xrightarrow[\text{CH}_2Cl_2, \text{Py}]{H_2O_2}$

C_6H_5 ⟶ OH   (22) (Ref. 14)

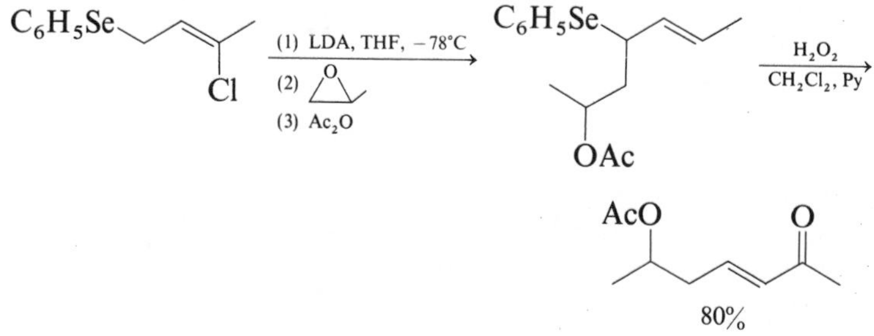

80%

C_6H_5Se ... Cl $\xrightarrow[\substack{\text{(2) } \triangle O \\ \text{(3) } Ac_2O}]{\text{(1) LDA, THF, } -78°C}$ C_6H_5Se ... OAc $\xrightarrow[\text{CH}_2Cl_2, \text{Py}]{H_2O_2}$

AcO ... O   (23) (Ref. 14)

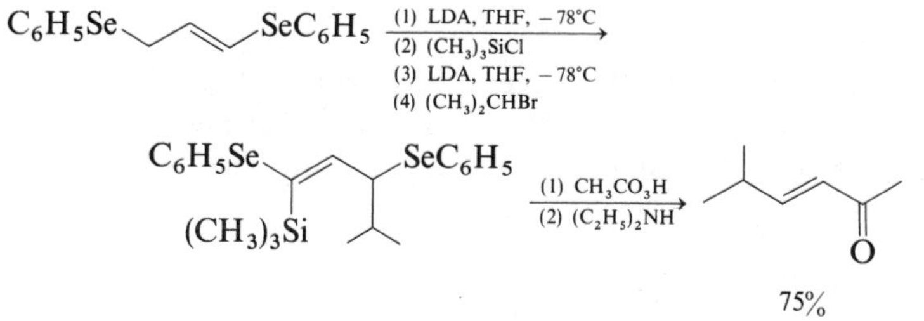

80%

C_6H_5Se ... SeC_6H_5 $\xrightarrow[\substack{\text{(2) } (CH_3)_3SiCl \\ \text{(3) LDA, THF, } -78°C \\ \text{(4) } (CH_3)_2CHBr}]{\text{(1) LDA, THF, } -78°C}$

C_6H_5Se ... SeC_6H_5
$(CH_3)_3Si$ ... $\xrightarrow[\text{(2) } (C_2H_5)_2NH]{\text{(1) } CH_3CO_3H}$ ... O   (24) (Ref. 25)

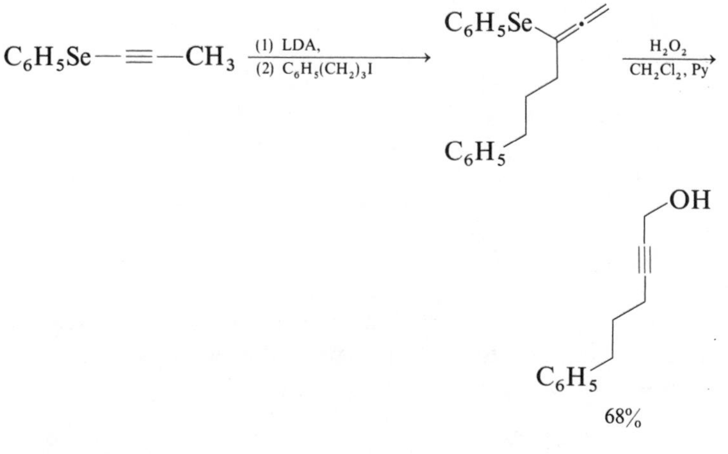

75%

C_6H_5Se—≡—CH_3 $\xrightarrow[\text{(2) } C_6H_5(CH_2)_3I]{\text{(1) LDA,}}$ C_6H_5Se ... C_6H_5 $\xrightarrow[\text{CH}_2Cl_2, \text{Py}]{H_2O_2}$

OH ... C_6H_5 (25) (Ref. 36)

68%

Conversion of an initially formed allyllithium reagent to another metal can be useful in achieving control of reactivity and regioselectivity.[33,68b] In the transformation of Eq. (26) the conversion of an allyllithium to an allyltin reagent allows an *in situ* generated iminium salt to be used as electrophile, and attack occurs exclusively γ to the tin substituent. The phenylseleno group of allyl selenides can also be displaced by trimethyltin lithium (Section 4.8), allowing a second aminomethylation of the same allyl fragment to be carried out.[32] If reaction conditions are carefully controlled, it is possible to convert α-lithio selenides to the corresponding copper derivatives, and these can be alkylated.[68b]

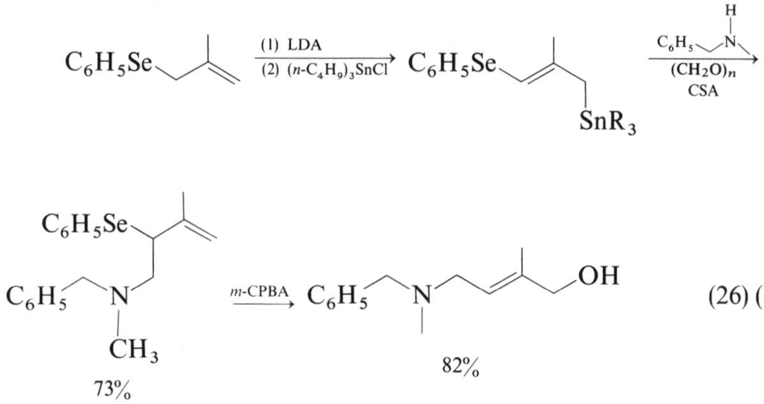

(26) (Ref. 32)

During the oxidation of α-silyl selenides a sila-seleno Pummerer reaction competes to varying degree with the selenoxide *syn* elimination[28,29] forming carbonyl compounds and vinyl selenides. Hence phenylselenotrimethylsilylmethane can serve as a formyl anion equivalent [Eq. (27)].

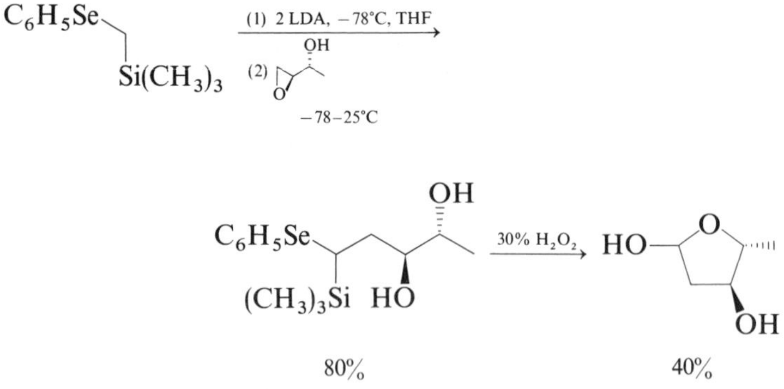

(27) (Ref. 30)

Curiously, if the α-silyl selenide is converted to an *N*-tosylselenimide instead of a selenoxide the balance appears to shift in favor of the elimination reaction over the possible aza-Pummerer process [Eq. (28)].[95]

$$
\underset{\substack{\\ n\text{-}C_9H_{19}}}{C_6H_5Se} \diagdown Si(CH_3)_3 \quad \xrightarrow[\text{THF}]{CH_3-\!\!\!\langle\bigcirc\rangle\!\!\!-\overset{\overset{O}{\|}}{\underset{\underset{O}{\|}}{S}}-\overset{Cl}{\underset{Na}{N}}}
$$

82%

$$
\left[\underset{\substack{\\ n\text{-}C_9H_{19}}}{\underset{\|}{\overset{\overset{Ts}{N}}{C_6H_5Se}}} \diagdown Si(CH_3)_3 \right] \xrightarrow{20^\circ C} \underset{\substack{\\ n\text{-}C_9H_{19}}}{} \diagup^{Si(CH_3)_3} \qquad (28)\ (Ref.\ 95)
$$

70%

4.2. Reductive Elimination of β-Hydroxy Selenides

Treatment of β-hydroxy selenides with many types of hydroxyl activating reagents results in reductive elimination to give olefins. The reagents that have been used include TsOH/pentane,[33b,65b] HClO$_4$/ether,[65b] CF$_3$CO$_2$H/N(C$_2$H$_5$)$_3$,[65b] 1,2-phenylenephosphorochloridite,[17] (CH$_3$)$_3$SiCl/NaI,[96] t-C$_4$H$_9$-(CH$_3$)$_2$SiCl/Na–NH$_3$,[97] SOCl$_2$/N(C$_2$H$_5$)$_3$,[65c] POCl$_3$/N(C$_2$H$_5$)$_3$,[42] P$_2$I$_4$/N(C$_2$H$_5$)$_3$,[98] PI$_3$/N(C$_2$H$_5$)$_3$,[80a,83,99] carbonyl diimidazole,[80a] and CH$_3$SO$_2$Cl/N(C$_2$H$_5$)$_3$.[10a,100] Of these, the last four or five are probably the most effective, since they have been found to work in difficult eliminations such as those leading to α-vinylketones,[100b,100c] alkylidene cyclopropanes,[80a,99] and allenes.[83]

Since the reaction of α-lithio selenides with aldehydes and ketones gives β-hydroxy selenides, a versatile olefin synthesis related to the Wittig and Peterson olefinations is available. The elimination proceeds with anti stereochemistry[10a,65c] and presumably involves an episelenonium ion as intermediate. Typical applications are exemplified in Eqs. (29), (30), (31), (32) and (33).

$$
(C_6H_5Se)_2CH_2 \xrightarrow[\substack{(2)\ \overset{O}{\underset{H}{\diagdown}}}]{(1)\ LDA,\ THF,\ -78^\circ C} \underset{\underset{OH}{}}{\overset{C_6H_5Se}{\underset{C_6H_5Se}{}}} \xrightarrow[N(C_2H_5)_3]{SOCl_2} C_6H_5Se\diagup\!\!\!\diagdown
$$

77%   81%

(29) (Ref. 10)

$$
C_6H_5Se\diagdown\!\!\diagup\!\!\diagdown \xrightarrow[\substack{(2)\ (C_2H_5)_3Al\\ (3)\ AcO-(CH_2)_8-CHO}]{(1)\ LDA} \underset{\underset{HO}{}}{\overset{C_6H_5Se}{\diagdown}}\!\!\diagup^{(CH_2)_8OAc} \xrightarrow{TsOH}
$$

70%

$$
AcO\diagdown\!\!\diagup\!\!\diagdown \qquad (30)\ (Ref.\ 33a)
$$

(E)/(Z)   84/16

77%

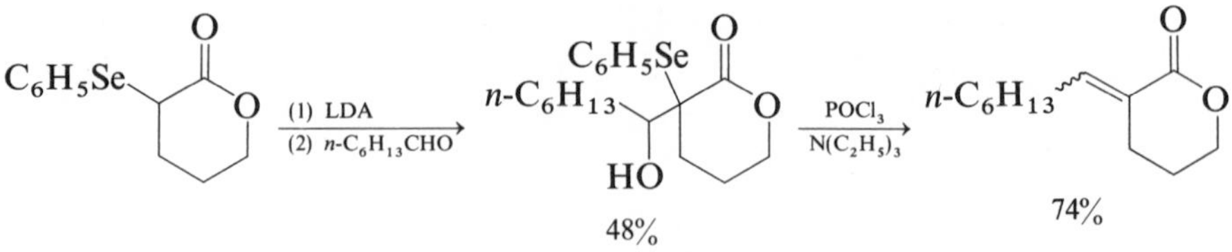

(31) (Ref. 42)

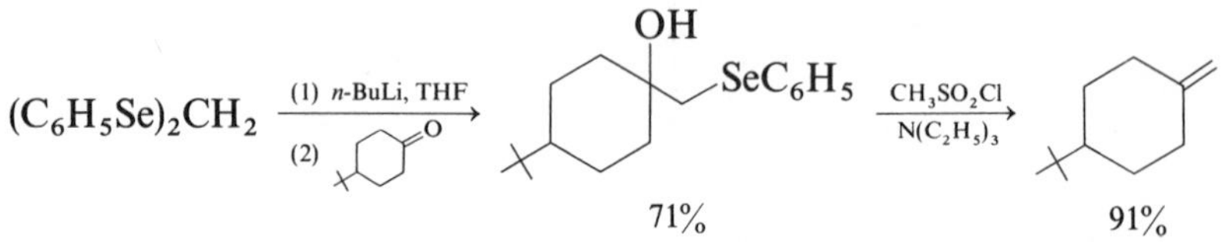

(32) (Ref. 10a)

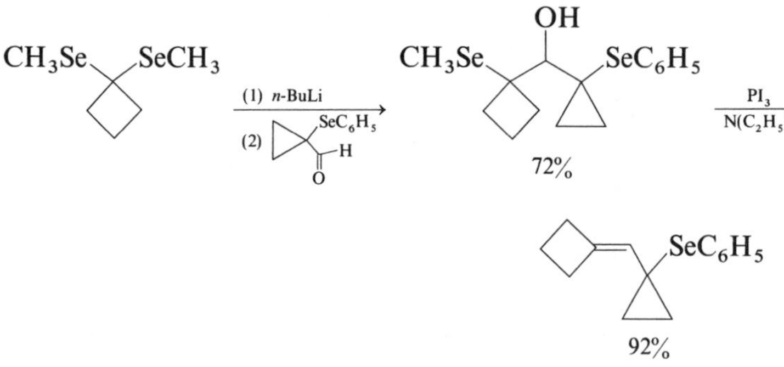

(33) (Ref. 73c)

4.3. Conversion to Lithium Reagents by a Second Lithium/Selenium Exchange Reaction

As outlined in Section 3.2, selenoacetals, selenoketals, and selenoorthoesters can be converted to α-lithio selenides by treatment with n-butyllithium. Other types of selenides are also cleaved, provided that at least one carbanion stabilizing group (e.g., RS, RSe, R_3Si, etc.) remains in the molecule. Some examples are shown in Eqs. (34)–(36). These show how selenoacetals or seleno ortho-esters can be used to generate one, two, or even more carbanion consecutively at the same carbon atom.

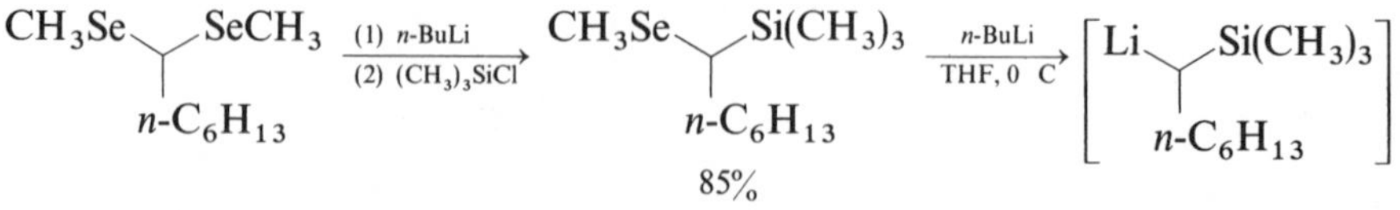

(34) (Ref 72)

$$C_6H_5Se\diagdown\diagup SeC_6H_5 \xrightarrow[\text{(2) } (C_6H_5S)_2]{\text{(1) } n\text{-BuLi}} C_6H_5Se\diagdown\diagup SC_6H_5 \xrightarrow{n\text{-BuLi}} \left[Li\diagdown\diagup SC_6H_5 \right]$$

$$70\%$$

$$\text{(35) (Ref 8)}$$

$$(C_6H_5Se)_2CH_2 \xrightarrow[\substack{\text{(2)} \\ \text{(3) } CH_3SO_2Cl}]{\text{(1) LDA}} \begin{array}{c} C_6H_5Se \\ \diagup \\ C_6H_5Se \end{array}\diagdown\diagup OMs \xrightarrow[\substack{\text{(2) LDA} \\ \text{(3) } HSO_3^-, I^-}]{\text{(1) } m\text{-CPBA}}$$

$$\begin{array}{c} C_6H_5Se \\ C_6H_5Se \end{array}\triangleright \xrightarrow{n\text{-BuLi}} \begin{array}{c} C_6H_5Se \\ Li \end{array}\triangleright \quad \text{(36) (Ref. 26)}$$

$$57\%$$

4.4 Hydrolysis of Selenoketals and Vinyl Selenides to Carbonyl Compounds

Both selenoketals[24,59,76] and vinyl selenides,[101] appear to hydrolyze somewhat more slowly than their sulfur analogs, and thus this technique has not been widely used for the preparation of carbonyl compounds. Two examples illustrating the use of α-lithio selenides as acyl anion equivalents[24,102] are shown in the following equations [Eqs. (37) and (38)].

$$\xrightarrow[\substack{CH_3Se \diagdown \diagup Li \\ CH_3Se \diagdown CH_3 \\ \text{THF, HMPA, } -78°C}]{} \xrightarrow[\text{acetone, 20°C}]{CuCl_2, 4\ CuO}$$

$$65\% \qquad\qquad\qquad 70\%$$

$$\text{(37) (Ref. 102)}$$

$$\begin{array}{c} C_6H_5Se \\ C_6H_5Se \end{array}\diagdown CH_3 \xrightarrow[\text{(2) } C_6H_5CH_2Br]{\text{(1) KDA, THF}} \begin{array}{c} C_6H_5Se \diagdown \diagup CH_3 \\ C_6H_5Se \diagup \diagdown CH_2C_6H_5 \end{array} \xrightarrow[\text{acetone}]{2\ CuCl_2, 4\ CuO} O{=}\begin{array}{c} CH_3 \\ \diagdown C_6H_5 \end{array}$$

$$88\% \qquad\qquad\qquad 86\%$$

$$\text{(38) (Ref. 24)}$$

4.5. Acid and Base Catalyzed Substitutions and Eliminations of Selenium Groups

Selenides are easily converted to selenonium salts. These form olefins on treatment with base[69a,81,82b] [Eq. (39)], and epoxides if a β-hydroxy group is present[18a,18b,65a,66,67,69a,82a] [Eqs. (40) and (41)].

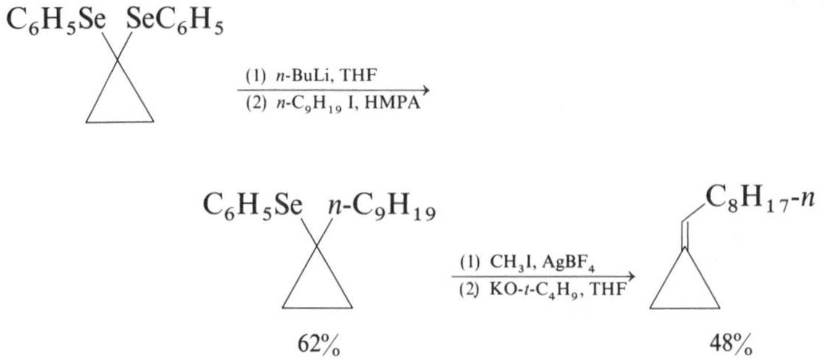

$C_6H_5Se \quad SeC_6H_5$

(1) *n*-BuLi, THF
(2) *n*-C_9H_{19} I, HMPA

(39) (Ref. 81)

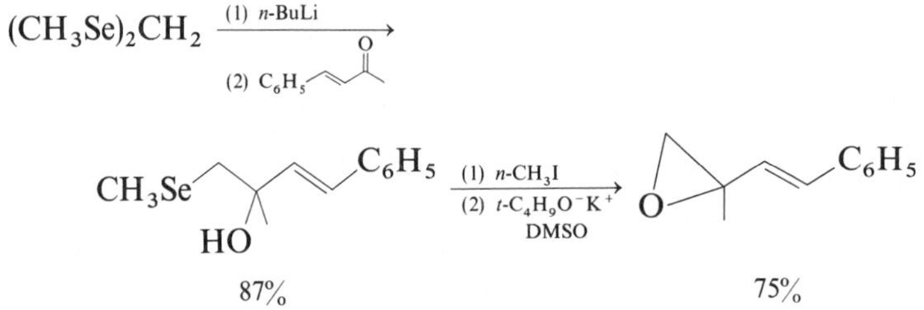

$C_6H_5Se \quad n-C_9H_{19}$

$C_8H_{17}-n$

(1) CH_3I, $AgBF_4$
(2) KO-*t*-C_4H_9, THF

62%   48%

$(CH_3Se)_2CH_2$ (1) *n*-BuLi

(2) C_6H_5

(40) (Ref. 67)

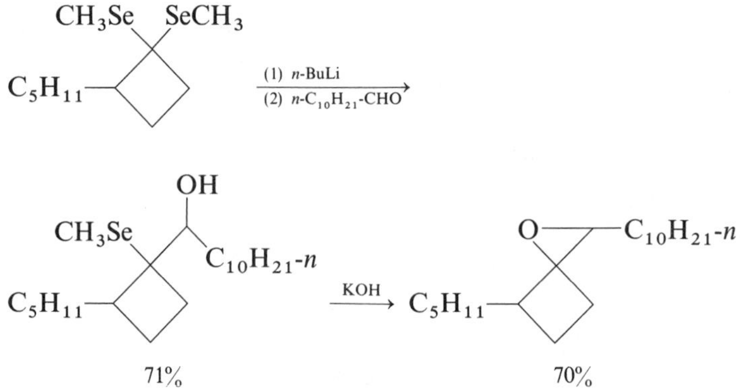

CH_3Se C_6H_5
                                            (1) *n*-CH_3I
                                            (2) *t*-$C_4H_9O^-K^+$
HO   DMSO

C_6H_5

87%   75%

$CH_3Se \quad SeCH_3$

C_5H_{11}

(1) *n*-BuLi
(2) *n*-$C_{10}H_{21}$-CHO

(41) (Ref. 82a)

OH

CH_3Se
                        $C_{10}H_{21}-n$

C_5H_{11}

KOH

O $C_{10}H_{21}-n$

C_5H_{11}

71%   70%

Phenylseleno groups can also be activated towards substitution by positive halogen [Eq. (42)]. In the absence of nearby functional groups, the treatment of selenides with bromine,[75,103] or selenoxides with HX[104] leads to the formation of alkyl halides [Se/Br exchange, Eq. (43)].

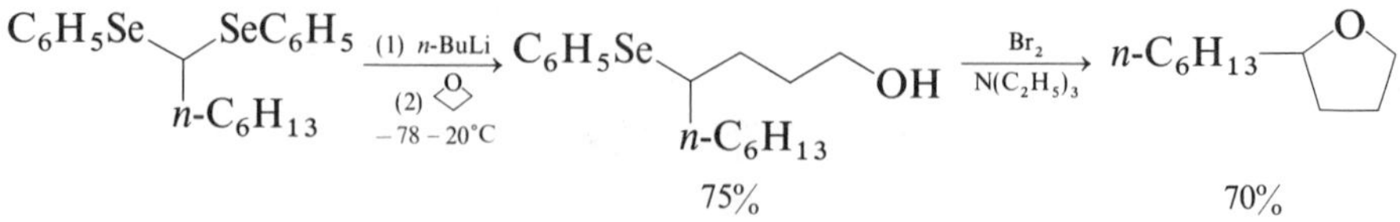

$C_6H_5Se \quad SeC_6H_5$ (1) *n*-BuLi
                                            (2)
n-C_6H_{13} $-78 - 20°C$

C_6H_5Se

$n-C_6H_{13}$ OH

Br_2
$N(C_2H_5)_3$

n-C_6H_{13} O

75%   70%

(42) (Ref. 69b)

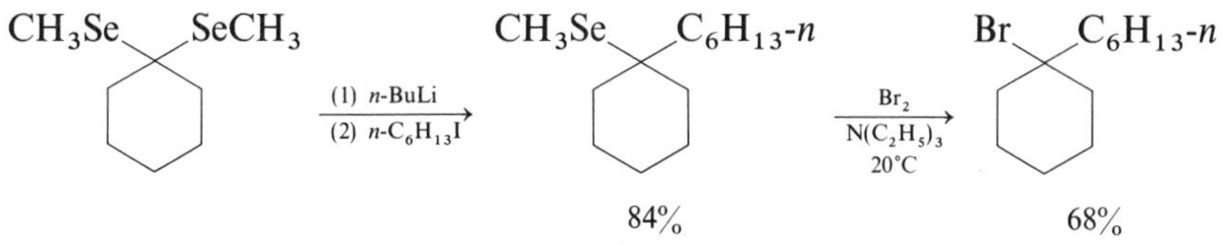

$$CH_3Se \diagdown \diagup SeCH_3$$

(1) n-BuLi
(2) n-$C_6H_{13}I$
→

$$CH_3Se \diagdown \diagup C_6H_{13}\text{-}n$$

84%

$\dfrac{Br_2}{\begin{array}{c}N(C_2H_5)_3\\20°C\end{array}}$ →

$$Br \diagdown \diagup C_6H_{13}\text{-}n$$

68%

(43) (Ref. 75)

An example of a pinacol rearrangement of an α-alkoxy selenoxide has been presented earlier [Section 3.3, Eq. (6)].[85] Similar ring expansions can be initiated by onium salt formation,[72a] with AgBF$_4$ [Eq. (44)], Tl(OEt),[77b] or with p-toluenesulfonic acid [Eq. (45)].

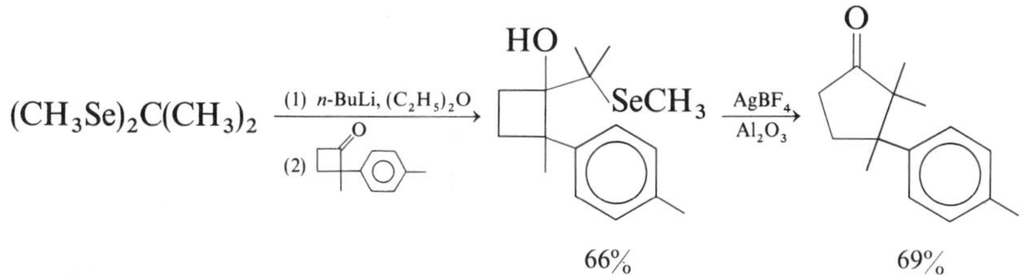

$$(CH_3Se)_2C(CH_3)_2$$

(1) n-BuLi, $(C_2H_5)_2O$
(2)
→

66%

$\dfrac{AgBF_4}{Al_2O_3}$ →

69%

(44) (Refs. 66, 105)

$$C_6H_5Se \diagup \diagdown SeC_6H_5$$

(1) n-BuLi, $(C_2H_5)_2O$
(2)
→

81%

$\dfrac{p\text{-TsOH}}{\begin{array}{c}C_6H_6\\80°C\end{array}}$ →

70%

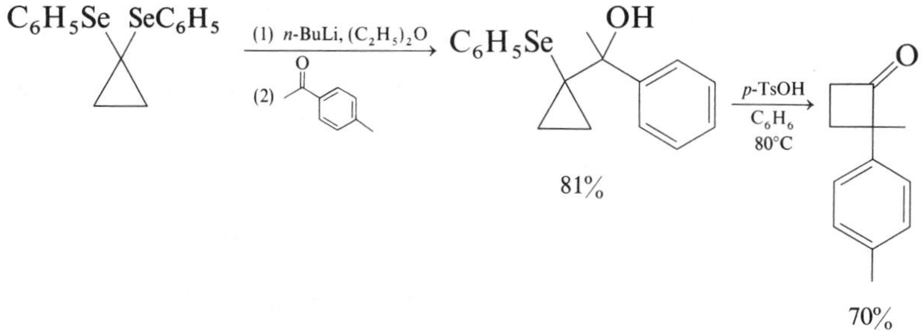

(45) (Ref. 66)

4.6. Replacement of RSe Groups by Hydrogen

The reductive removal of selenium functions can be carried out via the lithium reagent,[68] or by Raney nickel[51,65b] dissolving metal[65b,106] tri-n-butyltin hydride,[70,107] or palladium catalyzed triethylborohydride[108] reductions [Eqs. (46) and (47)].

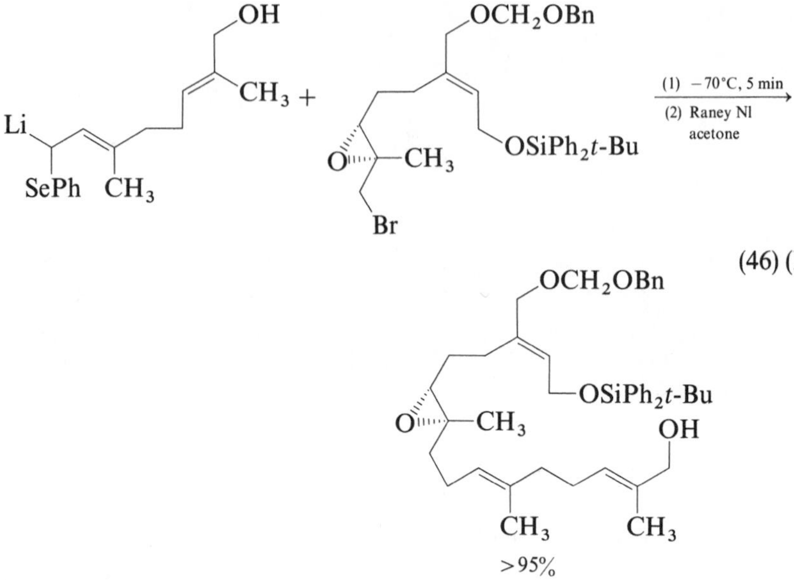

(46) (Ref. 51)

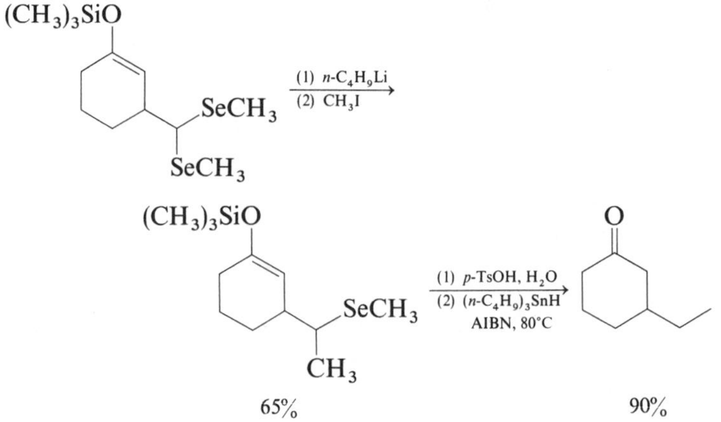

>95%

(47) (Ref. 70)

65%   90%

α-Phenylselenocarbonyl compounds are easily deselenated on treatment with soft nucleophiles (RS⁻, RSe⁻)[23,39,52,84a], in protic media. The selenium can thus be used as an easily removable activating group [Eq. (48)].

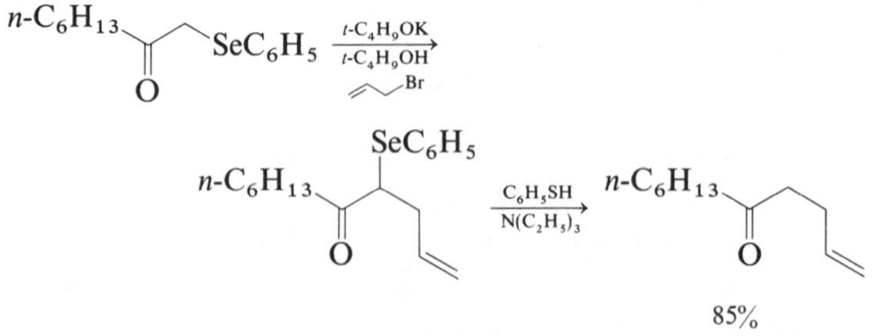

(48) (Ref. 39)

85%

4.7. Replacement of RSe Groups by Alkyl Groups

Only a few cases of carbanionic displacement of phenylseleno groups have been reported. Allyl selenides are displaced by dialkyl cuprates [Eq. (49)].

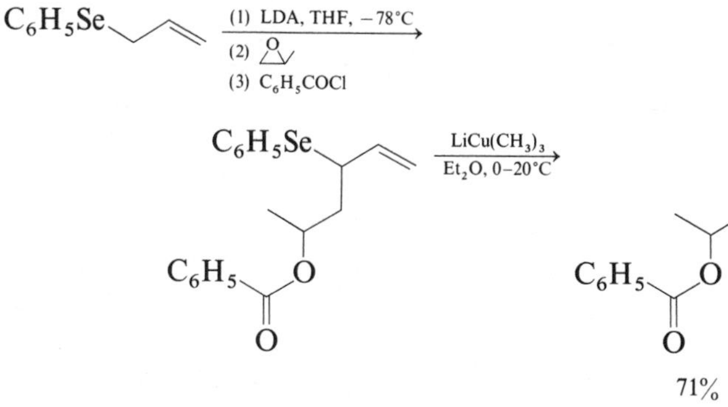

C_6H_5Se ⟶ (1) LDA, THF, $-78°C$ (2) ⟶ (3) C_6H_5COCl

C_6H_5Se ⟶ $\dfrac{LiCu(CH_3)_2}{Et_2O,\ 0\text{-}20°C}$ ⟶ (49) (Ref. 14)

C_6H_5—O—O C_6H_5—O—O

71%

An interesting synthetic procedure involving transfer of alkyl from boron to carbon has been reported. Here excellent regiocontrol ($>10/1$) is achieved by proper timing of the second electrophilic reaction [Eq. (50)].

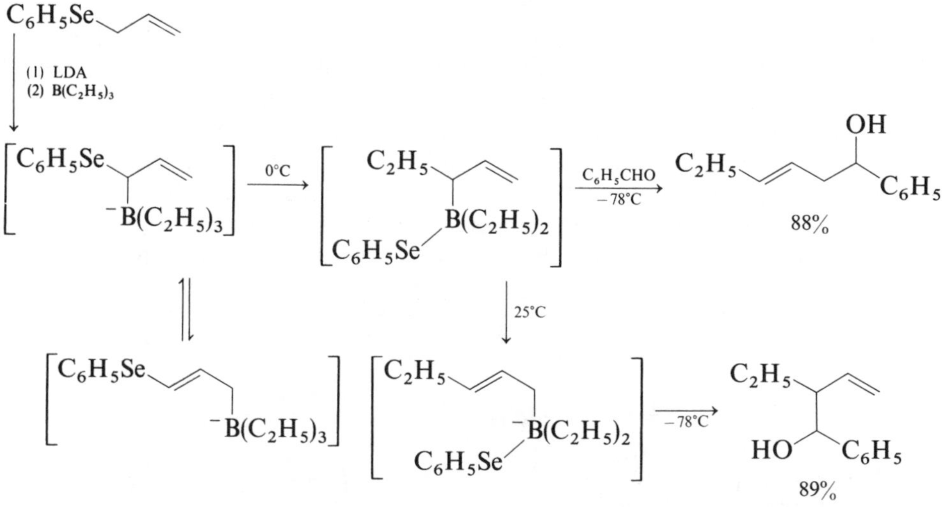

C_6H_5Se ⟶

(1) LDA
(2) $B(C_2H_5)_3$

$\left[C_6H_5Se \diagdown \underset{^-B(C_2H_5)_3}{}\right]$ $\xrightarrow{0°C}$ $\left[\underset{C_6H_5Se}{\overset{C_2H_5}{}} B(C_2H_5)_2\right]$ $\xrightarrow[-78°C]{C_6H_5CHO}$ C_2H_5 ⟶ OH C_6H_5

88%

$\left[C_6H_5Se \diagdown \underset{^-B(C_2H_5)_3}{}\right]$ $\left[\underset{C_6H_5Se}{\overset{C_2H_5}{}} \diagdown ^-B(C_2H_5)_2\right]$ $\xrightarrow{-78°C}$ C_2H_5 HO C_6H_5

25°C

89%

(50) (Ref. 33b)

4.8. Replacement of RSe Groups by Trialkylstannyl Groups

Allyltin reagents can be easily prepared by treatment of allyl selenides with trialkylstannyllithium. The reaction results in loss of regiochemical control, with the trialkylstannyl group appearing predominantly at the less substituted position [Eqs. (51) and (52)]. The analogous sulfur compounds do not react under similar conditions.

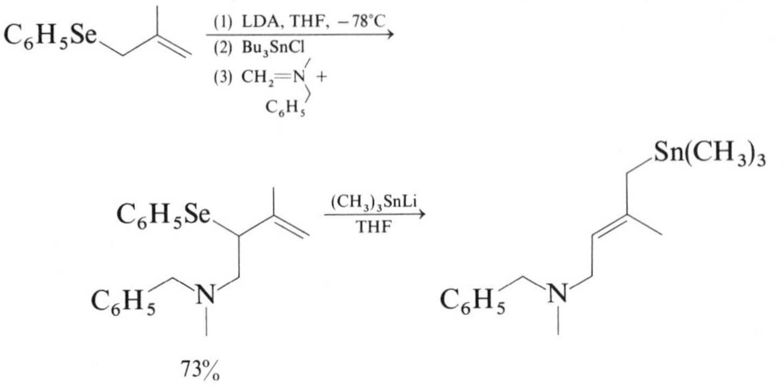

(51) (Ref. 32)

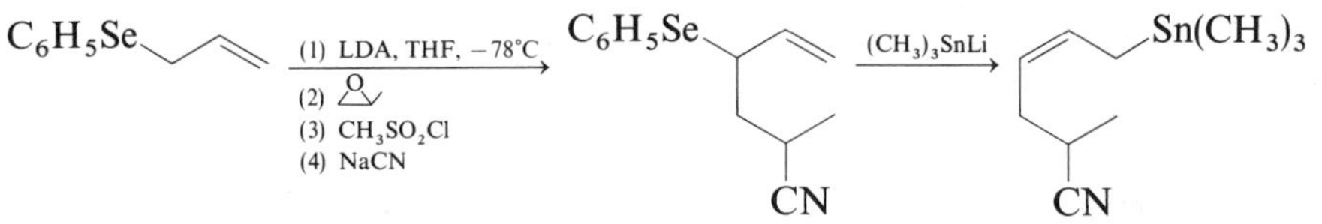

73%

(52) (Ref. 109)

The chemistry described above illustrates a variety of applications of [alpha]-lithio selenium compounds in organic chemistry. Although there are both advantages and disadvantages in comparison to the analogous sulfur compounds, it is clear that for at least some applications the selenium compounds are the reagents of choice because the selenium precursor is more easily prepared, or because the selenium functionality can be more easily removed at the end of the process. The discovery of new reactions and further development of the topics discussed in this review will certainly add to the value of metallo organoselenium compounds.

REFERENCES

1. H. Gilman and F. J. Webb, *J. Am. Chem. Soc.*, **71**, 4062 (1949); H. Gilman and R. L. Bebb, *J. Am. Chem. Soc.*, **61**, 109 (1939).

2. It was subsequently reported that some metalation of $C_6H_5SeCH_3$ does occur. (See Ref. 7.)

3. W. von E. Doering and A. K. Hoffmann, *J. Am. Chem. Soc.*, **77**, 521 (1955).

4. (a) A. I. Shatenshtein and H. A. Gvozdeva, *Tetrahedron*, **25**, 2749 (1969); (b) A. I. Shatenshtein, N. N. Magdesieva, Yu. I. Ranneva, I. O. Shapiro, and A. I. Serebryanskaya, *Teor. Eksp. Khim.*, **3**, 343 (1967); *Chem. Abstr.*, **68**, 58799n (1968).

5. E. G. Kataev, L. M. Kataeva, and G. A. Chmutova, *Zh. Org. Khim.* **2**, 2244 (1966); E. G. Kataev, G. A. Chmutova, A. A. Musina, and E. G. Yarkova, *Dokl. Akad. Nauk. SSSR*, **187**(6), 1308 (1969).

6. B. R. Muth and A. I. Kiss, *J. Org. Chem.*, **21**, 576 (1956): M. Renson and P. Pirson, *Bull. Soc. Chim. Belg.*, **75**, 456 (1966); L. Christiaens and M. Renson, *Bull. Soc. Chim. Belg.*, **77**, 153 (1968).

7. D. Seebach and N. Peleties, *Chem. Ber.*, **105**, 511 (1972); *Angew. Chem. Int. Ed. Engl.*, **8**, 450 (1969).

8. D. Seebach and A. K. Beck, *Angew. Chem. Int. Ed. Engl.*, **13**, 806 (1974).

9. (a) F. G. Bordwell, J. E. Bares, J. E. Bartmess, G. E. Drucker, J. Gerhold, G. M. McCollum, M. Van Der Puy, N. R. Vanier, and W. S. Matthews, *J. Org. Chem.*, **42**, 326 (1977); (b) F. G. Bordwell, M. Van Der Puy, and N. R. Vanier, *J. Org. Chem.*, **41**, 1883 (1976).

10. (a) H. J. Reich, F. Chow, and S. K. Shah, *J. Am. Chem. Soc.*, **101**, 6638 (1979); (b) H. J. Reich, W. W. Willis, Jr., and P. D. Clark, *J. Org. Chem.*, **46**, 2775 (1981); (c) D. J. Saez, "The chemistry of selenones" unpublished doctoral dissertation, University of Wisconsin, Madison, 1984.

11. G. Pourcelot and C. Georgoulis, *Bull. Soc. Chim. Fr.*, 866 (1964); G. Pourcelot and J.-M. Cense, *Bull. Soc. Chim. Fr.*, 1578 (1976).

12. J.-M. Lehn, G. Wipff, and J. Demuynck, *Helv. Chim. Acta*, **60**, 1239 (1977).

13. D. Seebach, J. Gabriel, and R. Haessig, *Helv. Chim. Acta*, **67**, 1083 (1984).

14. H. J. Reich, *J. Org. Chem.*, **40**, 2570 (1975).

15. R. H. Mitchell, *Chem. Commun.*, 990 (1975).

16. B.-T. Gröbel and D. Seebach, *Chem. Ber.*, **110**, 867 (1977).

17. D. Seebach, N. Meyer, and A. K. Beck, *Lieb. Ann. Chem.*, 846 (1977).

18. (a) W. Dumont, P. Bayet, and A. Krief, *Angew. Chem. Int. Ed. Engl.*, **13**, 804 (1974); (b) D. Van Ende, W. Dumont, and A. Krief, *Angew. Chem.*, **87**, 709 (1975); (c) D. Van Ende, A., Cravador, and A. Krief, *J. Organomet. Chem.*, **177**, 1 (1979).

19. (a) L. Brandsma, *Rec. Trav. Chim.*, **83**, 307 (1964). Ya. L. Gol'dfarb, V. P. Litvinov, and A. N. Sukiasyan, *Dokl. Akad. Nauk SSSR*, **182**, 340 (1968). V. P. Litvinov, I. P. Konyaeva, and Ya. L. Gol'dfarb, *Izv. Akad. Nauks SSSR, Ser. Khim.*, **2**, 477 (1976); (b) L. Christiaens, R. Dufour, and M. Renson, *Bull. Soc. Chim. Belg.*, **79**, 143 (1970); J. Morel, C. Paulmier, D. Semard, and P. Pastour, *C. R. hebd. Séances Acad. Sci., Ser. C*, **270**, 825 (1970); Ya. L. Gol'dfarb, V. P. Litvinov, and I. P. Konyaeva, *Khim. Geterotsikl Soed.*, **15**, 1072 (1979).

20. H. J. Reich and S. K. Shah, *J. Am. Chem. Soc.*, **97**, 3250 (1975); S. K. Shah, "Preparation, Reactions and Further Transformations of Alpha Lithio Selenides and Selenoxides. Synthetic Transformations Based on Propargyl Selenides." Unpublished doctoral dissertation, University of Wisconsin, Madison, Wisconsin 1977.

21. J. V. Comasseto and N. Petragnani, *J. Organomet. Chem.*, **152**, 295 (1978).

22. R. Holtan, Ph.D. Thesis, "Stereocontrolled Synthesis of Silyl Enol Ethers Using Alpha-Phenylthio silyl Ketones." Unpublished doctral dissertation, University of Wisconsin, Madison, Wisconsin, 1984.

23. H. J. Reich and M. L. Cohen, *J. Am. Chem. Soc.*, **101**, 1307 (1979); M. L. Cohen, "Organoselenium Chemistry. Some Mechanistic, Synthetic and Stereochemical Considerations." Unpublished doctoral dissertation, University of Wisconsin, Madison, Wisconsin, 1979.

24. S. Raucher and G. A. Koolpe, *J. Org. Chem.*, **43**, 3794 (1978).

25. B.-T. Gröbel and D. Seebach, *Chem. Ber.*, **110**, 852 (1977).

26. H. J. Reich, S. K., Shah and F. Chow, *J. Am. Chem. Soc.*, **101**, 6648 (1979).

27. C. A. Brandt, J. V. Comasseto, W. Nakamura, and N. Petragnani, *J. Chem. Res. (S)*, 156 (1983).

28. H. J. Reich and S. K. Shah, *J. Org. Chem.*, **42**, 1773 (1977).

29. K. Sachdev and H. S. Sachdev, *Tetrahedron Lett.*, 4223 (1976); 814 (1977).

30. J. D. White, M. Kang, and B. G. Sheldon, *Tetrahedron Lett.*, **24**, 4539 (1983).

31. M. R. Binns and R. K. Haynes, *J. Org. Chem.*, **46**, 3790 (1981).

274 **HANS J. REICH**

32. H. J. Reich, M. C. Schroeder, and I. L. Reich, *Israel J. Chem.*, **24**, 157 (1984).

33. (a) Y. Yamamoto, Y. Saito, and K. Maruyama, *Tetrahedron Lett.*, **23** 4597 (1982). Y. Yamamoto, H. Yatagai, Y. Saito, and K. Maruyama, *J. Org. Chem.*, **49**, 1096 (1984); (b) Y. Yamamoto, Y. Saito, and K. Maruyama, *J. Org. Chem.*, **48**, 5408 (1983).

34. H. Wetter, *Helv. Chim. Acta*, **61**, 3072 (1978).

35. H. J. Reich, M. C. Clark, and W. W. Willis, Jr., *J. Org. Chem.*, **47**, 1618 (1982).

36. H. J. Reich and S. K. Shah, *J. Am. Chem. Soc.*, **99**, 263 (1977); H. J. Reich, S. K. Shah, P. M. Gold, and R. E. Olson, *J. Am. Chem. Soc.*, **103**, 3112 (1981).

37. H. J. Reich and W. W. Willis, Jr., *J. Am. Chem. Soc.*, **102**, 5967 (1980).

38. M. Sevrin, J. N. Denis, and A. Krief, *Angew. Chem.*, **90**, 550 (1978).

39. T. Takahashi, H. Nagashima, and J. Tsuji, *Tetrahedron Lett.*, 799 (1978).

40. P. A. Grieco, M. Nishizawa, T. Oguri, S. D. Burke, and N. Marinovic, *J. Am. Chem. Soc.*, **99**, 5773 (1977).

41. D. Liotta, C. S. Barnum, and M. Saindane, *J. Org. Chem.*, **46**, 4301 (1981).

42. J. Lucchetti and A. Krief, *Tetrahedron Lett.*, 2693 (1978).

43. J. Lucchetti and A. Krief, *Tetrahedron Lett.*, 2697 (1978).

44. K. B. Sharpless, R. F. Lauer, and A. Y. Teranishi, *J. Am. Chem. Soc.*, **95**, 6137 (1973).

45. N. Petragnani and H. M. C. Ferraz, *Synthesis*, 476 (1978).

46. P. A. Grieco and M. Miyashita, *J. Org. Chem.*, **39**, 120 (1974).

47. Y. Masuyama, Y. Ueno, and M. Okawara, *Chem. Lett.*, 835 (1977).

48. P. A. Grieco and Y. Yokoyama, *J. Am. Chem. Soc.*, **99**, 5210 (1977).

49. T. Sakakibara, S. Ikuta, and R. Sudoh, *Synthesis*, 261 (1982).

50. A. Streitwieser, Jr., and S. P. Ewing, *J. Am. Chem. Soc.*, **97**, 190 (1975).

51. W. C. Still and D. Mobilio, *J. Org. Chem.*, **48**, 4785 (1983).

52. G. Zima, C. Barnum, and D. Liotta, *J. Org. Chem.*, **45**, 2736 (1980).

53. A. P. Batalov and L. A. Pogodina, *Zh. Obshch. Khim.*, **47**, 834 (1977); C. A. **87**,52401x (1977). A. P. Batalov, and G. A. Rostokin, *Zh. Obschch. Khim.*, **41**, 1735, 1738 (1971). A. P. Batalov, G. A. Rostokin, and I. A. Korshunov, *Tr. Khim. Tekhnol.*, 7 (1968); CA **71**, 60331 (1969). A. P. Batalov, G. A. Rostokin, and M. A. Skvortsova, *Zh. Obshch. Kim.*, **39**, 1840 (1969).

54. H. J. Reich, N. H. Phillips and I. L. Reich, J. Am. Chem. Soc., 107, 4101 (1985); I. L. Reich, unpublished results.

55. W. Dumont, M. Sevrin, and A. Krief, *Angew. Chem. Int. Ed. Engl.*, **16**, 541 (1977).

56. N. Petragnani and G. Schill, *Chem. Ber.*, **103**, 2271 (1970).

57. R. Pellicciari, M. Curini, P. Ceccherelli, and R. Fringvelli, *Chem. Commun.*, 440 (1979).

58. A. Anciaux, A. Eman, W. Dumont, D. Van Ende, and A. Krief, *Tetrahedron Lett.*, 1613 (1975).

59. W. Dumont and A. Krief, *Angew. Chem. Int. Ed. Engl.*, **16**, 540 (1977); A. Burton, L. Hevesi, W. Dumont, A. Cravador, and A. Krief, *Synthesis*, 877 (1979).

60. D. L. J. Clive and S. M. Menchen, *J. Org. Chem.*, **44**, 4279 (1979).

61. P. J. Giddings, D. I. John, and E. J. Thomas, *Tetrahedron Lett.*, **21**, 399 (1980); K. Hirai, Y. Iwano, and K. Fujimoto, *Tetrahedron Lett.*, **23**, 4021 (1982).

62. S. Halazy, J. Lucchetti, and A. Krief, *Tetrahedron Lett.*, 3971 (1978).

63. P. J. Stang, K. A. Roberts, and L. E. Lynch, *J. Org. Chem.*, **49**, 1653 (1984).

64. J. N. Denis and A. Krief, *Tetrahedron Lett.*, **23**, 3407 (1982).

65. (a) W. Dumont and A. Krief, *Angew. Chem.*, **87**, 347 (1975); (b) J. Rémion, W. Dumont, and A. Krief, *Tetrahedron Lett.*, 1385 (1976); (c) J. Rémion and A. Krief, *Tetrahedron Lett.*, 3743 (1976).

66. S. Halazy, F. Zutterman, and A. Krief, *Tetrahedron Lett.*, **23**, 4385 (1982).

67. D. Van Ende and A. Krief, *Tetrahedron Lett.*, 457 (1976).

68. (a) M. Sevrin, D. Van Ende, and A. Krief, *Tetrahedron Lett.*, 2643 (1976); (b) M. Clarembeau, J. L. Bertrand, and A. Krief, *Isr. J. Chem.*, **24**, 125 (1984).

69. (a) M. Sevrin and A. Krief, *Tetrahedron Lett.*, 187 (1978); (b) M. Sevrin and A. Krief, *Tetrahedron Lett.*, **21**, 585 (1980).

70. J. Lucchetti and A. Krief, *Tetrahedron Lett.*, **22**, 1623 (1981).

71. W. Dumont, D. Van Ende, and A. Krief, *Tetrahedron Lett.*, 485 (1979).

72. W. Dumont and A. Krief, *Angew. Chem.*, **88**, 184 (1976).

73. (a) D. Labar, L. Hevesi, W. Dumont, and A. Krief, *Tetrahedron Lett.*, 1141 (1978): (b) D. Labar, W. Dumont, L. Hevesi, and A. Krief, *Tetrahedron Lett.*, 1145 (1978); (c) S. Halazy and A. Krief, *Tetrahedron Lett.*, **22**, 1833 (1981).

74. J. N. Denis, W. Dumont, and A. Krief, *Tetrahedron Lett.*, 453 (1976).

75. M. Sevrin, W. Dumont, L. Hevesi, and A. Krief, *Tetrahedron Lett.*, 2647 (1976).

76. J. Lucchetti, W. Dumont, and A. Krief, *Tetrahedron Lett.*, 2695 (1979).

77. (a) D. Labar, J. Laboureur, and A. Krief, *Tetrahedron Lett.*, **23**, 983 (1982); (b) C. Schmit, S. Sahraoui-Taleb, E. Differding, C. G. D. Lombaert, and L. Ghosez, *Tetrahedron Lett.* **25**, 5043 (1984).

78. H. M. J. Gillissen, P. Schipper, P. J. J. M. van Ool, and H. M. Buck, *J. Org. Chem.*, **45**, 319 (1980).

79. S. Halazy, W. Dumont, and A. Krief, *Tetrahedron Lett.*, **22**, 4737 (1981).

80. (a) S. Halazy and A. Krief, *Chem. Commun.*, 1136 (1979); (b) S. Halazy and A. Krief, *Tetrahedron Lett.*, **22**, 1829 (1981).

81. S. Halazy and A. Krief, *Tetrahedron Lett.*, 4233 (1979).

82. (a) S. Halazy and A. Krief, *Chem. Commun.*, 1200 (1982); (b) S. Halazy and A. Krief, *Tetrahedron Lett.*, **21**, 1997 (1980).

83. J. N. Denis and A. Krief, *Tetrahedron Lett.*, **23**, 3411 (1982).

84. (a) H. J. Reich, J. M. Renga, and I. L. Reich, *J. Am. Chem. Soc.*, **97**, 5434 (1975); (b) S. J. Falcone and M. E. Munk, *Synth. Commun.*, **9**, 719 (1979); (c) D. Liotta, M. Saindane, and D. Brothers, *J. Org. Chem.*, **47**, 1598 (1982).

85. R. C. Gadwood, *J. Org. Chem.*, **48**, 2098 (1983).

86. A. Krief, W. Dumont, and J.-N. Denis, *J. Chem. Soc. Chem. Commun.*, 571 (1985).

87. (a) S. Raucher and G. A. Koolpe, *J. Org. Chem.*, **43**, 4252 (1978); (b) T. Kauffmann, H. Ahlers, H.-J. Tilhard, and A. Woltermann, *Angew. Chem. Int. Ed. Engl.*, **16**, 710 (1977).

88. T. Kauffmann, H. Ahlers, A. Hamsen, H. Schulz, H.-J. Tilhard, and A. Vahrenhorst, *Angew Chem. Int. Ed. Engl.*, **16**, 119 (1977)

89. (a) M. Shimizu and I. Kuwajima, *J. Org. Chem.*, **45**, 2921 (1980); (b) M. Shimizu and I. Kuwajima, *J. Org. Chem.*, **45**, 4063 (1980).

90. R. Ando, T. Sugawara, and I. Kuwajima, *Chem. Commun.*, 1514 (1983).

91. M. Shimizu, R. Ando, and I. Kuwajima, *J. Org. Chem.*, **46**, 5246 (1981).

92. P. Fuchs, private communication.

93. (a) J. Lucchetti and A. Krief, *J. Organomet. Chem.*, **194**, C49 (1980); (b) J. Lucchetti and A. Krief, *Chem. Commun.*, 127 (1982).

94. H. J. Reich, S. Wollowitz, J. E. Trend, F. Chow, and D. F. Wendelborn, *J. Org. Chem.*, **43**, 1697 (1978).

95. F. Ogura, T. Otsubo, and N. Ohira, *Synthesis*, 1006 (1983).

96. D. L. J. Clive and V. N. Kale, *J. Org. Chem.*, **46**, 231 (1981).

97. K. C. Nicolaou, W. J. Sipio, R. L. Magolda, and D. A. Claremon, *Chem. Commun.*, 83 (1979).

98. J. N. Denis, W. Dumont, and A. Krief, *Tetradehron Lett.*, 4111 (1979).

99. J. N. Denis and A.Krief, *Chem. Commun.*, 229 (1983).

100. (a) H. J. Reich and F. Chow, *Chem. Commun.*, 790 (1975); (b) D. L. J. Clive, C. G. Russell, and S. C. Suri, *J. Org. Chem.*, **47**, 1632 (1982); (c) C. J. Kowalski and J.-S. Dung, *J. Am. Chem. Soc.*, **102**, 7950 (1980).

101 (a) L. Hevesi, J. L. Piquard, and H. Wantier, *J. Am. Chem. Soc.*, **103**, 870 (1981); (b) R. A. McClelland and M. Leung, *J. Org. Chem.*, **45**, 187 (1980).

102. J. Lucchetti and A. Krief, *Synth. Commun.*, **13**, 1153 (1983).

103. M. Sevrin and A. Krief, *Chem. Commun.*, 656 (1980).

104. L. Hevesi, M. Sevrin, and A. Krief, *Tetrahedron Lett.*, 2651 (1976).

105. L. Fitjer, H.-J. Shevermann, and D. Wehle, *Tetrahedron Lett.*, **25**, 2329 (1984).

106. L. B. Agenas, *Ark. Kemi*, **23**, 463 (1965).

107. D. L. J. Clive, G. Chittattu, and C. K. Wong, *Chem. Commun.*, 41 (1978).

108. R. O. Hutchins, and K. Learn, *J. Org. Chem.*, **47**, 4380 (1982).

109. H.J. Reich, J. Ringer, I. L. Reich, and C. A. Hoeger, unpublished results.

6

The Chemistry of Selenocarbonyl Compounds

FRANK S. GUZIEC, JR.

Chemistry Department, New Mexico State University,
Las Cruces, New Mexico

CONTENTS

1. GENERAL CONSIDERATIONS

Studies of the chemistry of the carbon–selenium double bond have until recently been relatively limited. Selenocarbonyl compounds are much less well known than their thiocarbonyl analogs. There are a number of reasons for this obscurity. First of all, in many cases, the early literature describing the chemistry of selenocarbonyl compounds is fraught with confusion.[a] Many supposed selenocarbonyl compounds were subsequently shown not to contain a carbon–selenium double bond. Secondly, and perhaps a cause for this confusion, is the fact that although selenium compounds in many cases closely resemble their sulfur analogs, some significant differences do occur. While carbon disulfide and phosphorus (V) sulfide have been widely used in the synthesis of thiocarbonyl compounds, the selenium analogs prove to be much less useful in the preparation of selenocarbonyl compounds. For example, carbon diselenide is an intensely unpleasant compound that should be handled with great care.[1] This compound also exhibits a great tendency to polymerize under nucleophilic reaction conditions, limiting its utility relative to carbon disulfide. Phosphorus (V) selenide is an extremely insoluble material that is very difficult to characterize and has rarely been used successfully in selenations.[2] In addition, in many cases selenocarbonyl compounds are quite sensitive to light or moisture, liberating finely divided red selenium upon handling.

Finally, while most organic chemists have some experience in handling sulfur-containing compounds, selenium-containing reagents have until recently not been widely used in synthesis. The resulting "selenophobia" is perhaps not totally without reason. Many selenium compounds are known to be highly toxic. Hydrogen selenide, which was widely used in many attempted selenocarbonyl preparations, is extremely toxic. The acute unpleasant effects of even a small dose of hydrogen selenide are well known.[3] Still, many recently developed procedures avoid the hazards encountered by early researchers, and because of

---

[a] These early reports on the chemistry of the selenocarbonyl group have been extremely well documented.[4]

their unique reactivity in many cases selenocarbonyl compounds have proved to be valuable intermediates in organic synthesis.

A number of reviews on selenocarbonyl chemistry have appeared.[1,4-8] This discussion will concentrate on general preparations, reactions, and spectroscopic properties of well-characterized selenocarbonyl compounds, with special emphasis on the more recent literature.

2. STRUCTURE AND BONDING

The nature of carbon–selenium bonding in selenocarbonyl compounds has been a matter of question for some time. The inability initially to prepare selenium analogs of ketones that were not stabilized by resonance or by ligands strengthened the view that a *pure* carbon–selenium double bond would not be stable.[9] Indeed, due to less favorable overlap, the 2p–4p π bond would be expected to be less stable than the 2p–3p π bond of thiones and much less stable than the 2p–2p π bond of ketones. The recent preparation of stable selones whose spectroscopic and chemical properties closely resemble those of thiones (with a well-established carbon–sulfur double bond) strongly suggests the double bond nature of a selenocarbonyl. Recent calculations based on[77]Se and [17]O nuclear magnetic resonance (nmr) data on selones and ketones indicate that the bond orders of the selenocarbonyl and the carbonyl groups in these molecules are very similar.[10] Only a limited number of X-ray crystal structures of resonance stabilized selenocarbonyl compounds have been determined.[11] These are consistent with the supposed double bond structure of the selenocarbonyl. It is worth noting that most selenocarbonyl reactions involve the ready conversion of a carbon–selenium double bond into a carbon–selenium single bond.

Another important consideration is the polarity of the carbon–selenium double bond. While no data on dipole moments of the *pure* selenocarbonyl of selones has been reported, some useful information can be obtained by examination of orbital electronegativities (Table 1). A simple comparison of the σ and π components of the carbonyl, thiocarbonyl, and selenocarbonyl would lead to the prediction that the properties of selenocarbonyl compounds would much more closely resemble those of their sulfur analogs than the properties

TABLE 1

ORBITAL ELECTRONEGATIVITIES IN CARBONYL, THIOCARBONYL, AND SELENOCARBONYL COMPOUNDS, PAULING SCALE[a]

	C	O	S	Se
$\sigma\ sp^2$	2.75	5.54	3.46	3.29
$\pi\ p$	1.68	3.19	2.40	2.31

[a] Data from J. Hinze and H. H. Jaffe, *J. Am. Chem. Soc.*, **84**, 540 (1962); *J. Phys. Chem.*, **67**, 1501 (1963).

TABLE 2
STERIC CONSIDERATIONS IN CARBONYL, THIOCARBONYL,
AND SELENOCARBONYL COMPOUNDS[a]

Covalent Radii	0 (Å)	S (Å)	Se (Å)
Single bonded	0.74	1.04	1.17
Double bonded	0.62	0.94	1.07

[a] Data from L. Pauling, *The Nature of the Chemical Bond*, 3rd ed., Cornell University Press, Ithaca, 1960, pp. 224–228.

of thiocarbonyl compounds resemble those of the coresponding carbonyl compounds. This is in fact the case.

Finally, the greater steric bulk of selenium relative to sulfur (Table 2) and its more ready polarizability may lead in some cases to significant differences between the properties of selenocarbonyl- and the corresponding thiocarbonyl-containing molecules. All of these aspects of the carbon–selenium double bond lead to a diverse and occasionally unpredictable chemistry.

3. SELONES AND SELENOALDEHYDES

3.1. Early Reports

Although there are numerous reports of the preparations of simple monomeric selones (selenium analogs of ketones) and selenoaldehydes prior to 1975, there is good evidence that these compounds did not, in fact, contain a selenocarbonyl moiety.[12] Many such attempts involved the reaction of an aldehyde or ketone with hydrogen selenide or its salts. The analogous procedure using hydrogen sulfide has proved useful for the preparation of a variety of thiones. The products of the early hydrogen selenide reactions with ketones and moderately hindered aldehydes were often reported to red oils, and were generally characterized as dimeric. Subsequently a number of these materials were shown to be diselenides.[12] While disulfides are typically not observed in the thione preparations, hydrogen selenide is a significantly better reducing agent than hydrogen sulfide. Presumably the initially formed selenocarbonyl compound, 1, is reduced by hydrogen selenide to the unstable selenol, 2, generating red selenium, and under work-up conditions air oxidation to the deselenide, 3, occurs [Eq. (1)]. With unhindered aldehydes the hydrogen selenide reaction takes a different course, affording true trimeric selenoaldehyde derivatives 4 [Eq. (2)].[13,14]

Since no selenium is generated in these reactions, presumably no free selenoaldehyde is formed as an intermediate. In the case of benzaldehyde both reduction and a trimerization appear to occur simultaneously.[15,16] The reported thermal conversion of the benzaldehyde trimer 5 to *trans*-stilbene, 6 [Eq. (3)][17]

$$\underset{R}{\overset{R}{>}}C=O + H_2Se \longrightarrow \left[\underset{R}{\overset{R}{>}}=Se\right]$$
$$\underline{1}$$

$$\Bigg\downarrow {\scriptstyle H_2Se} \tag{1}$$

$$\left(\underset{R\ H}{\overset{R}{>}}-Se\overline{}\right)_2 \longleftarrow \left[\underset{R}{\overset{R}{>}}\underset{SeH}{\overset{H}{<}}\right] + Se$$
$$\underline{3} \qquad\qquad \underline{2}$$

$$H_2C=O + H_2Se \xrightarrow{\text{HCl}} \underset{Se\ \ \ Se}{\overset{Se}{\bigcirc}} \tag{2}$$
$$\underline{4}$$

(and presumably the trimer of the selenoaldehyde of 11-phenylundecapentaenal to 1,30-diphenyltricontapentadecaene)[18] perhaps deserves further attention. An interesting discussion of these early attempted selenocarbonyl preparations has been previously related.[4]

$$\underset{Se\ \ \ Se}{\overset{R\ \ Se\ \ R}{\bigcirc}}_{R} \xrightarrow{\ \Delta\ } \underset{H}{\overset{R}{>}}C=C\underset{R}{\overset{H}{<}} \tag{3}$$
$$\underline{5} \qquad\qquad\qquad \underline{6}$$
$$R = Ph, Ph(CH=CH)_5$$

3.2. Preparation

The first well-characterized preparation of a simple monomeric selone un-stabilized by vinylogy or metal ligands involved the pyrolysis of a phospho-ranylidene hydrazone 7 of a hindered ketone in the presence of selenium powder affording the selone, 8, nitrogen, and triphenylphosphine selenide, 9, [Eq. (4)].[19,20] This method proved to be very useful for the preparation of a variety of very sterically hindered selones (and thiones)[21] in good yield. Sulfur appears to be more reactive than selenium in this reaction.[21a] One problem with this method is the difficulty in controlling the temperature of the pyrolysis; the reaction is also quite sensitive to reagent purity. An additional shortcoming of this procedure is the fact that only very hindered selones can be prepared from the phosphoranylidene hydrazone. Attempts to prepare selenobenzophenone or selenocamphor under these conditions led only to the corresponding dimeric olefins. Pyrolysis of phosphoranylidene hydrazones is known to afford diazo compounds in good yield; these diazo compounds can react with the liberated selones affording olefins [see Eq. (9)]. While the mechanism of this selone preparation remains unclear, it is likely that the diazo compound is an intermediate in this reaction.[22,23]

$$
\underset{\underset{7}{}}{\text{N}\!\!=\!\!\text{PPh}_3} + \text{Se} \xrightarrow{120°\text{C}} \underset{\underset{8}{35\%}}{=\text{Se}} + \text{Ph}_3\text{P}\!\!\rightarrow\!\!\text{Se} + \text{N}_2 \qquad (4)
$$

Fenchone **10**, is reported to react with a mixture of bis-tricyclohexyltin selenide, **11**,–boron trichloride to afford selenofenchone, **12**, in excellent yield [Eq. (5)].[24] As yet no useful selenium analog of dimeric p-methoxyphenylthionophosphine sulfide, **13**, which is widely used as a thionating reagent for carbonyl compounds,[25] has been described.

$$
\underset{\underset{10}{}}{\text{O}} + [(\text{C}_6\text{H}_{12})_3\text{Sn}]_2\text{Se} \xrightarrow{\text{BCl}_3} \underset{\underset{12}{90\%}}{\text{Se}} \qquad (5)
$$

$$
\text{CH}_3\text{O}\!\!-\!\!\underset{\underset{13}{}}{\overset{\overset{\text{S}}{\parallel}}{\text{P}}\underset{\text{S}}{\overset{\text{S}}{\diagup}}\overset{\overset{\text{S}}{\parallel}}{\text{P}}}\!\!-\!\!\text{OCH}_3
$$

Perhaps the most convenient method to date for the preparation of sterically hindered selones involves the reaction of a hydrazone, **14**, with selenium (I) bromide,[23] analogous to the thionation using sulfur (I) chloride.[23a] Yields of 65–80% are typically obtained. Presumably the reaction proceeds through a N-selenonitrosimine, **15**, which cyclizes to the selenadiazetine, **16**. Extrusion of nitrogen affords the selone [Eq. (6)]. It is possible that these intermediates are also involved in the phosphoranylidene hydrazone reaction. The selenium (I) bromide procedure also does not afford the less hindered selenobenzophenone, selenofluorenone, or selenocamphor.

As yet, there are no reports of preparations of simple selenoaldehydes, however, the recent generation and trapping of unstable thioaldehydes,[26,27] and the isolation of the hindered 2,4,6-tri-*tert*-butylthiobenzaldehyde[28] suggests that routes to simple selenoaldehydes may become available.

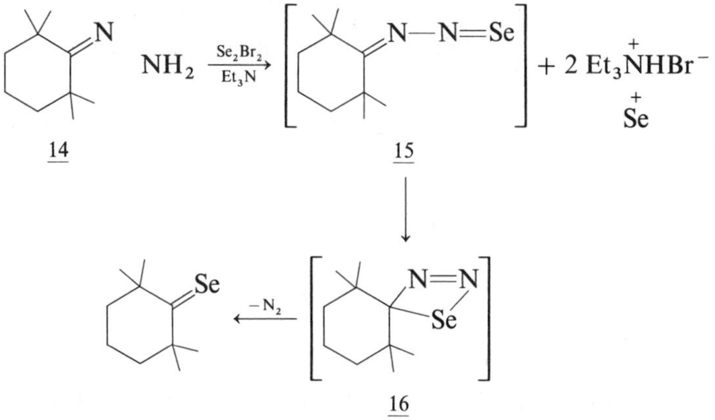

$$+ 2 \text{ Et}_3\overset{+}{\text{N}}\text{HBr}^- \qquad (6)$$

3.3. Reactions

Sterically hindered selones are typically deep-blue materials, stable for prolonged periods at 150°C under an inert atmosphere.[20] They are stable to visible light under nitrogen, but undergo rapid photochemical oxidation to the ketone and selenium in air. In the dark no decomposition of the selone is noted over extended periods of time.

Selones can be quantitatively reduced with sodium borohydride to the air-sensitive selenols, 17, which upon work-up afford diselenides, 18.[20] Lower yields of the same products are obtained with sodium or sodium amalgam [Eq. (7)].

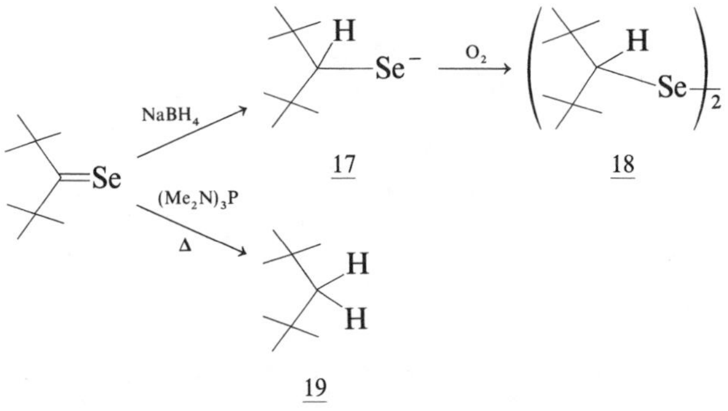

$$(7)$$

Treatment of a selone with tri-n-butylphosphine or tris(dimethylamino) phosphine at elevated temperatures affords the corresponding alkane, 19, in modest yield, as does reduction with a sodium–potassium alloy (1:5).[20] Reaction of the selone with pentacarbonyliron also led to selenium extrusion.

Low temperature oxidation of di-tert-butyl selone with m-chloroperoxy-benzoic acid afforded a thermally unstable compound, presumably the selenine,

20, which rapidly decomposed at $-20°C$ affording the ketone and selenium [Eq. (8)]. This selenine is considerably less stable than its sulfine analog, which can be easily isolated and is stable at room temperature.[20]

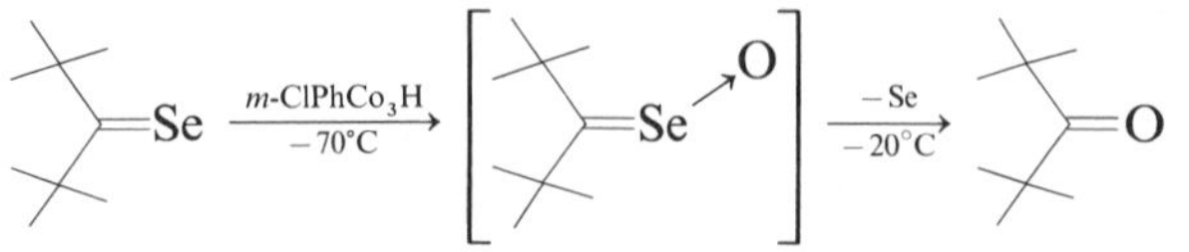

$$\text{(8)}$$

Selones, like thiones, readily undergo dipolar cycloaddition reactions. The reaction of a selone with a sterically hindered diazo compound has provided the best route for the preparation of extremely sterically hindered alkenes [Eq. (9)].[20,29-32] In this "twofold extrusion" reaction the initially formed cycloaddition product, a 1,3,4-selenadiazoline, 21, upon heating, extrudes nitrogen affording a thermally unstable episelenide, 22, as an intermediate. Thermal extrusion of selenium then affords the olefin, 23. In a number of cases symmetrical selenadiazolines have proved remarkably stable. Photolysis of the selenadiazoline affords the corresponding azine, 24, in good to excellent yield [Eq. (10)].[33] Selones appear to be more reactive than thiones in cycloadditions. In a competition experiment, an equimolar solution of di-*tert*-butyl selone and di-*tert*-butyl thione was treated with one equivalent of diphenyldiazomethane at $0°C$ [Eq. (11)]. Selenadiazoline, 25, was immediately formed to the exclusion of thiadiazoline, indicating the greater reactivity of the selone in the cycloaddition.[20] The successful use of the selone in the twofold extrusion preparation of the biindanylidene, 26, is reported when the corresponding reaction of the thione fails [Eq. (12)].[34] In addition use of the selone instead of a thione in the olefin preparation avoids the necessity of desulfurization with a phosphine, as the intermediate episelenide thermally extrudes selenium directly.

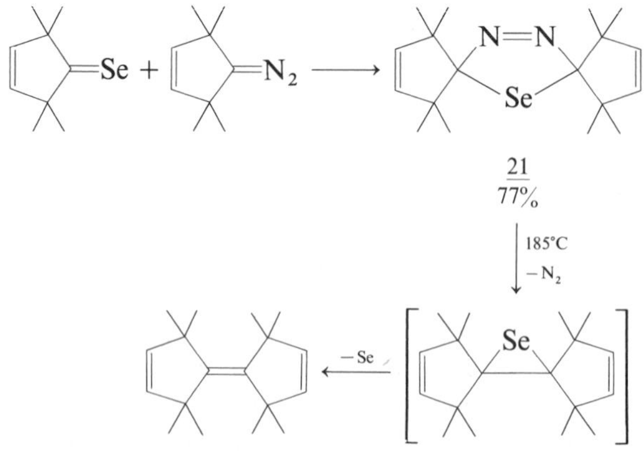

$$\text{(9)}$$

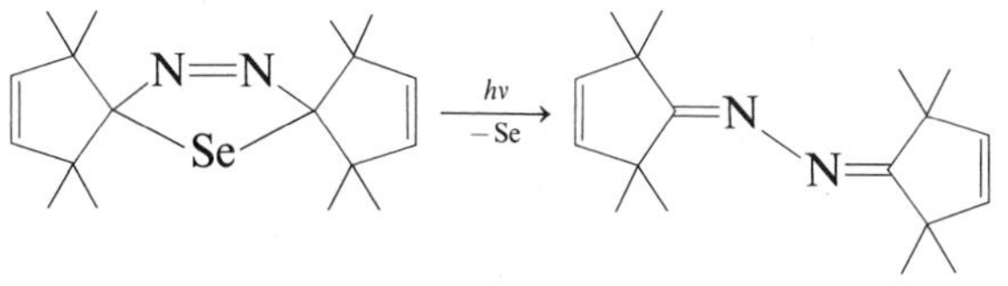

$$(10)$$

$$\underline{24}$$
$$80\%$$

$$(11)$$

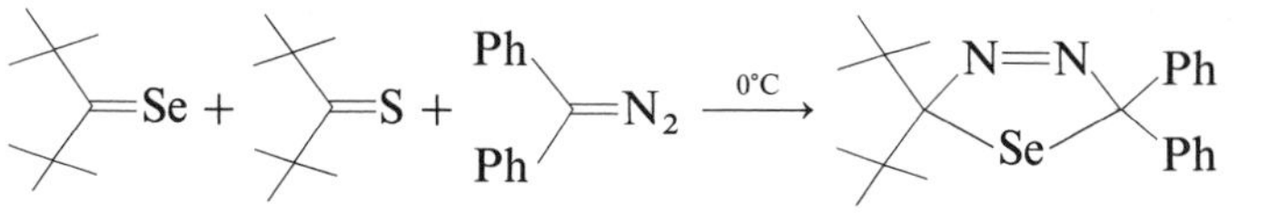

1 eq   1 eq   1 eq $\underline{25}$

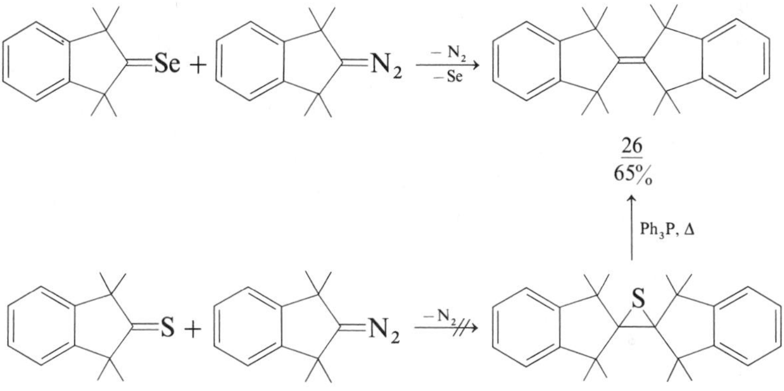

$$\underline{26}$$
$$65\%$$

$$(12)$$

One limitation of the twofold extrusion method with selones is its inability to prepare the extremely hindered tetra-*tert*-butylethylene, $\underline{27}$, or many unsymmetrical olefins containing geminal *tert*-butyl groups, $\underline{29}$.[20] In the case of $\underline{27}$ no selenadiazoline forms; in the second case the selenadiazoline $\underline{28}$ forms, but thermally retrocyclizes rather than extrudes nitrogen [Eq. (13)].[33]

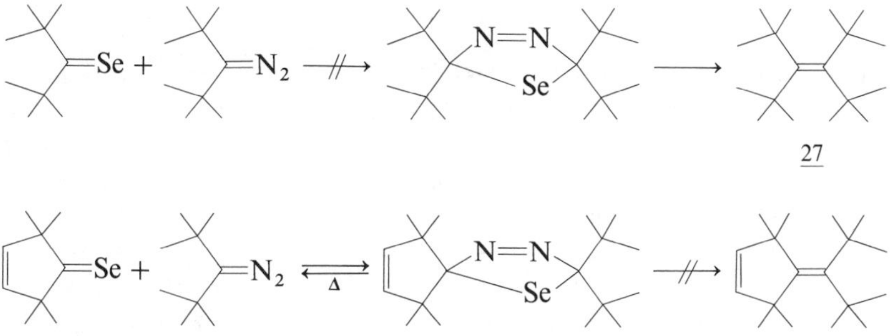

$$\underline{27}$$

$$(13)$$

$$\underline{28} \qquad\qquad \underline{29}$$

Selones also react with phenyl azide, 30, affording N-phenylimines, 32, in good to excellent yield [Eq. (14)].[35] By analogy to the diazo compound reaction, it is assumed that this process involves a 1,2,3-selenatriazoline intermediate, 31, which can, upon heating, extrude nitrogen and selenium affording the imine. While thiones react similarly, selones again are more reactive, and the products of the selone reaction are formed in higher yields.

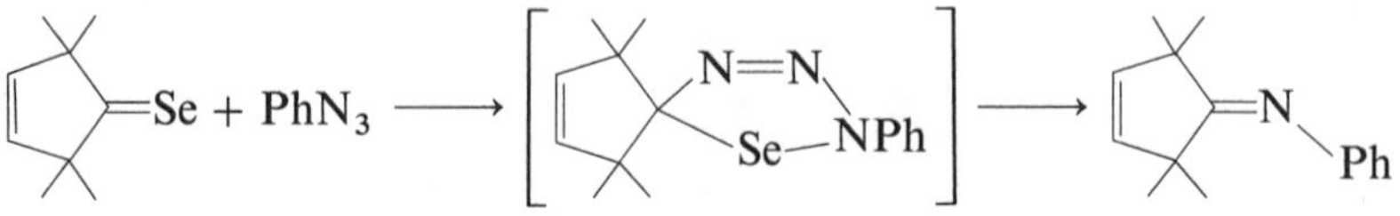

Sterically hindered selones react with organometallic compounds affording products of reduction as well as "selenophilic" addition, analogous to the related "thiophilic" reactions of thiones [Eq. (15)]. Di-*tert*-butyl selone reacts with phenyllithium to afford only the selenophilic product, di-*tert*-butylmethyl phenyl selenide 33.[36] In contrast the corresponding thione affords a mixture of thiophilic and carbophilic products.[36a] The stability of persistent selenoalkyl radicals [see Eq. (9)] is consistent with an electron transfer mechanism that has been similarly proposed for the thiophilic reaction.

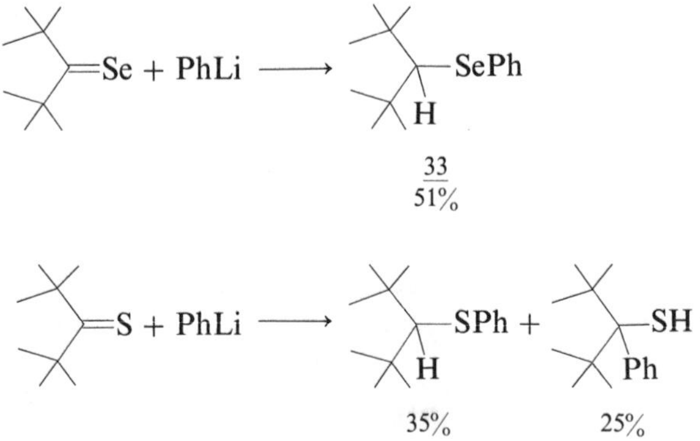

Heating a selone with a slight excess of sulfur affords the corresponding thione in good yield [Eq. (16)].[37] Selenoesters and selenoamides react similarly. This method appears to be very useful for introducing isotopically labeled sulfur into a thiocarbonyl moiety. A related conversion of the vinylogous selenocarbonate, 34, utilizes phosphorus pentasulfide [Eq. (17)].[38]

The detailed photochemistry of two selones has been investigated [Eq. (18)].[39,40] Like thiones, selones exhibit a weak intensity band in the visible region of the absorption spectrum. Unlike thiones, which upon radiation in the presence of olefins afford thietanes, the selones studied do not produce cyclo-

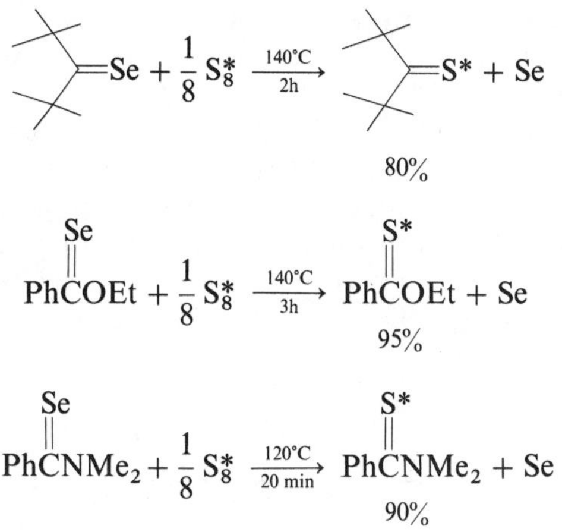

$$\text{t-Bu}_2C{=}Se + \frac{1}{8}S_8^* \xrightarrow[\text{2h}]{140°C} \text{t-Bu}_2C{=}S^* + Se$$

80%

$$
\underset{\text{Se}}{\overset{\text{Se}}{\text{PhCOEt}}} + \frac{1}{8}S_8^* \xrightarrow[\text{3h}]{140°C} \underset{\text{S*}}{\overset{\text{S*}}{\text{PhCOEt}}} + Se
\tag{16}
$$

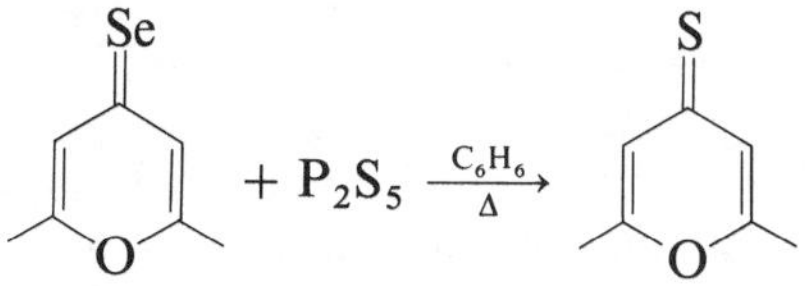

95%

$$
\underset{\text{Se}}{\overset{\text{Se}}{\text{PhCNMe}_2}} + \frac{1}{8}S_8^* \xrightarrow[\text{20 min}]{120°C} \underset{\text{S*}}{\overset{\text{S*}}{\text{PhCNMe}_2}} + Se
$$

90%

$$\text{Se-pyranone} + P_2S_5 \xrightarrow[\Delta]{C_6H_6} \text{S-pyranone} \tag{17}$$

<u>34</u>

adducts. The main product of irradiating di-*tert*-butyl selone in a variety of solvents is the diselenide. The intermediate radical has been detected in a low temperature irradiation. Irradiation of selenofenchone affords a mixture of selenium containing compounds, including products of intermolecular hydrogen transfer. No related intermolecular hydrogen transfer reaction is observed upon photolysis of thiofenchone, the products being the thiol and disulfide.

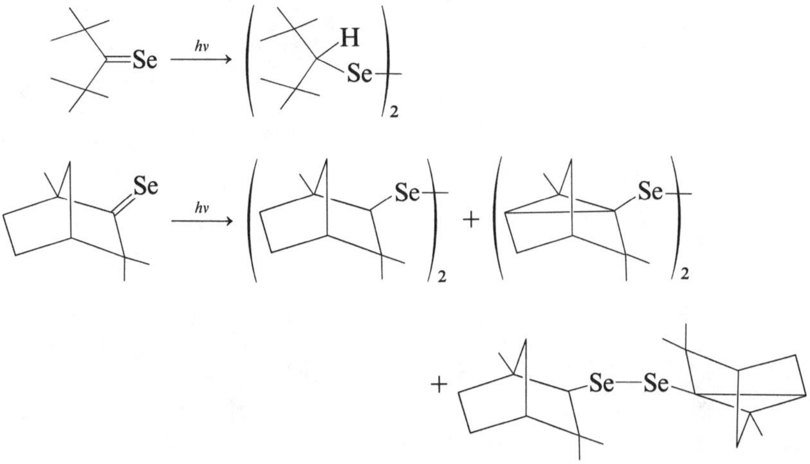

$$\text{t-Bu}_2C{=}Se \xrightarrow{h\nu} \left(\underset{\text{Se}}{\overset{\text{H}}{\text{t-Bu}_2C}} \right)_2$$

$$\text{selenofenchone} \xrightarrow{h\nu} \left(\text{—Se} \right)_2 + \left(\text{—Se} \right)_2 \tag{18}$$

$$ + \text{—Se—Se—} $$

The reactions of di-*tert*-butyl selone with radical species have been studied in detail. This selone reacts with a number of photochemically generated transient radicals (R_nM, where $M = C$, Se, O, Si, S, Sn, Ge) affording persistent selenoalkyl radicals [Eq. (19)].[41,42] The electron spin resonance (esr) parameters of these selenoalkyl radicals are sensitive to the nature of R_nM. Low spin density was observed on selenium in the selenoalkyl radicals suggesting an eclipsed geometry for these species. The selone proved to be a better radical trap than the corresponding thione, and has been used to trap 1,4-biradicals formed in photochemical Norrish type II reactions.[43] Selones may also be useful as radical traps in other applications.

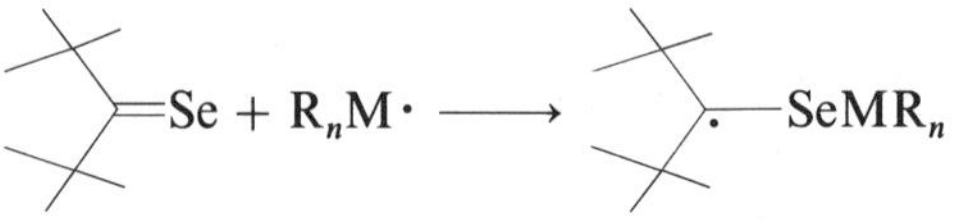

$$\text{\LARGE $\bigtimes$}{=}Se + R_nM\cdot \longrightarrow \text{\LARGE $\bigtimes$}\cdot{-}SeMR_n \qquad (19)$$

3.4. Stabilized Selones

Many selenocarbonyl compounds can be greatly stabilized by conjugation or ligand coordination. A number of selenium analogs of β-diketones have been prepared. These compounds are stabilized by intramolecular hydrogen bonding or by metal ligands. While the diselenoacetylacetone nickel complex, <u>35</u>, is

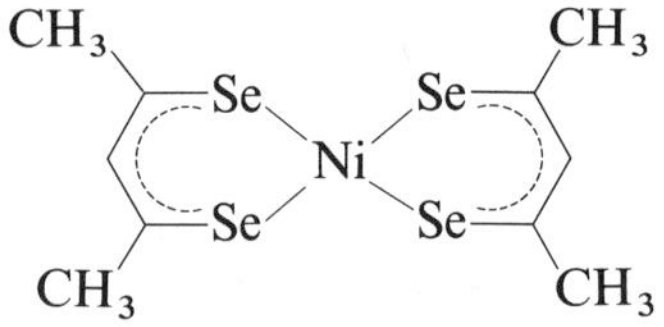

<u>35</u>

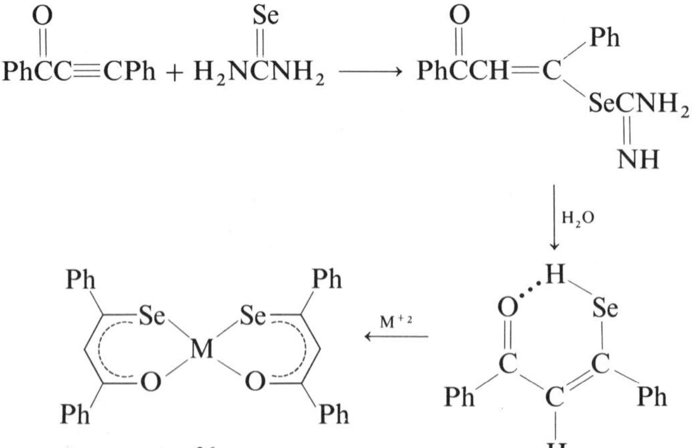

$$\underset{\text{PhCC}\equiv\text{CPh}}{\overset{\displaystyle O}{\overset{\|}{}}} + \text{H}_2\text{NCNH}_2 \longrightarrow \ldots \qquad (20)$$

<u>36</u>

relatively unstable in solution,[44] a variety of stable metal complexes, <u>36</u>, of mono- and diselenobenzoylacetophenone have been prepared [Eq. (20)].[45-48].

The bromocyclobutenedione, <u>37</u>, reacts with hydrogen selenide in pyridine to afford the resonance stabilized selone enolate, <u>38</u> [Eq. (21)].[49] This compound is protonated and alkylated on selenium, and reacts with aniline to afford the internal salt. These reactions and the spectral properties of the anion indicate there is little selenocarbonyl character associated with this enolate. Vinylogous stabilization of other selenocarbonyl compounds is discussed later.

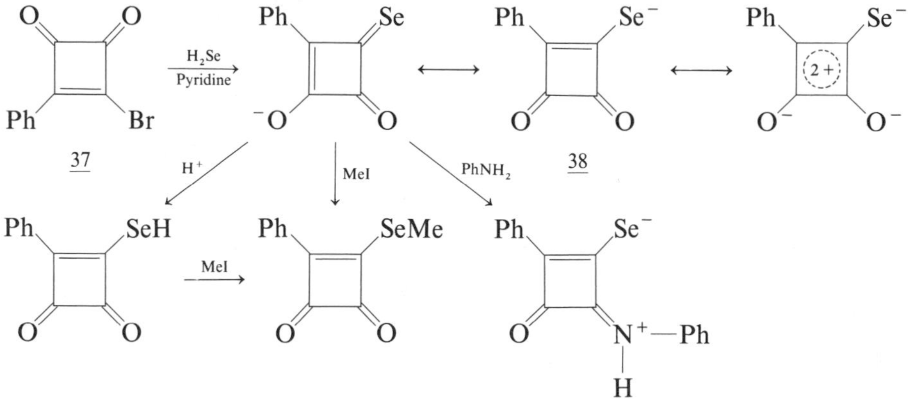

$$(21)$$

Other less general preparations and reactions of stabilized selones have been previously reported.[4]

4. SELENOESTERS

4.1. Preparation

A number of general methods for the preparation of selenoesters have recently been reported. Perhaps the most convenient method involves the reaction of dialkyliminium esters with sodium hydrogen selenide [Eq. (22)]. The iminium esters can be readily prepared by the reaction of a dialkylamide with phosgene, followed by the reaction of the intermediate iminium chloride, <u>39</u>, with the desired alcohol.[50] This procedure avoids the handling of the highly toxic hydrogen selenide by using an ethanolic solution of sodium hydrogen selenide prepared by borohydride reduction of selenium powder.[51] This method can be used for the preparation of a great variety of selenoesters <u>40</u>, affording selenobenzoates in typical yields of near 90%, and selenoformates and aliphatic selenoesters in about 70% yields. It is worth noting that when sodium hydrogen telluride was substituted for sodium hydrogen selenide, the product was the corresponding ether.[50]

$$
\underset{RCNR'_2}{\overset{\overset{\displaystyle O}{\|}}{}} \quad \xrightarrow{\text{COCl}_2} \quad R-\underset{\underset{39}{+}}{\overset{\overset{\displaystyle Cl}{|}}{C}}=NR'_2 Cl^-
$$

$$\downarrow R''OH \qquad\qquad (22)$$

$$
\underset{40}{\underset{R}{\overset{\overset{\displaystyle Se}{\|}}{\diagdown}} \diagup^{OR''}} \quad \xleftarrow[\text{EtOH}]{\text{NaHSe}} \quad R-\underset{\underset{+}{}}{\overset{\overset{\displaystyle OR''}{|}}{C}}=NR'_2 Cl^-
$$

R = aryl, alkyl, H
R' = alkyl
R'' = alkyl, phenyl

While phenylselenoacetate could be prepared in 65% yield using this, the sodium hydrogen selenide procedure, only the selenoamide, 41, could be isolated (77% yield) starting with the more hindered isobutyramide [Eq. (23)].[50]

$$
\underset{Cl^-}{\overset{\displaystyle OPh}{\underset{N}{\overset{+}{\diagup}}}} \quad \xrightarrow[\text{EtOH}]{\text{NaHSe}} \quad \underset{41}{\overset{\displaystyle Se}{\underset{N}{\diagup}}} \qquad\qquad (23)
$$

Related procedures involve the reaction of an imidate ester, 42,[52,53] readily available by acid catalyzed reaction of a nitrile with a low molecular weight alcohol, with hydrogen selenide [Eq. (24)]. Typical yields are 60–90%.

$$
RCN \xrightarrow[\text{HCl}]{\text{EtOH}} \underset{R}{\overset{\overset{\displaystyle \overset{+}{N}H_2 Cl^-}{\|}}{C}}\diagup^{OEt} \xrightarrow[-15°C]{\text{NH}_3} \underset{R}{\overset{\overset{\displaystyle N}{\|}}{C}}\diagup^{OEt} \xrightarrow{\text{H}_2\text{Se}} \underset{}{\overset{\overset{\displaystyle Se}{\|}}{RCOEt}} \qquad (24)
$$

$$42$$

This procedure also has been used for the preparation of selenolactones, 43 [Eq. (25)].[54] The diselenolactone, 43c, proved to be too unstable to be isolated.

Selenoesters, unsubstituted at the α position can be prepared by base catalyzed fragmentation of 4-substituted-1,2,3,-selenadiazolines, 44, presumably through the intermediacy of a selenoketene, 45 [Eq. (26)].[55]

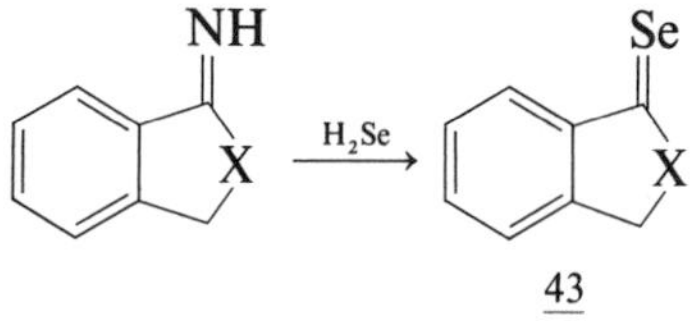

$$(25)$$

(a) X = O
(b) X = S
(c) X = Se

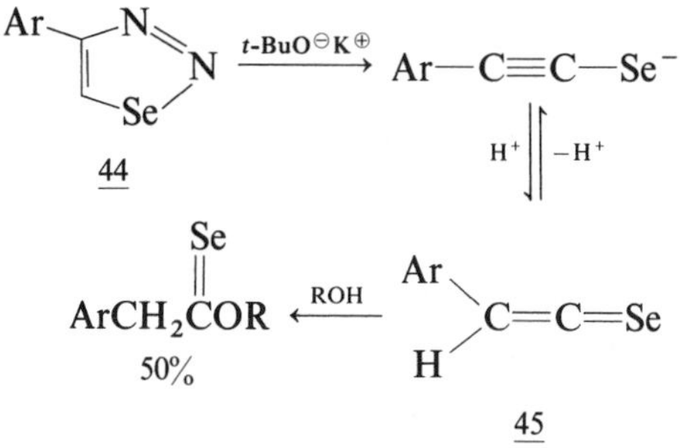

$$(26)$$

A less general procedure involves the reaction of pentacarbonylchromium (0) arylalkoxycarbenes, 46, with selenium to afford selenoesters in low yield [Eq. (27)].[56]

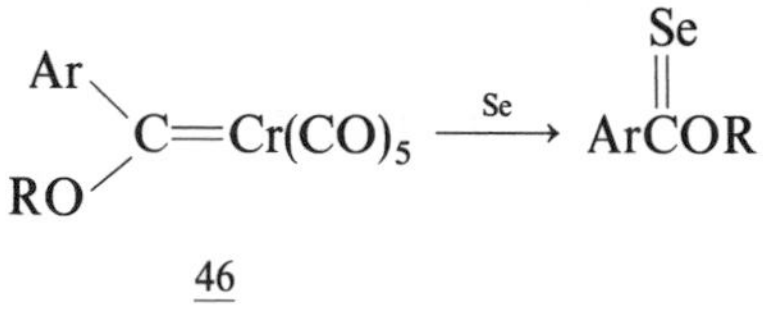

$$(27)$$

4.2. Reactions

Aliphatic selenoesters are typically yellow liquids while aromatic selenoesters are typically red oils.[53] The aromatic compounds in general are more stable than the aliphatic. They slowly deposit red selenium at room temperature but appear to be relatively stable under refrigeration.

Reduction of a selenoester with W-2 Raney nickel affords the corresponding ether, 47, in good yield [Eq. (28)].[57] This procedure is an extremely mild method for the preparation of ethers and has been applied to complex molecules such as amino glycoside antibiotics. Borohydride reduction of a selenoester followed by reaction with triethylphosphine also affords the ether, presumably due to the affinity of selenium for phosphorus. Quenching of the intermediate by alkylation affords the alkoxyalkyl alkylselenide, 48, while oxidation of this intermediate affords the bisalkoxyalkyl diselenide, 49.[57]

Treatment of the selenobenzoate of cholesterol, 50, with tributylstannane in refluxing toluene led to deoxygenation of the sterol [Eq. (29)].[50] While this

$$
\underset{\substack{R = H, CH_3 \\ R' = Cholesteryl}}{\overset{\overset{\displaystyle Se}{\overset{\|}{C}}}{R \diagdown OR'}}
\begin{array}{l}
\xrightarrow{\text{Raney Ni}} RCH_2OR' \quad \underline{47} \\[2mm]
\xrightarrow[\text{(2) Et}_3\text{P}]{\text{(1) NaBH}_4} RCH_2OR' \\[2mm]
\xrightarrow{\text{NaBH}_4}
\end{array}
\left[\underset{\substack{\\ H}}{\overset{Se^-}{\underset{|}{R-C-OR'}}}\right]
\begin{array}{l}
\xrightarrow{\text{MeI}} \underset{\underline{48}}{\overset{SeMe}{\underset{|}{RCHOR'}}} \\[3mm]
\xrightarrow{O_2} \underset{\underline{49}}{\overset{OR'}{\underset{|}{(RCHSe)_{\overline{2}}}}}
\end{array}
\qquad (28)
$$

reaction appears to be more rapid than that of the corresponding thiobenzoate, the yield of deoxygenated product is lower and more by-products are formed in this selenium reaction. The method does appear to be general for deoxygenations of secondary alcohols.[50]

$$
\underset{\underline{50}}{\overset{\overset{\displaystyle Se}{\overset{\|}{PhCO}}}{\text{(structure)}}} \xrightarrow[\substack{\text{toluene} \\ \Delta}]{\text{Bu}_3\text{SnH}} \underset{60\%}{\text{(structure)}} \qquad (29)
$$

Reaction of the selenobenzoates with methylenetriphenylphosphorane affords the unstable enol ethers, $\underline{51}$, in moderate yield [Eq. (30)].[57]

$$
\underset{\substack{\\ \\}}{\overset{\overset{\displaystyle Se}{\overset{\|}{PhCOR}}}{}} + Ph_3P{=}CH_2 \longrightarrow \underset{\substack{\underline{51} \\ 58-79\%}}{\overset{\overset{\displaystyle CH_2}{\overset{\|}{PhCOR}}}{}} \qquad (30)
$$

Treatment of O-cholesteryl selenoacetate, $\underline{52}$, with bis-(trimethylsilyl)-amide afforded the crotonate, $\underline{53}$ [Eq. (31)].[57] This is in marked contrast to O-alkyl thioesters, which are known to give normal Claisen products under similar conditions. Presumably in the selenium case expulsion of KSe^- is favored over loss of alkoxide. When ethyl selenoisobutyrate, $\underline{54}$, was treated under similar conditions and the intermediate alkylated with methyl iodide the dimethylketene monoselenoacetal, $\underline{55}$, was isolated in moderate yield [Eq. (32)].[47]

The reaction of a selenoester with a tertiary phosphine in the presence of oxygen leads to deselenation [Eq. (33)].[58] Aromatic selenoesters are typically more reactive then the corresponding aliphatic compounds. If the phosphine reaction is run in an inert atmosphere, a number of interesting compounds can be isolated depending on reaction conditions. Presumably nucleophilic attack of the phosphine occurs at the selenocarbonyl carbon and the resulting re-

$$2 \ CH_3\overset{\overset{\displaystyle Se}{\|}}{C}OR \xrightarrow{(Me_3Si)_2N^-K^+} RO\overset{\overset{\displaystyle Se^-K^+}{|}}{\underset{\underset{\displaystyle CH_3}{|}}{C}}-CH_2\overset{\overset{\displaystyle Se}{\|}}{C}OR \tag{31}$$

52

$$\downarrow {}_{-KSe^-}$$

$$\overset{\displaystyle RO}{\underset{\displaystyle CH_3}{}}C=C\overset{\displaystyle H}{\underset{\underset{\underset{\displaystyle Se}{\|}}{\displaystyle COR}}{}}$$

53
59%

$$(CH_3)_2CH\overset{\overset{\displaystyle Se}{\|}}{C}OEt \xrightarrow[\text{(2) MeI}]{\text{(1) } (Me_3Si)_2N^-K^+} \overset{\displaystyle CH_3}{\underset{\displaystyle CH_3}{}}C=C\overset{\displaystyle SeCH_3}{\underset{\displaystyle OCH_2CH_3}{}} \tag{32}$$

54

55
54%

latively stable tetrahedral intermediate, 55, expels selenium affording the ylid. This very reactive ylid can itself react normally or fragment to the α-alkoxy-carbene affording the observed products.[58]

$$Ph\overset{\overset{\displaystyle Se}{\|}}{C}OR \xrightarrow{Et_3P} \begin{bmatrix} Ph\overset{\overset{\displaystyle Se^-}{|}}{\underset{\underset{\displaystyle +PEt_3}{|}}{C}}OR \\ 55 \end{bmatrix} \tag{33}$$

Reactions:

$O_2 \rightarrow$ PhCOR + Et$_3$P → Se

PhCHO → $\overset{\displaystyle Ph}{\underset{\displaystyle H}{}}C=C\overset{\displaystyle OR}{\underset{\displaystyle Ph}{}}$

⬡ → bicyclic with Ph, OR

PhH → $\overset{\displaystyle Ph}{\underset{\displaystyle RO}{}}C=C\overset{\displaystyle OR}{\underset{\displaystyle Ph}{}}$ + $RO\overset{\displaystyle Ph}{\underset{\displaystyle H}{}}C-C\overset{\displaystyle Ph}{\underset{\displaystyle H}{}}OR$

MeI → $\begin{bmatrix} \overset{\displaystyle Me}{}\underset{\displaystyle Ph-C-OR}{\overset{\displaystyle \overset{+}{Se}}{|}}\underset{\underset{\displaystyle PEt_3}{|}}{} \overset{\displaystyle Me}{} \end{bmatrix} \xrightarrow{} \overset{\displaystyle +PEt_3}{Ph\overset{|}{C}-OR}$

Selenoesters can also be used as intermediates in the preparation of seleno-amides [see Eq. (39)].

5. DISELENOESTERS AND SELENOTHIOESTERS

Treatment of a Se-alkyl selenopiperidide, $\underline{57}$, (readily available by alkylation of the selenoamide) with hydrogen selenide affords the very unstable diseleno-esters, $\underline{58}$ [Eq. (34)].[59] These compounds are highly colored and difficult to characterize. Reaction of methyl diselenoacetate, $\underline{59}$, with dimethylamine afforded N,N-dimethylselenoacetamide, $\underline{60}$ (Eq. (35)].

$$\underset{\underline{57}}{\overset{\overset{\displaystyle Se}{\|}}{RC-N}}\bigcirc \xrightarrow{R'X} \underset{X^-}{R-\overset{\overset{\displaystyle SeR'}{|}}{C}=\overset{+}{N}}\bigcirc \xrightarrow{H_2Se} \underset{\underline{58}}{\overset{\overset{\displaystyle Se}{\|}}{RCSeR'}} \qquad (34)$$

$$\underset{\underline{59}}{\overset{\overset{\displaystyle Se}{\|}}{CH_3CSeCH_3}} \xrightarrow{(CH_3)_2NH} \underset{\underline{60}}{\overset{\overset{\displaystyle Se}{\|}}{CH_3CN(CH_3)_2}} \qquad (35)$$

The corresponding reaction on the S-alkylated amide afforded the S-alkyl selenothioesters, $\underline{61}$ [Eq. (36)].[59] These compounds are also highly colored and are more stable than their diselenium analogs. The isomeric Se-alkylated selenoesters have been previously reviewed.[5]

$$\underset{}{\overset{\overset{\displaystyle S}{\|}}{RC-N}}\bigcirc \xrightarrow{R'X} \underset{X^-}{R-\overset{\overset{\displaystyle SR'}{|}}{C}=\overset{+}{N}}\bigcirc \xrightarrow{H_2Se} \underset{\underline{61}}{\overset{\overset{\displaystyle Se}{\|}}{RCSR'}} \qquad (36)$$

6. SELENOAMIDES

6.1. Preparation

Primary selenoamides, $\underline{61}$, were initially prepared by the addition of hydrogen selenide to nitriles [Eq. (37)].[60,61] A much improved procedure uses aluminum selenide for this transformation [Eq. (38)].[62]

$$RCN \xrightarrow{H_2Se} \underset{\underline{61}}{\overset{\overset{\displaystyle Se}{\|}}{RCNH_2}} \qquad (37)$$

$$RCN \xrightarrow[\substack{pyridine- \\ H_2O,\ Et_3N}]{Al_2Se_3} \underset{70-98\%}{\overset{\overset{\displaystyle Se}{\|}}{RCNH_2}} \qquad (38)$$

Selenoamides could be prepared in poor yield by the reaction of the amide with phosphorus pentaselenide[63] or with mixtures of red phosphorus and selenium in refluxing xylene.[64] [The *in situ* generation procedure avoids the problem of characterizing phosphorus (V) selenide.]

Tertiary selenoamides, 62, can be prepared in moderate to good yields by reaction of a selenoester with a secondary amine [Eq. (39)].[53] In the reaction of a selenoester with a primary amine, loss of hydrogen selenide and imido ester formation competes with the formation of the desired selenoamide. This can be circumvented by using the magnesium halide salts of primary amines, affording the secondary selenoamides, 63, in high yields.[53]

$$
\begin{array}{c}
\underset{\displaystyle \underset{RCOEt}{\parallel}}{Se}
\end{array}
\quad
\xrightarrow[R']{\underset{R}{\displaystyle \overset{R'}{\diagdown}}NH}
\quad
\underset{\substack{62\\30-70\%}}{RCN\underset{R'}{\overset{R'}{\diagup\diagdown}}\overset{\displaystyle \overset{Se}{\parallel}}{}}
\tag{39}
$$

$$
\xrightarrow{R'NHMgBr}\quad \underset{\substack{63\\60-90\%}}{RCNHR'}\overset{\displaystyle \overset{Se}{\parallel}}{}
$$

The reaction of chlorimidates, 64, and iminium salts, 65, with sodium hydrogen selenide has also been used to prepare selenoamides[65] and vinylogous selenoamides [Eqs. (40) and (41)].[66] S-alkylthioamides have been similarly used.[67]

$$
\underset{64}{Ph-\underset{}{\overset{\displaystyle \overset{Cl}{|}}{C}}=N\diagdown_{Ph}}
\xrightarrow{HSe^-}
\underset{\underset{H}{\displaystyle |}}{Ph\overset{\displaystyle \overset{Se}{\parallel}}{C}NPh}
\quad
\underset{89\%}{}
\tag{40}
$$

$$
\underset{65}{\overset{Ar\diagdown}{\underset{Cl\diagup}{C}}=\overset{H}{\underset{H}{C}}\diagdown\overset{+}{\underset{R}{\overset{R}{\diagup}}C=N}}
\xrightarrow{NaHSe}
\underset{20-42\%}{\overset{Ar\diagdown}{\underset{Se\diagup}{C}}=\overset{H}{\underset{}{C}}\diagdown\overset{H}{\underset{H}{\underset{}{C-NR_2}}}}
\tag{41}
$$

Secondary and tertiary selenoamides, unsubstituted at the α-position, 66, can be readily prepared in good to excellent yields by a basic fragmentation of 1,2,3-selenadiazolines in the presence of an amine [Eq. (42)].[68] The analogous procedure has been used for selenoester formation, and the selenoketene is again the presumed intermediate. A related procedure generates the intermediate acetylenic selenolate by reaction of an acetylene with selenium.[69]

$$\text{R}\overbrace{}^{\text{N}}_{\underset{\text{Se}}{\text{N}}} \xrightarrow{\text{\textit{t}-BuO}^{\ominus}\text{K}^{\oplus}} \text{R—C}\equiv\text{C—Se}^- \xleftarrow{\text{Se}} \text{R—C}\equiv\text{C}^-$$

$$\Big\downarrow\Big\uparrow \text{H}^+$$

(42)

$$\underset{\underset{70-100\%}{\underline{66}}}{\text{RCH}_2\overset{\overset{\text{Se}}{\|}}{\text{C}}\text{N}\Big\langle\begin{smallmatrix}\text{R}\\\text{R}'\end{smallmatrix}} \xleftarrow{\text{RR'NH}} \begin{smallmatrix}\text{R}\\\text{H}\end{smallmatrix}\Big\rangle\text{C}=\text{C}=\text{Se}$$

R = H, alkyl, aryl

Tertiary selenoformamides, 69, can be prepared in moderate yield by reduction of dibenzyltriselenocarbonate, 67, with benzylselenol anion in the presence of triethylamine, followed by reaction of the resulting presumed polymeric diselenoformate, 68, with a secondary amine [Eq. (43)].[70] The corresponding reaction with a primary amine apparently also afforded selenoformamides, but these were destroyed upon attempted isolation.

$$\underset{\underline{67}}{\text{PhCH}_2\text{Se}\overset{\overset{\text{Se}}{\|}}{\text{C}}\text{SeCH}_2\text{Ph}} \xrightarrow[\text{Et}_3\text{N}]{\text{PhCH}_2\text{Se}^-} \underset{\underline{68}}{(\text{PhCH}_2\text{SeCHSe})_n}$$

$$\Big\downarrow \begin{smallmatrix}\text{R}\\\\\text{R}'\end{smallmatrix}\text{NH}$$

(43)

$$\underset{\underset{28-69\%}{\underline{69}}}{\begin{smallmatrix}\text{R}\\\text{R}'\end{smallmatrix}\text{N}-\overset{\overset{\text{Se}}{\|}}{\text{C}}-\text{H}}$$

Heterocyclic selenoformamides, 70, can be conveniently prepared by reaction of the corresponding amine with a dimethylformamide acetal followed by hydrogen selenide treatment [Eq. (44)].[70a]

$$\text{ArNH}_2 + (\text{CH}_3)_2\text{N—CH(OR)}_2 \longrightarrow \text{ArN}=\text{CHN(CH}_3)_2$$

$$\Big\downarrow \text{H}_2\text{Se}$$

(44)

$$\underset{\underset{12-79\%}{\underline{70}}}{\text{ArNH}\overset{\overset{\text{Se}}{\|}}{\text{C}}\text{H}}$$

A number of selenoformyl vinylogous amides, 71–73, have also been prepared, either directly using phosphorus pentaselenide [Eq. (45)],[71] or by
Vilsmeier–Haack formylation with hydrogen selenide treatment [Eq. (46)].[72]
These compounds are readily reduced to the corresponding methylated compounds with lithium aluminum hydride.

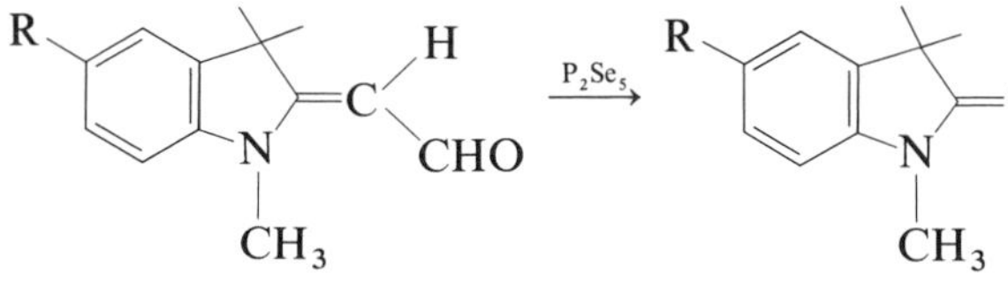

$$\text{(45)}$$

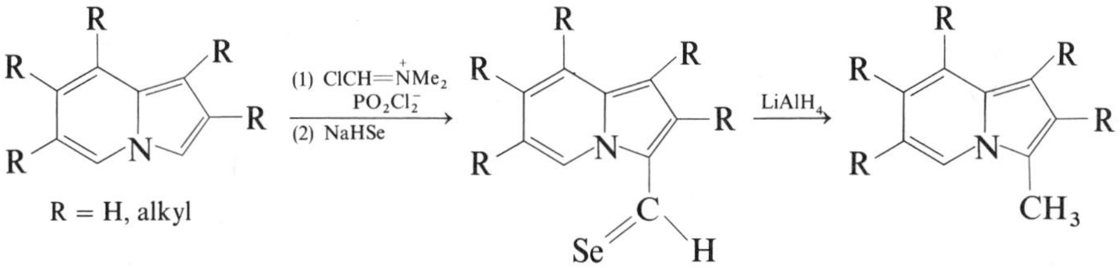

71

$$\text{(46)}$$

72
28–46%

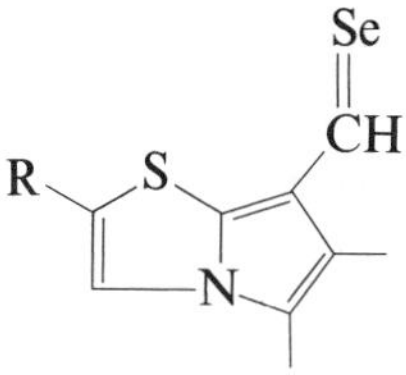

73a 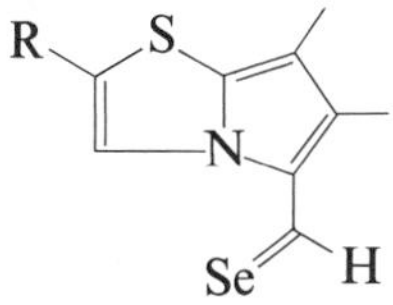   73b

6.2. Reactions

Selenoamides are generally colorless or slightly yellow compounds and are more
stable than the corresponding esters.[53] Hydrolysis of a selenoamide affords the
carboxylic acid [Eq. (47)].[73] Oxidation of a primary selenoamide with iodine
affords the selenadiazole, 74 [Eq. (48)].[61] Reaction of diselenomalonamide, 75,
with iodine affords the diselenolylium compound, 76 [Eq. (49)].[74]

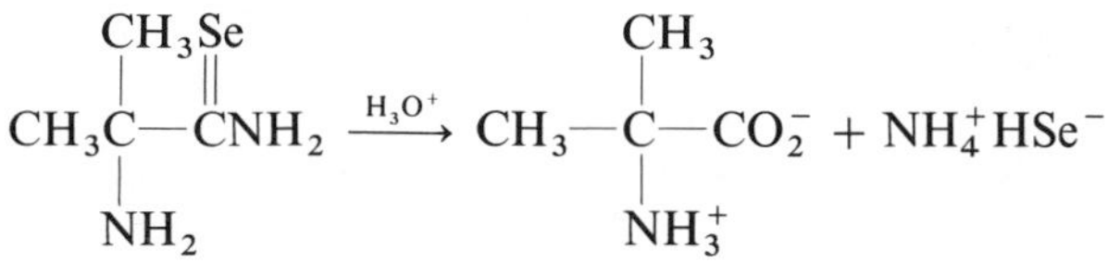

$$\text{(47)}$$

$$\text{PhCNH}_2 \ (\overset{\text{Se}}{\|}) \xrightarrow{\text{I}_2} \quad \underset{74}{\text{Ph}\overset{\text{N}-\text{N}}{\underset{\text{Se}}{\diagup}}\text{Ph}} + \text{Se} + \text{HI} \tag{48}$$

$$\underset{75}{\text{H}_2\text{NCCH}_2\text{CNH}_2} \ (\overset{\text{Se}\quad\text{Se}}{\|\quad\|}) \xrightarrow{\text{I}_2} \quad \underset{76}{\overset{\text{H}_2\text{N}}{\underset{\text{NH}_2}{\diagup}}\overset{\text{Se}}{\underset{\text{Se}}{(+)}}} \tag{49}$$

Alkylation of selenoamides occurs in high yield.[63,75] The alkylated products can in some cases be useful intermediates in the preparation of selenium-containing heterocycles [Eqs. (50) and (51)].[65,66,76]

$$\underset{\text{H}}{\overset{\text{Se}}{\text{PhCNPh}}} + \underset{\text{H}}{\overset{\text{Cl}}{\text{PhCCO}_2\text{H}}} \longrightarrow \text{Ph}\underset{}{\overset{\overset{\text{Ph}}{\underset{\|}{\text{N}}}}{}}\text{Se}-\underset{\text{H}}{\overset{\text{CO}_2\text{H}}{\text{C}}}-\text{Ph}$$

$$\downarrow \text{Et}_3\text{N}-\text{Ac}_2\text{O} \tag{50}$$

$$\underset{88\%}{\overset{\text{Ph}}{\underset{\text{Ph}}{\text{N}}}\overset{\text{O}}{\underset{\text{Se}}{\|}}{\text{Ph}}}$$

$$\underset{\overset{\text{O}}{\underset{\text{EtO}^-}{\text{XCH}_2\text{CR}'}}\downarrow}{\underset{\text{ArCCH}=\text{CHNR}_2}{\overset{\text{Se}}{\|}}} \nearrow^{\overset{\overset{\text{O}}{\|}}{\text{PhCH}_2\text{C}-\text{Cl}}} \ \underset{16-64\%}{\text{Ar}\overset{}{\overset{\text{Ph}}{\diagdown}}\text{Se}\overset{\text{O}}{=}} + \text{R}_2\text{NH}$$

$$\xrightarrow[\text{(2) H}^+]{\overset{\text{(1) PhCH}_2\text{CN,}}{\text{base}}} \ \underset{12-28\%}{\text{Ar}\overset{\text{Ph}}{\underset{\overset{+}{\text{Se}}}{\diagdown}}\text{NH}_2} \tag{51}$$

$$\underset{49-91\%}{\text{Ar}\overset{}{\underset{\text{Se}}{\diagup}}\overset{\text{R}'}{\underset{\overset{\|}{\text{O}}}{}}}$$

Tertiary selenoamides with α-hydrogens may be conveniently trimethyl-silyated in good yield [Eq. (52)].[77]

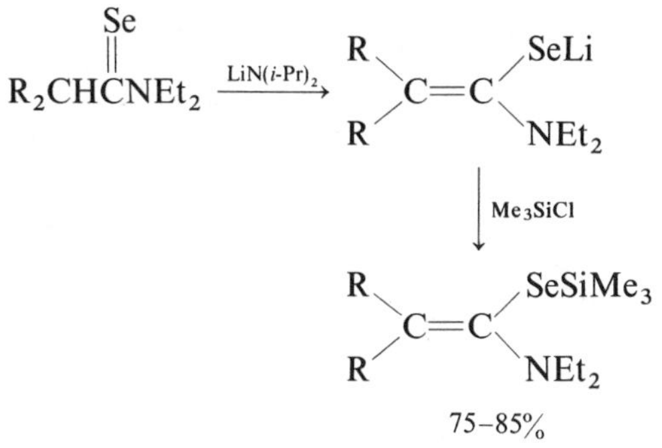

$$
\underset{R_2CHCNEt_2}{\overset{Se}{\overset{\|}{}}} \xrightarrow{\text{LiN}(i\text{-Pr})_2} \underset{R}{\overset{R}{\diagdown}} C=C \underset{NEt_2}{\overset{SeLi}{\diagup}}
$$

Me$_3$SiCl

$$
\underset{R}{\overset{R}{\diagdown}} C=C \underset{NEt_2}{\overset{SeSiMe_3}{\diagup}}
$$

75–85%

(52)

Other less general preparations and reactions of selenoamides, and especially their use in heterocyclic synthesis, have been previously reviewed.[6,78]

7. SELENOCARBOXYLIC ACIDS

Although thiocarboxylic acids can be prepared from acid chlorides and hydrogen sulfide salts, the corresponding reaction with hydrogen selenide typically affords diacylselenides, 78, by elimination of hydrogen selenide from the intermediate thermally unstable selenocarboxylic acids, 77. The diacylselenides themselves can undergo further reactions with the selenoacids affording diacyl-diselenides, 79 [Eq. (53)].[79]

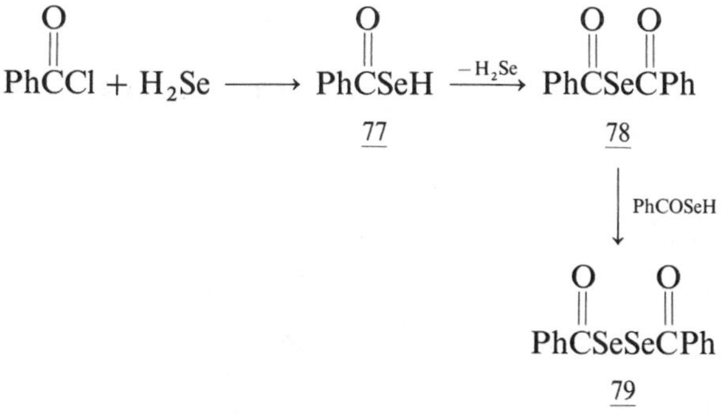

$$
\underset{}{\overset{O}{\overset{\|}{\text{PhCCl}}}} + H_2Se \longrightarrow \underset{77}{\overset{O}{\overset{\|}{\text{PhCSeH}}}} \xrightarrow{-H_2Se} \underset{78}{\overset{O\ \ O}{\overset{\|\ \ \|}{\text{PhCSeCPh}}}}
$$

PhCOSeH

$$
\underset{79}{\overset{O\ \ \ \ O}{\overset{\|\ \ \ \ \|}{\text{PhCSeSeCPh}}}}
$$

(53)

Although it undergoes similar reactions, selenobenzoic acid, 77, has been isolated as an unstable pink oil.[79] Infrared and nmr analysis of this compound indicates that it exists almost exclusively as the selenium protonated species, analogous to the thiocarboxylic acids. Selenobenzoic acid is soluble in aqueous alkali, and the resulting anion is easily Se-alkylated. This anion can most easily be prepared by alkaline cleavage of the diacylselenide, 78. Oxidation of the anion affords the diacyldiselenide, 79 [Eq. (54)].[80]

$$
\begin{array}{c}
\underset{77}{\text{PhCSeH}} \searrow \\
\end{array}
\quad
\underset{}{\text{PhC}\text{---}\text{Se} \ \text{K}^{+}} \xrightarrow{\text{RX}} \text{PhCSeR}
\qquad (54)
$$

$$
\underset{78}{\text{PhCSeCPh}} \xleftarrow[95\%]{\text{KOH}} \qquad \searrow{}^{\text{I}_2} \qquad \underset{79}{\text{PhCSeSeCPh}}
$$

Reaction of selenostearate, 80, with dichloromethane affords the diseleno-stearoyl methane, 81 [Eq. (55)].[80]

$$
2 \ \underset{80}{C_{17}H_{35}\overset{O}{\overset{\|}{C}}Se^{-}} \xrightarrow{\text{CH}_2\text{Cl}_2} \underset{81}{(C_{17}H_{35}\overset{O}{\overset{\|}{C}}Se)_2CH_2}
\qquad (55)
$$

The reaction of selenobenzoate with trimethylsilylchloride affords the very water sensitive trimethylsilylselenobenzoate, 82 in 68% yield.[81] It is interesting to note the corresponding reactions with trimethyltin chloride and trimethyl-germanium chloride afford the relatively more stable Se-metalated compounds [Eq. (56)]. This behavior parallels that of thiobenzoate in the corresponding reactions.[82]

$$
\text{PhC}\begin{array}{c} \nearrow \text{Se} \\ \searrow \text{O} \end{array}^{-}
\quad
\begin{array}{c}
\xrightarrow{\text{Me}_3\text{SiCl}} \underset{82}{\overset{\overset{\text{Se}}{\|}}{\text{PhCOSiMe}_3}} \\[2em]
\xrightarrow{\text{Me}_3\text{MCl}} \underset{(\text{M = Si, Ge})}{\overset{\overset{\text{O}}{\|}}{\text{PhCSeMMe}_3}}
\end{array}
\qquad (56)
$$

Dipropionyl selenide, 83, reacts with two equivalents of aniline indicating that selenoacids like the thioacids can act as acylating agents [Eq. (57)].[79]

$$
\underset{83}{\text{CH}_3\text{CH}_2\overset{O}{\overset{\|}{C}}\text{---}\text{Se}\text{---}\overset{O}{\overset{\|}{C}}\text{CH}_2\text{CH}_3} \xrightarrow{2 \ \text{PhNH}_2} 2 \ \text{PhNH}\overset{O}{\overset{\|}{C}}\text{CH}_2\text{CH}_3 + \text{H}_2\text{Se} \qquad (57)
$$

Attempts to prepare diselenocarboxylic acids by reaction of organometallic reagents with carbon diselenide have been unsuccessful.[83]

8. SELENOCARBONATES

8.1. Preparation

Reviews of selenocarbonic acids and derivatives have appeared.[84] The acids are not well characterized and readily decompose with liberation of hydrogen selenide.

Dimethyl triselenocarbonate **84** can be prepared by alkylating barium triselenocarbonate [Eq. (58)].[85]

$$Ba^{2+}CSe_3^{2-} \xrightarrow{\text{2 MeI}} \overset{\overset{\displaystyle Se}{\|}}{MeSeCSeMe} \tag{58}$$

$$\underset{20-30\%}{\underline{84}}$$

Diselenoxanthates, **85**, can be prepared from alkoxides and carbon diselenide followed by alkylation, but appear to be less stable than xanthates [Eq. (59)].[86] S-alkyl diselenothiocarbonates, **86**, can also be prepared in low yield from the corresponding thiolates [Eq. (60)].[87]

$$RO^- + CSe_2 \longrightarrow \overset{\overset{\displaystyle Se}{\|}}{ROCSe^-} \xrightarrow{\text{ClCH}_2\text{CO}_2^-} \overset{\overset{\displaystyle Se}{\|}}{ROCSeCH_2CO_2^-} \tag{59}$$

$$\underline{85}$$

$$RS^- + CSe_2 \longrightarrow \overset{\overset{\displaystyle Se}{\|}}{RSCSe^-} \xrightarrow{\text{BrCH}_2\text{CO}_2^-} \overset{\overset{\displaystyle Se}{\|}}{RSCSeCH_2CO_2^-} \tag{60}$$

$$\underline{86}$$

While the reaction of a selenol salt with carbon diselenide followed by alkylation is reported not to give dialkyl triselenocarbonates, **87**, these compounds can be prepared by the reaction of carbon diselenide, alkyl halide, and base in dimethyl sulfoxide (DMSO) [Eq. (61)].[88]

$$CSe_2 + RX + KOH \xrightarrow[\text{H}_2\text{O}]{\text{DMSO}} \overset{\overset{\displaystyle Se}{\|}}{RSeCSeR} \tag{61}$$

$$\underline{87}$$

8.2. Reactions

Dialkyl triselenocarbonates react with secondary amines to form diseleno-carbamates, 88, and with primary amines to form selenoureas.[89] Presumably the second reaction occurs via conversion of an intermediate unstable seleno-carbamate to the selenoisocyanate, which reacts further to form the selenourea [Eq. (62)].

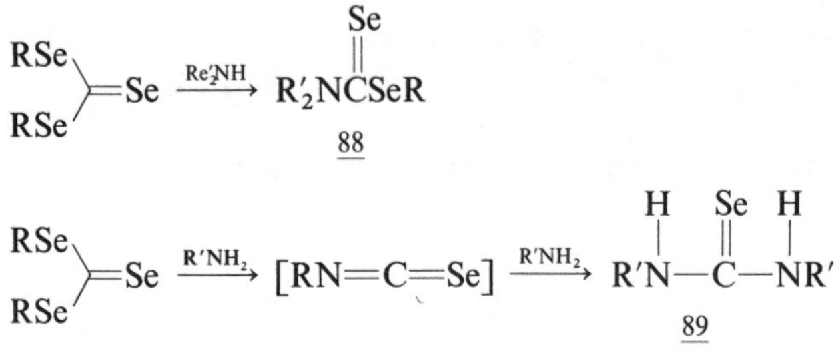

$$
\begin{array}{c}
\text{RSe} \\
\diagdown \\
\diagup \text{=Se} \xrightarrow{\text{Re}_2'\text{NH}} \text{R}_2'\text{N}\overset{\overset{\text{Se}}{\|}}{\text{C}}\text{SeR} \\
\text{RSe} \qquad\qquad\quad \underline{88}
\end{array}
\tag{62}
$$

$$
\begin{array}{c}
\text{RSe} \\
\diagdown \\
\diagup \text{=Se} \xrightarrow{\text{R}'\text{NH}_2} [\text{RN}=\text{C}=\text{Se}] \xrightarrow{\text{R}'\text{NH}_2} \text{R}'\text{N}\overset{\overset{\text{H}}{|}}{\underset{}{}}\,\overset{\overset{\text{Se}}{\|}}{\text{C}}\,\overset{\overset{\text{H}}{|}}{\text{N}}\text{R}' \\
\text{RSe} \qquad\qquad\qquad\qquad\qquad\qquad\qquad \underline{89}
\end{array}
$$

The contrast between these reactions and the previously discussed reaction of a dialkyl triselenocarbonate with an amine in the presence of a selenol salt to form a selenoformamide [see Eq. (43)][70] has been explained in terms of the reactions of hard versus soft nucleophiles with an ambident triseleno-carbonate electrophile.[89]

9. CYCLIC SELENOCARBONATES

Cyclic selenocarbonate derivatives have recently proved to be very important in the preparation of organic semiconductors and are therefore treated sepa-rately. A variety of cyclic triselenocarbonates, 91, have been prepared by reaction of α-haloketones with selenocarbamate anions, 90. Acid catalyzed cyclization, followed by reaction with hydrogen selenide led to the cyclic triselenocarbonate [Eq. (63)].[90]

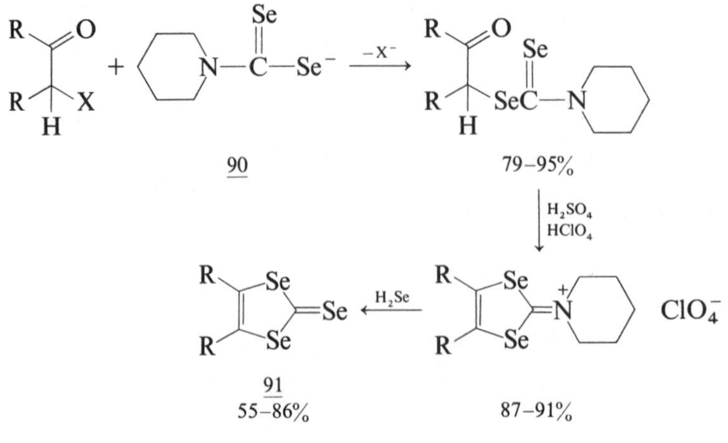

$$\tag{63}$$

The reaction of sodium acetylide with carbon diselenide and selenium affords the unsubstituted selenocarbonate, 92 [Eq. (64)].[91]

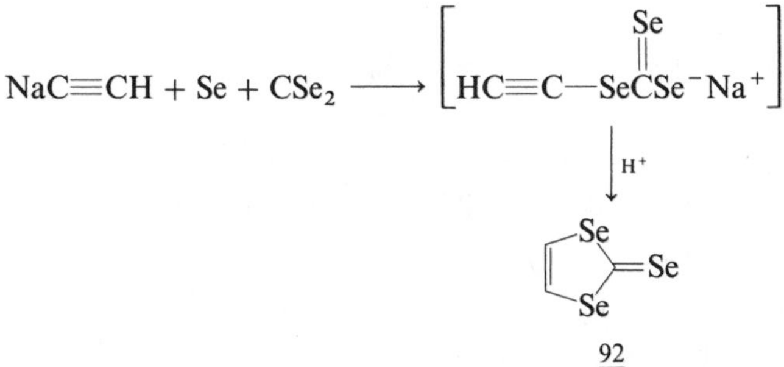

$$
NaC\equiv CH + Se + CSe_2 \longrightarrow \left[HC\equiv C-Se\overset{\overset{\displaystyle Se}{\|}}{C}Se^-Na^+ \right]
$$

$$\downarrow H^+ \tag{64}$$

92

Treatment of the 1,2-dilithiobenzene with selenium, followed by addition of thiocarbonyldiimidazole affords the thiocarbonyl derivative. Methylation with dimethoxycarbenium fluoroborate and hydrogen selenide addition affords the triselenocarbonate, 93 [Eq. (65)].[92] Alternatively this compound has been prepared by thermolysis of 1,2,3-benzoselenadiazole, 94, in the presence of excess carbon diselenide [Eq. (66)].[93]

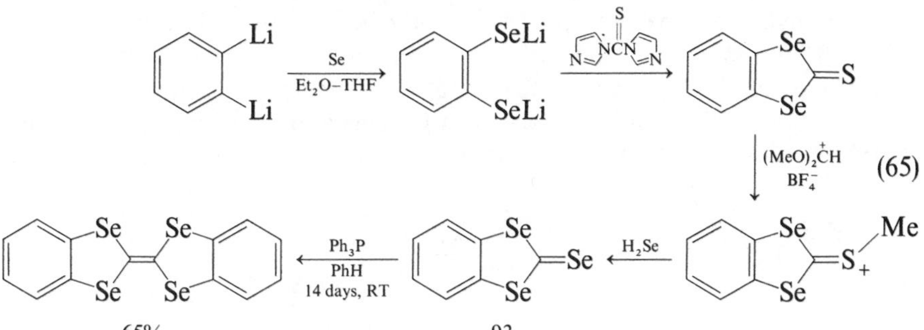

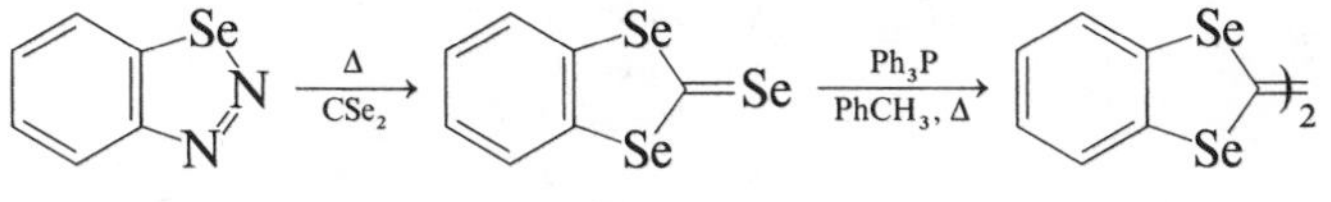

(65)

65%   93

(66)

94   69%   66%

Electrochemical reduction of carbon diselenide at 40°C affords 4,5-diselenolate-1,3-diselenole-2-selone, 95, which upon alkylation with 1,2-dibromoethane affords the cyclic triselenocarbonate, 96 [Eq. (67)].[94]

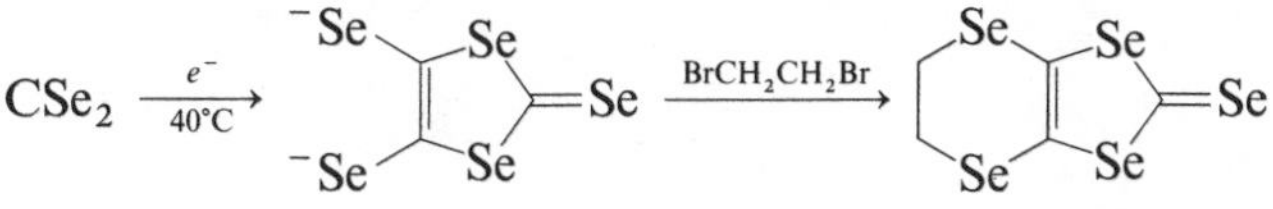

(67)

95   96
10%

Formation of the vinylogous cyclic selenocarbonates, 97 and 98, by procedures closely related to those already discussed have also been reported [Eqs. (68) and (69)].[95-97]

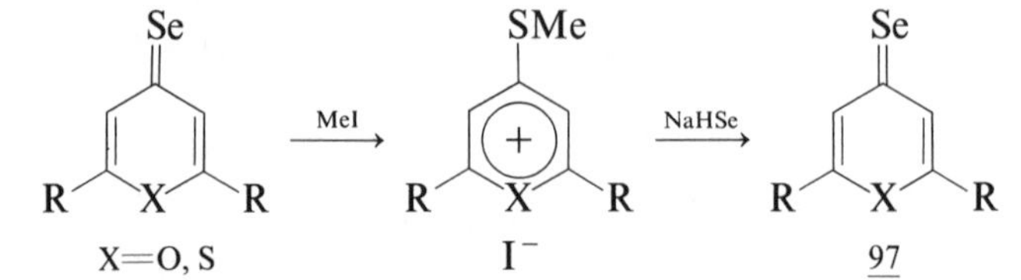

Treatment of a selenocarbonate with trialkyl phosphite[91] or triphenylphosphine[92,93] typically affords the fulvalene [Eqs. (65) and (66)]. In this reaction the selenocarbonyl compound is much preferred over the corresponding thiocarbonyl, and appears to be the method of choice for such transformations [Eq. (70)] (although an exception has been noted).[91,94] Alternative related conversions of vinylogous selenocarbonates to fulvalenes, 99 and 100, include simple pyrolysis,[95,96] and pyrolysis in a hydrogen atmosphere [Eqs. (71) and (72)].[95]

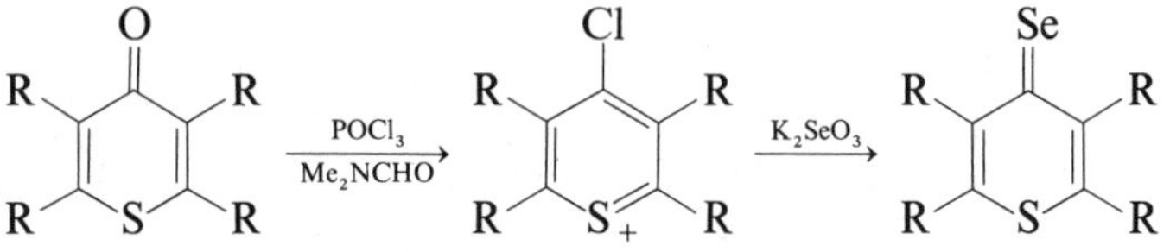

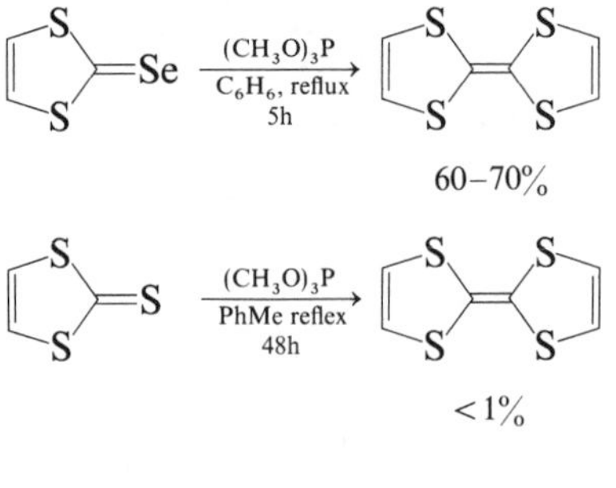

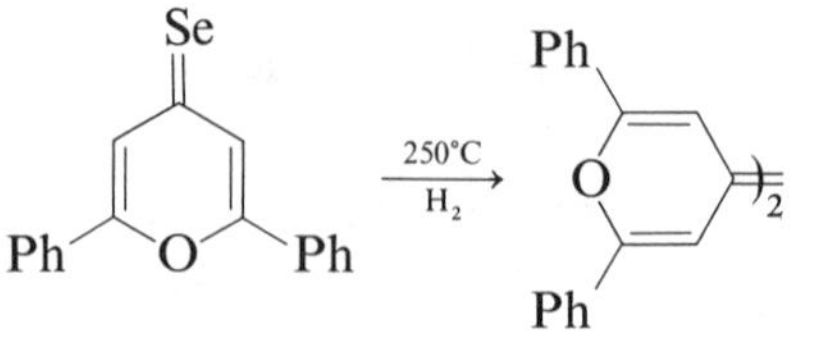

The selenocarbonates can be readily alkylated, and the alkylated products reduced with sodium borohydride [Eq. (73)]. Peracid oxidation, which can be

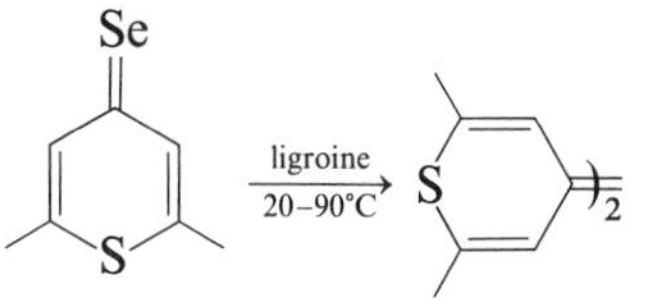

(72)

98 100

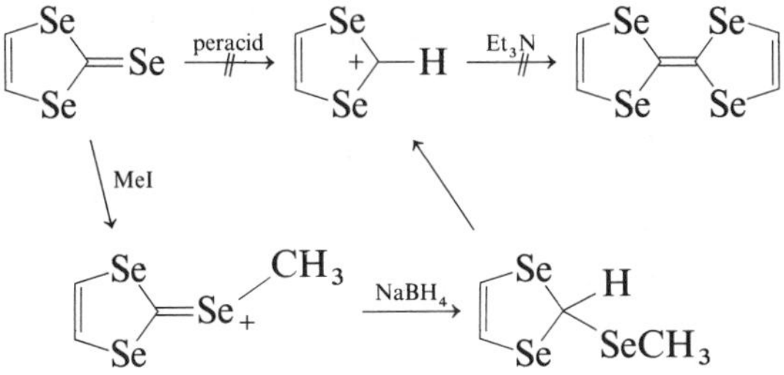

(73)

used for the formation of fulvalenes from trithiocarbonates, fails in the case of triselenocarbonates apparently due to oxidation of the ring seleniums.[90,91]

10. SELENOUREAS

10.1. Preparation

Selenoureas are probably the most stable and best studied selenocarbonyl molecules. A variety of methods are currently available for the preparation of this class of compounds. The most general method involves the conversion of thioureas to their selenium analogs. In this procedure the thiourea is alkylated, then treated with sodium hydrogen selenide under alkaline conditions [Eq. (74)]. The selenoureas, 101, are isolated in good yield. This procedure is especially useful for preparing tetrasubstituted selenoureas that cannot be readily prepared by other methods.[98]

$$\mathrm{R_2N\overset{S}{\overset{\|}{C}}NR_2 \xrightarrow{MeI} R_2\underset{+}{N}{=}\overset{SMe}{\overset{|}{C}}{-}NR_2 \xrightarrow[pH\ 8-9]{HSe^-} R_2N\overset{Se}{\overset{\|}{C}}NR_2} \qquad (74)$$

101

The reaction of an isoselenocyanate, 102, with ammonia or amines is also a relatively general method for the preparation of mono-, di-, and trisubstituted selenoureas [Eq. (75)].[99] The isoselenocyanate can also be generated *in situ.*[100] Selenocyanate ion also reacts with amines to afford selenoureas.[101]

Carbodiimides, 103, react with hydrogen selenide to give 1,3-disubstituted selenoureas [Eq. (76)].[102] Reaction of cyanamide, 104, or substituted cyanamide with hydrogen selenide provides a route to the parent selenourea, and mono- and 1,1-disubstituted selenoureas [Eq. (77)].[103,104]

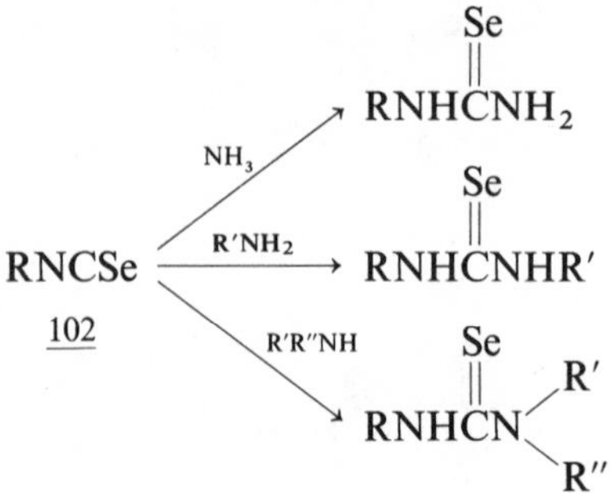

$$RNCSe \quad \begin{cases} \xrightarrow{NH_3} & RNHCNH_2 \\\\ \xrightarrow{R'NH_2} & RNHCNHR' \\\\ \xrightarrow{R'R''NH} & RNHCN\begin{smallmatrix}R'\\R''\end{smallmatrix} \end{cases} \quad (75)$$

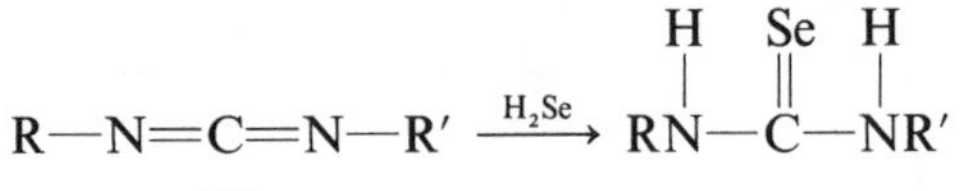

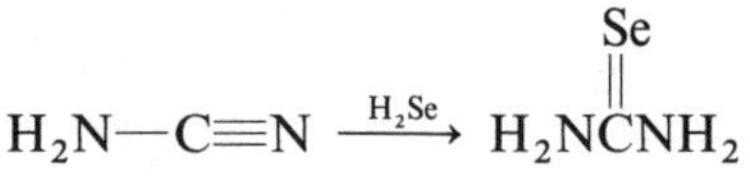

102

with Se double bonds above each product carbon.

$$R-N=C=N-R' \xrightarrow{H_2Se} RN\underset{H}{-}\overset{Se}{C}\underset{H}{-}NR' \quad (76)$$

103

$$H_2N-C\equiv N \xrightarrow{H_2Se} H_2N\overset{Se}{C}NH_2 \quad (77)$$

104

Other methods for the preparation of selenoureas involve the reaction of carbon diselenide with excess primary amine [Eq. (78)][105] and the reaction of triselenocarbonic acid esters with primary amines [see Eq. (62)].[89] Phosphorus pentaselenide converts urea into selenourea in poor yield.[106]

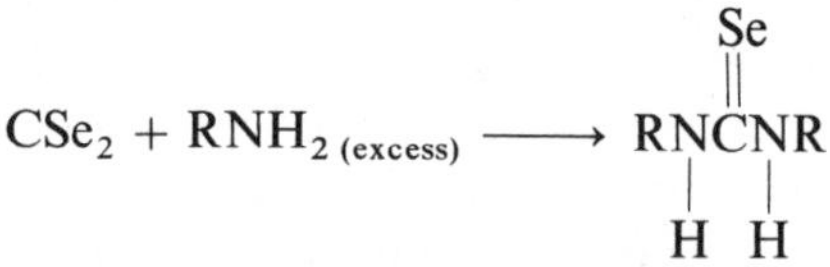

$$CSe_2 + RNH_{2\,(excess)} \longrightarrow RN\underset{H}{\overset{Se}{C}}NR \quad (78)$$

The highly strained quasiselenourea 1,2-bis-diisopropylaminocyclopropene-selone, 105, could be prepared in quantitative yield by reaction of cyclo-propenium salts with sodium hydrogen selenide, and could be readily alkylated.[107] The spectral properties of the molecule closely resemble those

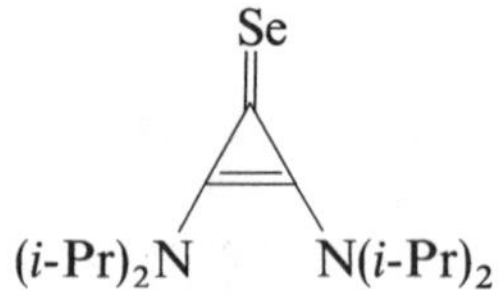

$(i\text{-Pr})_2N \qquad N(i\text{-Pr})_2$

105

$$H_2N\overset{Se}{C}NH\overset{X}{C}NH_2$$

106

(a) X = O, 83%
(b) X = S, 41%
(c) X = Se, 63%

of a selenourea rather than those of a selone, as would be expected because of strong conjugation through the cyclopropene ring.

The selenobiurets, 106, can be prepared from the corresponding thiobiurets by alkylation followed by sodium hydrogen selenide treatment.[108]

10.2. Reactions

Selenoureas have been widely used to introduce selenium into organic molecules,[6] and are especially useful in the preparation of selenium containing heterocycles.[78] Some recent examples include the preparation of the biologically interesting, 107[109] and 108,[110]

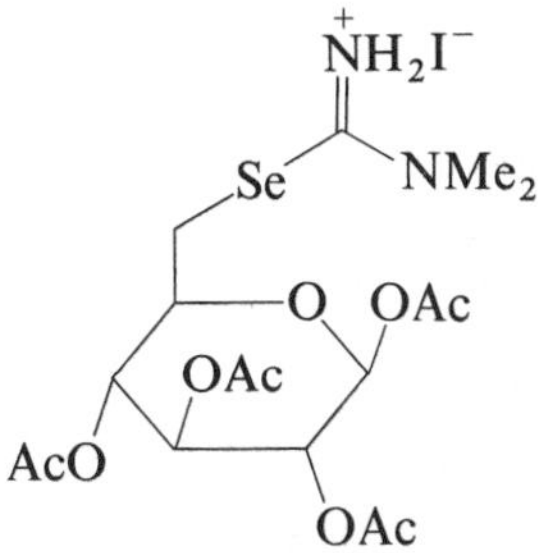

107

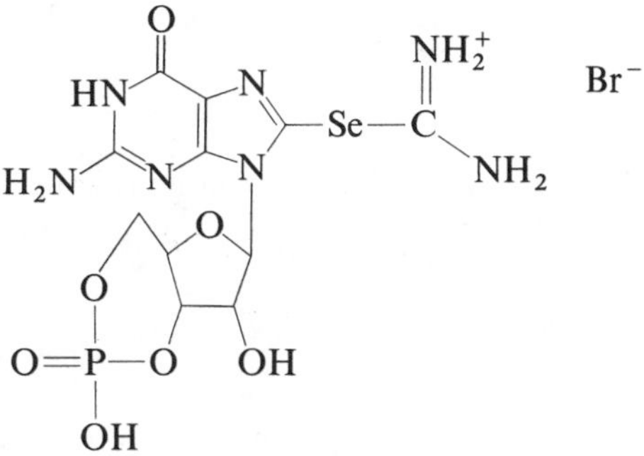

108

as well as the preparations of heterocycles, 109 and 110 [Eqs. (79) and (80)].[111]

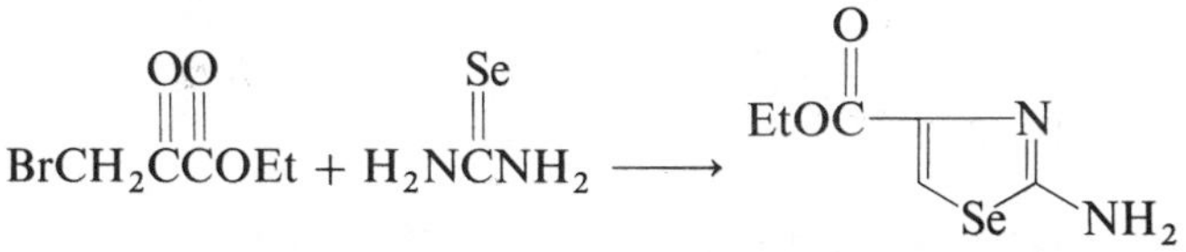

$$BrCH_2\overset{OO}{\underset{||\,||}{C}}COEt + H_2N\overset{Se}{\underset{||}{C}}NH_2 \longrightarrow \qquad (79)$$

109
82%

$$\underset{\underset{Cl}{|}}{H\overset{O}{\underset{||}{C}}CH\overset{O}{\underset{||}{C}}OMe} + H_2N\overset{Se}{\underset{||}{C}}NH_2 \longrightarrow \qquad (80)$$

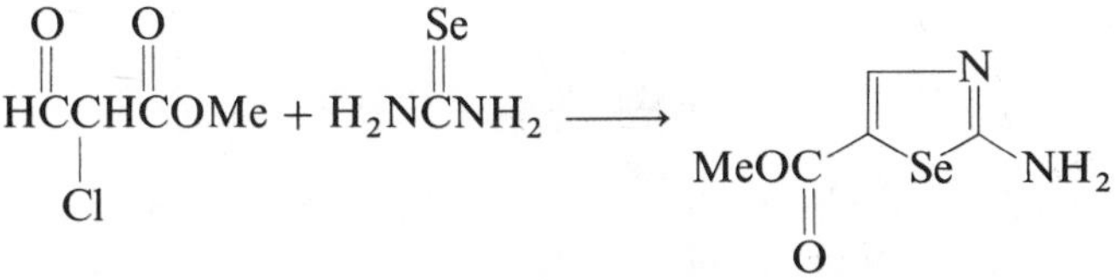

110
82%

Treatment of dicyclohexylselenourea with dimethyl sulfoxide in the presence of sulfuric acid affords the urea, selenium, and dimethyl sulfide in excellent yield [Eq. (81)].[112]

$$\text{cyclohexyl}-\overset{\overset{H}{|}}{N}\overset{\overset{H}{|}}{C}\overset{||}{N}-\text{cyclohexyl} \xrightarrow[\text{H}^+]{\text{Me}_2\text{SO}} \text{cyclohexyl}-\overset{\overset{H}{|}}{N}\overset{\overset{H}{|}}{C}\overset{||}{N}-\text{cyclohexyl} + \text{Se} + \text{Me}_2\text{S} \quad (81)$$

$$96\%$$

This deselenation is also potentially useful for other selenocarbonyl compounds. Other reported reactions include the oxidation of the selenourea with hydrogen peroxide,[113] ferricyanide,[114] p-benzoquinone, or electrochemically,[114] affording the α,α-diselenobisformamidium cation, 111 [Eq. (82)].

$$\overset{\overset{Se}{||}}{H_2N\overset{}{C}NH_2} \xrightarrow{[O]} \overset{\overset{+}{N}H_2}{H_2N\overset{}{C}}-Se-Se-\overset{\overset{+}{N}H_2}{\overset{}{C}NH_2} \quad (82)$$

$$\underline{111}$$

Selenoureas and related compounds have also found important uses as photographic sensitizers and in electrophotography.[115]

11. SELENOSEMICARBAZIDES

Selenosemicarbazides, 112, may be most conveniently prepared by the reaction of a isoselenocyanate with a hydrazine [Eq. (83)].[116,117]

$$\text{RNCSe} + \text{R'NHNH}_2 \longrightarrow \overset{\overset{Se}{||}}{\underset{\overset{|}{H}\ \overset{|}{H}}{RN\overset{}{C}NNHR'}} \quad (83)$$

$$\underline{112}$$

Aside from selenosemicarbazone formation,[116–120] these compounds have proved especially useful in heterocyclic synthesis.[6,78] Other reactions have been previously reviewed.[6]

12. SELENOCARBAMATES

The reaction of amines[121] or hydrazines[122] with carbon diselenide affords salts of diselenocarbamic acid, 113 [Eq. (84)].

$$\text{O}\diagdown\!\!\!\diagup\text{NH} \xrightarrow{\text{CSe}_2} \text{O}\diagdown\!\!\!\diagup\text{N}-\overset{\overset{Se}{||}}{C}Se^- \xrightarrow{\text{ClCH}_2\text{CO}_2\text{H}} \text{O}\diagdown\!\!\!\diagup\text{N}\overset{\overset{Se}{||}}{C}SeCH_2CO_2H \quad (84)$$

$$\underline{113} \qquad\qquad\qquad\qquad \underline{114}$$

These salts are readily alkylated to give the diselenourethanes, 114, in good yield.

Monoselenourethanes, 115, can be prepared by hydrogen selenide addition to alkyl cyanates [Eq. (85)],[123] or by ammonolysis of diselenocarbonates [Eq. (86)].

$$
\underset{}{\text{ROC}\equiv\text{N} + \text{H}_2\text{Se}} \longrightarrow \underset{115}{\text{RO}\overset{\overset{\displaystyle\text{Se}}{\|}}{\text{C}}\text{NH}_2} \tag{85}
$$

$$
\text{RO}\overset{\overset{\displaystyle\text{Se}}{\|}}{\text{C}}\text{SeCH}_2\text{CO}_2^- \quad
\begin{array}{c}
\xrightarrow{\text{NH}_3} \quad \text{RO}\overset{\overset{\displaystyle\text{Se}}{\|}}{\text{C}}\text{NH}_2 \\
\xrightarrow[\text{NH}_2\text{NH}_2]{} \quad \text{RO}\overset{\overset{\displaystyle\text{Se}}{\|}}{\text{C}}\text{NHNH}_2
\end{array} \tag{86}
$$

Selenothiocarbonates, 116, can be prepared by reaction of S-alkyl thiopseudoureas with hydrogen selenide ion at slightly acidic pH [Eq. (87)].[98]

$$
\text{R}_2\text{N}-\overset{\overset{\displaystyle\text{SCH}_3}{|}}{\underset{+}{\text{C}}}=\text{NR}_2 \xrightarrow[\text{pH 5-6}]{\text{HSe}^-} \underset{116}{\text{R}_2\text{N}\overset{\overset{\displaystyle\text{Se}}{\|}}{\text{C}}\text{SCH}_3} \tag{87}
$$

Under alkaline conditions the selenourea is formed.[98] Other less general procedures for the preparation of these and related compounds have been previously reported.[6]

Salts of diselenocarbamic acids afford a variety of complex polyselenides upon oxidation [e.g., Eq. (88)].[121,124]

$$
\text{R}_2\text{N}-\text{NHR} + \text{CSe}_2 \longrightarrow \text{R}_2\text{N}-\text{N}\overset{\displaystyle R}{\underset{\overset{\displaystyle\text{C}\text{SeH}}{\overset{\|}{\displaystyle\text{Se}}}}{}}
$$

$$
\text{I}_2 \qquad\qquad 40\text{-}82\% \qquad\qquad \Big\downarrow\text{O}_2 \tag{88}
$$

$$
\underset{\overset{\|}{\displaystyle\text{Se}}}{(\text{R}_2\text{NN}\overset{\displaystyle R}{\underset{}{-}}\text{CSe})_2\text{-Se}} \qquad\qquad \underset{\overset{\|}{\displaystyle\text{Se}}}{(\text{R}_2\text{N}-\overset{\displaystyle R}{\underset{}{\text{N}}}-\text{C}-\text{Se})_2^-}
$$

$$
\underline{78\%}
$$

13.   ISOSELENOCYANATES

13.1.   Preparation

The reaction of isonitriles with selenium appears to be a general method for
the preparation of isoselenocyanates [Eq. (89)].[125,126]

$$PhNC \xrightarrow{\text{Se}} PhNCSe \qquad (89)$$

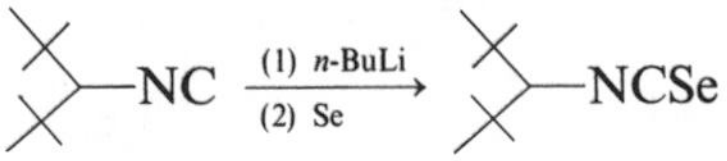

$$\chemfig{NC} \xrightarrow[\text{(2) Se}]{\text{(1) } n\text{-BuLi}} \chemfig{NCSe} \qquad (90)$$

<u>117</u>

The sterically hindered isoselenocyanate, <u>117</u>, could best be prepared by treatment of the isocyanide at $-80°C$ with n-butyllithium followed by addition of
selenium [Eq. (90)].[20]

   Displacements of activated alkyl halides by selenocyanate ion also provides
a route to isoselenocyanates [Eq. (91)].[126-128]

$$RX + SeCN^- \longrightarrow RN{=}C{=}Se + X^- \qquad (91)$$

A problem with this method is the formation of selenocyanates, <u>118</u>, as well as
isoselenocyanates in the displacement reaction due to the bidentate nature of
selenocyanate ion. Photochemical conversions of benzylic selenocyanates to the
corresponding isoselenocyanates have been reported [Eq. (92)].[129]

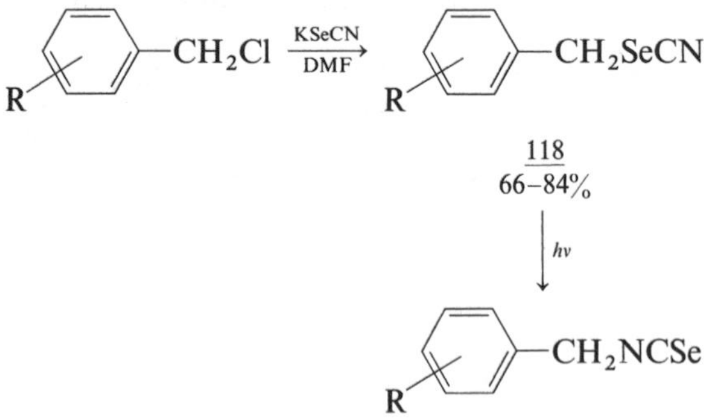

$$R{-}\!\!\!\bigcirc\!\!\!{-}CH_2Cl \xrightarrow[\text{DMF}]{\text{KSeCN}} R{-}\!\!\!\bigcirc\!\!\!{-}CH_2SeCN \qquad (92)$$

<u>118</u>
66–84%

$$\downarrow h\nu$$

$$R{-}\!\!\!\bigcirc\!\!\!{-}CH_2NCSe$$

26–40%

   The formation of intermediate isoselenocyanates has been postulated in the
reaction of carbon diselenide (or triselenocarbonates) with primary amines
to form selenoureas [see Eqs. (62) and (78)]. The use of mercuric chloride in
the carbon diselenide reaction allows for the isolation of the isoselenocyanate
in modest to good yield [Eq. (93)].[130]

$$RNH_2 + CSe_2 + HgCl_2 + 2Et_3N \longrightarrow R\text{---}NCSe + HgSe + 2Et_3N\cdot HCl$$

47–73%

(93)

Other less general methods for the preparation of these compounds have also been reviewed.[6]

13.2. Reactions

Isoselenocyanates are useful starting materials for the preparation of selenoureas and selenosemicarbazides [see Eq. (75)]. Heating trityl isoselenocyanate, 119, leads to extrusion of selenium and formation of the corresponding nitrile, 120 [Eq. (94)].[131]

$$Ph_3C\text{---}NCSe \xrightarrow[-Se]{\Delta} Ph_3CN \qquad (94)$$

119   120

Reduction of an isoselenocyanate with lithium aluminum hydride[131] or with zinc–hydrochloric acid[132] affords the amine. Treatment of the steroidal iso-selenocyanate, 121, with tri-*n*-butylstannane and azobisisobutyronitrile (AIBN) as a radical initiator affords the deaminated product in moderate yield [Eq. (95)].[133] This reaction works as well, however, with the isothiocyanate or the isocyanide.

(95)

121   55%

14. SELENOKETENES

These thermally unstable compounds have been postulated as intermediates in the photolysis and base catalyzed fragmentation of 1,2,3-selenadiazoles, 122. Treatment of 1,2,3-selenadiazoles with alkoxide affords an unstable acetylenic selenolate, which in dilute solution can react with alcohols or amines to form selenoesters or selenoamides, presumably through a selenoketene intermediate [see Eqs. (26) and (42)]. In more concentrated solution an isomeric mixture of dimeric 2, Ω-disubstituted-1,4-diselenafulvenes, 123, is formed.[134–136] The proposed mechanism of formation is the reaction of an intermediate selenoketene with the acetylenic selenolate followed by protonation [Eq. (96)]. The 1,4-diselenafulvenes are also formed upon photolysis of the 1,2,3-selenadiazole.[137]

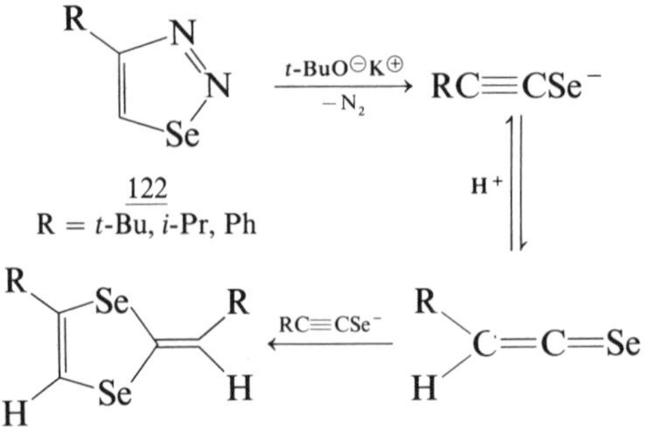

$$\begin{array}{c} R \\ \underset{Se}{\overset{N}{\diagdown}}N \end{array} \xrightarrow[-N_2]{t\text{-BuO}^{\ominus}K^{\oplus}} RC{\equiv}CSe^-$$

122
R = t-Bu, i-Pr, Ph

$$\underset{H}{\overset{R}{\diagdown}}\underset{Se}{\overset{Se}{\diagdown}}\underset{H}{\overset{R}{\diagdown}} \xleftarrow{RC{\equiv}CSe^-} \underset{H}{\overset{R}{\diagdown}}C{=}C{=}Se$$

123
72–90%

(96)

Vapor phase pyrolysis of 1,2,3-selenadiazoles at 500–600°C and trapping of intermediates at −196°C led to the isolation and characterization of three selenoketenes, 124 [Eq. (97)].[138,139] These highly colored compounds rapidly dimerized thermally. Trapping of the parent selenoketene at −80°C led to polymerization; reaction with dimethylamine vapor led to the expected N,N-dimethylselenoacetamide.

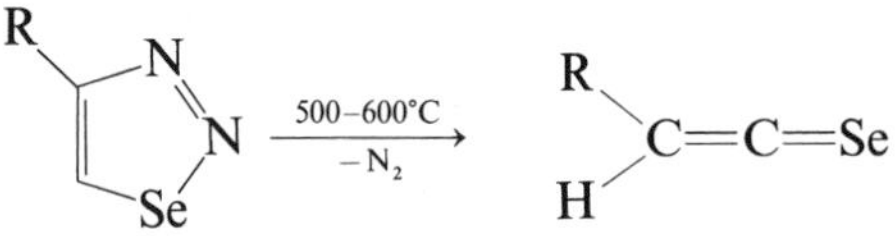

$$\underset{Se}{\overset{R}{\diagdown}}\underset{N}{\overset{N}{\diagdown}} \xrightarrow[-N_2]{500\text{–}600°C} \underset{H}{\overset{R}{\diagdown}}C{=}C{=}Se$$

R = H, Me, and t-Bu   124
(trapped at −196°C)

(97)

Bis-trimethylsilylselenoketene, 125, is a remarkably stable compound which can be prepared by trimethylsilylation of the acetylenic selenolate [Eq. (98)].[140] Reaction of 125 with diethylamine led to the desilylated selenoamides, 126 and 127 [Eq. (99)].

$$Me_3SiC{\equiv}CSeLi + Me_3SiCl \xrightarrow{-30°C} \underset{Me_3Si}{\overset{Me_3Si}{\diagdown}}C{=}C{=}Se$$

125

(98)

$$\underset{Me_3Si}{\overset{Me_3Si}{\diagdown}}C{=}C{=}Se \xrightarrow{Et_2NH} Me_3SiCH_2\overset{Se}{\overset{\|}{C}}NEt_2 + CH_3\overset{Se}{\overset{\|}{C}}NEt_2$$

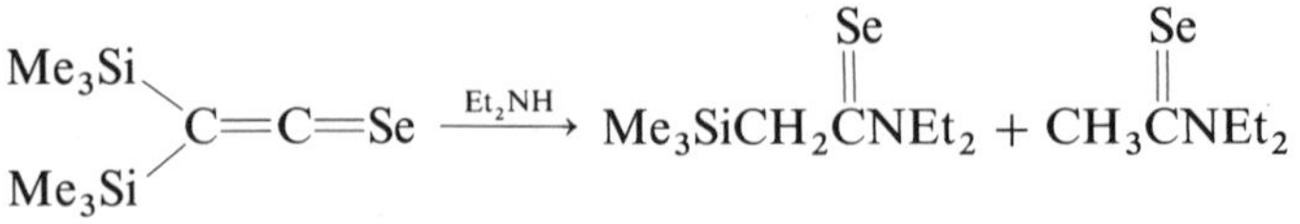

125   126   127

(99)

A number of other stable sterically hindered selenoketenes, 129, could be prepared through a selena–Cope reaction of the allylic acetylenic selenide, 128 [Eq.

(100)].[141] Aside from selenoamide formation, the selenoketenes formed (1:1) and (1:2) adducts, 130, 131, with 3,4-dihydroisoquinoline [Eq. (101)].

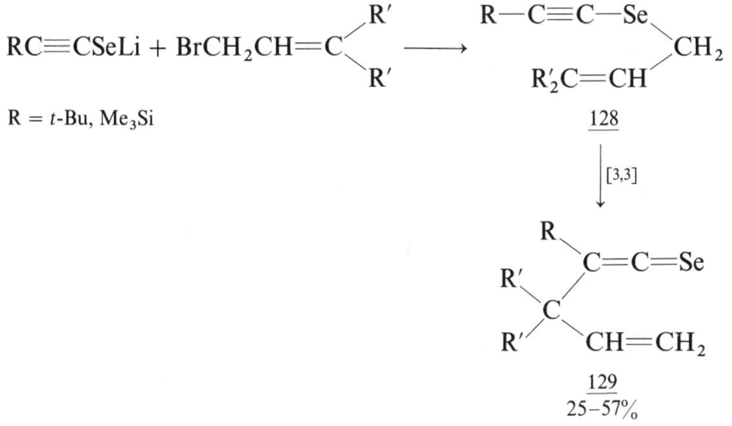
RC≡CSeLi + BrCH2CH=CR′2
R = t-Bu, Me3Si
R—C≡C—Se—CH2—CH=CR′2
128
[3,3]
R(C=C=Se)—CR′2—CH=CH2
129
25–57%

(100)

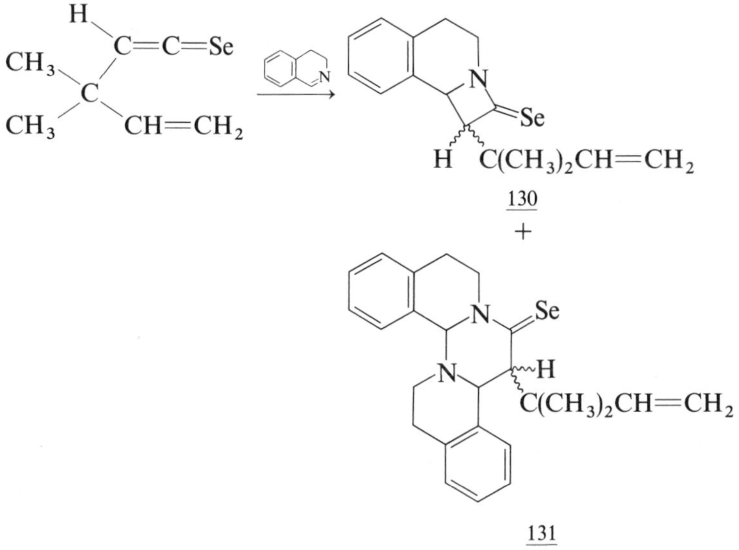
H
C=C=Se
CH3
C
CH3
CH=CH2
N
Se
H C(CH3)2CH=CH2
130
+
N
Se
N
H
C(CH3)2CH=CH2
131

(101)

15. CARBON DISELENIDE

The preparation and some reactions of carbon diselenide, 132, have been previously reivewed.[142] A more convenient preparation of this compound from dichloromethane and selenium has been recently reported [Eq. (102)].[142a]

$$CH_2Cl_2 + 2Se \xrightarrow{500°C} \underset{\substack{\underline{132} \\ 85\%}}{CSe_2} + 2HCl \qquad (102)$$

Carbon diselenide reacts with a variety of active methylene compounds affording 1,1-diselenolates, 133, which are readily alkylated [Eq. (103)].[143]

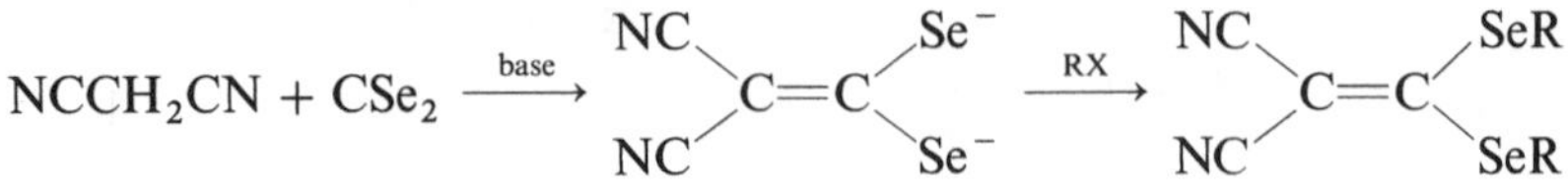

$$\text{NCCH}_2\text{CN} + \text{CSe}_2 \xrightarrow{\text{base}} \underset{\text{NC}}{\overset{\text{NC}}{\diagup}}\text{C}=\text{C}\underset{\text{Se}^-}{\overset{\text{Se}^-}{\diagdown}} \xrightarrow{\text{RX}} \underset{\text{NC}}{\overset{\text{NC}}{\diagup}}\text{C}=\text{C}\underset{\text{SeR}}{\overset{\text{SeR}}{\diagdown}} \quad (103)$$

133

The use of carbon diselenide in the preparations of selenoformamides, selenocarbonates, selenocarbamates, selenoureas, and related compounds have been described earlier.

16. SELENOCARBONYL COMPOUNDS OF BIOLOGICAL INTEREST

Selenopyrimidines and selenopurines have been prepared from the reaction of the corresponding chloropyrimidines or chloropurines and sodium hydrogen selenide or selenourea [Eq. (104)].[144] This method has also been used in the direct preparation of selenonucleosides.[145] Alternatively the alkylthio-substituted purines have been converted to the corresponding selenocarbonyl compounds using sodium hydrogen selenide.[146]

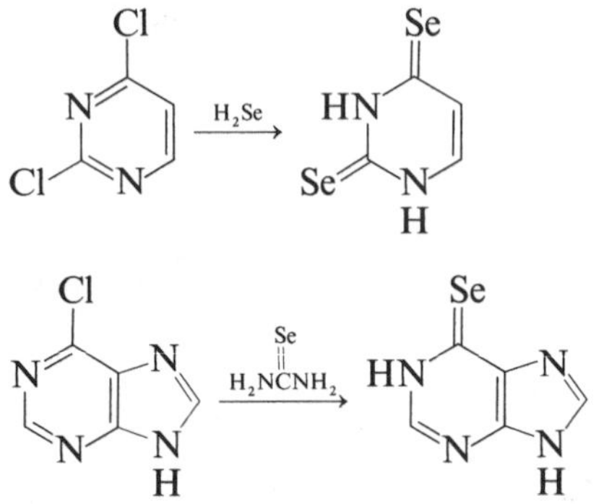

$$(104)$$

92%

Adenosine, 2-aminoadenosine and related compounds have been directly converted to the corresponding selenocarbonyl compounds 134 and 135 in good yield by treatment with hydrogen selenide in aqueous pyridine.[147] This reaction involves displacement of a heterocyclic amino group. The method can also be used for selenonucleotide preparations.

2-Selenobarbiturates 136,[148] 2-selenouracil, and 2-selenothymine 137[144] have been prepared by condensation of selenourea with β-dicarbonyl compounds [Eqs. (105) and (106)].

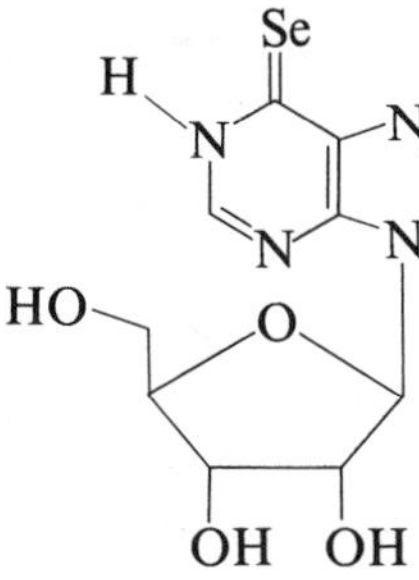

134

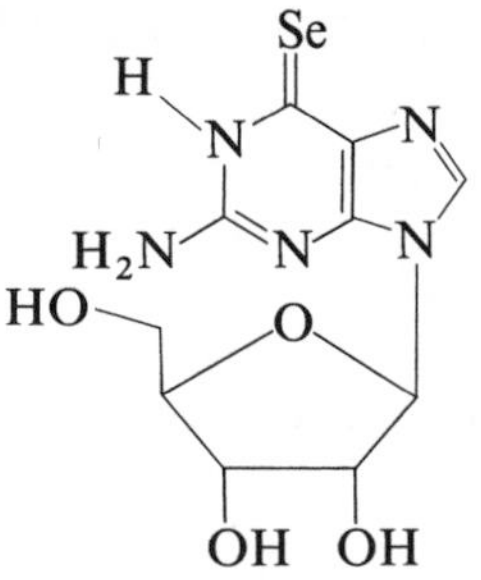

135

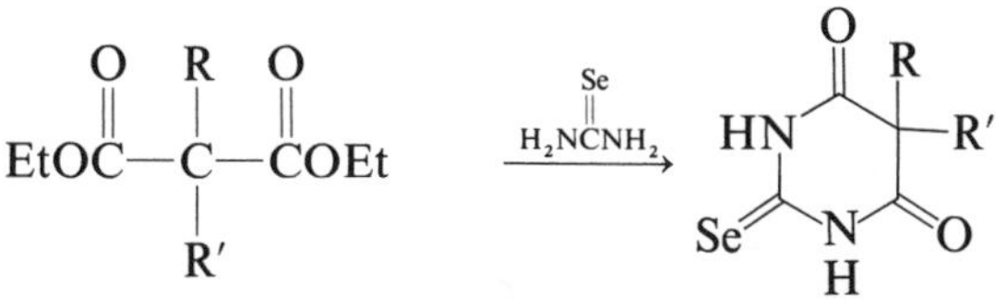

$$\underset{\substack{\text{EtOC}-\text{C}-\text{COEt}\\|\\R'}}{\overset{\substack{\text{O} \quad \text{R} \quad \text{O}\\||\qquad||}}{}} \xrightarrow{\underset{H_2NCNH_2}{\overset{Se}{||}}} 136 \qquad (105)$$

136

$$\underset{\substack{R \quad \text{ONa}}}{\overset{\substack{\text{O}\\||\\\text{EtOC}\qquad\text{H}}}{}} \xrightarrow{\underset{H_2NCNH_2}{\overset{Se}{||}}} 137 \qquad (106)$$

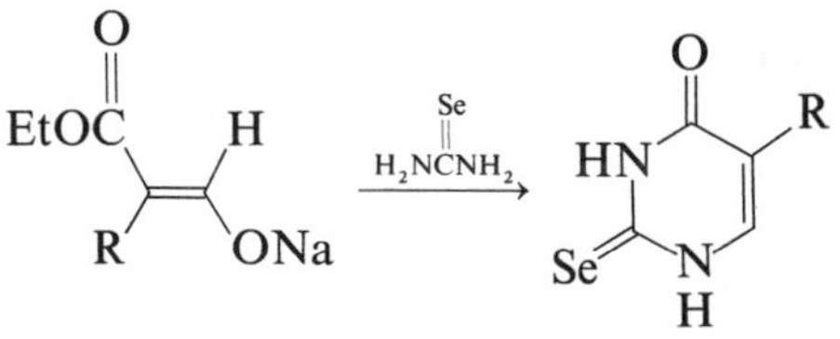

137

A modified selenonucleoside, 5-methylaminomethyl-2-selenouridine, 138, has been isolated from glutamate- and lysine-containing *t*-RNA from *E. coli* and *C. stricklandi* grown in the presence of selenite.[149,150] Reports of naturally occurring 4-selenouridine, 139, have also appeared.[151,152]

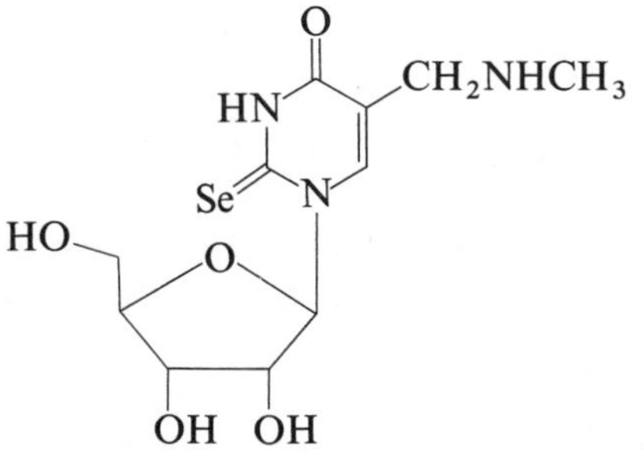

138

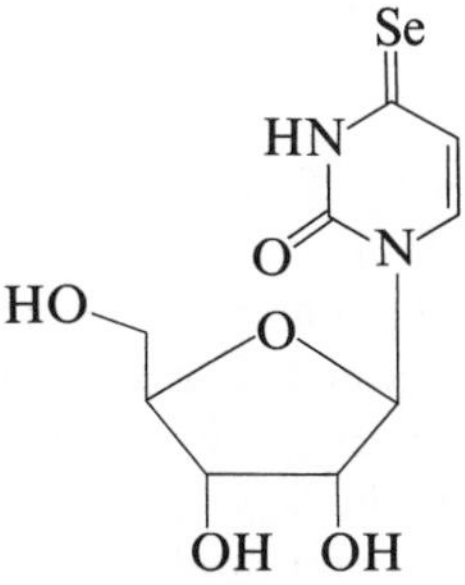

139

Selenosemicarbazones have also been evaluated as antineoplastic agents.[119]

17. OTHER SELENOCARBONYL-CONTAINING COMPOUNDS

Treatment of the vinylogous selenocarbonate, 140, with aqueous sodium selenide affords the diselone–dienediselenol tautomeric mixture 141 plus its air oxidation product 142 [Eq. (107)].[153] Acid-catalyzed hydrolysis or pyrolysis of 141 regenerates the selenocarbonate, 140. (It is worth noting that these are not the initially reported structures for the products of this reaction.)[38]

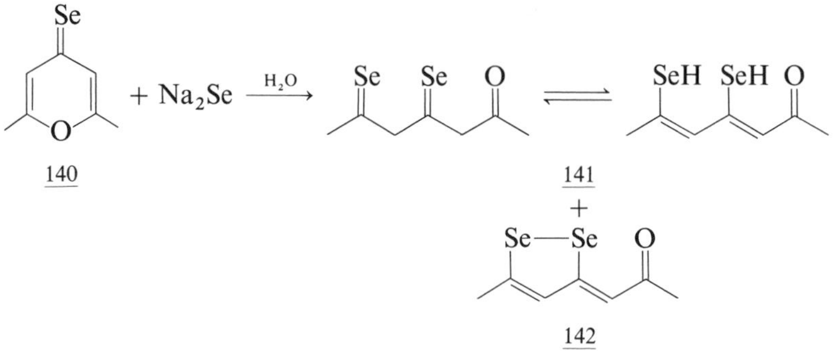

$$\text{(107)}$$

If the oxidation product, 142, is treated with phosphorus pentasulfide two interesting compounds, selenothiophthenes, 143 and 144, are obtained. These are formally selenocarbonyl compounds, and are stabilized by so-called "no-bond resonance" [Eq. (108)].[154]

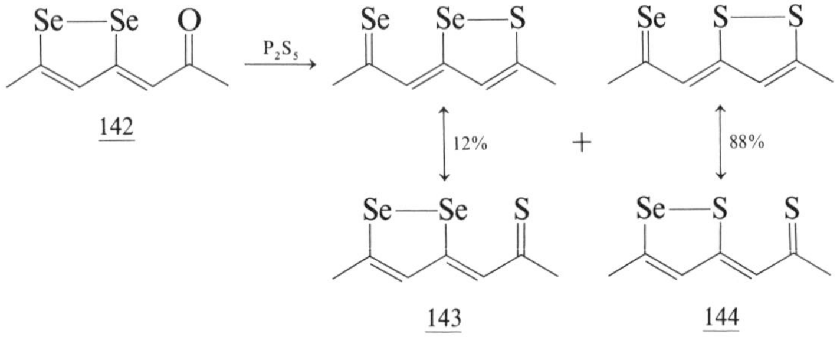

$$\text{(108)}$$

Treatment of the 1,2-diselenafulvene aldehyde, 145, with phosphorus pentaselenide afforded an analogous compound, 146 [Eq. (109)]. It is worth noting that an X-ray analysis of 146 showed a longer S—S bond than an S—Se bond suggesting a greater resonance contribution from 146b, and hence less selenocarbonyl character in this molecule.[155]

Perhaps the best way to envision molecules of this type is as the hypervalent selenium species, and not as true selenocarbonyl compounds. Related compounds have been prepared from the vinylogous thioselenocarbonates [Eq. (110)].[97]

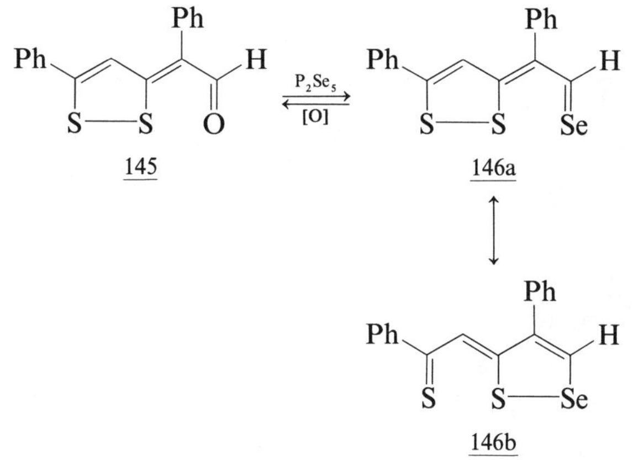

(109)

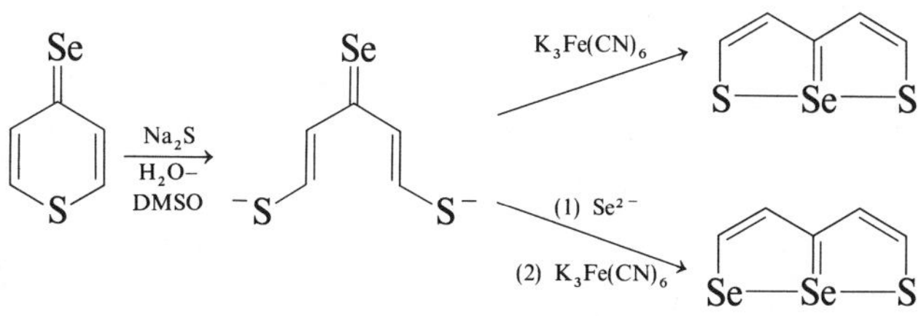

(110)

A variety of other heterocyclic selenocarbonyl-containing compounds have been recently prepared. Some interesting reactions include [Eqs. (111), (112), (113), (114), and (115)].

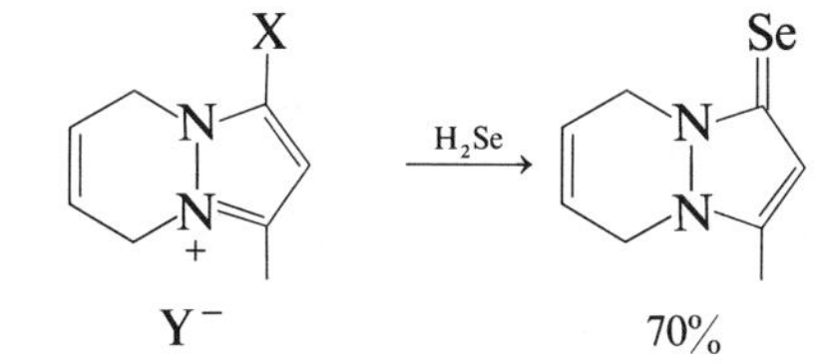

(111) (Ref. 156)

$$X = Cl, \quad Y = Cl$$
$$X = OEt, \quad Y = BF_4^-$$

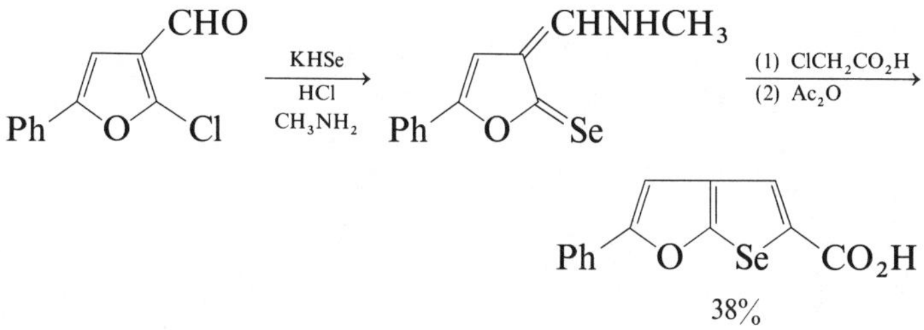

(112) (Ref. 157)

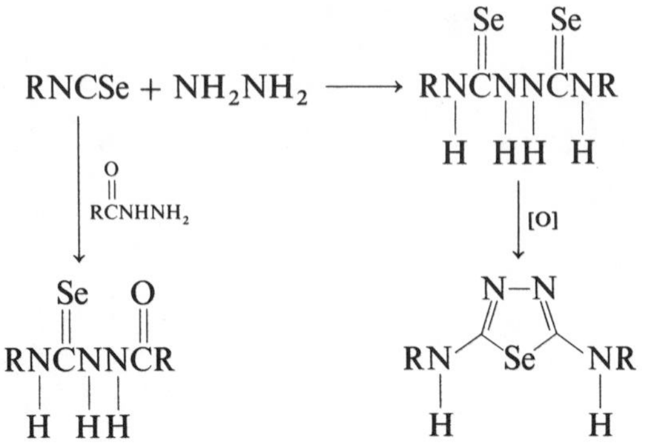

$$\text{RNCSe} + \text{NH}_2\text{NH}_2 \longrightarrow \overset{\text{Se}\ \ \ \text{Se}}{\underset{\text{H}\ \ \text{H}\text{H}\ \ \text{H}}{\text{RNCNNCNR}}}$$

$$\downarrow \overset{\text{O}}{\underset{\text{RCNHNH}_2}{\|}}$$

$$\overset{\text{Se}\ \ \ \text{O}}{\underset{\text{H}\ \ \text{H}\text{H}}{\text{RNCNNCR}}} \qquad\qquad \overset{\text{N--N}}{\underset{\underset{\text{H}}{\text{RN}}\ \ \text{Se}\ \ \underset{\text{H}}{\text{NR}}}{}}$$

$$\Big\downarrow [\text{O}] \qquad\qquad\qquad (113)\ (\text{Ref. 158})$$

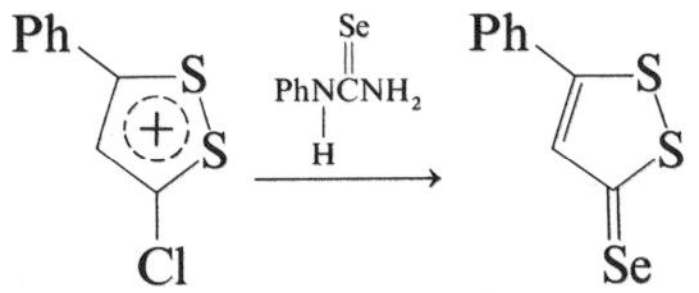

(114) (Ref. 159)

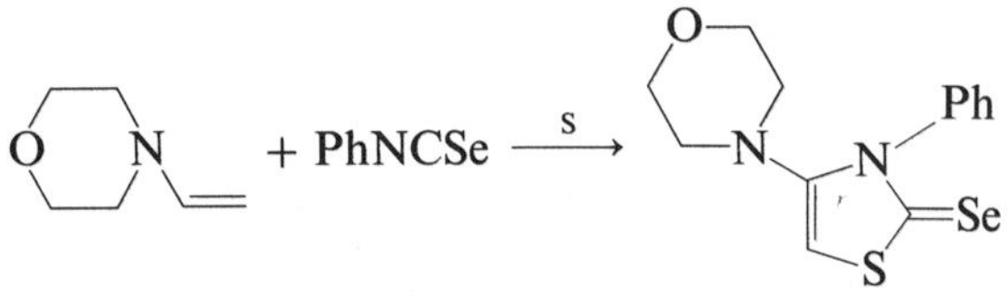

(115) (Ref. 160)

Transition metal stabilization of formally selenocarbonyl containing molecules has also been reported [Eq. (116)].[161,162]

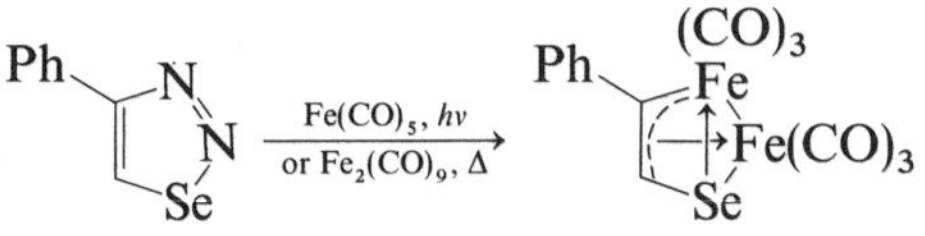

(116)

18. SPECTROSCOPIC STUDIES OF SELENOCARBONYL COMPOUNDS

Nuclear magnetic resonance (NMR) studies have proved to be especially useful in investigating the structure of selenocarbonyl-containing molecules. Selenium 77 NMR studies of a series of selenocarbonyl compounds show that the $\delta(^{77}\text{Se})$ chemical shifts are drastically shifted downfield (~ 100 ppm from dimethyl selenide), and are remarkably sensitive to small changes in the electronic structure of the selenium atom.[163] The ^{77}Se chemical shift closely parallels the λ_{max} of the $n \to \pi^*$ transitions in these compounds. The $^{77}\text{Se}-^{13}\text{C}$ coupling constants were also determined, and found to be much larger than had been previously observed (209–221 Hz).[163,164] Comparison of ^{13}C chemical shifts of selenocarbonyl compounds with their sulfur and oxygen analogs shows a significant downfield shift due to selenium, although the relative downfield shift on going from sulfur to selenium (3–13 ppm) is much less than that observed in going from the carbonyl to the thiocarbonyl compound (13–60 ppm).[165] The selenocarbonyl carbon is the most deshielded carbon observed in neutral molecules.[166] Comparison of [77]Se and [17]O chemical shifts has also been reported.[10] The ^{13}C isotope effect on [77]Se shielding has also been determined, and correlated with C—Se bond distances for a number of selenocarbonyl compounds.[167]

Nuclear magnetic resonance studies of amides, thioamides, and selenoamides show an increased barrier to rotation about the C—N bond going from oxygen, to sulfur, to selenium. This suggests an enhanced contribution of the dipolar resonance from 147b for selenoamides [Eq. (117)].[168] Chemical shifts[169] of ^{14}N labeled compounds and ^{13}C studies[165] are in agreement with this conclusion.

$$
\underset{\underline{147a}}{\overset{\displaystyle \overset{X}{\underset{\displaystyle \|}{}}}{R-C-N{<}}} \longleftrightarrow \underset{\underline{147b}}{\overset{\displaystyle \overset{X^{-}}{\underset{\displaystyle |}{}}}{RC{=}\underset{+}{N}{<}}}
\tag{117}
$$

X = O, S, Se

Related nmr studies of selenoureas and selenosemicarbazides have also been carried out.[170–172] While the semicarbazides behave like the amides, free rotation about the C—N bond of tetramethylselenourea occurs even at −120°C. This has been explained by steric strain in the ground state due to selenium, lowering the barrier to rotation.

A detailed review describes a number of infrared spectral studies that have been carried out on selenocarbonyl compounds to determine the C=Se absorption frequency.[173] In general these studies involved comparing the selenocarbonyl compounds with thiocarbonyl analogs, and noting differences in the spectra. The ultraviolet–visible spectra of some series of selenocarbonyl compounds has appeared[163,174] as has some mass spectral data.[62,175] The chiroptical properties of (−) selenofenchone have been compared with its thione and ketone analogs.[176] The electrochemical properties of methyl selenobenzoate and selenobenzamide have been compared with the sulfur analogs.[177]

ACKNOWLEDGMENT

I gratefully acknowledge many talented coworkers in this area. I especially thank Lynn James San Filippo and Loide Mayer Wasmund for their help and suggestions in preparing this manuscript.

REFERENCES

1. W. H. H. Günther, in *Organic Selenium Compounds: Their Chemistry and Biology*, D. L. Klayman and W. H. H. Günther, Eds., Wiley-Interscience, New York, 1973, p. 59.

2. W. H. H. Günther, in *Organic Selenium Compounds: Their Chemistry and Biology*, D. L. Klayman and W. H. H. Günther, Eds., Wiley-Interscience, New York, 1973, p. 45.

3. W. H. H. Günther, in *Organic Selenium Compounds: Their Chemistry and Biology*, D. L. Klayman and W. H. H. Günther, Eds., Wiley-Interscience, New York, 1973, p. 35.

4. R. B. Silverman, in *Organic Selenium Compounds: Their Chemistry and Biology*, D. L. Klayman and W. H. H. Günther, Eds., Wiley-Interscience, New York, 1973, p. 245.

5. K. A. Jensen, in *Organic Selenium Compounds: Their Chemistry and Biology*, D. L. Klayman and W. H. H. Günther, Eds., Wiley-Interscience, New York, 1973, p. 263.

6. R. J. Shine, in *Organic Selenium Compounds: Their Chemistry and Biology*, D. L. Klayman and W. H. H. Günther, Eds., Wiley-Interscience, New York, 1973, p. 273.

7. H. G. Mautner, in *Organic Selenium Compounds: Their Chemistry and Biology*, D. L. Klayman and W. H. H. Günther, Eds., Wiley-Interscience, New York, 1973, p. 497.

8. K. J. Irgolic and M. V. Kudchadker, in *Selenium*, R. A. Zingaro and W. C. Cooper, Eds., Van Nostrand Reinhold, New York, 1974, p. 408.

9. R. B. Silverman, in *Organic Selenium Compounds: Their Chemistry and Biology*, D. L. Klayman and W. H. H. Günther, Eds., Wiley-Interscience, New York, 1973, p. 246.

10. T. C. Wong, F. S. Guziec, Jr., and C. A. Moustakis, *J. Chem. Soc. Perkin Trans. 2*, 1471 (1983).

11. I. L. Karle and J. Karle, in *Organic Selenium Compounds: Their Chemistry and Biology*, D. L. Klayman and W. H. H. Günther, Eds., Wiley-Interscience, New York, 1973, pp. 992–993.

12. D. S. Margolis and R. W. Pittman, *J. Chem. Soc.*, 799 (1957).

13. H. J. Bridger and R. W. Pittman, *J. Chem. Soc.*, 1371 (1950).

14. L. Mortillaro, L. Credali, M. Mammi, and G. Valle, *J. Chem. Soc.*, 807 (1965).

15. L. Szperl and W. Wiorogorski, *Rocz. Chem.*, **12**, 270 (1932).

16. R. B. Silverman, in *Organic Selenium Compounds: Their Chemistry and Biology*, D. L. Klayman and W. H. H. Günther, Eds., Wiley-Interscience, New York, 1973, p. 248.

17. L. Vanino and A. Schinner, *J. Prakt. Chem.*, **91**, 116 (1915).

18. R. Kuhn, *J. Chem. Soc.*, 605 (1938).

19. T. G. Back, D. H. R. Barton, M. R. Britten-Kelly, and F. S. Guziec, Jr., *Chem. Commun.*, 539 (1975).

20. T. G. Back, D. H. R. Barton, M. R. Britten-Kelly, and F. S. Guziec, Jr., *J. Chem. Soc. Perkin Trans. 1.*, **19**, 2079 (1976).

21. P. de Mayo, G. L. R. Petrasiunas, and A. C. Weedon, *Tetrahedron Lett.*, 4621 (1978).

21b. N. Y. M. Fung, unpublished masters thesis, University of Western Ontario, 1979.

22. F. S. Guziec, L. J. San Filippo, C. J. Murphy, C. A. Moustakis, and E. R. Cullen, *Tetrahedron*, **41**, 4843 (1985).

23. F. S. Guziec, Jr., and C. A. Moustakis, *J. Org. Chem.*, **49**, 189 (1984). (a) R. Okazàki, K. Inoue, and N. Inamoto, *Tetrahedron Letters*, 3673 (1979).

24. K. Steliou and M. Mrani, *J. Am. Chem. Soc.*, **104**, 3104–3106 (1982).

25. B. S. Pedersen, S. Scheibye, N. H. Nilsson, and S. O. Lawesson, *Bull. Soc. Chem. Belg.*, **87**, 223 (1978).

26. J. E. Baldwin and R. C. Gerald-Lopez, *Chem. Commun.*, 1029 (1982).

27. E. Vedejs, T. H. Eberlein, and D. J. Varie, *J. Am. Chem. Soc.*, **104**, 1445 (1982).

28. R. Okazaki, A. Ishii, N. Fukuda, H. Oyama, and N. Inamoto, *Chem. Commun.*, 1187 (1982).

29. D. H. R. Barton, F. S. Guziec, Jr., and I. Shahak, *J. Chem. Soc. Perkin Trans. 1.*, **15**, 1794 (1974).

30. F. S. Guziec, Jr., and C. J. Murphy, *J. Org. Chem.*, **45**, 2890 (1980).

31. E. R. Cullen, F. S. Guziec, Jr., M. I. Hollander, and C. J. Murphy, *Tetrahedron Lett.*, 4563 (1981).

32. E. R. Cullen, F. S. Guziec, Jr., and C. J. Murphy, *J. Org. Chem.*, **47**, 3563 (1982).

33. F. S. Guziec, Jr., C. J. Murphy, and E. R. Cullen, *J. Chem. Soc. Perkin Trans 1*, 107 (1985).

34. A. Krebs, W. Ruger, and W. U. Nickel, *Tetrahedron Lett*, 4937 (1981).

35. F. S. Guziec, Jr. and C. A. Moustakis, *Chem. Commun.*, 63 (1984).

36. F. S. Guziec, Jr., C. A. Moustakis, and L. J. San Filippo, to be published. (a) A. Ohno, K. Nakamura, M. Uohoma, S. Oka, T. Yamabe, and S. Nagata, *Bull. Chem. Soc. Jpn.*, **48**, 3718 (1975).

37. C.-P. Klages and J. Voss, *Angew. Chem. Int. Ed. Engl.*, **16**, 725 (1977).

38. G. Traverso, *Ann. Chim. (Rome)*, **47**, 3 (1957).

39. B. J. McKinnon, P. de Mayo, N. C. Payne, and B. Ruge, *Nouv. J. Chim.*, **2**, 91 (1978).

40. N. Y. M. Fung, P. de Mayo, B. Ruge, A. C. Weedon, and S. K. Wong, *Can. J. Chem.*, **58**, 6 (1980).

41. J. C. Scaiano and K. U. Ingold, *Chem. Comm.*, 205 (1976).

42. J. C. Scaiano and K. U. Ingold, *J. Phys. Chem.*, **80**, 1901 (1976).

43. J. C. Scaiano, *J. Am. Chem. Soc.*, **99**, 1494 (1977).

44. C. G. Barraclough, R. L. Martin, and I. M. Stewart, *Aust. J. Chem.*, **22**, 891 (1969).

45. G. Wilke and E. Uhlemann, *Z. Chem.*, **14**, 288 (1974).

46. G. Wilke and E. Uhlemann, *Z. Chem.*, **15**, 66 (1975).

47. V. A. Alekseevskii, O. A. Ramazanova, and L. A. Deryabina, *Zh. Neorg. Khim.*, **23**, 545 (1978). [*Chem. Abst.* **88** (1978), 142408]

48. E. Uhlemann, *Int. Symp. Specific Interact. Mol. Ions.* [Proc] 3rd. **2**, 539 (1976). [*Chem. Abst.* **88** (1978), 79947].

49. A. H. Schmidt, W. Ried, and P. Pustoslemsek, *Angew. Chem. Int. Ed. Engl.*, **15**, 704 (1976).

50. D. H. R. Barton and S. W. McCombie, *J. Chem. Soc. Perkin Trans. 1*, 1574 (1975).

51. D. L. Klayman and T. S. Griffin, *J. Am. Chem. Soc.*, **95**, 197 (1973).

52. C. Collard-Charon and M. Renson, *Bull. Soc. Chim. Belg.*, **71**, 563 (1962).

53. V. I. Cohen, *J. Org. Chem.*, **42**, 2645 (1977).

54. I. Wallmark, M. H. Krackov, S.-H. Chu, and H. G. Mautner, *J. Am. Chem. Soc.*, **92**, 4447 (1970).

55. F. Malek-Yazdi and M. Yalpani, *J. Org. Chem.*, **41**, 729 (1976).

56. E. O. Fischer and S. Riedmuller, *Ber.*, **107**, 915 (1974).

57. D. H. R. Barton, P.-E. Hansen, and K. Picker, *J. Chem. Soc. Perkin Trans. 1*, 1723 (1977).

58. P.-E. Hansen, *J. Chem. Soc. Perkin Trans. 1*, 1627 (1980).

59. K. A. Jensen, H. Mygind, and P. H. Nielsen, in *Organic Selenium Compounds: Their Chemistry and Biology*, D. L. Klayman and W. H. H. Günther, Eds., Wiley-Interscience, New York, 1973, p. 267.

60. F. von Dechend, *Ber.*, **7**, 1273 (1874).

61. W. Becker and J. Meyer, *Ber.*, **37**, 2550 (1904).

62. V. I. Cohen, *Synthesis*, 668 (1978).

63. K. A. Jensen and P. H. Nielsen, *Acta Chem. Scand.*, **20**, 597 (1966).

64. H. A. Hallam and C. M. Jones, *J. Chem. Soc. A.*, 1033 (1969).

65. M. P. Cava and L. E. Saris, *Chem. Commun.*, 617 (1975).

66. J. Liebscher and H. Hartmann, *Synthesis*, 521 (1976).

67. H. Hartmann, *Z. Chem.*, **11**, 60 (1971).

68. F. Malek-Yazdi and M. Yalpani, *Synthesis*, 328 (1977).

69. R. S. Sukhai, R. de Jong, and L. Brandsma, *Synthesis*, 888 (1977).

70. L. Henriksen, *Synthesis*, 501 (1974). (a) M. Tisler, B. Stenovnik, Z. Zrimsek, and C. Stropnik, *Synthesis*, 299 (1981).

71. Ger. Patent 910, 199/1954 (*Chem. Abst.* **53**, (1959) 936h).

72. D. H. Reid, R. G. Webster, and S. McKenzie, *J. Chem. Soc. Perkin Trans. 1*, 2334 (1979).

73. F. Asinger, H. Berding, and H. Offermanns, *Monatsh. Chem.*, **99**, 2072 (1968).

74. K. A. Jensen and U. Henriksen, *Acta Chem. Scand.*, **21**, 1991 (1967).

75. C. O. Maese, W. Walter, H. Mrotzek, and H. Miryai, *Chem. Ber.*, **109**, 956 (1976).

76. J. Liebscher and H. Hartmann, *Tetrahedron*, **33**, 731 (1977).

77. R. S. Sukhai and L. Brandsma, *Synthesis*, 455 (1979).

78. E. Bulka, in *Organic Selenium Compounds: Their Chemistry and Biology*, D. L. Klayman and W. H. H. Günther, Eds., Wiley-Interscience, New York, 1973, p. 459.

79. K. A. Jensen, L. Boje, and L. Henriksen, *Acta Chem. Scand.*, **26**, 1465 (1972).

80. H. Ishihara and Y. Hirabayashi, *Chem. Letters.*, 203 (1976).

81. H. Ishihara and S. Kato, *Tetrahedron Lett.*, 3751 (1972).

82. S. Kato, W. Akada, M. Mizuta, and Y. Ishii, *Bull. Chem. Soc. Jpn.*, **46**, 244 (1973).

83. K. A. Jensen, in *Organic Selenium Compounds: Their Chemistry and Biology*, D. L. Klayman and W. H. H. Günther, Eds., Wiley-Interscience, New York, 1973, p. 268.

84. M. Drager and G. Gattow, *Angew. Chem. Int. Ed. Engl.*, **11**, 868 (1968).

85. M. Drager and G. Gattow, *Ber.*, **104**, 1429 (1971).

86. K. A. Jensen, P. A. A. Frederiksen, and L. Henriksen, *Acta Chem. Scand.*, **24**, 2061 (1970).

87. K. A. Jensen and U. Anthoni, *Acta Chem. Scand.*, **24**, 2055 (1970).

88. L. Henriksen, *Acta Chem. Scand.*, **21**, 1981 (1967).

89. L. Henriksen and E. S. S. Kristiansen, *Ann. N.Y. Acad. Sci.*, 101 (1972).

90. K. Bechgaard, D. O. Cowan, A. N. Bloch, and L. Henriksen, *J. Org. Chem.*, **40**, 746 (1975).

91. E. M. Engler, B. A. Scott, S. Etemad, T. Penney, and V. V. Patel, *J. Am. Chem. Soc.*, **99**, 5909 (1977).

92. K. Lerstrup, M. Lee, F. M. Wiggel, T. J. Kistenmacher, and D. O. Cowan, *Chem. Commun.*, 294 (1983).

93. I. Johannsen, K. Bachgaard, K. Mortensen, and C. Jacobsen, *Chem. Commun.*, 295 (1983).

94. V. Y. Lee, E. M. Engler, R. R. Schumaker, and S. S. P. Parkin, *Chem. Commun.*, 234 (1983).

95. G. Traverso, *Ann. Chim. (Rome)*, **47**, 1244 (1957).

96. D. J. Sandman, A. J. Epstein, T. J. Holmes, and A. P. Fisher III, *Chem. Commun.*, 177 (1977).

97. D. H. Reid, *J. Chem. Soc. (C)*, 3187 (1971).

98. D. L. Klayman and R. J. Shine, *J. Org. Chem.*, **34**, 3549 (1969).

99. C. Collard-Charon and M. Renson, *Bull. Soc. Chim. Belg.*, **72**, 149 (1963).

100. M. Lipp, F. Dallacker, and I. Meier, *Monatsh. Chem.*, **90**, 41 (1959).

101. M. Giua and R. Bianco, *Gazz. Chim. Ital.*, **89**, 693 (1959).

102. R. A. Zingaro, F. C. Bennett, Jr., and G. W. Hammer, *J. Org. Chem.*, **18**, 292 (1953).

103. C. King and R. J. Hlavacek, *J. Am. Chem. Soc.*, **69**, 1833 (1951).

104. F. C. Bennett, Jr., and R. A. Zingaro, *Organic Syntheses*, Vol. 4, N. Rabjohn, Ed., Wiley, New York, 1963, p. 359.

105. H. G. Grimm and H. Metzger, *Ber.*, **69**, 1356 (1936).

106. K. A. Jensen, G. Felbert, and B. Kagi, *Acta. Chem. Scand.*, **20**, 281 (1966).

107. Z. I. Yoshida, H. Konishi, and H. Ogoshi, *Chem. Commun.*, 359 (1975).

108. T. S. Griffin and D. L. Klayman, *J. Org. Chem.*, **39**, 3161 (1974).

109. G. C. Chen, R. A. Zingaro, and C. R. Thomspon, *Carbohydr. Res.*, **39**, 61 (1975).

110. S. H. Chu, C.-T. Shiue, and M. Y. Chu., *J. Med. Chem.*, **18**, 559 (1975).

111. A. Shafiee and I. Lalezari, *J. Heterocycl. Chem.*, **12**, 675 (1975).

112. M. Mikolajczyk and J. Luczak, *Chem. Ind.*, 76 (1972).

113. A. Chiesi, G. Grossoi, M. Nardelli, and M. E. Vidoni, *Chem. Commun.*, 404 (1969).

114. P. W. Preisler and T. N. Scortia, *J. Am. Chem. Soc.*, **80**, 2309 (1958).

115. Gerald Lucovsky and Mark D. Tabak, in *Selenium* R. A. Zingaro and W. C. Cooper, Eds., Van Nostrand Reinhold, New York 1974, p. 788.

116. C. Collard-Charon, R. Huls, and M. Renson, *Bull. Soc. Chim. Belg.*, **71**, 541 (1962).

117. K. A. Jensen, G. Felbert, C. T. Pedersen, and U. Svanholm, *Acta Chem. Scand.*, **20**, 278 (1966).

118. H. G. Mautner and W. D. Kumler, *J. Am. Chem. Soc.*, **78**, 97 (1956).

119. K. C. Agrawal, B. A. Booth, R. L. Michand, E. C. Moore, and A. C. Santorelli, *Biochem. Phamacol.*, **23**, 2421 (1967).

120. A. A. Tsurkau and V. S. Basalitskaya, *Farm. Zh. (Kiev.)*, **30**, 45 (1975) [*Chem. Abstr.*, **84**, (1976) 164919].

121. D. Barnard and D. T. Woodbridge, *J. Chem. Soc.*, 2922 (1961).

122. U. Anthoni, *Acta Chem. Scand.*, **20**, 2742 (1966).

123. K. A. Jensen, M. Due, A. Holm, and C. Wentrup, *Acta Chem. Scand.*, **20**, 2091 (1966).

124. U. Anthoni, B. M. Dahl, C. Larsen, and P. H. Nielsen, *Acta Chem. Scand.*, **24**, 959 (1970).

125. E. Bulka and K.-D. Ahlers, *Z. Chem.*, **3**, 348 (1963).

126. C. Collard-Charon and M. Renson, *Bull. Soc. Chim. Belg.*, **71**, 531 (1962).

127. C. T. Pedersen, *Acta Chem. Scand.*, **17**, 1459 (1963).

128. I. B. Douglass, *J. Am. Chem. Soc.*, **59**, 740 (1937).

129. H. Suzuki, M. Usuki, and T. Hanafusa, *Synthesis*, 705 (1979).

130. L. Hendriksen and J. Ehrbor, *Synthesis*, 519 (1976).

131. T. Tarantelli and C. Pecile, *Ann. Chim. (Rome)*, **52**, 75 (1962).

132. W. J. Franklin and R. L. Werner, *Tetrahedron Lett.*, 3003 (1965).

133. D. H. R. Barton, G. Bringmann, G. Lamotte, W. B. Motherwell, R. S. H. Motherwell, and A. E. A. Porter, *J. Chem. Soc. Perkin Trans. 1.*, 2657 (1980).

134. I. Lalezari, A. Shafiee, and M. Yalpani, *Tetrahedron Lett.*, 5105 (1969).

135. I. Lalazari, A. Shafiee, and M. Yalpani, *J. Org. Chem.*, **38**, 338 (1973).

136. M. H. Ghadehari, D. Davalian, M. Yalpani, and M. H. Partovi, *J. Org. Chem.*, **39**, 3906 (1974).

137. H. Meier and I. Menzel, *Tetrahedron Lett.*, 445 (1972).

138. B. Bak, O. J. Nielsen, H. Svanholt, and A. Holm, *Chem. Phys. Lett*, **53**, 374 (1978); **55**, 36 (1978).

139. A. Holm, C. Berg, C. Bjerre, B. Bak, and H. Svanholt, *Chem. Commun.*, 99 (1979).

140. R. S. Sukhai and L. Brandsma, *Rec. Trav. Chim.*, **98**, 55 (1979).

141. E. Schaumann and F. F. Grabley, *Tetrahedron Lett.*, 4251 (1980).

142. K. A. Jensen, *Quart. Rept. Sulfur Chem.*, **5**, 45 (1970).

142a. L. Henriksen and E. S. Kristiansen, *Int. J. Sulfur Chem.*, **A2**, 133 (1972).

143. K. A. Jensen and L. Henriksen, *Acta Chem. Scand.*, **34**, 3213 (1970).

144. H. G. Mautner, *J. Am. Chem. Soc.*, **78**, 5293 (1956).

145. L. B. Townsend and G. H. Milne, *J. Heterocycl. Chem.*, **7**, 753 (1970).

146. F. Bergmann and M. Rashi, *Isrl. J. Chem.*, **7**, 63 (1969).

147. C.-Y. Shiue and S.-H. Chu, *Chem. Commun.*, 319 (1975).

148. H. G. Mautner and E. M. Clayton, *J. Am. Chem. Soc.*, **81**, 6270 (1959).

149. W.-M. Ching, *Fed. Proc., Fed. Am. Soc. Exp. Biol.*, **42**, 2238 (1983).

150. A. J. Wittwer and L. Tsai, *Fed. Proc., Fed. Am. Soc. Exp. Biol.*, **42**, 2238 (1983).

151. J. H. Hoffman and K. P. McConnell, *Biochem. Biophys. Acta*, **366**, 109 (1974).

152. Y. S. Prasado Rao and J. D. Cherayil, *Life Sci.*, **14**, 2051 (1974).

153. S. Bezzi, *Gazz. Chim. Ital.*, **92**, 859 (1962).

154. S. Pietra, C. Garbuglio, and M. Mammi, *Gazz. Chim. Ital.*, **94**, 48 (1964).

155. J. H. van den Hende and E. Klingsberg, *J. Am. Chem. Soc.*, **88**, 5045 (1966).

156. D. Farnum, A. T. Au, and K. Rasheed, *J. Heterocycl. Chem.*, **8**, 25 (1971).

157. I. Y. Kvito, N. B. Sokolova, and L. S. Etros, *Khim. Geterotsikl. Soedin.*, 715 (1973).

158. E. Bulka and D. Ehlers, *J. Prakt. Chem.*, **315**, 155 (1973).

159. H. Spies, K. Gewald, and R. Mayer, *J. Prakt. Chem.*, **313**, 804 (1971).

160. H. Spies, K. Gewald, and R. Mayer, *J. Prakt. Chem.*, **314**, 646 (1972).

161. G. N. Schrauzer and H. J. Kisch, *J. Am. Chem. Soc.*, **95**, 2501 (1973).

162. K. H. Pannell, A. J. Mayr, R. Hoggard, and R. C. Pettersen, *Angew. Chem. Int. Ed. Engl.*, **19**, 632 (1980).

163. E. R. Cullen, F. S. Guziec, Jr., C. J. Murphy, T. C. Wong, and K. K. Andersen, *J. Am. Chem. Soc.*, **103**, 7055 (1981).

164. W. Gombler, *Z. Naturforsch B.*, **36B**, 1561 (1981).

165. E. R. Cullen, F. S. Guziec, Jr., C. J. Murphy, T. C. Wong, and K. K. Andersen, *J. Chem. Soc. Perkin Trans. 2* 473 (1982).

166. G. A. Olah, T. Nakajima, and G. K. Surya Prakash, *Angew. Chem. Int. Ed. Engl.*, **19**, 811 (1980).

167. W. Gombler, *J. Am. Chem. Soc.*, **104**, 6616 (1982).

168. G. Schwenker and H. Rosswag, *Tetrahedron Lett.*, 4237 (1967).

169. P. Hampson and A. Mathias, *Mol. Phys.*, **13**, 361 (1967).

170. K. A. Jensen and J. Sandstron, *Acta Chem. Scand.*, **23**, 1911 (1969).

171. U. Svanholm, *Ann. N.Y. Acad. Sci.*, **192**, 124 (1972).

172. W. Walter, E. Schaumann, and H. Rose, *Tetrahedron*, 3233 (1972).

173. K. A. Jensen, K. Henriksen, and P. H. Nielsen, *Organic Selenium Compounds: Their Chemistry and Biology*, D. L. Klayman and W. H. H. Günther, Eds., Wiley-Interscience, New York, 1973, p. 835.

175. A. M. Kirkien, R. J. Shine, and J. R. Plimmer, *Org. Mass Spectrom.*, **7**, 233 (1973).

175a. F. S. Guziec, Jr., and C. A. Moustakis, *Phosphorus and Sulfur*, **21**, 189 (1984).

176. K. K. Andersen, D. M. Gash, J. D. Robertson, and F. S. Guziec, Jr., *Tetrahedron Lett.*, 911 (1982).

177. R. Mayer, S. Schiethauer, and D. Kunz. *Ber.*, **99**, 1393 (1966).

7

Radical Reactions
of Selenium Compounds

THOMAS G. BACK

Department of Chemistry, University of Calgary,
Calgary, Alberta, Canada

CONTENTS

1. INTRODUCTION

Early work on the subject of organoselenium radicals was reviewed in 1973 by Shine,[1a] who remarked on the paucity of well-documented reactions of such species. Despite a growing interest in this branch of organoselenium chemistry, much territory remains uncharted, especially when compared with the extensive literature dealing with organosulfur radicals. Progress has, however, been made on several fronts in the past decade and the present review covers the literature from 1973 to early 1983.

The free-radical reactions of common classes of compounds such as diselenides, selenides, and selenols have been investigated by several groups and are now reasonably well understood. Electron spin resonance (esr) techniques have been particularly valuable for gaining insight into the mechanisms of such processes, despite the fact that selenium centered radicals are often difficult to detect because of severe line broadening caused by large spin-orbit coupling. Although mechanistic information gleaned from esr experiments will be included where appropriate, it is not the intent of this chapter to review the spectroscopic properties of organoselenium radicals in detail.

A number of free-radical selenium reactions are proving to be of general synthetic utility, particularly when employed in conjunction with other, non-radical selenium-based functional group transformations. These include deselenizations effected by tin hydrides, carbon–carbon bond forming methods based on selenium extrusion processes, photoreductions and $S_{RN}1$ reactions of selenols and their derivatives, and the selenosulfonation of olefins, allenes, and acetylenes.

Apart from purely chemical considerations, the radical reactions of selenium are of importance in other domains. There is evidence that selenium compounds play an important role as antioxidants in living organisms. The selenoprotein glutathione peroxidase is believed to suppress damage to cell membranes from free-radical processes associated with polyunsaturated lipid peroxides.[2] The possible use of selenium compounds as radioprotective agents against ionizing radiation has been investigated.[1b] The patent literature reveals that there is current interest in the industrial sector in techniques for the deposition of elemental selenium by means of the homolytic decomposition of compounds such as dibenzyl diselenide. Applications exist in microimaging processes and in the production of photoconducting materials. Although the biochemical and industrial aspects of the subject are beyond the scope of this review, they provide additional impetus for seeking a better understanding of the free-radical chemistry of selenium.

A sizable number of photochemical reactions of organoselenium compounds have been reported and are the topic of a recent review by Martens and Praefcke.[3] Some of these are clearly free-radical processes, although many others proceed via mechanisms that are uncertain. Only reactions of the former type are included herein.

2. REACTIONS OF DISELENIDES

2.1. Photolysis

Despite two early reports to the contrary,[4,5] it is now generally accepted that the principal primary processes in the photolysis of diselenides are the cleavage of the Se—Se and C—Se bonds. The former is reversible and so does not necessarily lead to the formation of new products. The latter process is of particular importance when the resulting alkyl radical is relatively stable, as in the case of $PhCH_2 \cdot$.

The photolysis of diphenyl diselenide at low temperatures was studied by esr spectroscopy.[6] Both the olive-green phenylselenyl radical PhSe· and a species identified as the PhSeSe· radical were detected. Subsequent spin-trapping experiments with nitrosodurene indicated the formation of the nitroxide adduct, 1 [Eqs. (1) and (2)], but not of 2, thus suggesting that Se—Se, but not C—Se cleavage is the favored process in diaryl diselenides.[7] In a more recent study, Ito[8] produced the PhSe· radical by flash photolysis of the diselenide and monitored its disappearance by means of its absorption at 490 nm. Its decay via dimerization was slower than that of the analogous sulfur radical PhS· and followed the expected second-order kinetics. The phenylselenyl radical displays low reactivity towards oxygen, and hydrogen or halogen donors. It adds reversibly to activated olefins [Eq. (3)] to produce the alkyl radical, 3, which can be trapped with oxygen. The rate constants for the addition were measured and found to vary from 2.9×10^6 mole^{-1} s^{-1} for α-methylstyrene to $\sim 10^3$ mole^{-1} s^{-1} for vinyl acetate. These values are ca. 10–50 times slower than those for PhS·, indicating the greater stability of the selenyl radical.

$$PhSeSePh \underset{}{\overset{h\nu}{\rightleftharpoons}} 2\ PhSe\bullet \tag{1}$$

$$PhSe\bullet \ +\ ArN{=}O \longrightarrow PhSe{-}N\overset{\bullet O}{\underset{Ar}{\diagdown}} \qquad Ph{-}N\overset{\bullet O}{\underset{Ar}{\diagdown}} \tag{2}$$

Ar= duryl

1   2

$$PhSe\bullet \ +\ \diagup\!\!\diagup_{R} \rightleftharpoons PhSe\diagdown\!\!\diagup\overset{\bullet}{\diagdown}_R \tag{3}$$

3

Dimethyl diselenide undergoes similar photolytic Se—Se cleavage. Although the resulting MeSe· radicals could not be directly observed by esr spectroscopy, their presence was confirmed by spin trapping with t-butyl phenyl nitrone, di-t-butyl selenoketone (see Section 7) or the corresponding thioketone to afford adducts 4–6, respectively.[9] A weak signal attributed to 7 suggests that some C—Se cleavage also occurs. No spin adducts were detected when MeSe·

was generated in the presence of 2-methyl-2-nitrosopropane or 1,1-di-*t*-butyl-ethylene.

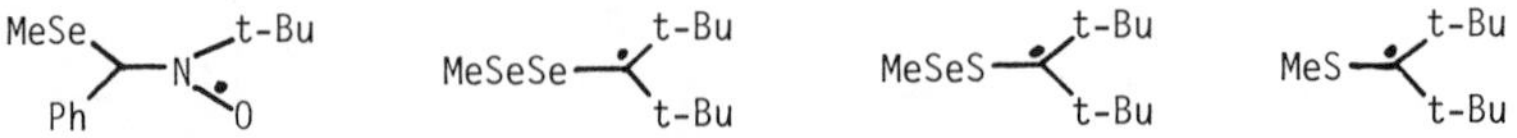

$$
\begin{array}{cccc}
\underset{4}{\text{MeSe}\ \diagdown_{\text{Ph}}\ \diagup^{\text{t-Bu}}_{\text{N}\cdot\text{O}}} & \underset{5}{\text{MeSeSe}-\overset{\bullet}{\underset{\text{t-Bu}}{\overset{\text{t-Bu}}{<}}}} & \underset{6}{\text{MeSeS}-\overset{\bullet}{\underset{\text{t-Bu}}{\overset{\text{t-Bu}}{<}}}} & \underset{7}{\text{MeS}-\overset{\bullet}{\underset{\text{t-Bu}}{\overset{\text{t-Bu}}{<}}}}
\end{array}
$$

In contrast to PhSeSePh, the C—Se bond of dibenzyl diselenide is readily cleaved by irradiation with wavelengths in the range 280–350 nm in degassed benzene[10] or acetonitrile.[11] High yields of dibenzyl selenide and elemental selenium are formed when the reaction is carried out at 0°C to prevent thermal side reactions (see Section 2.2). The mechanism in Scheme 1 was proposed by Chu et al.[11] and the intermediacy of benzyl radicals was confirmed by Franzi and Geoffroy[7] by spin-trapping experiments.

$$
\begin{aligned}
\text{PhCH}_2\text{SeSeCH}_2\text{Ph} &\xrightarrow{h\nu} \text{PhCH}_2\text{\Large.} \ + \ \text{PhCH}_2\text{SeSe\Large.} \\
\text{PhCH}_2\text{SeSe\Large.} &\longrightarrow \text{PhCH}_2\text{Se\Large.} \ + \ \text{Se} \\
\text{PhCH}_2\text{\Large.} \ + \ \text{PhCH}_2\text{SeSeCH}_2\text{Ph} &\longrightarrow \text{PhCH}_2\text{SeCH}_2\text{Ph} \ + \ \text{PhCH}_2\text{Se\Large.} \\
\text{PhCH}_2\text{\Large.} \ + \ \text{PhCH}_2\text{Se\Large.} &\longrightarrow \text{PhCH}_2\text{SeCH}_2\text{Ph}
\end{aligned}
$$

SCHEME 1.

Bis(9-anthrylmethyl)diselenide, 8, displays yet another mode of behavior. Photolysis in toluene solution results in the extrusion of both selenium atoms to afford a mixture of the hydrocarbons biplanene, 9, and lepidopterene, 10.[12] The products are formed by the symmetrical or unsymmetrical coupling of 9-anthrylmethyl radicals to give 11 and 12, respectively, followed by appropriate pericyclic reactions as indicated in Scheme 2.

When oxygen is present during the photolysis of dibenzyl diselenide,[10,13] dibenzyl ditelluride,[13] or 8,[12] the corresponding aldehydes and alcohols are produced along with selenium [Eq. (4)].

$$
\text{ArCH}_2\text{SeSeCH}_2\text{Ar} \xrightarrow[\text{O}_2]{h\nu} \text{ArCHO} \ + \ \text{ArCH}_2\text{OH} \ + \ \text{Se} \tag{4}
$$

The irradiation of diselenides in the presence of tertiary phosphines proceeds via a free-radical chain mechanism initiated by homolysis of the Se—Se linkage (Scheme 3).[14–16] The PhSe· radicals thus produced are effectively trapped by the phosphine and so are prevented from dimerizing back to the diselenide. The process resembles the photochemical desulfurization of disulfides with trivalent phosphorus compounds.[17–18] The corresponding selenide and phosphine selenide are the principal products. Bibenzyl is also formed from

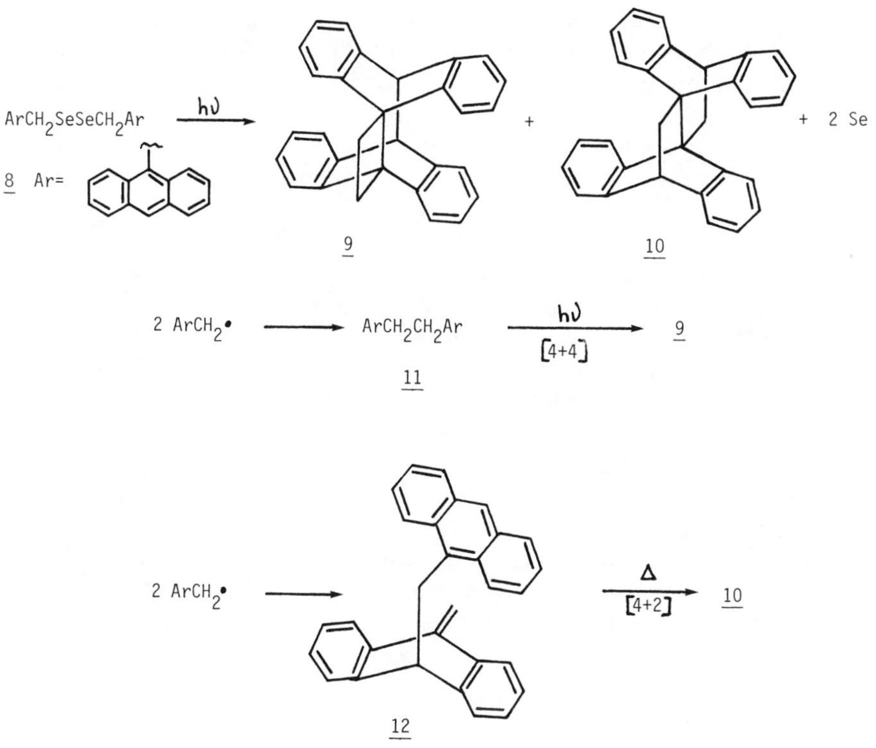

SCHEME 2.

$PhCH_2SeSeCH_2Ph$,[15,16,19] while diselenide, 8, produces hydrocarbons 9 and 10 in a 1:3 ratio.[12] The rate of disappearance of the diselenide is markedly increased by the presence of the phosphine[16] (i.e., the process in Scheme 3 is faster than that in Scheme 1). Furthermore, the quantum yield for the photolysis of dibenzyl diselenide at 313 nm was found to increase from 0.16–6.80 as the concentration of added triphenylphosphine was increased from 0–0.2 M.[15] These observations support the chain mechanism in Scheme 3. Initiation by

$$\mathrm{RSeSeR} \xrightleftharpoons{h\nu} 2\ \mathrm{RSe}\bullet$$

$$\mathrm{RSe}\bullet + \mathrm{R'_3P} \longrightarrow \mathrm{RSe{-}\dot{P}R'_3}$$

$$\mathrm{RSe{-}\dot{P}R'_3} \longrightarrow \mathrm{R}\bullet + \mathrm{Se{=}PR'_3}$$

$$\mathrm{R}\bullet + \mathrm{RSeSeR} \longrightarrow \mathrm{RSeR} + \bullet\mathrm{SeR}$$

$$2\ \mathrm{R}\bullet \longrightarrow \mathrm{R{-}R}$$

SCHEME 3.

C—Se bond cleavage is relatively unimportant in the photoreactions of diselenides with phosphines, even in the case of $PhCH_2SeSeCH_2Ph$,[15,16] but assumes greater prominence in the corresponding reactions of ditellurides.[20]
Several such processes have been investigated by Chemically Induced Dynamic
Nuclear Polarization (CIDNP) techniques.[19,21]

2.2. Pyrolysis

Diselenides display thermochromism, which at one time was attributed to reversible dissociation into selenyl radicals at higher temperatures.[22] This was
later proved in error by Kuder and Lardon,[23] who demonstrated that the
phenomenon was caused by closely spaced vibrational levels in the diselenide
chromophore, which resulted in a pronounced temperature dependence of the
width of the absorption.

Free-radical formation from diselenides does, however, occur at elevated
temperatures. Pyrolysis of dibenzyl diselenide at 150–170°C produced a mixture
of the selenide, diselenide, polyselenides, and free selenium.[24] At temperatures
above 200°C, toluene and bibenzyl were also observed [Eq. (5)]. The reactions
are complex and evidently involve the homolytic scission of both the C—Se
and Se—Se bonds. The organic products are in equilibrium with each other
and with selenium [Eq. (6)].

$$PhCH_2SeSeCH_2Ph \xrightarrow[\Delta]{150-170^\circ C} PhCH_2Se_nCH_2Ph + Se \quad (+ PhCH_2CH_2Ph + PhMe \qquad (5)$$
$$n= 1,2,3..... \qquad \text{at higher temperatures)}$$

$$PhCH_2Se_nCH_2Ph + x Se \rightleftharpoons PhCH_2Se_{n+x}CH_2Ph \qquad (6)$$

Bis(diphenylmethyl)diselenide, <u>13</u>, shows similar behavior when heated in
chlorobenzene at $100-120^\circ C$.[25] Its decomposition follows first-order kinetics
and has an activation energy of 20.9 kcal mole^{-1}. The products are the selenide,
<u>14</u>, diselenide, <u>13</u>, and polyselenides, along with 1,1,2,2-tetraphenylethane [Eq.
(7)]. At higher temperatures the latter product dominates; it is formed quantitatively at 210°C, along with elemental selenium.[25]

$$Ph_2CHSeSeCHPh_2 \xrightarrow[\Delta]{100-120^\circ C} Ph_2CHSeCHPh_2 + \underline{13} + Ph_2CHSe_nCHPh_2 \ (n>2)$$
$$\underline{14}$$
$$+ Ph_2CHCHPh_2 + Se$$
$$\underline{15}$$
$$\underline{13}$$
$$\xrightarrow[\Delta]{210^\circ C} \underline{15} + 2\ Se$$
$$99.4\% \quad 99.8\%$$

$$(7)$$

The pyrolysis of diselenide, 8, at 210°C produced a 3:1 mixture of hydrocarbons, 10 and 11, along with selenium [Eq. (8)], and with the complete exclusion of the corresponding selenide, diselenide, and polyselenides.[12] This reaction therefore resembles that of diselenide, 13 (at 210°C) more closely than that of dibenzyl diselenide. Biplanene, 9, is not produced under thermal conditions as its formation requires a photochemical [4 + 4] cycloaddition step (see Scheme 2).

$$8 \xrightarrow[\Delta]{210\,°C} \underset{3}{10} + \underset{1}{11} + Se \qquad (8)$$

Under some conditions, the product distributions from the pyrolysis of selenides (see Section 3.1) resemble those from their diselenide counterparts (e.g., 8 and 13). This suggests that in some circumstances selenides are produced as intermediates in the thermal decomposition of diselenides.

Several diselenides were subjected to flash vacuum pyrolysis at 600°C at 20 mm.[26] Diphenyl diselenide was recovered intact while the dibenzyl derivative was converted to bibenzyl (96% yield) and selenium. In contrast, dialkyl diselenides, 16a–16c, afforded the corresponding terminal olefins as the principal products [Eq. (9)].

$$RCH_2CH_2SeSeCH_2CH_2R \xrightarrow[\Delta]{600\,°C} RCH=CH_2 + Se$$

16a	R = Ph	95%
16b	R = n-Bu	90%
16c	R = n-C$_6$H$_{13}$	93%

$$(9)$$

The results of the pyrolyses of dimethyl and diisopropyl diselenides at 820 K and 680 K in a flow tube are provided in Eqs. (10) and (11).[27]

$$MeSeSeMe \xrightarrow[\Delta]{820\,K} MeSeMe + MeSeH + CH_4 + Se \qquad (10)$$

$$i\text{-PrSeSe-}i\text{-Pr} \xrightarrow[\Delta]{680\,K} CH_2=CHMe + i\text{-PrSeH} + H_2Se + Se \qquad (11)$$

2.3. Radical Substitution (S$_H$2) Processes

Diselenides, like disulfides, undergo facile homolytic substitution (S$_H$2) reactions in which a radical intermediate attacks one of the diselenide selenium atoms and displaces an RSe· moiety [Eq. (12)]. For instance, the chain propagation steps in Schemes 1 and 3 provide examples of such processes.

$$R\cdot + RSeSeR \longrightarrow RSeR' + RSe\cdot \qquad (12)$$

Organocobaloximes, 17, react with PhSeSePh according to Scheme 4.[28] The reaction is initiated by the thermal or photochemical rupture of the R—Co bond, followed by S_H2 attack of the resulting radicals upon the diselenide. The observation that optically active *sec*-octylcobaloximes produce racemic *sec*-octyl phenyl selenide is consistent with the intermediacy of the free *sec*-octyl radical in this process. Product formation via the reaction of PhSe· radicals with the moiety R in the undissociated complex, 17, is not a major pathway.

$$RCo(dmgH)_2py \quad + \quad PhSeSePh \quad \xrightarrow[\Delta]{h\nu \atop or} \quad RSePh \quad + \quad PhSeCo(dmgH)_2py$$

17

$$17 \longrightarrow R\cdot \quad + \quad \cdot Co(dmgH)_2py$$

$$R\cdot \quad + \quad PhSeSePh \longrightarrow RSePh \quad + \quad \cdot SePh$$

$$\cdot Co(dmgH)_2py \quad + \quad PhSeSePh \longrightarrow PhSeCo(dmgH)_2py \quad + \quad \cdot SePh$$

$$R\cdot \quad + \quad \cdot SePh \longrightarrow RSePh$$

$$\cdot Co(dmgH)_2py \quad + \quad \cdot SePh \longrightarrow PhSeCo(dmgH)_2py$$

(py= pyridine; dmgH= dimethylglyoximato ligand)

SCHEME 4.

Organozirconium complexes, 18, also react with diphenyl diselenide as shown in Scheme 5.[29] Although the exact mechanism for this transformation is not known, it is believed to involve homolysis of the σ-Zr—C bond, followed by S_H2 displacement by the radicals thus formed upon the diselenide.

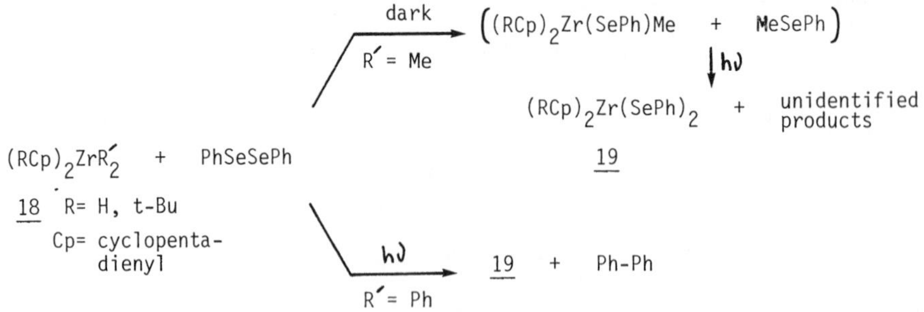

$$\xrightarrow[R' = Me]{dark} \left((RCp)_2Zr(SePh)Me \quad + \quad MeSePh \right)$$

$$\downarrow h\nu$$

$$(RCp)_2Zr(SePh)_2 \quad + \quad \text{unidentified products}$$

19

$(RCp)_2ZrR'_2 \quad + \quad PhSeSePh$

18 R= H, t-Bu
Cp= cyclopenta-
    dienyl

$$\xrightarrow[R' = Ph]{h\nu} \quad 19 \quad + \quad Ph\text{-}Ph$$

SCHEME 5.

Several photochemical reactions of diselenides with metal carbonyls have been reported [Eqs. (13)–(15)].[30,31] The authors have suggested that such processes proceed by way of the coupling of radical species $\cdot Mn(CO)_5$ or $\cdot Fe(CO)_4$ with $RSe\cdot$, followed by dimerization or tetramerization and loss of carbon monoxide. However, it appears that alternative mechanisms such as homolytic substitution (S_H2) by the photochemically generated metal carbonyl radical upon the diselenide cannot be ruled out on the basis of the available evidence.

$$CF_3SeSeCF_3 \quad + \quad Mn_2(CO)_{10} \xrightarrow{h\nu} \left(Mn(CO)_4SeCF_3\right)_2 \quad (13) \text{ (Ref. 30)}$$
$$67\%$$

$$CF_3SeSeCF_3 \quad + \quad Fe(CO)_5 \xrightarrow{h\nu} \left(Fe(CO)_3SeCF_3\right)_2 \quad (14) \text{ (Ref. 30)}$$
$$73\%$$

$$PhSeSePh \quad + \quad Mn_2(CO)_{10} \xrightarrow{h\nu} \left(Mn(CO)_3SePh\right)_4 \quad (15) \text{ (Ref. 31)}$$
$$35\text{-}40\%$$

1-Adamantyl, undecyl, and 5-hexenyl radicals react with PhSeSePh by means of S_H2 reactions to produce the corresponding selenides [Eqs. (16)–(18)].[32] The rate of the radical substitution step in the latter process was determined to be 5×10^7 L mole^{-1} s^{-1} at 80°C in benzene.[32] The preparation of the pyridyl selenide, <u>20</u>, was accomplished by the S_H2 transformation in [Eq. (19)].[33]

$$AdCO_2H \xrightarrow{Pb(OAc)_4} Ad\cdot \xrightarrow{PhSeSePh} AdSePh \qquad (16)$$
$$Ad = \text{1-adamantyl} \qquad\qquad\qquad\quad <20\%$$

$$(n\text{-}C_{11}H_{23}\overset{O}{C}O)_2 \xrightarrow{\Delta} n\text{-}C_{11}H_{23}\cdot \xrightarrow{PhSeSePh} n\text{-}C_{11}H_{23}SePh \qquad (17)$$

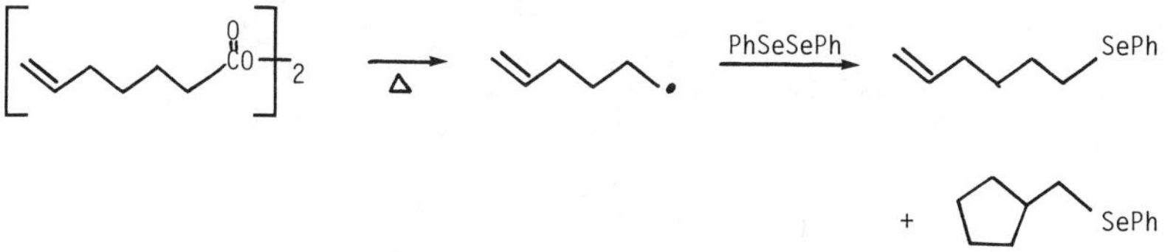

$$(18)$$

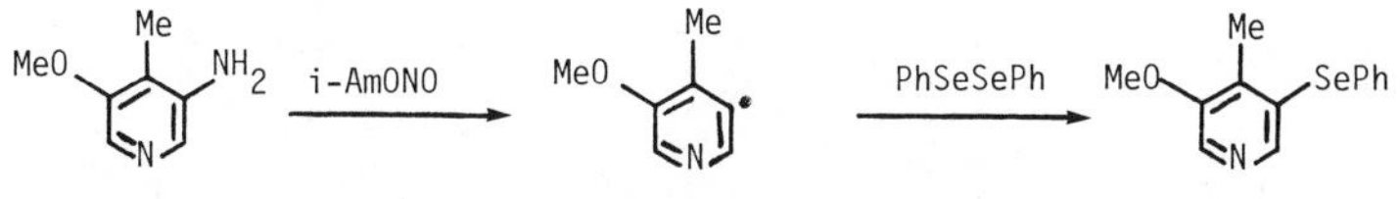

$$(19)$$

$$\underline{20} \quad 50\%$$

$$PhSeSePh \xrightarrow{h\nu} 2 \ PhSe\bullet$$

$$PhSe\bullet \ + \ CH_2N_2 \longrightarrow PhSeCH_2N=N\bullet$$

$$PhSeCH_2N=N\bullet \longrightarrow PhSeCH_2\bullet \ + \ N_2$$

$$PhSeCH_2\bullet \ + \ \bullet SePh \longrightarrow PhSeCH_2SePh$$

SCHEME 6.

2.4. Other Reactions

Disulfides, diselenides, and ditellurides are converted to bis(thio-, seleno-, or telluro)methane derivatives upon treatment with diazomethane[34] [Eq. (20)]. The ditellurides react in the dark, probably via an ionic pathway, whereas the other reactions require irradiation with sunlight. The latter processes may involve a carbene insertion reaction or, alternatively, the free-radical mechanism in Scheme 6. It appears possible that the $PhSeCH_2\bullet$ radical could also react with the diselenide instead of with $PhSe\bullet$ in the product forming step.

$$RYYR \ + \ CH_2N_2 \longrightarrow RYCH_2YR \ + \ N_2 \tag{20}$$
$$Y= S, \ Se, \ Te$$

Russell and Tashtoush[35] reported that alkylmercury halides participate in free-radical chain reactions when irradiated in the presence of diphenyl diselenide, Se-phenyl p-tolueneselenosulfonate, 21a, or various sulfur and tellurium compounds. Initiation occurs through photolytic cleavage of the diselenide or selenosulfonate linkage in PhSeSePh or 21a, respectively, and proceeds according to Scheme 7. Alkyl phenyl selenides are produced in generally high yields.

$$
RHgCl \ + \
\begin{matrix} PhSeSePh \\ or \\ ArSO_2SePh \end{matrix}
\ \xrightarrow{h\nu} \ RSePh \ + \
\begin{matrix} PhSeHgCl \\ or \\ ArSO_2HgCl \end{matrix}
$$

21a   Ar= p-tolyl

$$PhSeY \longrightarrow PhSe\bullet \ + \ Y\bullet$$
$$(Y= PhSe \ or \ ArSO_2)$$

$$Y\bullet \ + \ RHgCl \longrightarrow R\overset{\bullet}{H}gClY$$

$$R\overset{\bullet}{H}gClY \longrightarrow R\bullet \ + \ HgClY$$

$$R\bullet \ + \ PhSeY \longrightarrow RSePh \ + \ Y\bullet$$

SCHEME 7.

The rates of the product forming steps in the case of the 5-hexenyl radical and the diselenide or the selenosulfonate were determined to be 1.2×10^7 L mole^{-1} s^{-1} and 3.0×10^6 L mole^{-1} s^{-1}, respectively.[35] Vinyl mercurials afford vinyl selenides under similar conditions, although not through the intermediacy of free vinyl radicals.[36] A radical addition–elimination mechanism proceeding via adduct, 22, has been proposed (Scheme 8).

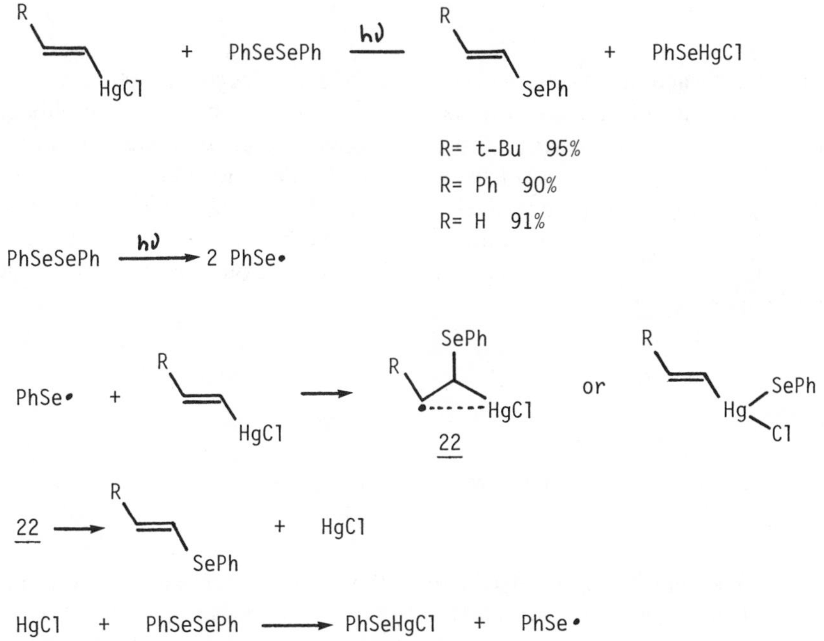

R= t-Bu 95%

R= Ph 90%

R= H 91%

SCHEME 8.

The pulse radiolysis of bis-(p-methoxyphenyl) diselenide in neutral or acidic methanol produced both ArSe· and ArSeSe· transients.[37] The former species was also detected from the corresponding selenide while the analogous telluride and ditelluride generated only the ArTe· radical under similar conditions.

3. REACTIONS OF SELENIDES

3.1. Selenium Extrusion Reactions

As mentioned in the previous section, the free-radical decomposition of selenides sometimes produces similar products to those obtained from diselenides. Thus, pyrolysis of bis(diphenylmethyl) selenide, 14, at 140°C afforded a mixture of compounds, 13–15, as well as polyselenides and selenium [Eq. (21); cf. Eq. (7)].[25]

Moreover, bis(9-anthrylmethyl) selenide formed the same products as its di-selenide analog when subjected to the conditions of Scheme 2 or Eq. (8).[12]

$$Ph_2CHSeCHPh_2 \xrightarrow[\Delta]{140\,^\circ C} Ph_2CHSe_nCHPh_2 + Se$$

$$\underline{14} \qquad\qquad \underline{15} \ n=0 \ \ 55.2\% \qquad 16\%$$
$$\underline{14} \ n=1 \ \ 15.2\%$$
$$\underline{13} \ n=2 \ \ 15.8\%$$
$$n>2 \ \ 13.8\%$$

$$(21)$$

The flash vacuum pyrolysis of a variety of selenides at ca. 600°C has been studied by Misumi and co-workers.[26,38-40] Under such conditions dibenzyl selenide and symmetrical ring substituted derivatives undergo cleavage of both C—Se bonds, with smooth extrusion of selenium and formation of the corresponding bibenzyls in preparatively useful yields [Eq. (22)].[26,38] The unsymmetrical selenide, 23, produced all three possible bibenzyl products [Eq. (23)], thus ruling out a concerted mechanism and supporting one involving free-radical intermediates.[38]

$$ArCH_2SeCH_2Ar \xrightarrow[\Delta]{600\,^\circ C} ArCH_2CH_2Ar + Se$$
$$63-89\%$$

$$(22)$$

$$PhCH_2SeCH_2Ar \xrightarrow[\Delta]{600\,^\circ C} PhCH_2CH_2Ph + PhCH_2CH_2Ar + ArCH_2CH_2Ar + Se$$
$$\underline{23} \ \ Ar=p\text{-MeOPh} \qquad 13\% \qquad\qquad 24\% \qquad\qquad 8\%$$

$$(23)$$

Phenyl selenides of general structure RSePh, where R· is a relatively stable radical, react exclusively via cleavage of that C—Se linkage, as shown in Eq. (24).[26,39] In those cases where a stable, carbon centered radical cannot be

$$RSePh \xrightarrow[\Delta]{600\,^\circ C} \left(R\bullet + \bullet SePh \right) \longrightarrow R-R + PhSeSePh$$

$$R= ArCH_2 \qquad 66-92\%$$
$$R= CH_2CH=CH_2 \qquad 67\% \qquad 88\%$$
$$R= CH_2CN \qquad 87\% \qquad 90\%$$
$$R= CH_2COMe \qquad 52\% \qquad 93\%$$

$$(24)$$

$$R= \qquad 76\%$$

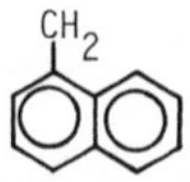

$$R= \qquad 91\%$$

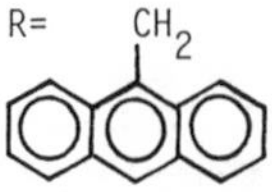

$$\underline{11}$$

produced by fission of either C—Se bond, the selenides were recovered intact (e.g., $PhCH_2CH_2SePh$, $PhCH_2CH_2SeCH_2CH_2Ph$, and $n\text{-}C_6H_{13}Se\text{-}n\text{-}C_6H_{13}$).[26] The product, <u>11</u>, obtained in 91% yield from the flash vacuum pyrolysis of 9-anthrylmethyl phenyl selenide [Eq. (24), bottom entry] at 600°C, reacted further in solution at 150–170°C to afford lepidopterene, <u>10</u>, as the principal product.[26] Styrene was produced nearly quantitatively from benzyl 2-phenylethyl selenide [Eq. (25)], and compounds, <u>24–26</u>, were similarly obtained from the corresponding bis- or tris(benzyl selenides).[26]

$$PhCH_2SeCH_2CH_2Ph \xrightarrow[\Delta]{600\,^\circ C} \left(PhCH_2{}^\bullet \;+\; {}^\bullet SeCH_2CH_2Ph \right) \longrightarrow \underset{96\%}{PhCH=CH_2} \;+\; \underset{48\%}{PhMe}$$

$$+\; \underset{37\%}{PhCH_2CH_2Ph} \;+\; Se$$

$$(25)$$

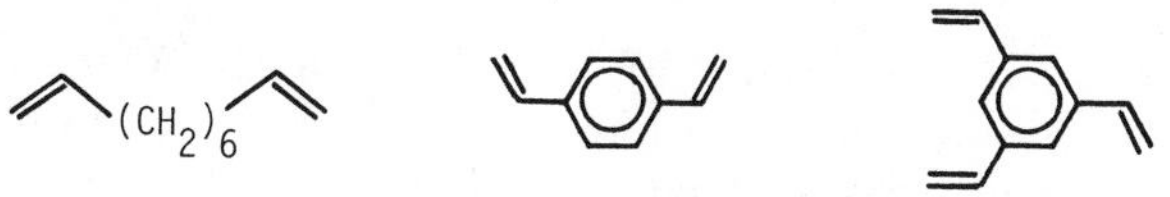

<u>24</u> 93%   <u>25</u> 88%

<u>26</u> 48%

Highly strained molecules such as benzocyclobutane, <u>27</u>, and [2.2]p-cyclophane, <u>28</u> [Eq. (27)] have been prepared by the pyrolysis of their appropriate bisselenide precursors.[26,39] A number of other cyclophanes, including three-layered ones, have been obtained by selenium extrusion methods using pyrolytic,[40] photolytic [e.g., Eq. (28)],[41] and other methods.[40] Farnesyl phenyl selenide affords dihydrofarnesene upon irradiation.[42]

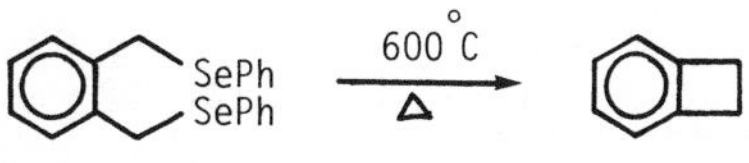

$$(26)$$

<u>27</u> up to 40%

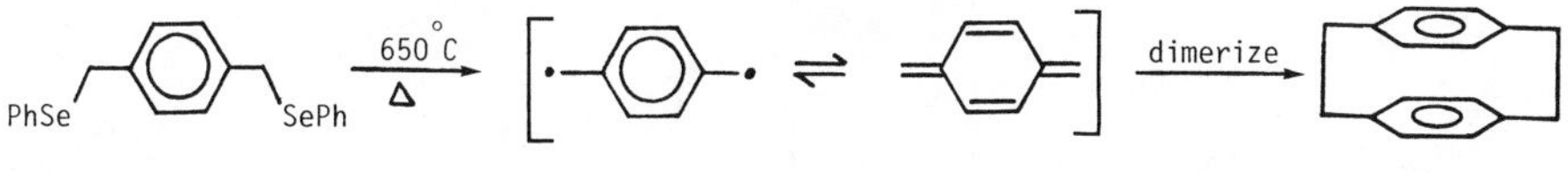

$$(27)$$

<u>28</u> 23%

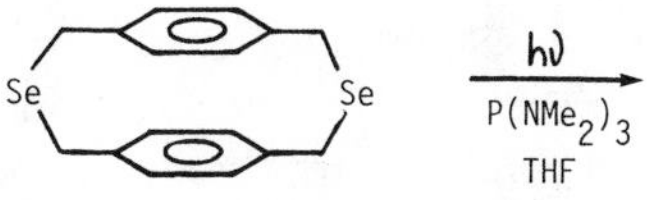

$$(28)$$

<u>28</u> 93%

3.2. Reduction with Tin Hydrides

Alkyl selenides undergo facile S_H2 cleavage of the alkyl–selenium bond when attacked by trialkyl- or triaryltin radicals. Such processes were first investigated by Ingold and co-workers,[43] who treated several sulfides, selenides, and tellurides with n-$Bu_3Sn\cdot$, in turn generated from the photolysis of hexa-n-butyldistannane [Eq. (29)]. Rate measurements indicated that the reactivity decreases in the order $R_2Te > R_2Se > R_2S$ and increases as more stable alkyl radicals are displaced (t-$Bu\cdot > Et\cdot > Me\cdot$).

$$RXR \quad + \quad n\text{-}Bu_3Sn\bullet \quad \longrightarrow \quad n\text{-}Bu_3SnXR \quad + \quad R\bullet$$

$$X = S,\ Se,\ Te \tag{29}$$

The reaction in Eq. (29) suggests a protocol for the free-radical deselenization of aryl alkyl selenides, where cleavage of the alkyl–Se linkage by $R_3Sn\cdot$ is expected to be favored. The use of tin hydride reagents in this context was first independently reported by the groups of Clive et al.,[44-47] Corey et al.,[48] and Nicolaou et al.[49-53] An extensive study by Clive et al.[47] demonstrated that a large variety of phenyl selenides (and tellurides) were smoothly reduced to selenium-free products with triphenyltin hydride in refluxing toluene. The reaction is initiated by the homolysis of the tin hydride, which proceeds at elevated temperatures even in the absence of added initiator. Product formation occurs via the chain mechanism in Scheme 9. The possibility that the formation of selenuranyl radical intermediates, 29, or that electron transfer from $Ph_3Sn\cdot$ to the selenide precedes C—Se scission was also considered and cannot be ruled out.

$$RSePh \quad + \quad Ph_3SnH \quad \xrightarrow{\Delta} \quad RH \quad + \quad Ph_3SnSePh$$

$$Ph_3SnH \quad \longrightarrow \quad Ph_3Sn\bullet \quad + \quad H\bullet$$

$$Ph_3Sn\bullet \quad + \quad RSePh \quad \longrightarrow \quad R\bullet \quad + \quad Ph_3SnSePh$$

$$R\bullet \quad + \quad Ph_3SnH \quad \longrightarrow \quad RH \quad + \quad Ph_3Sn\bullet$$

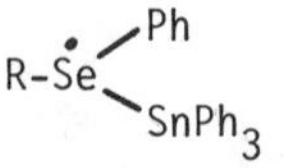

$$R\text{-}\overset{\bullet}{Se}\overset{\diagup Ph}{\diagdown SnPh_3}$$

29

SCHEME 9.

Clive et al.[44,47] also reported that triphenyltin deuteride provides access to deuterated products. Furthermore, selenides are readily reduced in the presence of sulfides, as expected from their previously determined relative rates of reaction in Eq. (29). Yields of deselenized products are often superior to those obtained with classical reducing agents such as Raney nickel or alkali metals. The groups of Corey et al.[48] and Nicolaou et al.[49–53] employed tri-n-butyltin hydride under similar conditions, generally in the presence of the radical initiator azobis(isobutyronitrile) (AIBN). Both tin hydride reagents have been subsequently employed by numerous other workers.

Tin hydride reductions can be used in conjunction with other selenium-based reactions to provide several types of overall transformations of wide applicability in synthesis. Thus, the deoxygenation of alcohols can be accomplished through their conversion to selenides with aryl selenocyanates and tri-n-butylphosphine via the method of Grieco et al.[54] (see Chapter 1, Section 5.4), followed by tin hydride mediated deselenization [Eq. (30)].[44,47] Similarly, diselenoketals can be efficiently reduced to the corresponding hydrocarbons.[44,47,55] Since the former compounds are readily available from ketones, this provides an attractive alternative to the Wolff–Kishner reaction [Eq. (31)].[44,47] Tin hydrides have also been frequently used to remove selenium-containing residues from the products of cyclization reactions mediated by selenenic electrophiles (see Chapter 2).[44–53,56–60]

$$ ROH \xrightarrow[\text{n-Bu}_3\text{P}]{\text{ArSeCN}} RSeAr \xrightarrow{\text{Ph}_3\text{SnH}} RH \qquad (30) $$

$$ \begin{array}{c} R \\ \diagdown \\ R \end{array}\!\!=\!\!O \xrightarrow[\substack{\text{or} \\ \text{B(SePh)}_3}]{\text{PhSeH,H}^+} \begin{array}{c} R \\ R \end{array}\!\!\diagup\!\!\begin{array}{c} \text{SePh} \\ \text{SePh} \end{array} \xrightarrow{\text{Ph}_3\text{SnH}} \begin{array}{c} R \\ R \end{array}\!\!\diagup\!\!\begin{array}{c} H \\ H \end{array} \qquad (31) $$

The reductive cleavage of arylseleno groups with tin hydrides has been exploited in the synthesis of steroids,[61] prostacyclin derivatives,[48,49,53] penicillins,[62] necines,[63] lauraceae lactones,[60] macrolides,[50,64] 2′-deoxy disaccharides,[65] caparrapi oxide,[66,67] and other products.[68,69] Several illustrative examples are provided in Eqs. (32)–(37).

$$ (32) \text{ (Ref. 47)} $$

77%

$$ (33) \text{ (Ref. 49)} $$

80%

The carbon centered radical intermediates in such processes can be captured by appropriate olefins in an inter-[70] or intramolecular[71–73] fashion, leading to

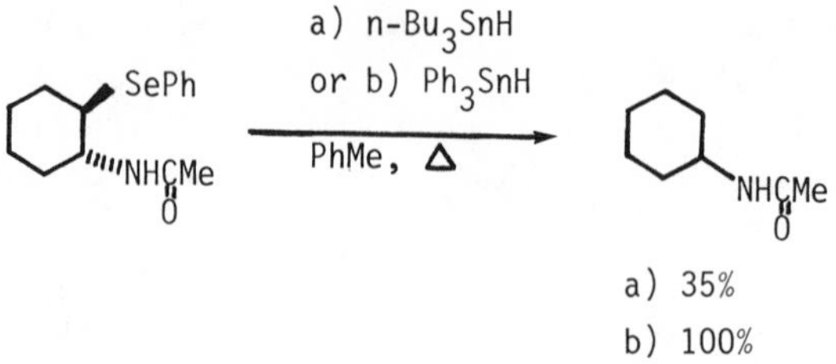

a) n-Bu$_3$SnH

or b) Ph$_3$SnH

PhMe, Δ

SePh

NHCMe
    O
(IIINHCMe)

(34) (Ref. 68)

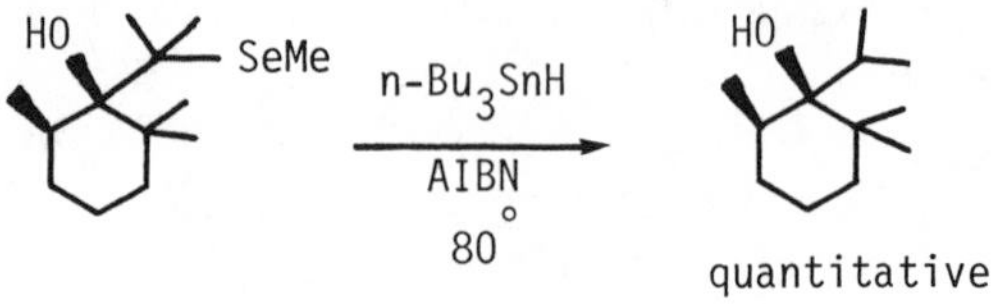

a) 35%
b) 100%

HO — SeMe

n-Bu$_3$SnH

AIBN

80°

HO

quantitative

(35) (Ref. 69)

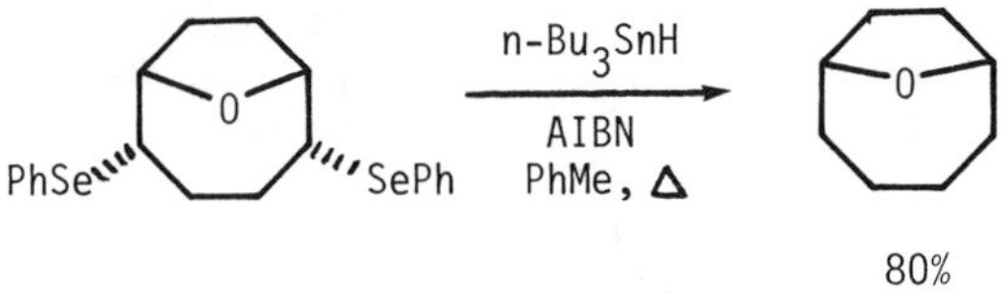

n-Bu$_3$SnH

AIBN

PhMe, Δ

PhSe   SePh

O

80%

(36) (Ref. 56)

PhSe

Ph$_3$SnH

AIBN

PhMe, Δ

O   O

91%

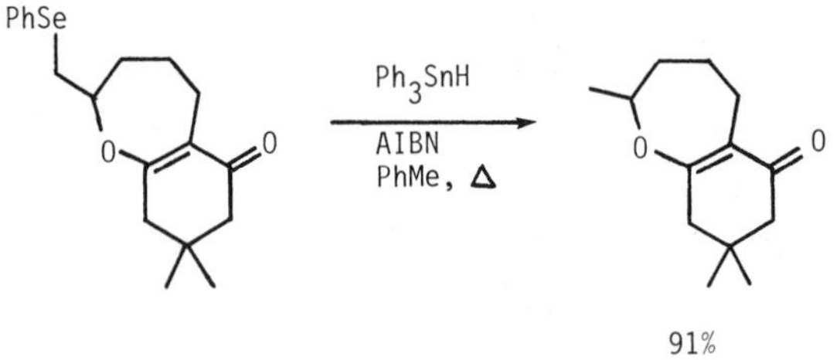

(37) (Ref. 57)

new carbon–carbon bond formation [e.g., Eqs. (38)–(41)]. Vinyl selenides [Eqs. (42) and (43)],[74,75] and alkyl selenides possessing β-chloro-[47] or β-sulfonyl[76] substituents afford the corresponding olefins [Eqs. (44) and (45)]. Selenoesters are reduced to aldehydes, or to hydrocarbons by decarbonylation of the intermediate acyl radicals [Eq. (46)].[77] Selenocarbonates produce formates, as well as alcohols by decarbonylation, and hydrocarbons by decarboxylation [Eq. (47)].[77] The product distributions are sensitive to the conditions and the nature

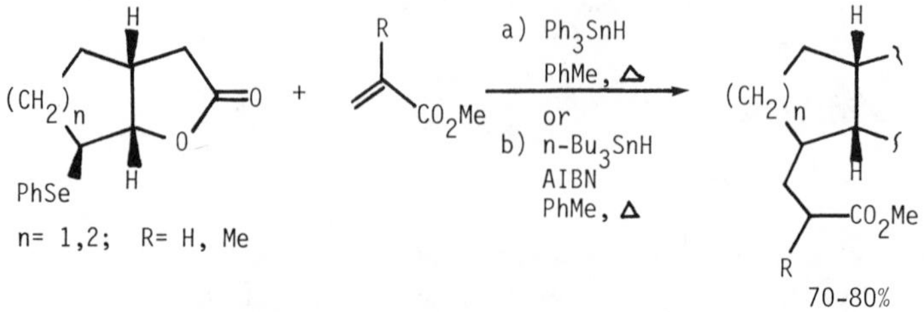

(CH$_2$)$_n$

O   O

PhSe   H

n= 1,2; R= H, Me

+

R

CO$_2$Me

a) Ph$_3$SnH

PhMe, Δ

or

b) n-Bu$_3$SnH

AIBN

PhMe, Δ

(CH$_2$)$_n$

CO$_2$Me

R

70-80%

(38) (Ref. 70)

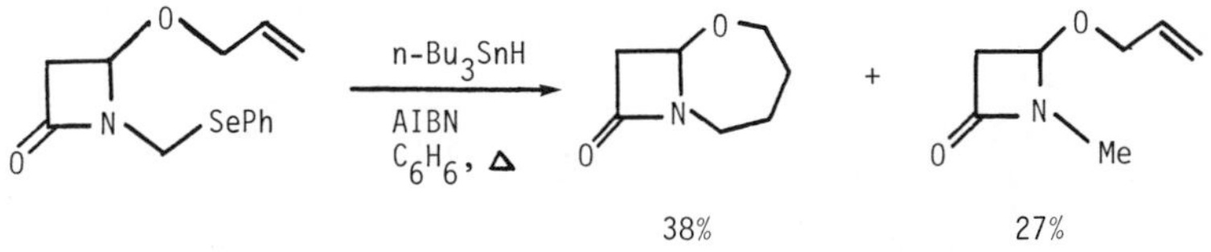

(39) (Ref. 72)

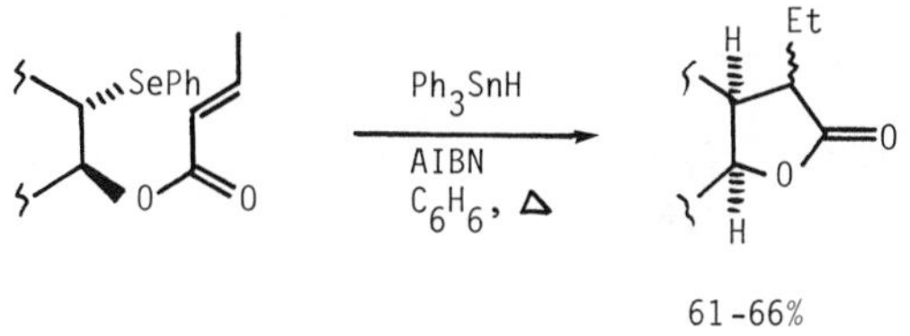

(40) (Ref. 73)

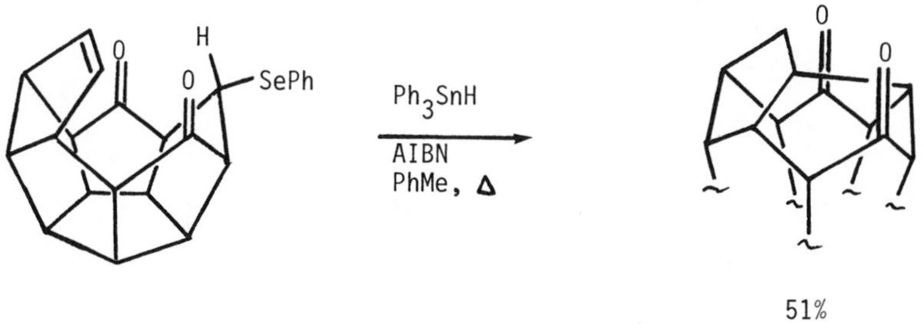

(41) (Ref. 71)

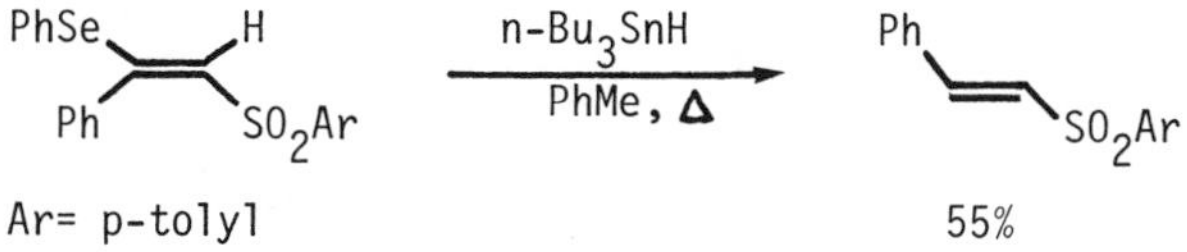

Ar= p-tolyl   55%

(42) (Ref. 74)

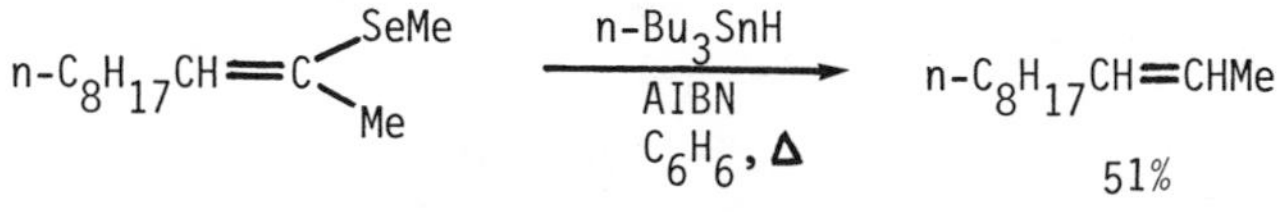

(43) (Ref. 75)

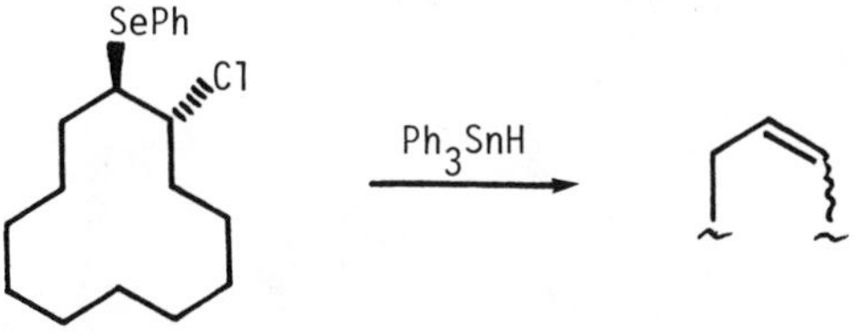

(44) (Ref. 47)

of the substrate. The reduction of selenoester, $\underline{30}$, to noralkane, $\underline{31}$, was a key step in a synthetic approach to quassins [Eq. (48)].[78] Compounds $\underline{32}$ and $\underline{33}$

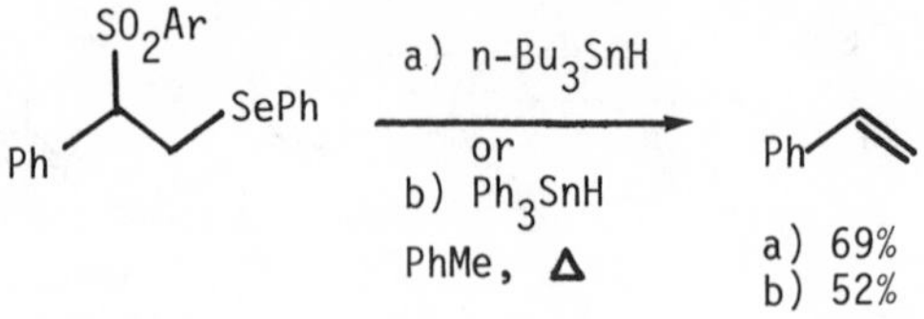

(45) (Ref. 76)

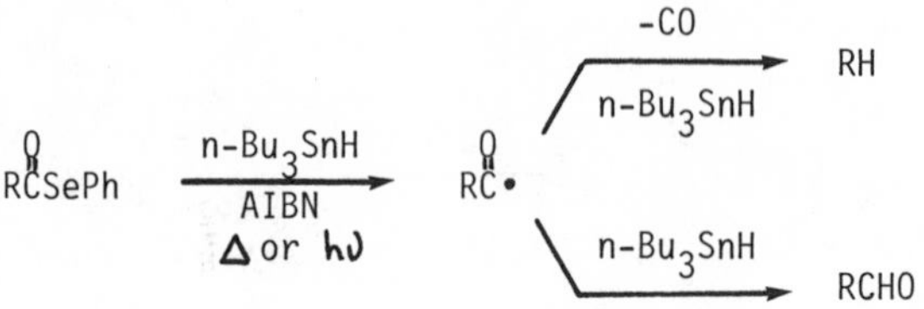

(46) (Ref. 77)

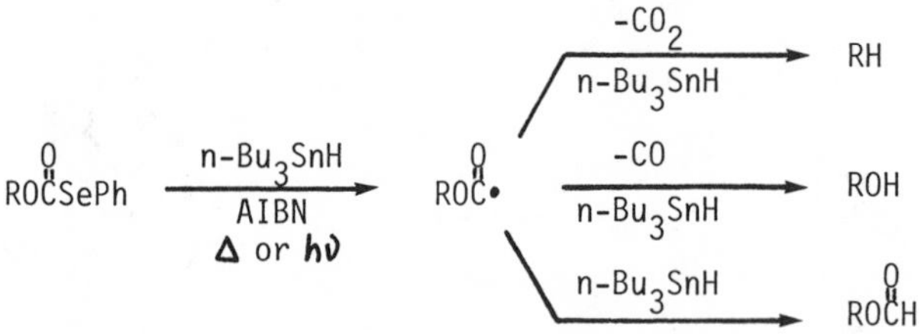

(47) (Ref. 77)

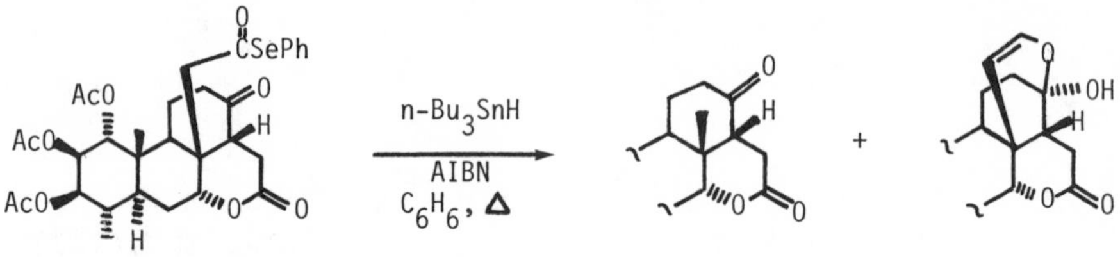

30 31

(48) (Ref. 78)

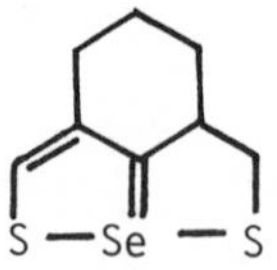

32

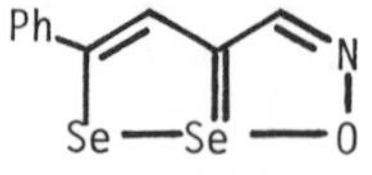

33

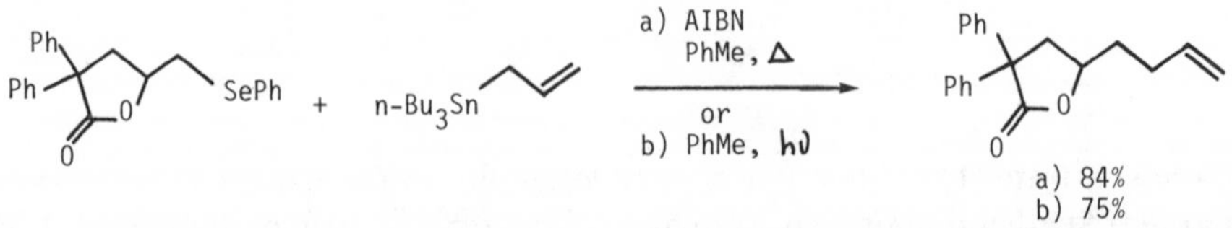

(49) (Ref. 80)

form adduct radicals when treated with n-Bu$_3$Sn$\cdot$.[79] The use of allyl tri-n-butylstannane instead of tin hydrides results in the replacement of the phenyl-seleno group in an alkyl phenyl selenide with an allyl substituent [e.g., Eqs. (49) and (50)].[80,81]

$$(50) \ (Ref. \ 81)$$

3.3. Other Reactions

Windle et al.[4,5] investigated the products of the low-temperature photolysis of several selenides by means of esr spectroscopy. Di-n-octadecyl and di-n-dodecyl selenide each produced signals from RSe$\cdot$ as well as an eight line spectrum attributed to secondary alkyl radicals —CH$_2\dot{C}$HCH$_2$—. The PhCH$_2$Se$\cdot$ radical was observed during the irradiation of benzylselenol (PhCH$_2$SeH), but not during the photolysis of dibenzyl selenide. In the latter instance, the only detectable radical was the species PhĊHSeCH$_2$Ph. The failure to observe the primary products of C—Se cleavage from this selenide is somewhat surprising in view of their relative stability. Other α-carbon centered radicals, 34 and 35, were generated by Scaiano and Ingold[9] by hydrogen abstraction from the α position of dimethyl and diethyl selenides with t-BuO$\cdot$ [Eq. (51)]. Although 34 and 35 could not be detected by esr, they were readily observed as their spin adducts, 36 and 37, in the presence of di-t-butyl thioketone. The rate constant for the formation of 34 from MeSeMe was determined to be 2.5×10^4 mole^{-1} s^{-1}.

$$(51)$$

Several homolytic substitution (S$_H$2) processes between alkyl or aryl radicals and selenides have been reported [Eq. (52)].[9] Relative rates were measured in several instances.

$$RSeR + R'\cdot \longrightarrow RSeR' + R\cdot$$

$$R = Me, Et; \qquad R' = Me, Ph, C_6F_5$$

$$(52)$$

Selenuranyl radicals, 38, were detected by esr spectroscopy when various radicals were generated in the presence of selenides [Eq. (53)].[82] This implies that in at least some cases substitution reactions of selenides may be two-step processes in which $\underline{38}$ act as intermediates, as opposed to concerted displacements.

$$RSeR \quad + \quad X\bullet \quad \longrightarrow \quad R_2Se\overset{\bullet}{-}X$$

$$\underset{\underline{38}}{}$$ (53)

$$X= Me_3SiO, \ t\text{-}BuO, \ EtCS, \ t\text{-}BuCS, \ CF_3S$$
$$\qquad\qquad\qquad \overset{\|}{O} \qquad \overset{\|}{O}$$

The γ-irradiation of MeSeMe was investigated by Williams and co-workers.[83,84] The selenide produces the corresponding radical cation, $\underline{39}$, by positive charge transfer in a Freon matrix,[84] but forms the dimer radical cation, $\underline{40}$, when irradiated in a single crystal [Eq. (54)].[83] Pulse radiolysis studies of a diaryl selenide and related compounds have also been reported.[37]

$$MeSeMe
\begin{array}{l}
\xrightarrow[\text{Freon matrix}]{\gamma\text{-irradiation}} \quad \overset{+\bullet}{MeSeMe} \quad \underline{39} \\
\\
\xrightarrow[\text{single crystal}]{\gamma\text{-irradiation}} \quad (Me_2Se\text{-}SeMe_2)^{+\bullet} \quad \underline{40}
\end{array}$$ (54)

4. REACTIONS OF SELENOLS, SELENOLATES, AND RELATED SPECIES

4.1. $S_{RN}1$ Reactions

Pierini and Rossi[85] reported that the photochemical reactions of the selenolate anion PhSe$^-$ with aryl chlorides or bromides in liquid ammonia produce aryl phenyl selenides via the $S_{RN}1$ mechanism displayed in Scheme 10. The reaction fails in the dark and the relative reactivities of the aryl halides (naphthyl > biphenyl > Ph) are consistent with their expected ease of reduction in the initial electron transfer step.

Dihaloarenes afford bisselenides[86] and reactions employing PhTe$^-$ instead of PhSe$^-$ give analogous telluride products.[86] The photochemical reaction of the

$$ArX \quad + \quad PhSe^- \quad \xrightarrow[\text{NH}_3(\text{liq.})]{h\nu} \quad ArX^{\overline{\bullet}} \quad + \quad PhSe\bullet$$

$$ArX^{\overline{\bullet}} \quad \longrightarrow \quad Ar\bullet \quad + \quad X^-$$

$$Ar\bullet \quad + \quad PhSe^- \quad \longrightarrow \quad ArSePh^{\overline{\bullet}}$$

$$ArSePh^{\overline{\bullet}} \quad + \quad ArX \quad \longrightarrow \quad ArSePh \quad + \quad ArX^{\overline{\bullet}}$$

SCHEME 10.

selenolate anion with perfluoroalkyl iodides results in the formation of per-fluoroalkyl phenyl selenides.[87]

Benzyl halides react with methaneselenol (or with thiols) in the presence of triethylamine to afford both substitution and reduction products, 42 and 43, respectively [Eq. (55)].[88] The ratio of 43:42 is increased by the presence of nitro substituents on the aryl group, as well as by the use of benzyl iodides instead of chlorides or bromides. The mechanism is uncertain, but may involve electron transfer from MeSe⁻ to the halide to generate the radical anion intermediate, 41.

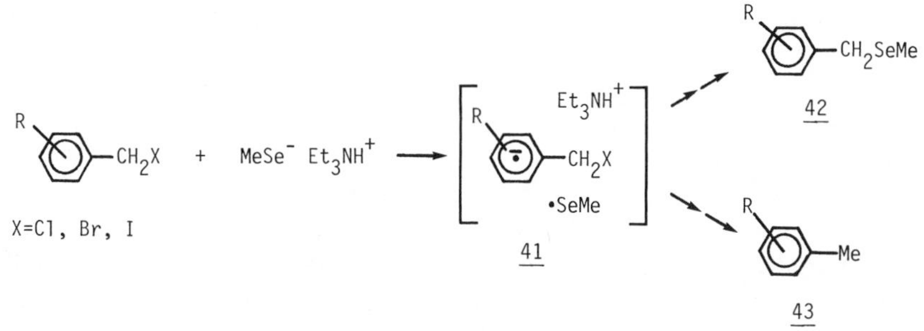

$$(55)$$

The photochemical reaction of iodobenzene (or other aryl halides) with selenide anion (Se^{2-}) produces PhSe⁻ via the $S_{RN}1$ process in Scheme 11.[89] The selenolate anion can in turn be transformed to diphenyl selenide by further reaction with the aryl halide (cf. Scheme 10), to the diselenide by air oxidation, or to alkyl phenyl selenides by quenching with an alkyl halide. Sulfide and telluride ions are less effective in this process. The formation of organic selenide and tellurides has also been observed from the thermal reaction of sodium or

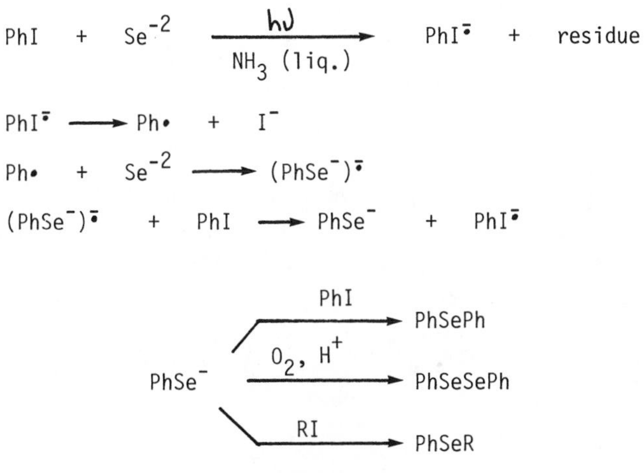

SCHEME 11.

potassium selenide or telluride with aryl halides in polar, aprotic solvents.[90] Poly(p-phenylene selenide) was thus obtainned from p-dihalobenzenes.

Sodium and potassium selenocyanate undergo photochemical reactions with aryl halides in aqueous t-butanol according to Eq. (56).[91] An $S_{RN}1$ process is plausible. Radical intermediates have also been implicated in the reactions of thallous selenolate (T1SePh) or thiolate (T1SPh) with various organic halides.[92]

$$ArX + {}^-SeCN \xrightarrow[h\upsilon]{t\text{-}BuOH\text{-}H_2O} ArSeCN \qquad (56)$$
$$X = Cl, Br$$

4.2. Radical Reductions

Selenols act as reducing agents towards a number of organic functionalities. Radical pathways have been suggested in several such processes, although precise mechanistic evidence is generally unavailable. Perkins et al.[93] converted arenediazonium fluoroborates to mixtures of aryl phenyl selenides and arenes upon treatment with PhSeH in methylene chloride–acetone. The mechanism in Scheme 12 was suggested to rationalize the observed products. Interestingly, the use of pure methylene chloride as the solvent produced arylhydrazine fluoroborates in lieu of the previous products.

$$ArN_2^+ BF_4^- + PhSeH \xrightarrow[Me_2C=O]{CH_2Cl_2-} ArSePh + ArH$$

$$ArN_2^+ + PhSeH \longrightarrow Ar\bullet + PhSe\bullet + N_2 + H^+$$

$$Ar\bullet + PhSeH \longrightarrow ArH + PhSe\bullet$$

$$2 PhSe\bullet \longrightarrow PhSeSePh$$

$$Ar\bullet + PhSeSePh (or PhSe\bullet) \longrightarrow ArSePh$$

SCHEME 12.

The same group[94] also employed PhSeH to reduce α,β-unsaturated carbonyl compounds, activated olefins, β-phenylseleno ketones, and benzaldehyde hydrazones, -oxime, and -anilide to saturated carbonyl compounds, alkanes, ketones, and benzylhydrazine, -hydroxylamine, and -aniline, respectively [Eq. (57)]. The observation that these reactions generally require irradiation with a sunlamp is more consistent with free-radical processes than with ionic ones. Benzyl phenyl selenide produced toluene under similar conditions [Eq. (58)].[94]

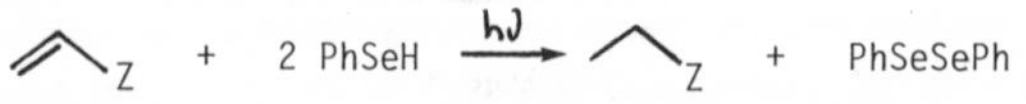

$$\diagup\!\!\!\diagdown_Z + 2 PhSeH \xrightarrow{h\upsilon} \diagup\!\!\!\diagdown_Z + PhSeSePh \qquad (57)$$

$$PhCH_2SePh \quad + \quad PhSeH \xrightarrow{h\nu} PhMe \quad + \quad PhSeSePh \tag{58}$$

The reductions of various nitrogenous compounds to primary amines with benzeneselenol were investigated by Oae and co-workers.[95] The reactions were accelerated by 1,4-diazabicyclo[2.2.2]octane (DABCO) and the substrates displayed the following order of reactivity:

$$
\begin{array}{c}
O \\
\uparrow \\
RNO \gg RN{=}NR > RNHOH > RN{=}NR \simeq RNO_2 \gg RNHNHR
\end{array}
$$

Such processes may involve a one-electron transfer step from the corresponding anion PhSe$^-$ to the nitrogen compound. When benzylselenol was used instead of PhSeH, the formation of toluene occurred, suggesting the intermediacy of benzyl radicals.

The photoreduction of ketones with hydrogen selenide has been reported[96] and probably proceeds via hydrogen abstraction from the latter reagent by the triplet excited state of the ketone [Eq. (59)].

$$R_2C{=}O \quad + \quad H_2Se \xrightarrow{h\nu} R_2CHOH \quad + \quad Se \tag{59}$$

The two-step reduction of methylene blue (MB) to leukomethylene blue (LMB) with thiolate ion (Scheme 13) is catalyzed by elemental selenium.[97] This is apparently due to the formation of more effective one-electron reducing species such as $RSSe_2^-$ from RS^- and Se^0.

$$RS^- \quad + \quad MB \xrightarrow{Se} MB^{\overline{\bullet}} \quad + \quad RS\bullet$$

$$MB^{\overline{\bullet}} \quad + \quad RS^- \xrightarrow{\text{buffer}} LMB \quad + \quad RS\bullet$$

$$2\ RS\bullet \longrightarrow RSSR$$

SCHEME 13.

5. SELENOSULFONATION

5.1. Decomposition of Selenosulfonates

The first manifestation of a homolytic reaction of a compound containing an Se—SO$_2$ linkage was observed in 1952 by Foss,[98] who noted that selenium dibenzenesulfonate (PhSO$_2$SeSO$_2$Ph) readily extrudes selenium when exposed to sunlight. However, such reactions were completely ignored for the next 30 years.

More recently, free-radical reactions of Se-phenyl areneselenosulfonates, 21, have come under scrutiny. These compounds are yellow or red in color and decompose upon exposure to uv or visible light, or when heated. Gancarz and Kice[99,100] investigated their photodecomposition and found that it proceeds via the mechanism in Scheme 14. When thermolyzed or photolyzed in the presence of unsaturated substrates, free-radical addition reactions termed selenosulfonations occur.

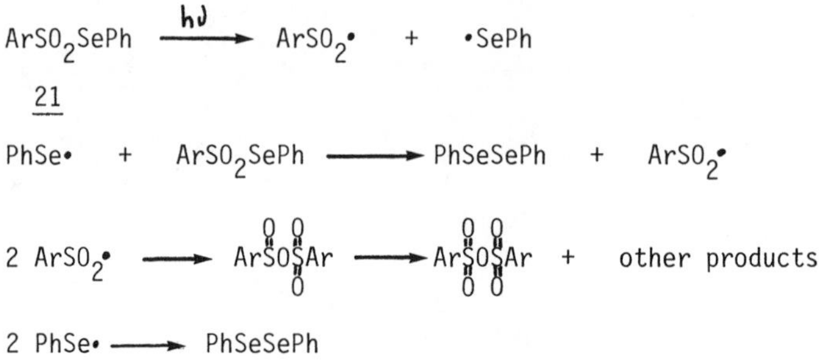

$$ArSO_2SePh \xrightarrow{h\nu} ArSO_2\bullet \ + \ \bullet SePh$$

21

$$PhSe\bullet \ + \ ArSO_2SePh \longrightarrow PhSeSePh \ + \ ArSO_2\bullet$$

$$2 \ ArSO_2\bullet \longrightarrow Ar\overset{O\ O}{\underset{O}{SOSAr}} \longrightarrow Ar\overset{O\ O}{\underset{O\ O}{SOSAr}} \ + \ \text{other products}$$

$$2 \ PhSe\bullet \longrightarrow PhSeSePh$$

SCHEME 14.

5.2. Selenosulfonation of Olefins and Allenes

Back and Collins,[76,101] and independently Gancarz and Kice,[99,100] reported that selenosulfonates, 21, add efficiently to olefins to produce adducts, 44 [Eq. (60)], which can in turn be converted to vinyl sulfones, 45, by oxidation–elimination. The former group employed thermal initiation, typically by refluxing the reactants in chloroform or benzene, while the latter employed photoinitiation. In either case the reaction proceeds via the same type of free-radical chain mechanism shown in Scheme 15. It is accelerated by initiators such as AIBN and suppressed by inhibitors such as 2,6-di-*t*-butyl-4-cresol.[76] Product formation occurs with a strong preference for anti-Markovnikov orientation and provides adducts of complementary regiochemistry to those observed when the selenosulfonation is performed under electrophilic conditions (see Chapter 1).[76] The additions to cyclic olefins are highly *anti* stereoselective, whereas acyclic substrates react nonstereospecifically. Thus, cyclohexene[100] and indene[76] give

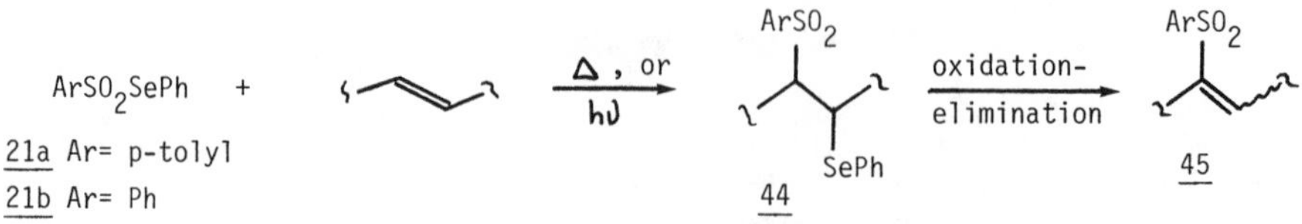

$$ArSO_2SePh \ + \ \text{(olefin)} \xrightarrow[h\nu]{\Delta, \ or} \ \underset{\underset{SePh}{44}}{\overset{ArSO_2}{\diagup}} \xrightarrow[\text{elimination}]{\text{oxidation-}} \underset{45}{\overset{ArSO_2}{\diagup}}$$

21a Ar= p-tolyl
21b Ar= Ph

(60)

only the trans adducts [Eqs. (61) and (62)] and (E)- and (Z)-5-decene form identical mixtures of threo and erythro diastereomers [Eq. (63)].[76] Yields are generally high with unhindered monosubstituted olefins [Eq. (64)], but are reduced significantly with increasing substitution [cf. Eq. (63)]. The conversion of the adducts to vinyl sulfones has been effected with m-chloroperbenzoic acid[76] or hydrogen peroxide.[100] The elimination step proceeds spontaneously and rapidly, giving exclusively vinyl (vs allyl) sulfones, frequently in quantitative yield. The products obtained from monosubstituted olefins are formed exclusively with the (E) configuration. Several vinyl sulfones obtained by the selenosulfonation method were recently reported by Paquette and Kinney[102] and Paquette and Crouse[103] and have found applications as dienophiles in Diels–Alder cycloadditions.

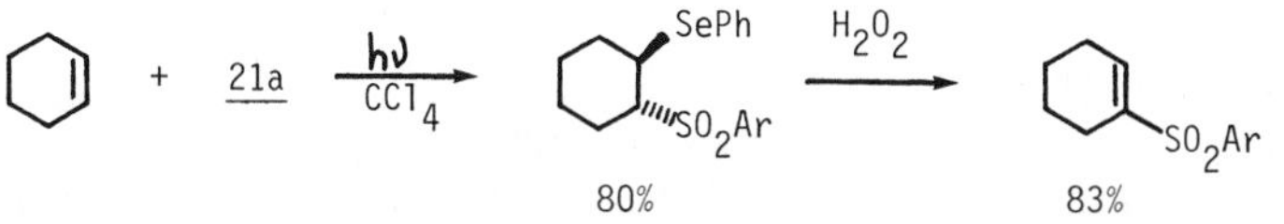

$$(61) \text{ (Ref. 100)}$$

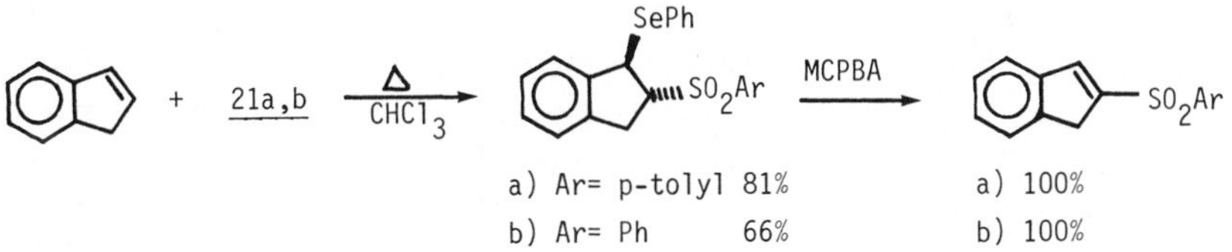

80%   83%

a) Ar= p-tolyl 81%   a) 100%
b) Ar= Ph   66%   b) 100%

$$(62) \text{ (Ref. 76)}$$

threo : erythro   E : Z
  4.5 : 1   4.3 : 1
  42-45%   100%

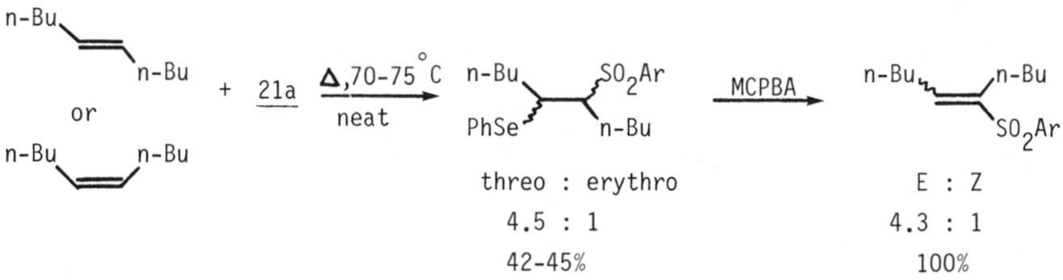

$$(63) \text{ (Ref. 76)}$$

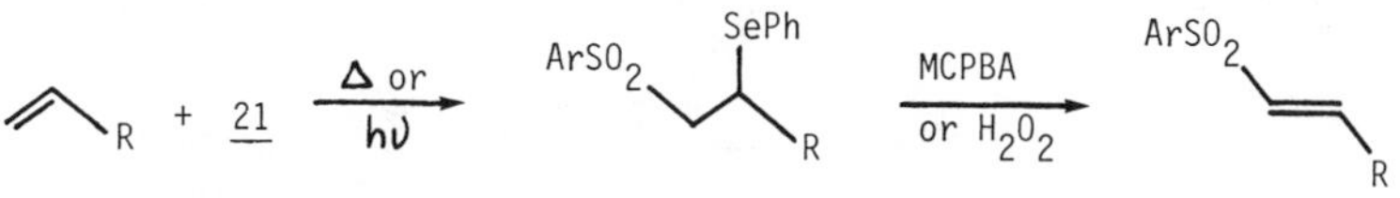

R= Ph, alkyl, substituted alkyl, CN, $SiMe_3$, $CH(OMe)_2$, CH_2OPh, CH_2OSiMe_2t-Bu, etc.

$$(64) \text{ (Refs. 76, 100, 102, 103)}$$

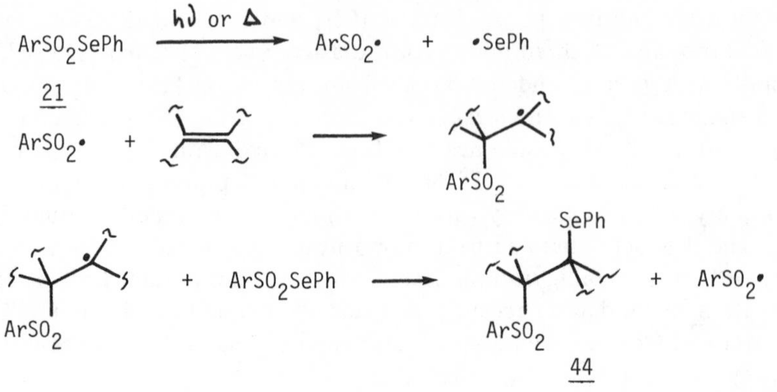

SCHEME 15.

Norbornadiene produces chiefly the nortricyclic products, 46, and only small amounts of the simple 1,2 adduct, 47 [Eq. (65)]. Furthermore, 1,5-cyclooctadiene affords roughly equal amounts of the 1,2 adduct, 48, and the corresponding *cis*-bicyclo[3.3.0]octane derivative, 49 [Eq. (66)]. Since the ratio of 1,2 addition to transannular addition is proportional to the efficiency of the product forming chain-transfer step in free-radical additions to such substrates, it is possible to conclude that selenosulfonates are comparable in reactivity to sulfonyl bromides, superior to carbon tetrachloride, and inferior to sulfonyl iodides, hydrogen bromide, or thioacetic acid in their role as chain-transfer agents.[100]

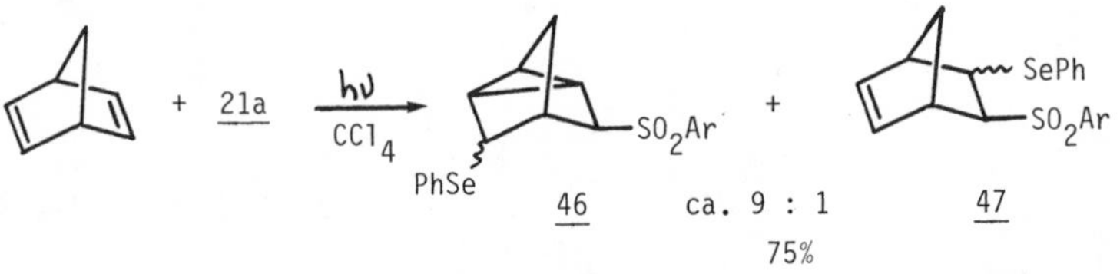

(65) (Ref. 100)

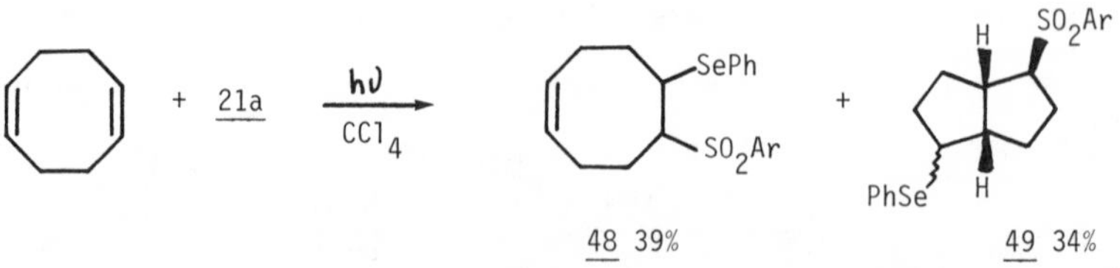

(66) (Ref. 100)

Alkaneselenosulfonates have been observed to undergo desulfonylation $(RSO_2\cdot \rightarrow R\cdot + SO_2)$ in several instances under either thermal[74] or photochemical[104] conditions.

Kang and Kice[105] investigated the photochemical selenosulfonation of allenes and demonstrated that addition of the sulfonyl moiety occurs regiospecifically to the central carbon atom and the phenylseleno group becomes incorporated at the less substituted terminal position. Oxidation of the products provides allylic alcohols, 50, via [2,3]sigmatropic rearrangements of the resulting selenoxides [Eq. (67)].

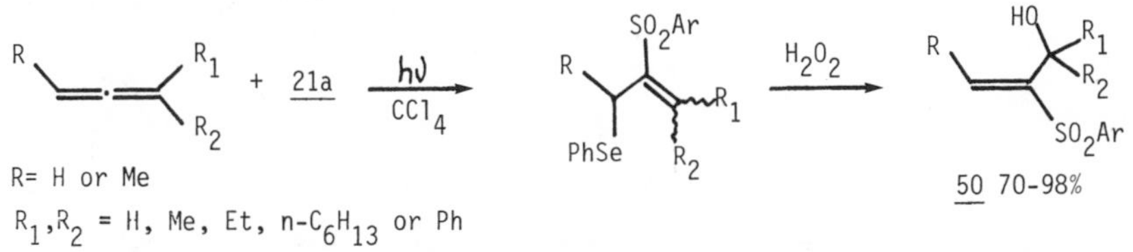

R
R1
R2
+ 21a
hν
CCl4
R= H or Me
R1,R2 = H, Me, Et, n-C6H13 or Ph
SO2Ar
R
PhSe
R1
R2
H2O2
HO
R
R1
R2
SO2Ar
50 70-98%

(67) (Ref. 105)

5.3. Selenosulfonation of Acetylenes

The thermally initiated selenosulfonation of acetylenes was independently reported by Back et al.[106–108] and by Miura and Kobayashi.[109] Terminal acetylenes produce excellent yields of anti-Markovnikov adducts, 51, which are readily oxidized to selenoxides, 52. The latter compounds eliminate spontaneously,[106,107,109] or under base catalyzed conditions[107] to afford acetylenic sulfones, 53 [Eq. (68)]. Disubstituted acetylenes form 1,2 adducts under similar conditions, but the corresponding selenoxides display more complex behavior,[107,108] as illustrated by the examples in Eqs. (69)–(71). The adducts, 51, have also been converted to various β-keto sulfone derivatives.[107]

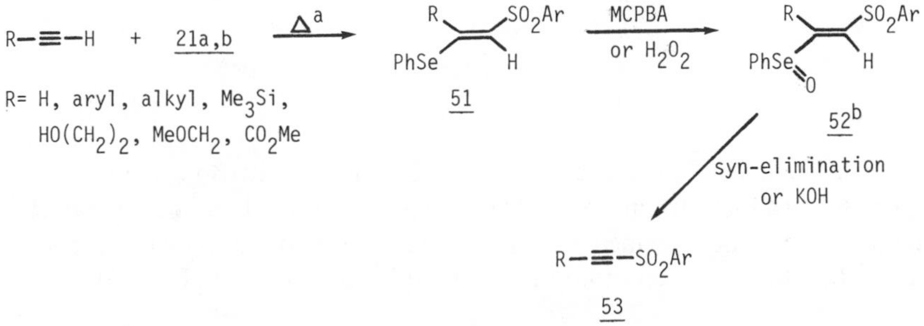

R—≡—H + 21a,b
Δ a
R= H, aryl, alkyl, Me3Si, HO(CH2)2, MeOCH2, CO2Me
R
SO2Ar
PhSe
H
51
MCPBA
or H2O2
R
SO2Ar
PhSe
O
H
52b
syn-elimination
or KOH
R—≡—SO2Ar
53

a. The addition of ca. 5 mol % of AIBN to the reaction is recommended[106,107]

b. Two of the selenoxides 52 studied (R= H and R= CO_2Me) underwent redox or other reactions and did not afford the acetylenic sulfones 53[107].

(68) (Refs. 106–109)

The addition process proceeds via a similar radical chain mechanism to that which occurs with olefins (cf. Scheme 15). The use of AIBN as an initiator permits shorter reaction times and sometimes improves yields.[106,107] On the other

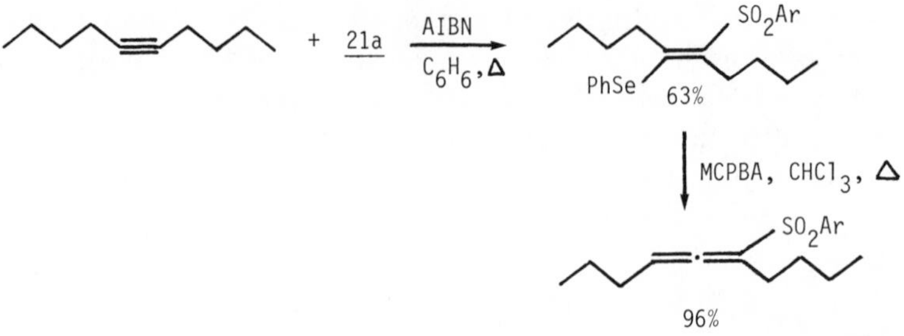

(69) (Ref. 108)

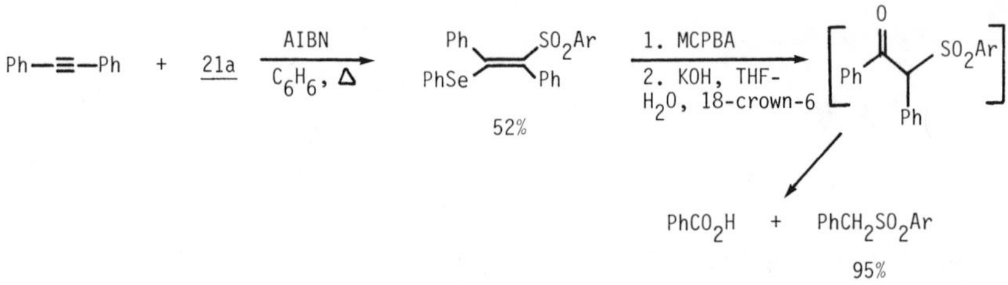

(70) (Ref. 107)

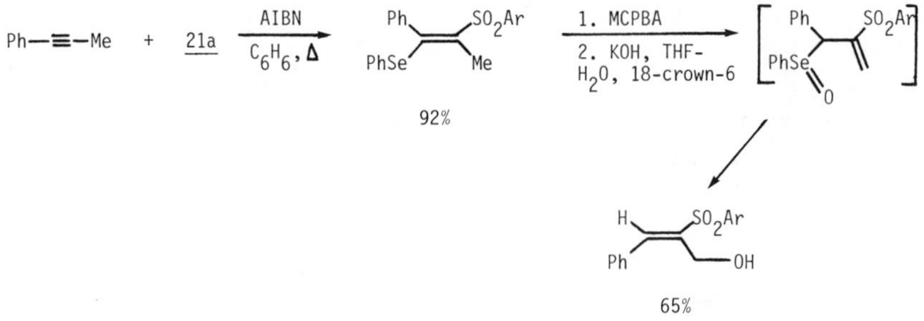

(71) (Ref. 108)

hand, the presence of the radical scavenger 2-methyl-2-nitrosopropane strongly suppresses product formation.[109] The additions are highly *anti* stereoselective, suggesting that the intermediate vinyl radical, 54, undergoes chain transfer in the product forming step more rapidly than inversion to 55 [Eq. (72)].[107]

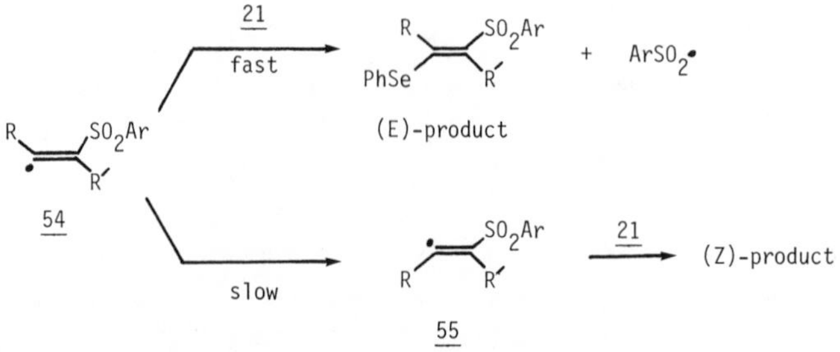

(72)

An anomalous reaction was observed when 1-trimethylsilylpropyne was treated with selenosulfonate, 21a, in a sealed glass tube at 120°C. The ketene diselenoacetal, 56, was formed in high yield [Eq.(73)] instead of the expected 1,2 adduct. The mechanism for its formation is not known.[108]

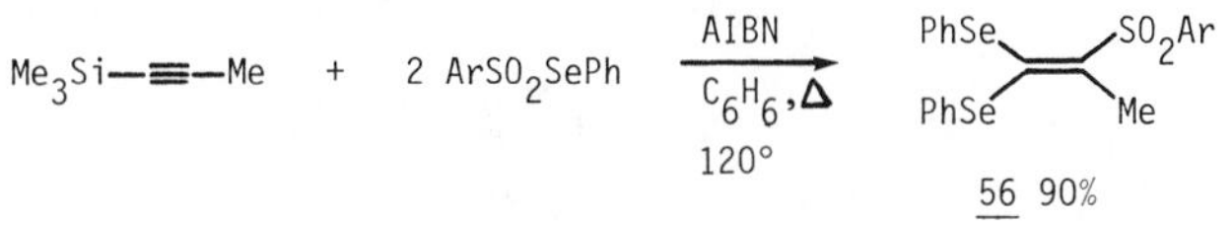

$$Me_3Si\!-\!\!\equiv\!\!-Me \quad + \quad 2\ ArSO_2SePh \quad \xrightarrow[\substack{C_6H_6,\Delta \\ 120°}]{AIBN} \quad \substack{PhSe \\ PhSe} \!\!\Big/\!\!=\!\!\Big\backslash\!\! \substack{SO_2Ar \\ Me} \tag{73}$$

$$\underline{56}\ 90\%$$

5.4. Reaction with Diazomethane

An unusual photochemical reaction between selenosulfonate, 21a, and excess diazomethane has been reported[110]. The product is the β-selenosulfone, 57, which is not produced via the intermediacy of the α-selenosulfone, 58. The radical chain mechanism in Scheme 16 has been proposed for this process. An interesting feature of this mechanism is that it requires the radical intermediates $ArSO_2\cdot$ and $ArSO_2CH_2\cdot$ to display contrasting behavior to that of the β-sulfonylalkyl radical $ArSO_2CH_2CH_2\cdot$. The former species are electrophilic and so react selectively with the nucleophilic substrate diazomethane. On the other hand, the latter intermediate is a more nucleophilic radical because the electron-withdrawing sulfonyl group is now removed from the radical center. It therefore selectively attacks the more electrophilic substrate, namely the selenosulfonate. This last step is analogous to the product forming step in the

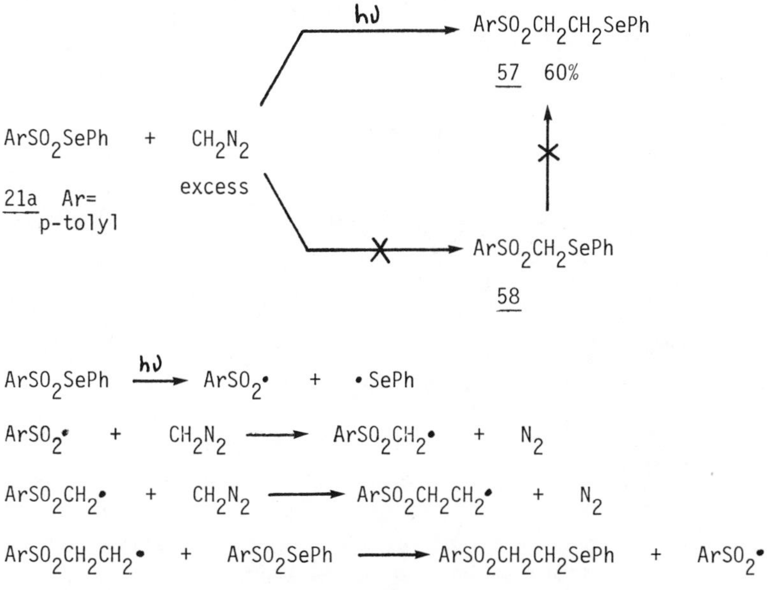

$$
\begin{array}{ccc}
& \xrightarrow{\ h\nu\ } & ArSO_2CH_2CH_2SePh \\
& & \underline{57}\ \ 60\% \\
ArSO_2SePh \ + \ CH_2N_2 & & \Big\updownarrow\!\!\!\times \\
\underline{21a}\ \ Ar= & \text{excess} & \\
\overline{\ }\ \text{p-tolyl} & & \\
& \xrightarrow{\ \ \times\ \ } & ArSO_2CH_2SePh \\
& & \underline{58}
\end{array}
$$

$$ArSO_2SePh \xrightarrow{\ h\nu\ } ArSO_2\cdot \ + \ \cdot SePh$$

$$ArSO_2\cdot \ + \ CH_2N_2 \longrightarrow ArSO_2CH_2\cdot \ + \ N_2$$

$$ArSO_2CH_2\cdot \ + \ CH_2N_2 \longrightarrow ArSO_2CH_2CH_2\cdot \ + \ N_2$$

$$ArSO_2CH_2CH_2\cdot \ + \ ArSO_2SePh \longrightarrow ArSO_2CH_2CH_2SePh \ + \ ArSO_2\cdot$$

SCHEME 16.

selenosulfonation of olefins (cf. Scheme 15). The corresponding dark reaction is slower and generates other products via an ionic mechanism (see Chapter 1).

6. RADICAL REACTIONS OF Se (IV) COMPOUNDS

According to Woodbridge,[111] the catalytic decomposition of t-butyl hydroperoxide by diselenides, seleninic acids, or anhydrides occurs via the free-radical pathway shown in Scheme 17. Formation of an intermediate peroxyseleninate 59 is followed by homolysis of either the O—O or Se—O bonds, thus generating t-BuO· or t-BuOO· radicals, respectively, and ultimately leading to the production of t-butanol and oxygen. The species RSe(O)O· or RSe(O) are reconverted to the seleninic acid or anhydride by an unknown mechanism to complete the catalytic cycle.

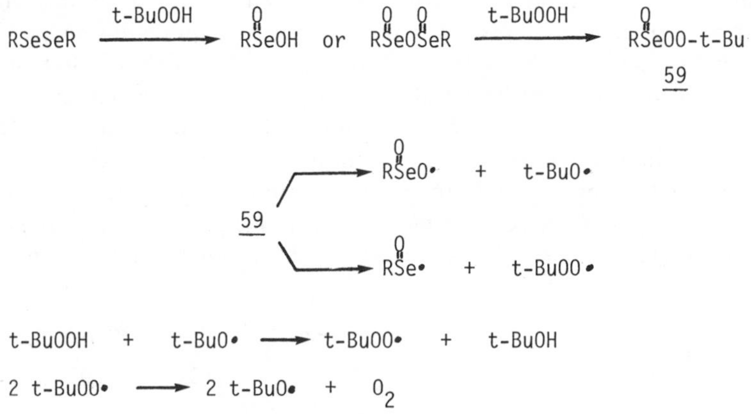

SCHEME 17.

More recently, t-butyl benzeneperoxyseleninate (59, R = Ph) was isolated and subjected to pyrolysis and photolysis.[112] The t-BuOO· radical was detected by esr spectroscopy during the photochemical experiment. This is consistent with the mechanism in Scheme 17, although its formation could also arise through hydrogen abstraction from t-butyl hydroperoxide, itself adventitiously generated by hydrolysis of the perester 59.[112]

Evidence has been presented for a free-radical mechanism in the selenium dioxide catalyzed oxidation of olefins with t-butyl hydroperoxide.[113] This involves C—Se cleavage of the allylic peroxyseleninate intermediate, 60, in competition with the more conventional [2,3]sigmatropic rearrangement (Scheme 18). The dissociative pathway is favored in cyclic systems where the olefin is unable to adopt the E-configuration. A similar homolytic decomposi-

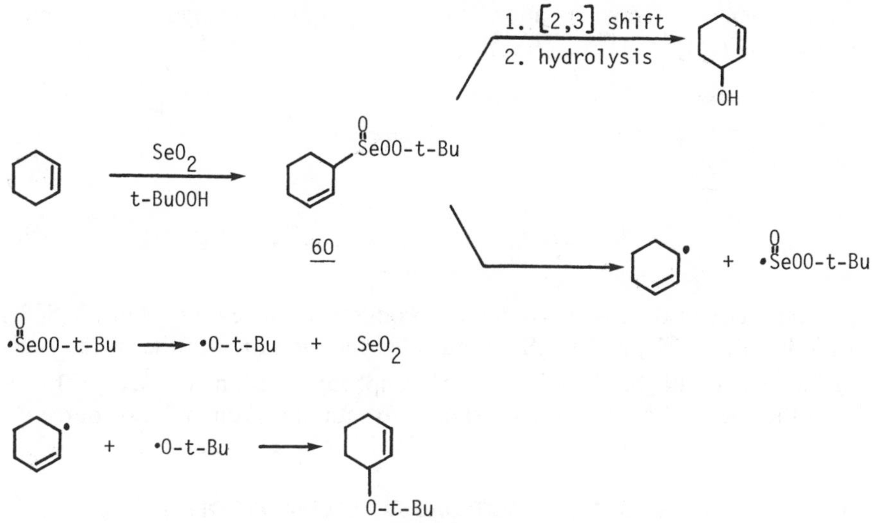

SCHEME 18.

tion of allylic seleninic acids, 61, has been suggested as occurring during the oxidation of some olefins with selenium dioxide [Eq. (74).[113]

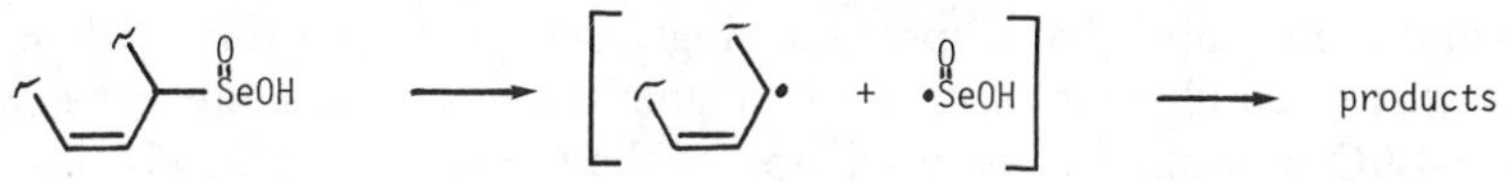

61 (74)

Thioseleninates, 62, are postulated intermediates in the oxidation of thiols with benzeneseleninic acid.[114] Their further decomposition to products possibly occurs via the radical pathway in Scheme 19.

$$\underset{\underline{62}}{\text{Ph}\overset{\text{O}}{\overset{\|}{\text{Se}}}\text{SR}} \longrightarrow \text{Ph}\overset{\text{O}}{\overset{\|}{\text{Se}}}\cdot + \cdot\text{SR}$$

$$2\ \text{Ph}\overset{\text{O}}{\overset{\|}{\text{Se}}}\cdot \longrightarrow 2\ \text{PhSe}\cdot + O_2$$

$$\text{PhSe}\cdot + \cdot\text{SR} \longrightarrow \text{PhSeSePh} + \text{PhSeSR} + \text{RSSR}$$

SCHEME 19.

Adamantyl phenyl selenoxide produces 1-adamantanol quantitatively upon pyrolysis.[32] The intermediacy of the adamantyloxy radical (Ad O·) was confirmed by spin-trapping experiments and is consistent with the mechanism in

Eq. (75). Benzyl selenoxides afford the corresponding aldehydes in a similar process [Eq. (76)].[115,116]

$$
\text{AdSePh} \xrightarrow{\Delta} (\text{Ad} \cdot + \cdot \text{SePh}) \longrightarrow (\text{AdOSePh}) \longrightarrow (\text{AdO} \cdot + \cdot \text{SePh}) \xrightarrow{\text{H}} \text{AdOH} \quad (75)
$$

Ad= 1-adamantyl

$$
\text{ArCH}_2\text{SeAr}' \xrightarrow{\Delta} (\text{ArCH}_2\text{OSeAr}') \longrightarrow \text{ArCHO} + \text{Ar}'\text{SeH} \quad (76)
$$

An esr study of the X-ray irradiation products of single crystals of Ph_3SeBr, Ph_2SeBr_2, Ph_2SeCl_2, and Ph_3SeCl indicated the formation of the corresponding $Ph_2\dot{Se}Br$ or $Ph_2\dot{Se}Cl$ radicals.[117,118] Diphenyl selenone ($PhSeO_2Ph$) produced the species $PhSeO_2\cdot$ when exposed to radiation from a ^{60}Co source.[119]

7. RADICAL REACTIONS OF SELENOKETONES

Scaiano and Ingold[9,120,121] demonstrated that di-t-butyl selenoketone, 63, functions as an effective spin-trapping agent for a variety of radicals of general structure $R_nM\cdot$, where M is C, O, S, Se (cf. 5), Si, Ge, P, or Sn [Eq. (77)]. The adduct radicals, 64, are relatively persistent and esr studies indicate that they adopt conformations where the spin-bearing carbon atom is planar or nearly planar and the group R_nM eclipses the singly occupied $2p_z$ orbital. Most such species exist in equilibrium with a diamagnetic dimer. The adduct, 64, where $R_nM = t$-BuO decomposes via γ scission of the C—O bond [Eq. (78)].

$$
\underset{63}{\overset{t\text{-Bu}}{\underset{t\text{-Bu}}{>}}}\!\!\text{=Se} \;+\; R_n\!M\cdot \longrightarrow \underset{64}{\overset{t\text{-Bu}}{\underset{t\text{-Bu}}{}}}\!\!\text{Se}^{MR_n} \rightleftharpoons \begin{array}{c}\text{diamagnetic}\\\text{dimer}\end{array} \quad (77)
$$

$$
t\text{-BuO-Se}\!\!\underset{t\text{-Bu}}{\overset{t\text{-Bu}}{<}} \longrightarrow t\text{-Bu}\cdot \;+\; O\text{=Se}\!\!\underset{t\text{-Bu}}{\overset{t\text{-Bu}}{<}} \longrightarrow O\!\!\underset{t\text{-Bu}}{\overset{t\text{-Bu}}{<}} \;+\; \text{Se} \quad (78)
$$

Ultraviolet irradiation of selenoketone, 63, results in its reduction to the diselenide, 65.[9,122,123] Furthermore, de Mayo et al.[123] reported that the selenoketone can be excited to either its S_1 or S_2 state, with the latter acting as a strong hydrogen abstracting species. The photoreduction is enhanced in the presence of a good hydrogen donor such as the selenol, 66, in which case the radical chain mechanism in Scheme 20 ensues. The reaction of the selenoketone and selenol in the presence of oxygen proceeds differently to afford triselenide, 67, as well as diselenide, 65.[124] The formation of 67 indicates a step involving C—Se cleavage, as shown in Scheme 21. Irradiation of selenofenchone, 68, produced a mixture of diselenides, 69–71 [Eq. (79)].[123]

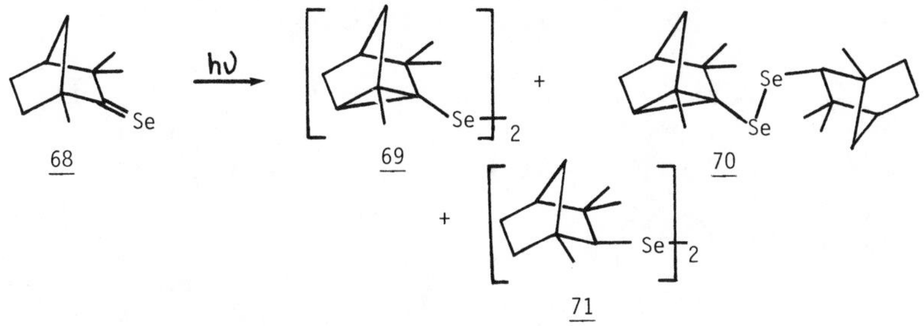

(79)

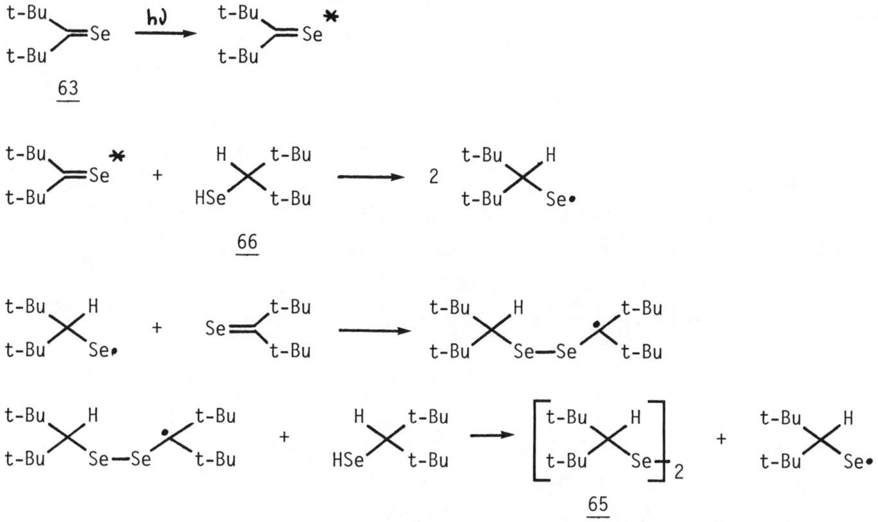

SCHEME 20.

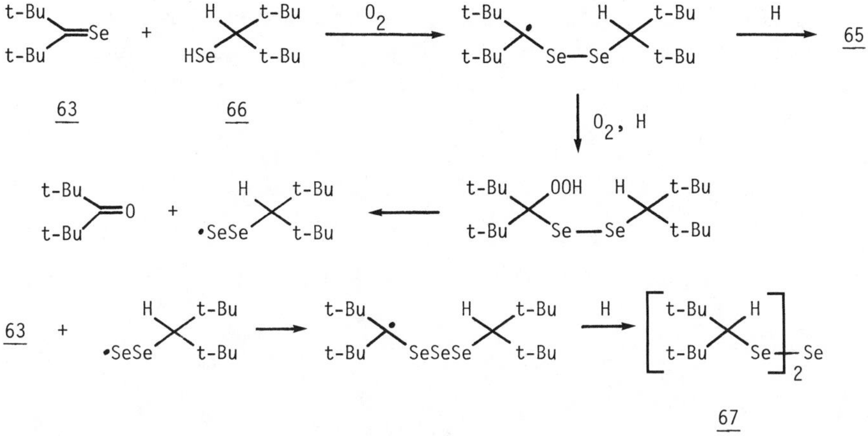

SCHEME 21.

8. MISCELLANEOUS

The formation of a diradical intermediate (or alternatively a carbene) has been suggested in the photolysis of selenadiazoles, 72.[125] Dimers, 73, are formed by the further attack of this intermediate on its selenoketene rearrangement product [Eq. (80)].

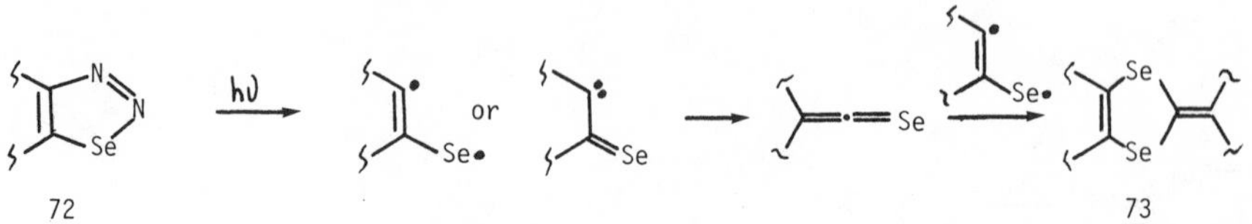

$$(80)$$

The photolysis of dibenzyl triselenocarbonate, 74, effects extrusion of the C=Se moiety to afford dibenzyl diselenide in 58% yield, providing that long wavelengths centered around the triselenocarbonate absorption ($\lambda_{max} = 540$ nm) are employed to avoid further decomposition of the product [Eq. (81)].[10]

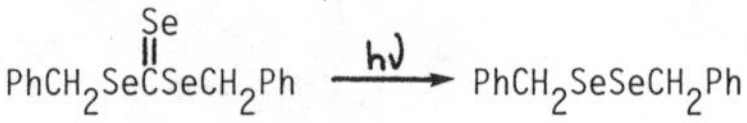

$$(81)$$

The phenylseleno metalloid derivatives $(PhSe)_4Sn$, $(PhSe)_3As$, and $(PhSe)_3Bi$, as well as several thio analogs, are stable at temperatures of $110-135°C$ under an inert atmosphere. In the presence of oxygen, however, the tin and bismuth reagents decompose to diphenyl diselenide by a free-radical chain process.[126]

An esr study of several radicals and ion radicals derived from heterocyclic precursors has been reported.[127] Thus, radical 75 and radical cation 76 were generated by oxidation of the parent compounds with lead dioxide and aluminum trichloride in nitroethane, respectively, while the radical anions, 77 and 78, were produced by reduction with potassium.

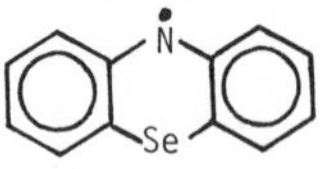

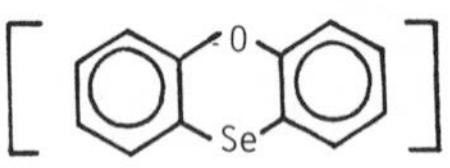

9. ADDENDUM

9.1. Reactions of Diselenides

Diselenides react with aluminum tribromide with homolytic cleavage of both the C—Se and Se—Se bonds. The radical complexes $(ArSe \cdot)AlBr_3$ and $(ArSeSe \cdot)AlBr_3$ were detected by esr, along with aryl radicals.[128,129] The above complexes were stable for several months in the absence of water and oxygen[128], but hydrolyzed to provide recovered diselenide along with ArSeAr and Ar—Ar.[129]

9.2. Reactions of Selenides

The reductive deselenization of an o-nitrophenyl selenide with triphenyltin hydride in 55% yield comprised the final step in a recent synthesis of dihydrocorynantheol.[130] Interestingly, the use of tri-n-butyltin hydride and AIBN resulted in reduction of the nitro group in lieu of deselenization.[130]

Pyrolysis of the 3-arylselenoclavulanic acid derivative $\underline{79}$ in the presence of AIBN resulted in C—Se cleavage followed by oxidation or rearrangement, depending on the presence or absence of oxygen[131] [Eq. (82)].

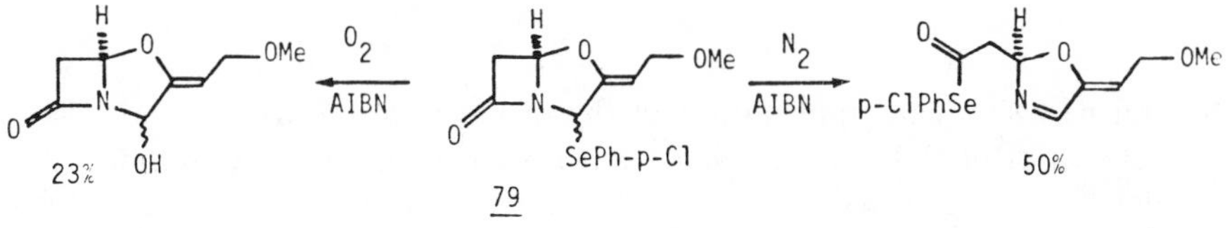

$$(82)$$

9.3. Reactions of Selenols, Selenolates and Related Species

Competition experiments performed in the photochemical $S_{RN}1$ reactions of the species PhX^- with 2-chloroquinoline in liquid ammonia (cf. Scheme 10) have established the following order of reactivity: $PhTe^- > PhSe^- > PhS^- \gg PhO^-$.[132]

9.4. Selenosulfonation

The free-radical selenosulfonation of β-pinene was accompanied by ring-opening of the four-membered ring[133] [Eq. (83)], while 1,6-heptadiene afforded the cyclized product $\underline{80}$ along with the expected products of 1,2-addition to either one or both of the double bonds.[133] Cyclohexene reacted with bis(p-tolysulfonyl) selenide $\underline{81}$ to produce adduct $\underline{82}$ and the diselenide $\underline{83}$ [Eq. (84)]. The latter was presumably formed by further S—Se cleavage and dimerization of 82. The analogous diselenide was favored when the olefin was styrene.[133]

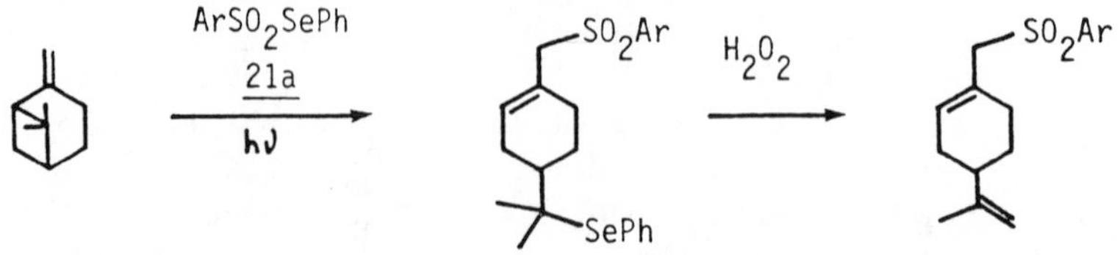

(83)

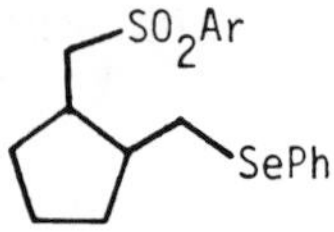

80

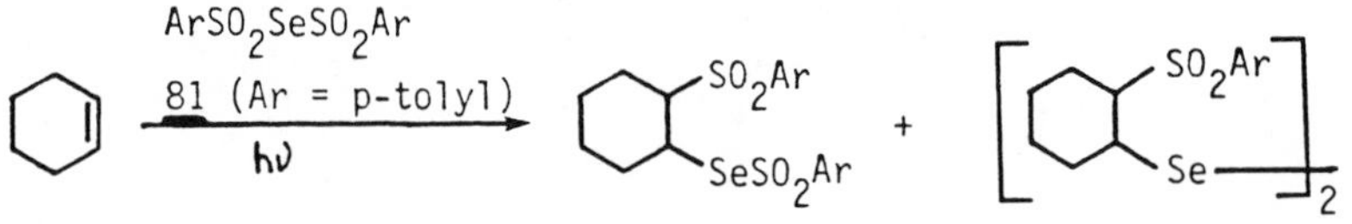

(84)

Several new synthetic applications of the products of the free-radical selenosulfonation of olefins and acetylenes have been reported. Vinyl sulfones thus obtained were employed in a new approach to the synthesis of substituted cyclohexenones and in a preparation of zingiberenol[134] (cf. refs. 102, 103).

The 1,2-addition products from the selenosulfonation of acetylenes undergo stereoselective substitution reactions of the phenylseleno group with organocuprates[135] [Eq. (85)], whereas enamine sulfones are produced when the corresponding selenoxides are treated with amines[136] [Eq. (85)].

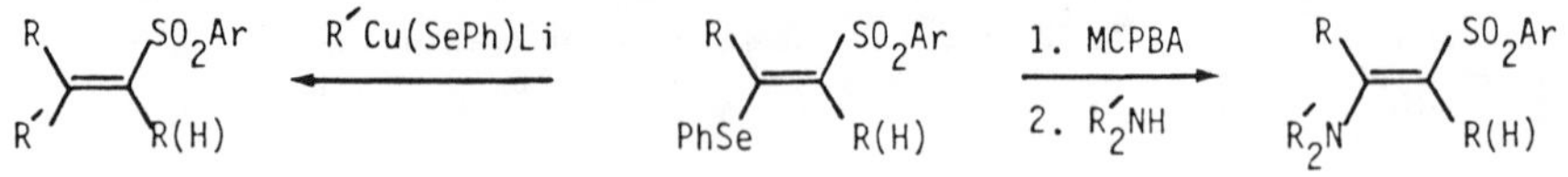

(85)

9.5. Miscellaneous

The trapping of alkyl radicals with phenylseleno-substituted captodative olefins 84 results in dimerization and spontaneous loss of PhSeSePh to afford 85[137] [Eq. (86)].

(86)

Z = electron-withdrawing
group

REFERENCES

1. *Organic Selenium Compounds: Their Chemistry and Biology*, D. L. Klayman and W. H. H. Günther, Eds., Wiley-Interscience, New York, 1973. (a) H. J. Shine, Chap. 15F; (b) D. L. Klayman, Chap. 13D, pp. 749–751.

2. J. R. Marier and J. F. Jaworski, *Interactions of Selenium*, Publication No. NRCC 20643 of the Environmental Secretariat, National Research Council of Canada, Ottawa, 1983.

3. J. Martens and K. Praefcke, *J. Organomet. Chem.*, **198**, 321 (1980).

4. J. J. Windle, A. K. Wiersema, and A. L. Tappel, *J. Chem. Phys.*, **41**, 1996 (1964).

5. J. J. Windle, A. K. Wiersema, and A. L. Tappel, *Nature*, **203**, 404 (1964).

6. U. Schmidt, A. Müller, and K. Markau, *Chem. Ber.*, **97**, 405 (1964).

7. R. Franzi and M. Geoffroy, *J. Organomet. Chem.*, **218**, 321 (1981).

8. O. Ito, *J. Am. Chem. Soc.*, **105**, 850 (1983).

9. J. C. Scaiano and K. U. Ingold, *J. Am. Chem. Soc.*, **99**, 2079 (1977).

10. W. Stanley, M. R. Van De Mark, and P. L. Kumler, *J. Chem. Soc. Chem. Commun.*, 700 (1974).

11. J. Y. C. Chu, D. G. Marsh, and W. H. H. Günther, *J. Am. Chem. Soc.*, **97**, 4905 (1975).

12. A. Couture, A. Lablache-Combier, R. Lapouyade, and G. Félix, *J. Chem. Res. (S)*, 258 (1979).

13. H. K. Spencer and M. P. Cava, *J. Org. Chem.*, **42**, 2937 (1977).

14. R. J. Cross and D. Millington, *J. Chem. Soc. Chem. Commun.*, 455 (1975).

15. J. Y. C. Chu and D. G. Marsh, *J. Org. Chem.*, **41**, 3204 (1976).

16. D. H. Brown, R. J. Cross, and D. Millington, *J. Chem. Soc. Dalton Trans.*, 159 (1977).

17. C. Walling and R. Rabinowitz, *J. Am. Chem. Soc.*, **79**, 5326 (1957).

18. C. Walling and R. Rabinowitz, *J. Am. Chem. Soc.*, **81**, 1243 (1959).

19. G. Vermeersch, N. Febvay-Garot, S. Caplain, A. Couture, and A. Lablache-Combier, *Tetrahedron Lett.*, 609 (1979).

20. D. H. Brown, R. J. Cross, and D. Millington, *J. Organomet. Chem.*, **125**, 219 (1977).

21. G. Vermeersch, N. Febvay-Garot, S. Caplain, A. Couture, and A. Lablache-Combier, *Proc. 7th IUPAC Symp. Photochem.*, 347 (1978); *Chem. Abstr.*, **90**, 120645U (1979).

22. T. W. Campbell, H. F. Walker, and G. M. Coppenger, *Chem. Rev.*, **50**, 279 (1952).

23. J. E. Kuder and M. A. Lardon, *Ann. N. Y. Acad. Sci.*, 192, 147 (1972).

24. M. A. Lardon, *Ann. N.Y. Acad. Sci.*, **192**, 132 (1972).

25. J. Y. C. Chu and J. W. Lewicki, *J. Org. Chem.*, **42**, 2491 (1977).

26. H. Higuchi, T. Otsubo, F. Ogura, H. Yamaguchi, Y. Sakata, and S. Misumi, *Bull. Chem. Soc. Jpn.*, **55**, 182 (1982).

27. T. Hirabayashi, S. Mohmand, and H. Bock, *Chem. Ber.*, **115**, 483 (1982).

28. J. Deniau, K. N. V. Duong, A. Gaudemer, P. Bougeard, and M. D. Johnson, *J. Chem. Soc. Perkin Trans. 2*, 393 (1981).

29. S. Pouly, G. Tainturier, and B. Gautheron, *J. Organomet. Chem.*, **232**, C65 (1982).

30. M. K. Chaudhuri, A. Haas, and A. Wensky, *J. Organomet. Chem.*, **116**, 323 (1976).

31. P. Jaitner, *J. Organomet. Chem.*, **210**, 353 (1981).

32. M. J. Perkins and E. S. Turner, *J. Chem. Soc., Chem. Commun.*, 139 (1981).

33. Y. Tamura, M. Fujita, L.-C. Chen, M. Inoue, and Y. Kita, *J. Org. Chem.*, **46**, 3564 (1981).

34. N. Petragnani and G. Schill, *Chem. Ber.*, **103**, 2271 (1970).

35. G. A. Russell and H. Tashtoush, *J. Am. Chem. Soc.*, **105**, 1398 (1983).

36. G. A. Russell and J. Hershberger, *J. Am. Chem. Soc.*, **102**, 7603 (1980).

37. J. Bergman, N. Eklund, T. E. Eriksen, and J. Lind, *Acta Chem. Scand., Ser. A*, **32**, 455 (1978).

38. T. Otsubo, F. Ogura, H. Yamaguchi, H. Higuchi, and S. Misumi, *Synth, Commun.*, **10**, 595 (1980).

39. H. Higuchi, Y. Sakata, S. Misumi, T. Otsubo, F. Ogura, and H. Yamaguchi, *Chem. Lett.*, 627 (1981).

40. H. Higuchi and S. Misumi, *Tetrahedron Lett.*, **23**, 5571 (1982).

41. H. Higuchi, M. Kugimiya, T. Otsubo, Y. Sakata, and S. Misumi, *Tetrahedron Lett.*, **24**, 2593 (1983).

42. J. Tanaka, T. Katagiri, M. Aoh, T. Ida, N. Morozumi, and K. Takabe, *Shizuoka Diagaku Kogakubu Kenkyu Hokoku*, **31**, 73 (1980); *Chem. Abstr.*, **96**, 26808b (1982).

43. J. C. Scaiano, P. Schmid and K. U. Ingold, *J. Organomet. Chem.*, **121**, C4 (1976).

44. D. L. J. Clive, G. Chittattu, and C. K. Wong, *J. Chem. Soc. Chem. Commun.*, 41 (1978).

45. D. L. J. Clive, C. K. Wong, W. A. Kiel, and S. M. Menchen, *J. Chem. Soc. Chem. Commun.*, 379 (1978).

46. D. L. J. Clive, G. Chittattu, and C. K. Wong, *J. Chem. Soc. Chem. Commun.*, 441 (1978).

47. D. L. J. Clive, G. J. Chittattu, V. Farina, W. A. Kiel, S. M. Menchen, C. G. Russell, A. Singh, C. K. Wong, and N. J. Curtis, *J. Am. Chem. Soc.*, **102**, 4438 (1980).

48. E. J. Corey, H. L. Pearce, I. Székely, and M. Ishiguro, *Tetrahedron Lett.*, 1023 (1978).

49. K. C. Nicolaou, W. E. Barnette, R. L. Magolda, P. A. Grieco, W. Owens, C.-L. J. Wang, J. B. Smith, M. Ogletree, and A. M. Lefer, *Prostaglandins*, **16**, 789 (1978).

50. K. C. Nicolaou, D. A. Claremon, W. E. Barnette, and S. P. Seitz, *J. Am. Chem. Soc.*, **101**, 3704 (1979).

51. K. C. Nicolaou, S. P. Seitz, W. J. Sipio, and J. F. Blount, *J. Am. Chem. Soc.*, **101**, 3884 (1979).

52. K. C. Nicolaou, R. L. Magolda, W. J. Sipio, W. E. Barnette, Z. Lysenko, and M. M. Joullie, *J. Am. Chem. Soc.*, **102**, 3784 (1980).

53. K. C. Nicolaou, W. E. Barnette, and R. L. Magolda, *J. Am. Chem. Soc.*, **103**, 3480 (1981).

54. P. A. Grieco, S. Gilman, and M. Nishizawa, *J. Org. Chem.*, **41**, 1485 (1976).

55. J. Lucchetti and A. Krief, *Tetrahedron Lett.*, **22**, 1623 (1981).

56. R. M. Scarborough, Jr., A. B. Smith III, W. E. Barnette, and K. C. Nicolaou, *J. Org. Chem.*, **44**, 1742 (1979).

57. W. P. Jackson, S. V. Ley, and J. A. Morton, *J. Chem. Soc. Chem. Commun.*, 1028 (1980).

58. W. P. Jackson, S. V. Ley, and A. J. Whittle, *J. Chem. Soc. Chem. Commun.*, 1173 (1980).

59. S. V. Ley, B. Lygo, H. Molines, and J. A. Morton, *J. Chem. Soc. Chem. Commun.*, 1251 (1982).

60. S. W. Rollinson, R. A. Ames, and J. A. Katzenellenbogen, *J. Am. Chem. Soc.*, **103**, 4114 (1981).

61. J. P. Konopelski, C. Djerassi, and J. P. Reynaud, *J. Med. Chem.*, **23**, 722 (1980).

62. P. J. Giddings, D. I. John, and E. J. Thomas, *Tetrahedron Lett.*, **21**, 399 (1980).

63. H. Rüeger and M. H. Benn, *Heterocycles*, **19**, 23 (1982).

64. P. A. Grieco, J. Inanaga, N.-H. Lin, and T. Yanami, *J. Am. Chem. Soc.*, **104**, 5781 (1982).

65. G. Jaurand, J.-M, Beau, and P. Sinaÿ, *J. Chem. Soc. Chem. Commun.*, 572 (1981).

66. T. Kametani, K. Fukumoto, H. Kurobe, and H. Nemoto, *Tetrahedron Lett.*, **22**, 3653 (1981).

67. T. Kametani, H. Kurobe, H. Nemoto, and K. Fukumoto, *J. Chem. Soc. Perkin Trans. 1*, 1085 (1982).

68. A. Toshimitsu, T. Aoai, H. Owada, S. Uemura, and M. Okano, *J. Org. Chem.*, **46**, 4727 (1981).

69. D. Labar and A. Krief, *J. Chem. Soc. Chem. Commun.*, 564 (1982).

70. S. D. Burke, W. F. Fobare, and D. M. Armistead, *J. Org. Chem.*, **47**, 3348 (1982).

71. R. L. Sobczak, M. E. Osborn, and L. A. Paquette, *J. Org. Chem.*, **44**, 4886 (1979).

72. M. D. Bachi and C. Hoornaert, *Tetrahedron Lett.*, **22**, 2693 (1981).

73. D. L. J. Clive and P. L. Beaulieu, *J. Chem. Soc. Chem. Commun.*, 307 (1983).

74. T. G. Back and S. Collins, unpublished results.

75. J. N. Denis and A. Krief, *Tetrahedron Lett.*, **23**, 3411 (1982).

76. T. G. Back and S. Collins, *J. Org. Chem.*, **46**, 3249 (1981).

77. J. Pfenninger, C. Heuberger, and W. Graf, *Helv. Chim. Acta*, **63**, 2328 (1980).

78. J. Pfenninger and W. Graf, *Helv. Chim. Acta*, **63**, 1562 (1980).

79. K. U. Ingold, D. H. Reid, and J. C. Walton, *J. Chem. Soc. Perkin Trans. 2*, 431 (1982).

80. G. E. Keck and J. B. Yates, *J. Am. Chem. Soc.*, **104**, 5829 (1982).

81. R. R. Webb II, and S. Danishefsky, *Tetrahedron Lett.*, **24**, 1357 (1983).

82. J. R. M. Giles, B. P. Roberts, M. J. Perkins, and E. S. Turner, *J. Chem. Soc. Chem. Commun.*, 504 (1980).

83. K. Nishikida and F. Williams, *Chem. Phys. Lett.*, **34**, 302 (1975).

84. J. T. Wang and F. Williams, *J. Chem. Soc. Chem. Commun.*, 1184 (1981).

85. A. B. Pierini and R. A. Rossi, *J. Organomet. Chem.*, **144**, C12 (1978).

86. A. B. Pierini and R. A. Rossi, *J. Org. Chem.*, **44**, 4667 (1979).

87. V. G. Voloshchuk, V. N. Boiko, and L. M. Yagupol'skii, *Zh. Org. Khim.*, **13**, 2008 (1977).

88. L. Hevesi, *Tetrahedron Lett.*, 3025 (1979).

89. R. A. Rossi and A. B. Peñéñory, *J. Org. Chem.*, **46**, 4580 (1981).

90. D. J. Sandman, J. C. Stark, L. A. Acampora, and P. Gagne, *Organomet.*, **2**, 549 (1983).

91. A. N. Frolov, A. V. El'tsov, E. V. Smirnov, and O. V. Kul'bitskaya, *Zh. Org. Khim.*, **13**, 2007 (1977).

92. M. R. Detty and G. P. Wood, *J. Org. Chem.*, **45**, 80 (1980).

93. F. G. James, M. J. Perkins, O. Porta, and B. V. Smith, *J. Chem. Soc. Chem. Commun.*, 131 (1977).

94. M. J. Perkins, B. V. Smith, and E. S. Turner, *J. Chem. Soc. Chem. Commun.*, 977 (1980).

95. K. Fujimori, H. Yoshimoto, and S. Oae, *Tetrahedron Lett.*, 4397 (1979).

96. N. Kambe, K. Kondo, S. Murai, and N. Sonoda, *Angew. Chem. Int. Ed., Engl.*, **19**, 1008 (1980).

97. W. J. Rhead and G. N. Schrauzer, *Bioinorg. Chem.*, **3**, 225 (1974).

98. O. Foss, *Acta Chem. Scand.*, **6**, 508 (1952).

99. R. A. Gancarz and J. L. Kice, *Tetrahedron Lett.*, **21**, 4155 (1980).

100. R. A. Gancarz and J. L. Kice, *J. Org. Chem.*, **46**, 4899 (1981).

101. T. G. Back and S. Collins, *Tetrahedron Lett.*, **21**, 2215 (1980).

102. L. A. Paquette and W. A. Kinney, *Tetrahedron Lett.*, **23**, 5127 (1982).

103. L. A. Paquette and G. D. Crouse, *J. Org. Chem.*, **48**, 141 (1983).

104. J. L. Kice, personal communication.

105. Y.-H. Kang and J. L. Kice, *Tetrahedron Lett.*, **23**, 5373 (1982).

106. T. G. Back and S. Collins, *Tetrahedron Lett.*, **22**, 5111 (1981).

107. T. G. Back, S. Collins, and R. G. Kerr, *J. Org. Chem.*, **48**, 3077 (1983).

108. T. G. Back, S. Collins, U. Gokhale, and K.-W. Law, *J. Org. Chem.*, **48**, 4776 (1983).

109. T. Miura and M. Kobayashi, *J. Chem. Soc. Chem. Commun.*, 438 (1982).

110. T. G. Back, *J. Org. Chem.*, **46**, 5443 (1981).

111. D. T. Woodbridge, *J. Chem. Soc. Part B*, 50 (1966).

112. A. J. Bloodworth and D. J. Lapham, *J. Chem. Soc. Perkin Trans. 1*, 471 (1983).

113. M. A. Warpehoski, B. Chabaud and K. B. Sharpless, *J. Org. Chem.*, **47**, 2897 (1982).

114. J. L. Kice and T. W. S. Lee, *J. Am. Chem. Soc.*, **100**, 5094 (1978).

115. I. D. Entwistle, R. A. W. Johnstone, and J. H. Varley. *J. Chem. Soc. Chem. Commun.*, 61 (1976).

116. B. E. Norcross, J. M. Lansinger, and R. L. Martin, *J. Org. Chem.*, **42**, 369 (1977).

117. M. Geoffroy, *J. Chem. Phys.*, **70**, 1497 (1979).

118. R. Franzi, M. Geoffroy, L. Ginet, and N. Leray, *J. Phys. Chem.*, **83**, 2898 (1979).

119. M. Geoffroy and N. Leray, *J. Chem. Phys.*, **72**, 775 (1980).

120. J. C. Scaiano and K. U. Ingold, *J. Chem. Soc. Chem. Commun.*, 205 (1976).

121. J. C. Scaiano and K. U. Ingold, *J. Phys. Chem.*, **80**, 1901 (1976).

122. T. G. Back, D. H. R. Barton, M. R. Britten-Kelly, and F. S. Guziec, Jr., *J. Chem. Soc. Perkin Trans. 1*, 2079 (1976).

123. N. Y. M. Fung, P. de Mayo, B. Ruge, A. C. Weedon, and S. K. Wong, *Can. J. Chem.*, **58**, 6 (1980).

124. B. J. McKinnon, P. de Mayo, N. C. Payne, and B. Ruge, *Nouv. J. Chim.*, **2**, 91 (1978).

125. H. Meier and I. Menzel, *Tetrahedron Lett.*, 445 (1972).

126. D. H. R. Barton, H. Dadoun, and A. Gourdon, *Nouv. J. Chim.*, **6**, 53 (1982).

127. M. F. Chiu and B. C. Gilbert, *J. Chem. Soc. Perkin Trans. 2*, 258 (1973).

128. I. P. Romm, I. V. Oliferenko, E. N. Gur'yanova, V. V. Troitskii, V. A. Chernoplekova, L. M. Kataeva, and E. G. Kataev, *Zh. Obshch. Khim.*, **53**, 477 (1983).

129. I. V. Olifirenko, I. P. Romm, E. N. Guryanova, V. V. Troitsky, L. M. Kataeva, V. A. Chernoplekova, and E. G. Kataev, *J. Organomet. Chem.*, **265**, 27 (1984).

130. B. Danieli, G. Lesma, G. Palmisano, and S. Tollari, *J. Chem. Soc., Perkin Trans. 1*, 1237 (1984).

131. G. Brooks and E. Hunt, *J. Chem. Soc., Perkin Trans. 1*, 2513 (1983).

132. A. B. Pierini, A. B. Penenory, and R. A. Rossi, *J. Org. Chem.*, **49**, 486 (1984).

133. Y.-H. Kang and J. L. Kice, *J. Org. Chem.*, **49**, 1507 (1984).

134. W. A. Kinney, G. D. Crouse, and L. A. Paquette, *J. Org. Chem.*, **48**, 4986 (1983).

135. T. G. Back, S. Collins and K.-W. Law, *Tetrahedron Lett.*, **25**, 1689 (1984).

136. T. G. Back, S. Collins, and K.-W. Law, *Can. J. Chem.*, **63**, 2313 (1985).

137. Z. Janousek, S. Piettre, F. Gorissen-Hervens, and H. G. Viehe, *J. Organomet. Chem.*, **250**, 197 (1983).

8

[2,3]Sigmatropic Rearrangements of Organoselenium Compounds

HANS J. REICH

Department of Chemistry, University of Wisconsin,
Madison, Wisconsin

CONTENTS

1. INTRODUCTION

The concept of indirect functionalization by transfer of a group from a hetero-atom to carbon has been widely used in many aspects of organometalloid and organometallic chemistry. In the area of sulfur and selenium chemistry two important applications are [2,3]sigmatropic rearrangements and Pummerer reactions.[1–4] The prototype [2,3]sigmatropic rearrangement [Eq. (1)] involves transfer of oxygen from selenium to carbon, although transfer of other groups (N,C) has been studied extensively for sulfur and to some extent also for selenium compounds. The advantages of techniques such as the [2,3]sigmatropic rearrangement for the formation of C—O bonds is that the oxidation itself is carried out at a relatively invariant site so that reproducibility and control of the oxidation is easily achieved. In contrast, direct oxidation of, for example, C—H bonds often requires specialized reagents and conditions for specific situations.

$$R{\diagdown}^{Se}{\diagup}{\diagdown}{\diagup} \xrightarrow{\text{[Ox]}} \underset{R}{\overset{O}{\diagdown}}^{Se}{\diagup}{\diagdown}{\diagup} \underset{\xleftarrow{}}{\xrightarrow{\text{[2, 3]}}} \underset{R}{\diagdown}^{Se}{\overset{O{\diagup}{\diagdown}}{\diagdown}} \xrightarrow{\text{H}_2\text{O}} HO{\diagup}{\diagdown}{\diagup} \tag{1}$$

The various types of [2,3]sigmatropic rearrangements developed in sulfur chemistry invariably have analogs in the selenium area, although these have been studied much less intensively. A comparison of the advantages and disadvantages of the sulfur and selenium versions of each reaction type is therefore not possible. Some general observations may be made, however:

1. Selenium compounds are more expensive and must be handled more carefully because of their toxicity.
2. If the S or Se function is to be carried through several reaction steps, or submitted to extreme reaction or purification conditions, the sulfur group is more likely to emerge unscathed.
3. The nucleophilicity of and ease of oxidation of divalent Se compounds exceeds that of sulfur compounds so the Se group can usually be introduced and manipulated under milder conditions than the sulfur analog. In contrast, the ease of oxidation and stability of Se (IV) compounds is less than that of S analogs.
4. The activation barrier and hence reaction temperature for the [2,3] sigmatropic rearrangement itself may be very much lower for selenium than sulfur.
5. Scission of the bond between Se and the rest of the molecule after the rearrangement is easier for Se than for S, and therefore the auxiliary Se group is easier to remove.

It is hoped that the results outlined in this chapter will lead to further exploration of a hitherto little developed area of selenium chemistry.

2. ALLYL SELENIDES AND SELENOXIDES

Allyl selenoxide [2,3]sigmatropic rearrangements have been the most widely studied and utilized of the group of reactions being considered here. Allyl selenides, which are invariably precursors to the allyl selenoxides, are readily available by a number of effective synthetic procedures. This section begins with a description of the kinetics and thermodynamics of the system and a discussion of the oxidation reaction. The remainder is a description of a number of applications and is divided into sections according to the method of preparation of the allyl selenide.

2.1. Kinetics and Thermodynamics of the Selenoxide [2,3]Sigmatropic Rearrangement

The first clear cut example of a [2,3]sigmatropic rearrangement of an allyl selenoxide was reported by Sharpless and Lauer[5] in connection with their studies of the selenium dioxide oxidation of olefins [Eqs. (2) and (3)]. They showed (1) rearrangements are rapid at room temperature (2) allylic transposition is the predominant process (3) trans double bonds are formed (4) a cheap and convenient oxidant such as hydrogen peroxide can be used without danger of overoxidation, and (5) no special reagent or conditions need be employed to achieve cleavage of the presumed intermediate selenenate ester as is the case with sulfur analogs.[6,7] In fact, selenenate esters have never been identified as intermediates in the phenylseleno series. They can, however, be observed if o-nitrophenyl allyl selenides are used.[8] Sharpless and Lauer[9] also identified a disadvantage of the selenium system: Allyl selenides more easily undergo [1,3]sigmatropic rearrangements, a process that results in loss of allyl regiospecifically. Subsequent work showed that isomerization was caused by acid catalysis and photolysis, as well as thermolysis.[10-14]

$$R \overset{SeC_6H_5}{\diagup\!\!\!\diagdown} \xrightarrow{\text{H}_2\text{O}_2} R \diagup\!\!\!\diagdown\!\!\!\diagup OH \qquad (2)$$

$$R \diagup\!\!\!\diagdown\!\!\!\diagup SeC_6H_5 \xrightarrow{\text{H}_2\text{O}_2} R \overset{OH}{\diagup\!\!\!\diagdown} \qquad (3)$$

Before discussing in detail the applications of selenoxide [2,3]sigmatropic rearrangements it is worthwhile to consider the kinetic and thermodynamic relationships between the two partners. Such studies have been reported by Tang and Mislow[6] for the sulfoxide–sulfenate equilibration. Free energies of activation were typically 19–22 kcal/mole (i.e., $t_{1/2} \approx$ 1–30 min at room temperature). In most cases the sulfoxide isomer was slightly favored over the sulfenate (1–3 kcal/mole), although in several cases the sulfenate was actually more stable [e.g., $2,4\text{-}(NO_2)_2C_6H_3\text{—}S\text{—}O\text{—}CH_2CH=CH_2$].

Simple bond strength arguments (i.e., C—Se is weaker than C—S bond) predict that in the selenium series the bias would be stronger in the direction of the selenenate, and that rearrangement of selenoxides should be faster than that of sulfoxides. The latter conclusion would also be reached on the basis that the selenoxide *syn* elimination, an electronically closely related process, occurs at temperatures about 100°C lower ($\sim$ 1000 times as fast) as similar sulfoxide eliminations.[15]

Kinetic measurements have been carried out on several allyl and propargyl selenoxide isomerizations (Table 1).[16–18] They occur with reasonable rates between −80 and −20°C, with the exception of the 3-(trimethylsilyl)-2-propynyl phenyl selenoxide (Entry 3),[18] which survives even at room temperature for a few minutes.

The rearrangement of selenenates to selenoxides, on the other hand, is not easily studied. In several of the examples of Table 1 (Entries 2 and 6) selenenate esters can be observed as products of the [2,3]sigmatropic rearrangement. No detectable amount of selenoxide remained at equilibrium, however. This is in contrast to the sulfur series, where the treatment of allyl and propargyl alcohols with sulfenyl halides is an important synthetic route for the preparaton of allyl and allenyl sulfoxides.[7]

The first demonstration that selenenates are in equilibrium with selenoxides was provided by the observation that when the selenenate ester, 1 (formed by treatment of alcohol with *o*-nitrobenzeneselenenyl chloride) was heated at 65°C it was slowly converted to the diene, 2, presumably by a [2,3]sigmatropic rearrangement and selenoxide *syn*-elimination sequence.[19–21]

$Ar = 2\text{-}NO_2\text{—}C_6H_4$

TABLE 1
RATES OF SELENOXIDE [2,3]SIGMATROPIC REARRANGEMENTS

Entry	Compound	Temperature (°C)	$t_{1/2}$ (min)	$\Delta G^{\ddagger}$ (kcal mole^{-1})	References
1	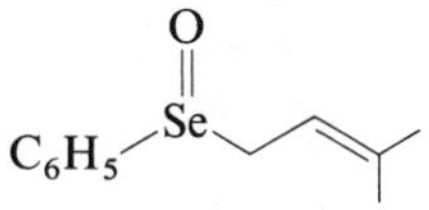	−50	44	16.5	16
2	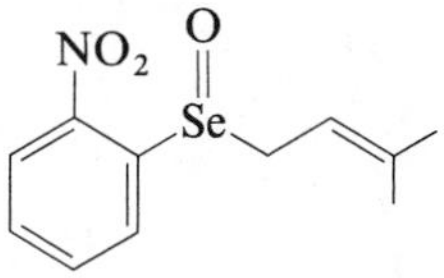	−80	∼6	∼13.5	17
3	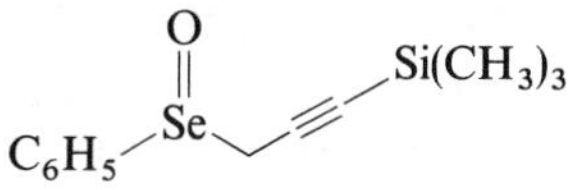	−31	33,000	21.2	18
4	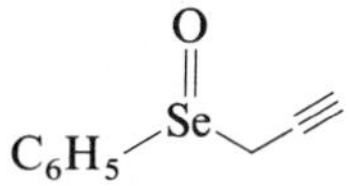	−31	60	18.7	18
5	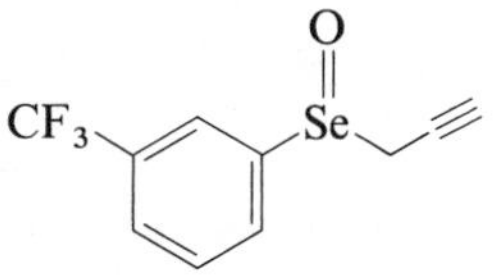	−31	116	18.5	18
6	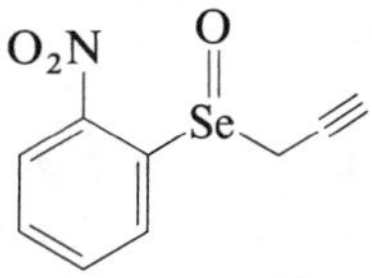	−50	33	16.5	18

The results of Eq. (4) suggested that the *o*-nitrophenylseleno group might provide a vehicle for a quantitative study of the [2,3]sigmatropic rearrangement energetics, and this was done using prenyl *o*-nitrophenyl selenoxide and sulfoxide [Eq. (5)].

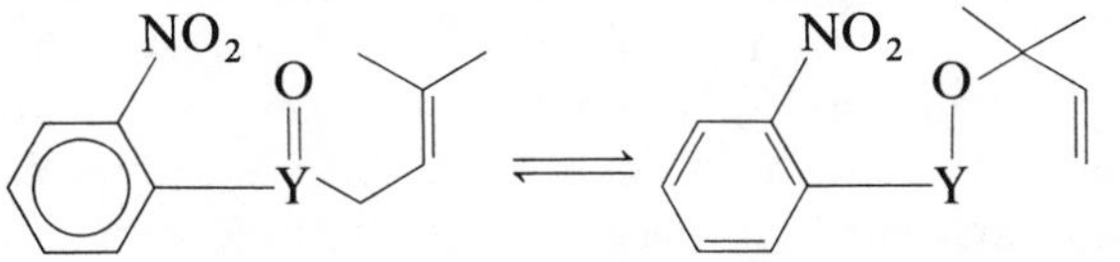

$$\text{(5) (Ref. 17)}$$

Y = S	Sulfoxide	Sulfenate
Y = Se	Selenoxide	Selenenate

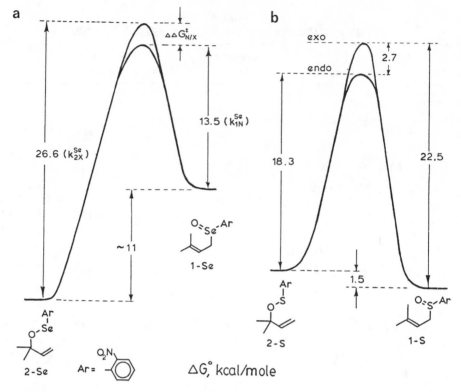

FIGURE 1. Energetics of the [2,3]sigmatropic rearrangements of allylic sulfoxides and selenoxides.

Figure 1 summarizes the results obtained. They follow earlier qualitative observations: The sulfoxide is more stable by 1.5 kcal/mole, whereas in the selenium system the selenenate is more stable by ~11 kcal/mole. This number is not very precise both because of the long temperature extrapolations involved, and because of uncertainty regarding the *exo/endo* ratio of the selenoxide [2,3]sigmatropic rearrangement that is an integral part of the analysis.[17] Thus in most situations the [2,3]sigmatropic rearrangement of selenoxides is effectively irreversible, in the sense that the forward reaction occurs at ~100°C lower temperature than the backward.

However, if there is a strong electronic or steric bias in favor of the selenoxide, it can become the more stable form. This is the case for benzyl selenoxides, which have not been shown to rearrange, as well as for the furylmethyl, phenanthrylmethyl, and 1-vinylcyclopropyl phenyl selenoxides shown in [Eq. (6)–(8)]. In each of these examples no selenenate ester is directly detectable, but it is present in low concentration since treatment with a secondary amine results in rapid conversion to the rearranged allyl alcohols as shown.

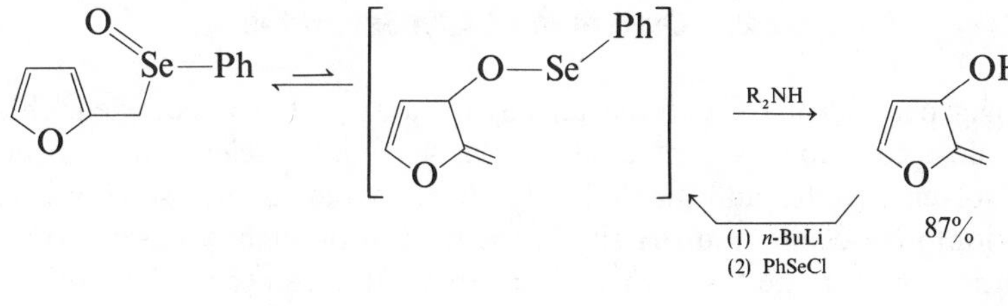

R$_2$NH

(1) n-BuLi
(2) PhSeCl

87%

(6) (Ref. 17)

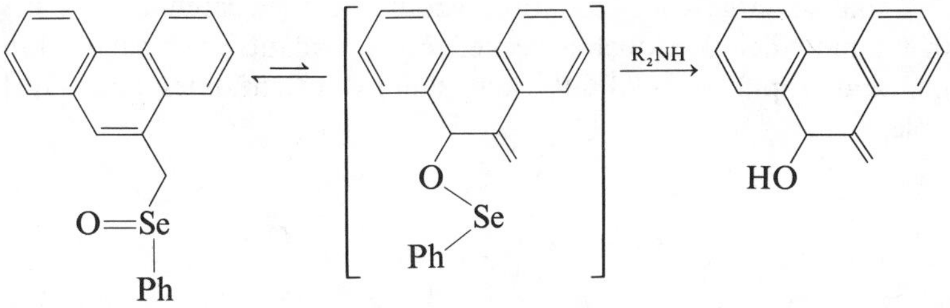

R$_2$NH

(7) (Ref. 17)

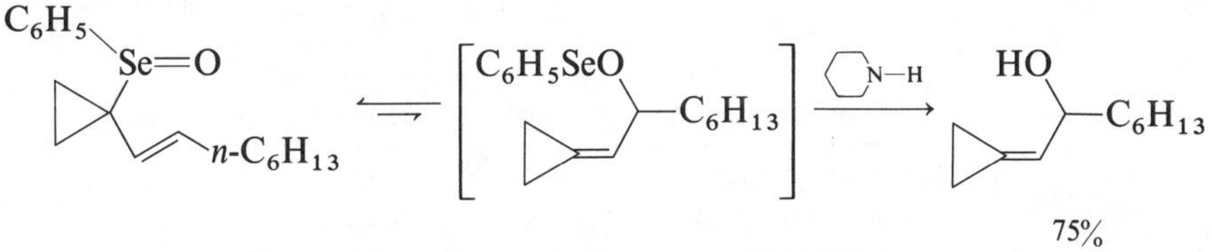

75%

(8) (Ref. 22)

The competition between [2,3]sigmatropic rearrangement and the selenoxide *syn* elimination in compounds where both are possible is generally in favor of the [2,3]sigmatropic rearrangement although a number of cases, usually involving a conjugated allyl selenide, have been reported where some[8,19,21,23,24] or all[25-27] of the product is a diene rather than an allyl alcohol. A particularly interesting example is depicted in Eq. (9).[24] The behavior of this system suggested that one selenoxide diastereomer gave allyl alcohol and the other diene.

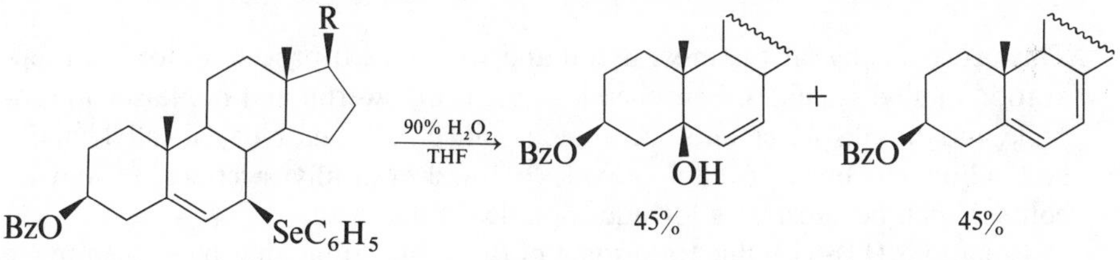

90% H$_2$O$_2$, THF

45%

+

45%

(9) (Ref. 24)

2.2. Oxidation of Allyl Selenides

Hydrogen peroxide in solvents like tetrahydrofuran (THF) or methylene chloride is the most common oxidant used for the selenide to selenoxide conversion. Polar solvents (water and alcohols) should be avoided since solvolysis (ether formation) may occur, and pH should be neutral or slightly basic. No special precautions need to be taken to avoid overoxidation, since the [2,3]rearrangement is invariably faster than oxidation of selenoxide to selenone. The hydrogen peroxide oxidation is autocatalytic and exothermic ($C_6H_5SeO_2H$ is a catalyst)[28,29] and so, large scale oxidations should be done carefully.[1]

When a more delicate touch is needed, other oxidants, such as m-chloroperbenzoic acid (rapid at $-78°C$)[17] and sulfonyl oxaziridines [Eq. (10)], are available.

$$(10)\ (\text{Ref. 30})$$

Conditions	% Yield	% Yield
THF/H$_2$O/25°C	25	49
Py/THF/H$_2$O	88	0

Equation (10) also illustrates the solution to a problem sometimes encountered when selenenic acids are generated in the presence of olefins: i.e., electrophilic additions leading to β-hydroxy selenide by products.[15,18]

Since oxaziridines are chiral, they offer the potential for asymmetric oxidation of selenides to optically active selenoxides. In this way nonracemic allyl alcohols can be prepared.[31] Enzymatic oxidations of allyl selenides have been reported, but products have not been analyzed for optical activity.[32]

2.3. Nucleophilic Selenenylation of Allyl Cation Equivalents

This process is by far the most useful and widely used procedure for the preparation of allyl selenides. Benzeneselenolate is a powerful and nonbasic nucleophile, and thus displacements occur under mild conditions. Not only allyl halides and sulfonates, but also allyl epoxides[5,33] and even allyl acetates[3,34] and alcohols[35] can be substrates for nucleophilic displacement.

Equations (11)–(13) illustrate some of the applications that have been made of the displacements of halides and mesylates. There seems to be a strong preference for S_N2 as opposed to S_N2' displacement [Eqs. (11) and (12)]. The double

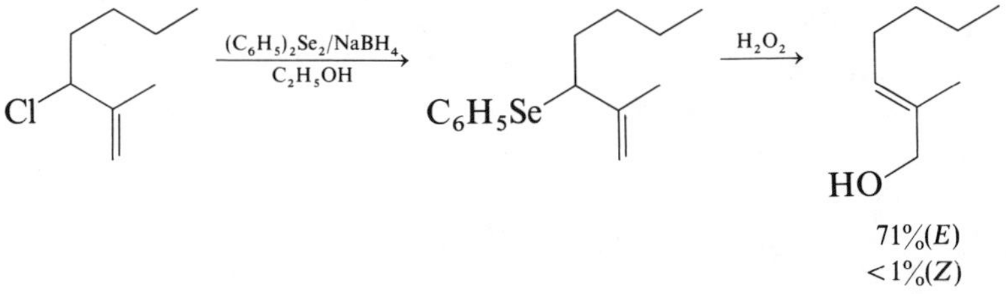

$$\xrightarrow[\text{C}_2\text{H}_5\text{OH}]{(\text{C}_6\text{H}_5)_2\text{Se}_2/\text{NaBH}_4} \qquad \text{C}_6\text{H}_5\text{Se} \qquad \xrightarrow{\text{H}_2\text{O}_2}$$

71% (E)
<1% (Z)

(11) (Ref. 5)

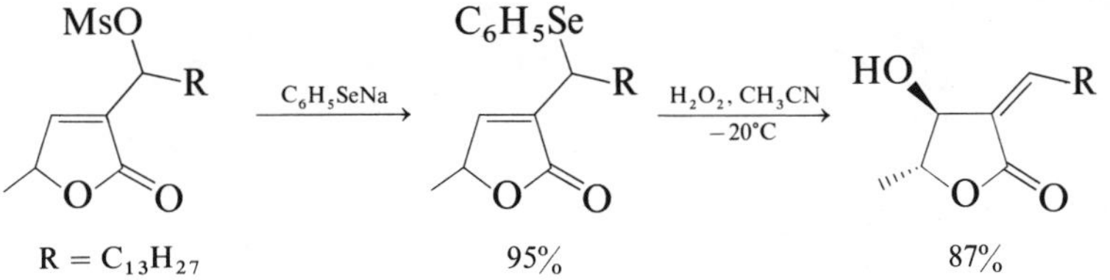

$$\text{MsO} \qquad \xrightarrow{\text{C}_6\text{H}_5\text{SeNa}} \qquad \text{C}_6\text{H}_5\text{Se} \qquad \xrightarrow[-20°C]{\text{H}_2\text{O}_2, \text{ CH}_3\text{CN}} \qquad \text{HO}$$

R = C$_{13}$H$_{27}$ 95%   87%

(12) (Refs. 36, 37)

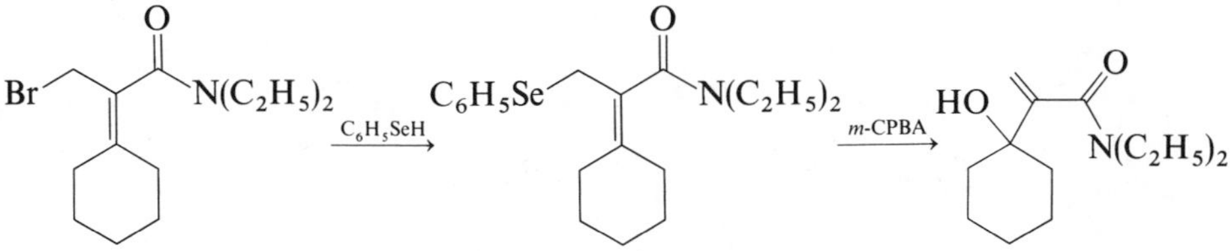

$$\text{Br} \qquad \text{N(C}_2\text{H}_5)_2 \qquad \xrightarrow{\text{C}_6\text{H}_5\text{SeH}} \qquad \text{C}_6\text{H}_5\text{Se} \qquad \text{N(C}_2\text{H}_5)_2 \qquad \xrightarrow{m\text{-CPBA}} \qquad \text{HO} \qquad \text{N(C}_2\text{H}_5)_2$$

(13) (Ref. 38)

bonds formed during the selenoxide [2,3]sigmatropic rearrangement usually show a strong preference for the (E)-configuration, an effect also seen for allyl sulfoxide rearrangements. Primary, secondary, and tertiary allyl alcohols can be prepared. Direct conversion of alcohols to selenides can be achieved using the oxidation–reduction condensation method.[28,39,40] This technique serves well for a two-step contrathermodynamic allylic isomerization of allyl alcohols [Eqs. (14) and (15)].[41,42] Care must be taken when hydrogen peroxide is used with trisubstituted olefins, since the areneseleninic acid serves as a catalyst for epoxidation.[28,29,42]

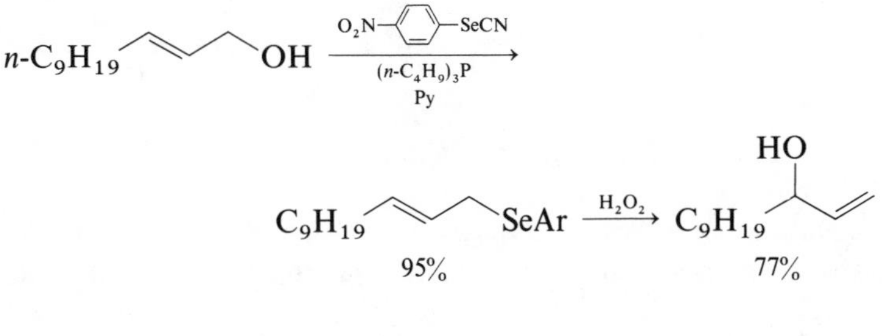

$$n\text{-C}_9\text{H}_{19} \qquad \text{OH} \qquad \xrightarrow[\substack{(n\text{-C}_4\text{H}_9)_3\text{P} \\ \text{Py}}]{\text{O}_2\text{N}-\!\!\!\!\bigcirc\!\!\!\!-\text{SeCN}}$$

(14) (Ref. 41)

$$\text{C}_9\text{H}_{19} \qquad \text{SeAr} \qquad \xrightarrow{\text{H}_2\text{O}_2} \qquad \text{C}_9\text{H}_{19} \qquad \text{HO}$$

95%   77%

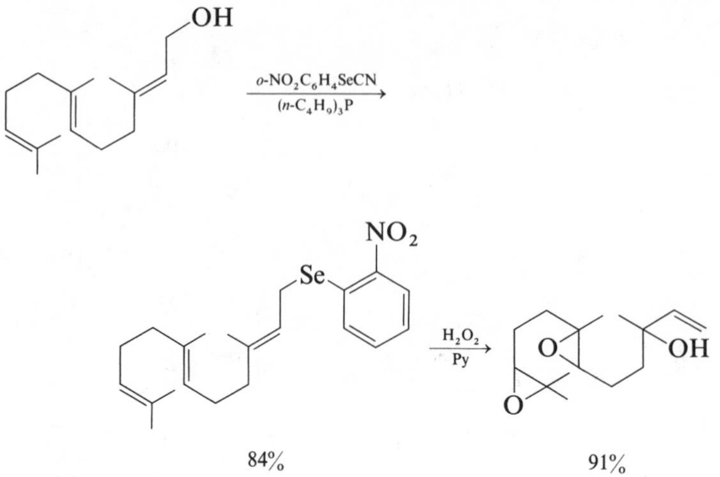

$$\text{(15) (Ref. 42)}$$

84%   91%

A procedure has been reported that permits replacement of allylic amino functions with selenium [Eq. (16)].

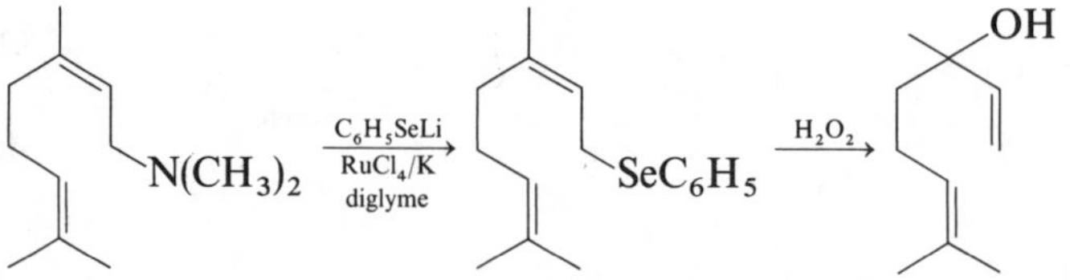

$$\text{(16) (Ref. 43)}$$

Dienes can also serve as precursors to allyl selenides, either via diene mono-epoxides[33,44] [Eq. (17)] or by selenenyl halide addition to the diene (Section 2.4).[45] The epoxide procedure leads to a specific 1,4-*trans*-bis-hydroxylation of the diene.

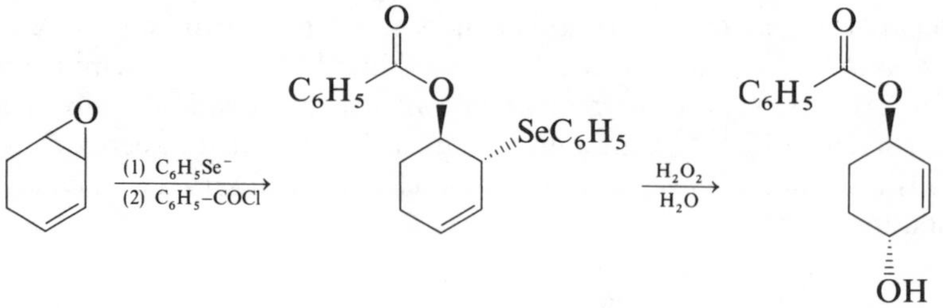

$$\text{(17) (Refs. 33, 44)}$$

2.4. Addition of PhSeX to Dienes

The electrophilic addition of PhSeX to dienes can occur in a 1,2 or a 1,4 fashion. In either case, allylic selenides can be products. Equation (18) illustrates the utilization of a 1,4-addition product, whereas Eq. (19) illustrates the preparation of allyl selenides using a selenolactonization process.[46,47]

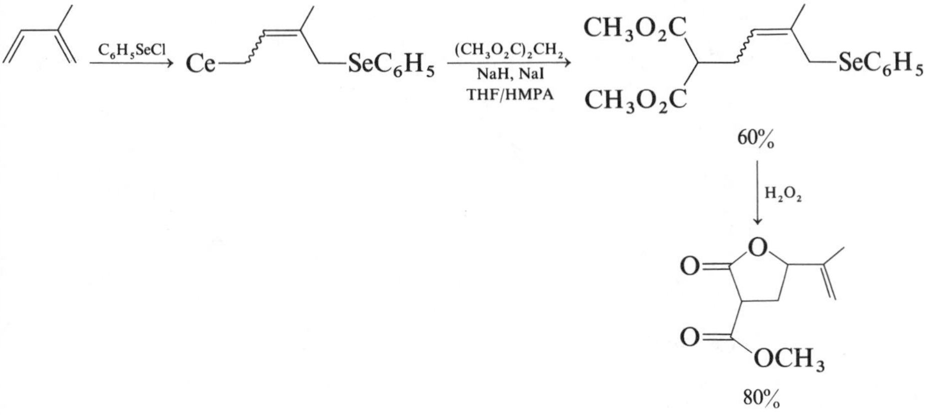

(18) (Ref. 45)

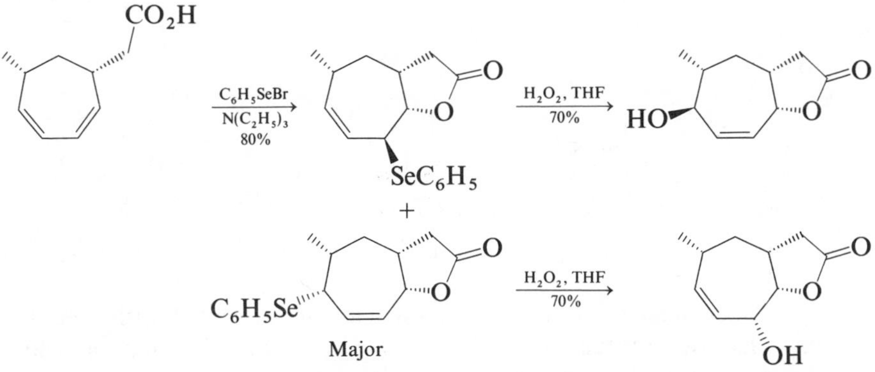

(19) (Ref. 46)

2.5. Electrophilic Selenenylation and Seleninylation of Allyl Anion Equivalents

Selenenyl halides, diselenides, and elemental selenium[48] are useful reagents for electrophilic introduction of selenium functions. In situations where allyl anion equivalents are available, this reaction can provide a facile synthesis of allyl selenides. After oxidation rearrangement the overall process corresponds to an allylic oxidation.

Dienolates and dienol silyl ethers are generally selenenylated at the γ position [Eqs. (20)–(22)] although it is not known whether these are the kinetic products.[51,52] The selenoxides derived from conjugated allyl selenides like these show some tendency to undergo *syn* elimination in competition with [2,3]sigma-tropic rearrangement.[25,27]

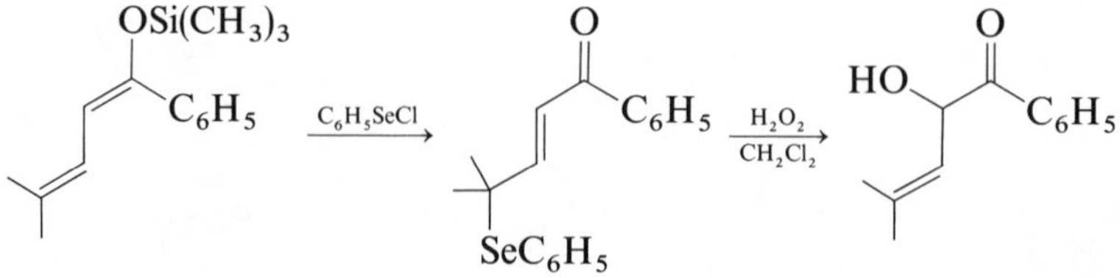

(20) (Ref. 49)

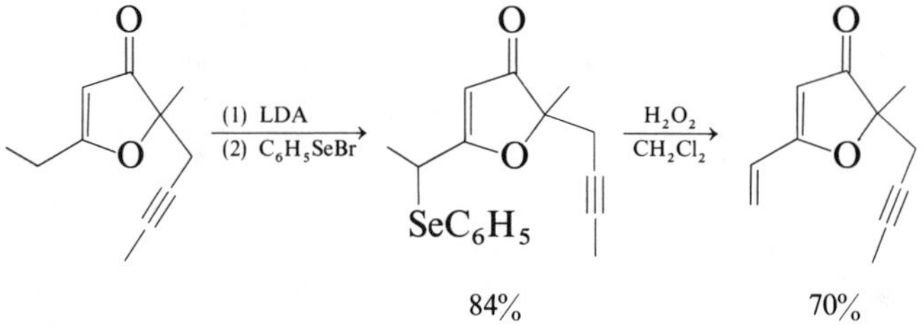

84%   70%

(21) (Ref. 26)

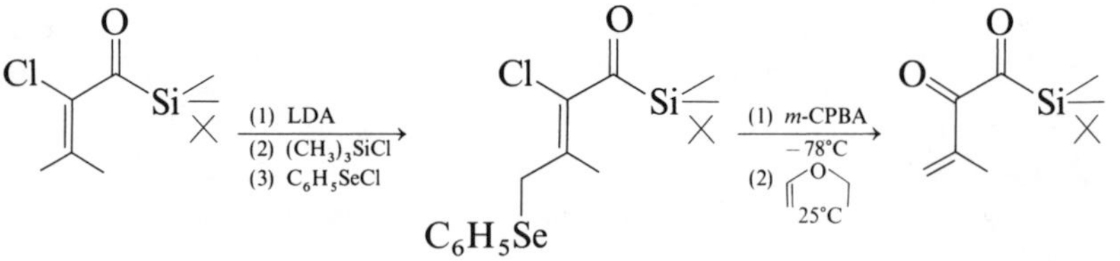

48%   (22) (Ref. 50)

Allyl silanes are converted to allyl selenides on treatment with benzene-selenenyl chloride and stannous chloride.[11–13,53,54] The acidic conditions cause equilibration of the phenylseleno group to the thermodynamically more stable location [Eq. (23)].

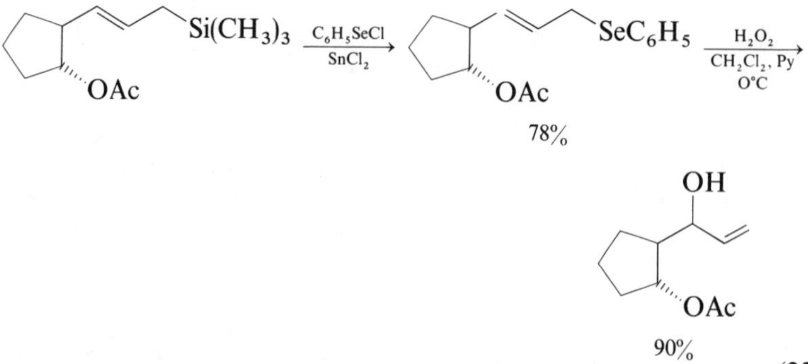

78%

90%

(23) (Ref. 12)

Treatment of allyl silanes with benzeneseleninic anhydride directly forms the allyl selenoxide which then rearranges to allyl alcohol [Eq. (24)].

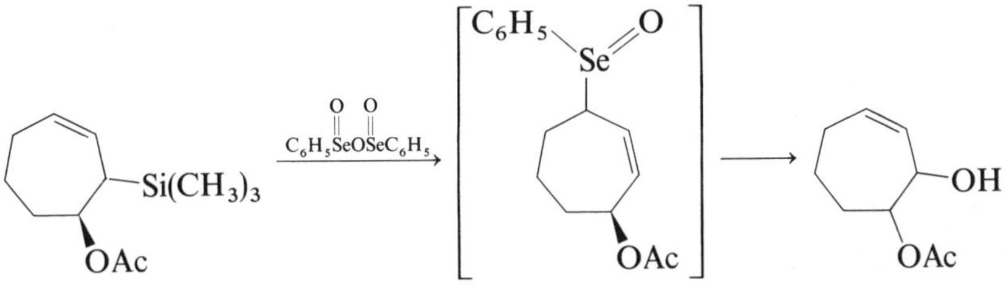

(24) (Ref. 55)

2.6. Alkylation of Allyl Selenide Anions

Allyl selenides are deprotonated on treatment with lithium diisopropylamide or other similar bases.[23] These lithium reagents show excellent nucleophilic properties, and react smoothly with primary and secondary halides, epoxides, and halosilanes to give (in most cases) good yields of substituted allyl selenides with favorable α/γ ratios. Carbonyl compounds tend to react at the γ position, although this can sometimes be controlled by use of metal cations other than lithium (e.g., Zn,[19] B,[56] Al,[57] Sn[10]). Small amounts of γ-substitution products are not a serious problem since oxidation converts them to stable vinyl selenoxides that are easily removed during work-up. Depending on the substitution pattern of the α-substituted allyl selenide and the electrophile chosen, oxidation leads to allyl alcohols, enones, or enals [Eqs. (25)-(28)].

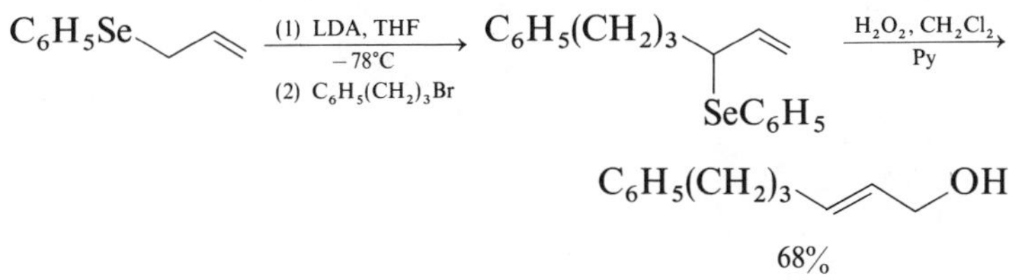

68%

(25) (Ref. 23)

74%

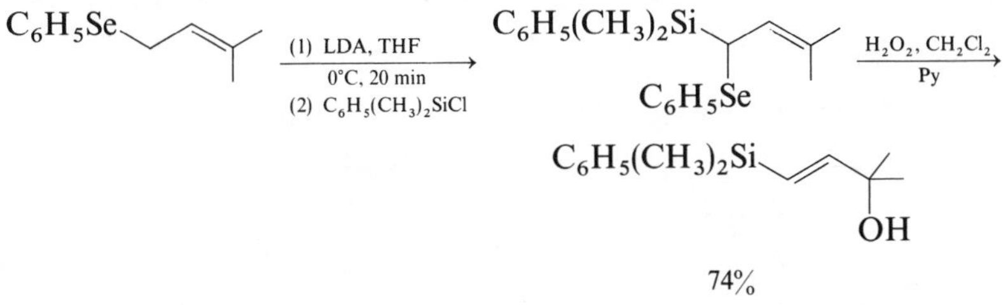

(26) (Ref. 23)

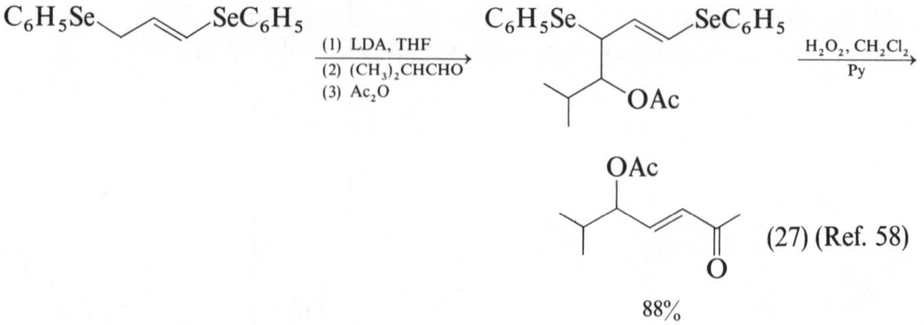

(27) (Ref. 58)

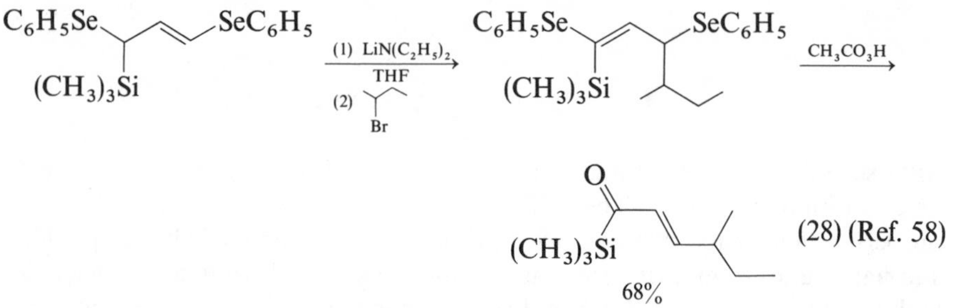

88%

(28) (Ref. 58)

68%

In some situations conversion of lithium to another metal serves to modify the regioselectivity or reactivity.[56,57] In Eq. (29) the normal tendency for reactions with aldehydes and ketones to occur γ to selenium is reversed if lithium

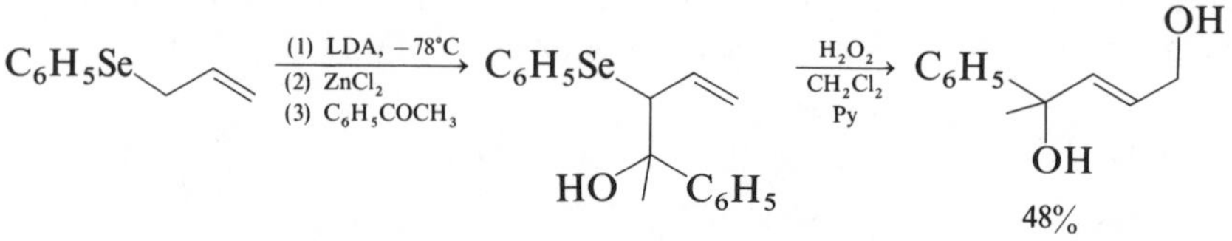

48%

(29) (Ref. 19)

51%

84%

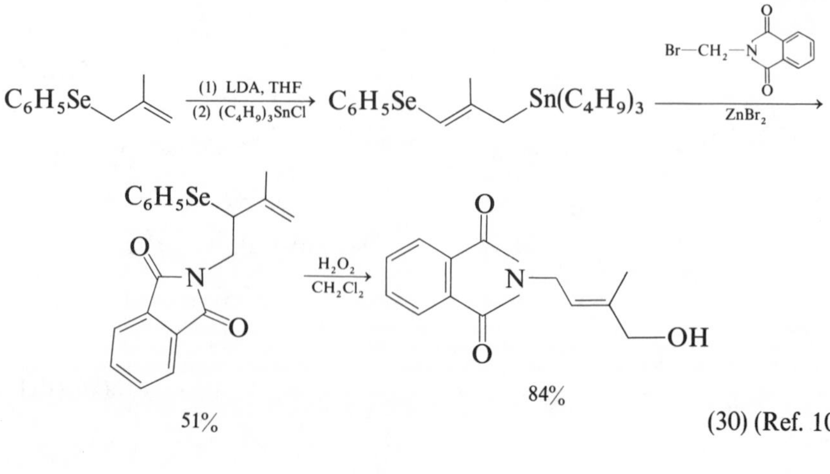

(30) (Ref. 10)

is replaced by zinc. A similar effect has been reported for aluminum.[57] Stan-nylated allyl selenides also have useful reactivity, which complements that of the lithium reagent. Stannylation occurs quantitatively γ to selenium, so that electrophilic substitution is again directed α [Eqs. (30) and (31)]. The latter case also illustrates how the [1,3]shift of allyl selenides[9,14] can be used to control the regiochemistry of the allyl alcohol formed after oxidation.

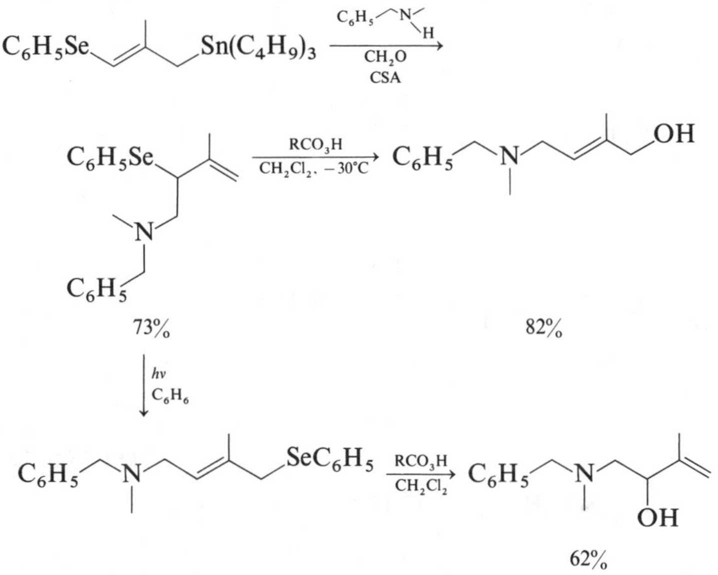

(31) (Ref. 10)

73%   82%

62%

2.7. Conversion of α-Seleno Carbonyl Compounds to Allyl Selenides

α-Phenylseleno aldehydes, which can be prepared by the selenation of alde-hydes or aldehyde derivatives[1,59-61] or by the formylation of α-lithiosele-nides,[62] are converted to selenides by carbonyl olefination procedures such as the Wittig[14,22,64] [Eqs. (32) and (33)], Wadsworth-Emmons,[51,64,65] Pe-terson,[13,53,54] [Eq. (34)], or β-hydroxy selenide reductive elimination[66,67] [Eq.

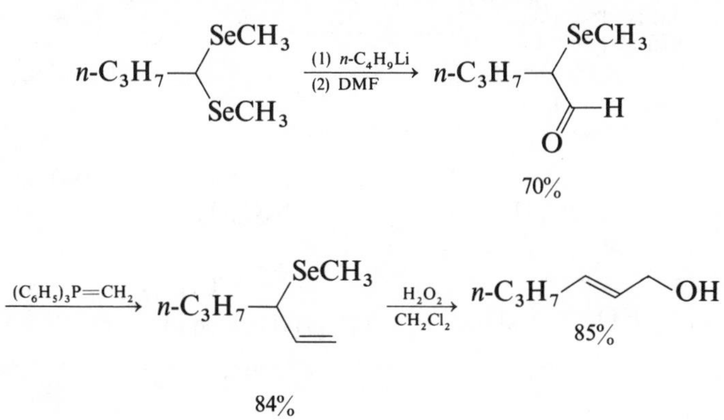

70%

(32) (Ref. 14)

84%   85%

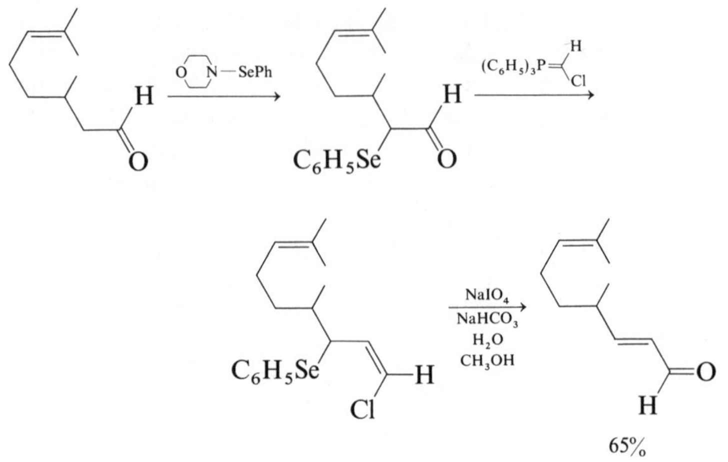

(33) (Ref. 64)

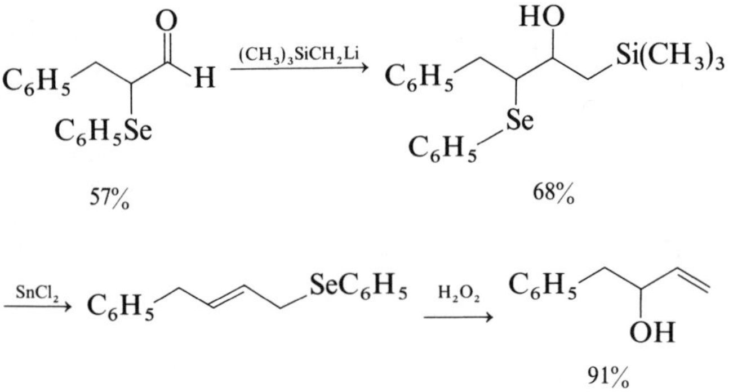

65%

57%   68%

(34) (Ref. 13)

91%

35, see Eq. 8]. The Wittig reaction in particular is well suited for the preparation of allyl selenides in a regioselective fashion. The less stable of two isomers related by a [1,3] PhSe shift can be prepared provided that care is taken to avoid thermal or catalyzed isomerization of the phenylseleno group.[14]

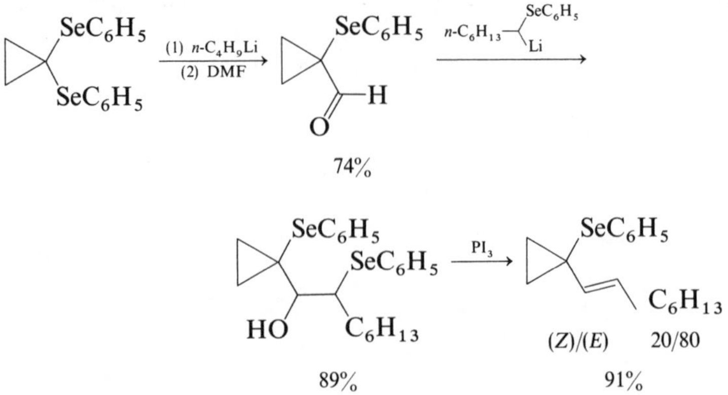

74%

(35) (Ref. 67)

89%   (Z)/(E)   20/80

91%

Under the conditions of the acid catalyzed Peterson elimination[13,54] equilibration to the more stable isomer occurs.

2.8. Allylic Selenoxides by Deconjugation of Vinyl Selenoxides

Two examples in which vinyl selenoxides were converted to allyl alcohols under basic conditions have been reported. In each case, the presence of polar and conjugating groups facilitates base catalyzed prototropic isomerization [Eqs. (36) and (37)]. Since there are many procedures for the preparation of vinyl selenides.[69] this reaction deserves further study.

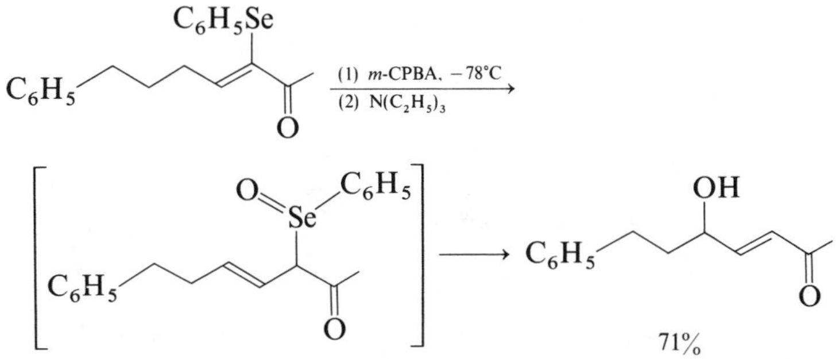

(36) (Ref. 8)

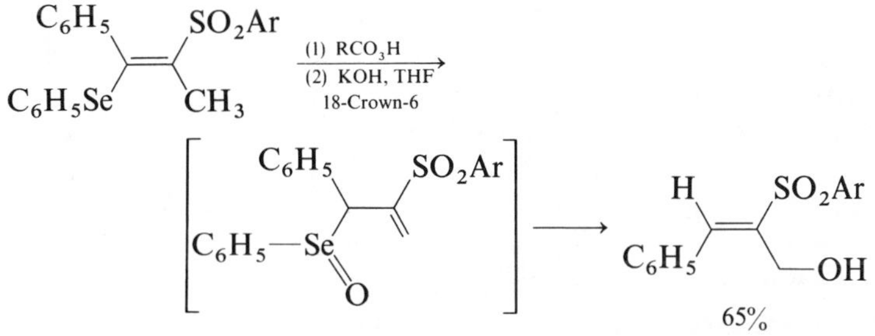

(37) (Ref. 68)

2.9. Miscellaneous Preparation of Allyl Selenides

The procedures for the preparation of ally selenides and selenoxides discussed in Sections 2.3–2.8 will undoubtedly be augmented by many others as organoselenium chemistry matures. Two more esoteric approaches are outlined in Eqs. (38) and (39). In the first a radical chain addition to allene provides a functionalized allyl selenide, whereas in the second, attack by an electrochemically

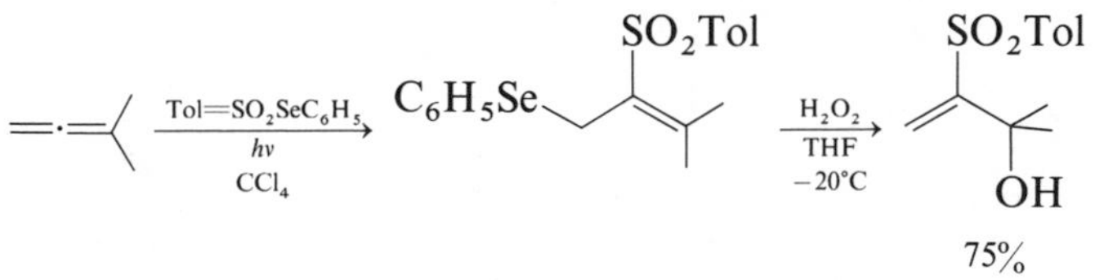

(38) (Ref. 70)

generated electrophilic selenium species induces fragmentation of β-pinene. Further electrochemical oxidation produces the allyl selenoxide, which rearranges and releases the selenium fragment for reuse.

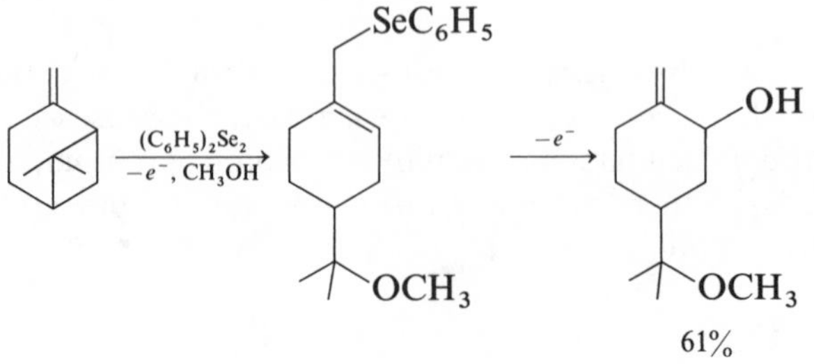

$$(39) \text{ (Ref. 71)}$$

3. ALLYL SELENINIC ACIDS—THE SELENIUM DIOXIDE OXIDATION

A delightful example of how the concepts of orbital symmetry control served to help unravel the mechanism of a reaction that had long been in dispute is provided by the examination of the selenium dioxide oxidation of olefins by Sharpless and co-workers.[5,33,72,73] Two of the most perplexing features of the oxidation were that the reaction usually occurred without allylic rearrangement of the double bond and that only the methyl group trans to the alkyl group was attacked when a 1-alkyl-2,2-dimethylethylene was oxidized.[74] These effects are nicely explained by a mechanism involving two consecutive pericyclic reactions: An initial ene reaction by selenium dioxide, or some other related electrophilic selenium species, to give an allyl seleninic acid [Eq. (40)], which undergoes [2,3]sigmatropic rearrangement and hydrolysis to the allyl alcohol. Detailed studies of these processes have shown that there is a 7/3 preference for attack of the selenium dioxide on the (E) versus (Z) methyl groups[75] and that stepwise polar processes can intervene to a greater or lesser extent de-

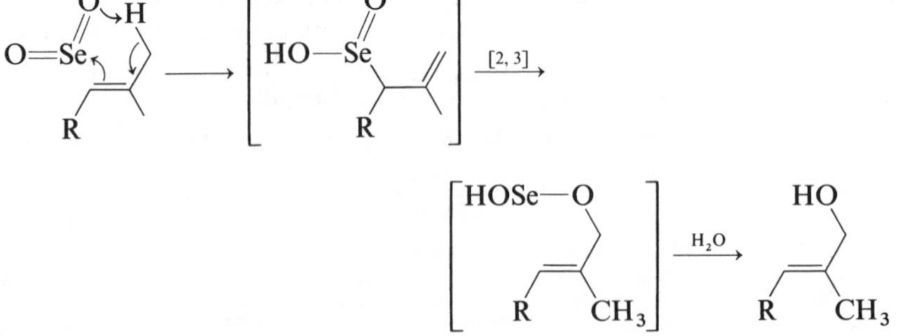

$$(40)$$

pending on conditions and substrates in both the ene reaction step[72] and the [2,3] sigmatropic rearrangement step.[76] However, the basic features of the mechanism have survived careful scrutiny.

No allyl seleninic acids have been isolated or detected, and this is a weak feature of the mechanism. However, intermediate seleninic acids have been trapped in the form of a seleninolactone during selenium dioxide oxidation [Eq. (41)]. Nuclear magnetic resonance observation of two allyl seleninic acid esters prepared by the reaction of allyltrimethyltin with selenium dioxide in methanol solution has been achieved [Eq. (42)]. These slowly decompose at room temperature to a mixture of allyl alcohol and allyl selenide. Thus the isolation of allyl seleninic acids themselves should pose no insurmountable obstacles, although it is not likely to be possible during selenium dioxide oxidation of olefins since reaction conditions are generally too vigorous to allow buildup of the intermediate.

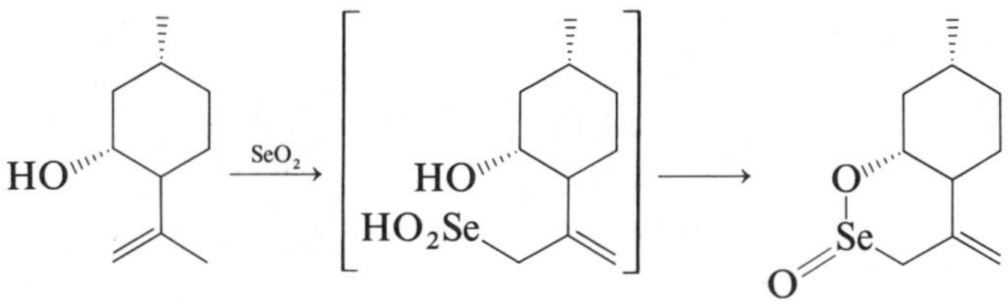

(41) (Ref. 73)

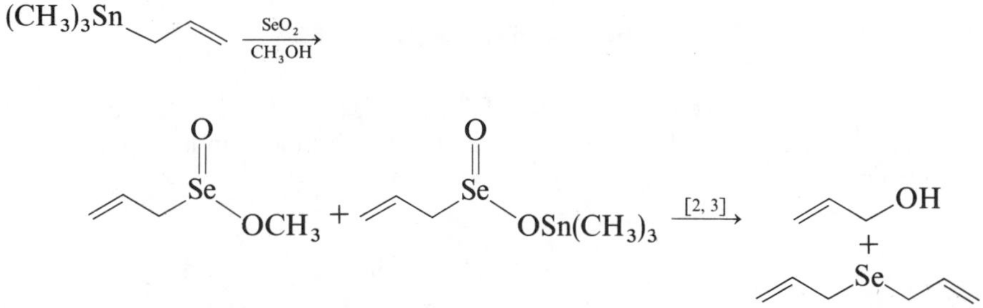

(42) (Ref. 77)

The selenium dioxide oxidation has been improved and extended in two ways:

1. The use of catalytic amounts of selenium dioxide with *t*-butylhydroperoxide as oxidant results in suppression of side reactions and improvement in yields.[78]

2. Imide analogs of selenium dioxide can be used to perform allylic aminations [Eq. (43)], and diene bisaminations [Eq. (44)]. In each case a sulfonamide function is introduced by the [2,3]sigmatropic rearrangement of an intermediate seleninimidic amide.

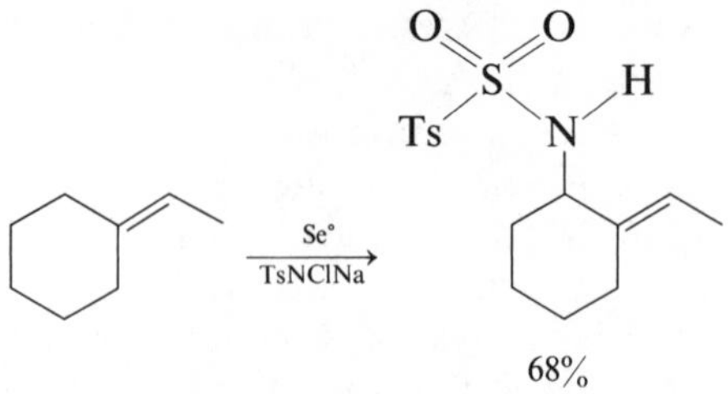

(43) (Ref. 79)

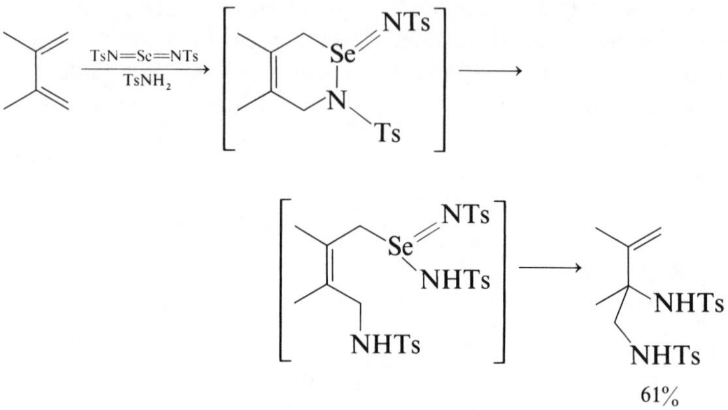

68%

(44) (Ref. 80)

61%

4. BUTADIENYL SELENOXIDES

The [2,3]sigmatropic rearrangement of 2-phenylselenino-1,3-butadienes [Eq. (45)] occurs with difficulty, but good yields are obtained under appropriate conditions.

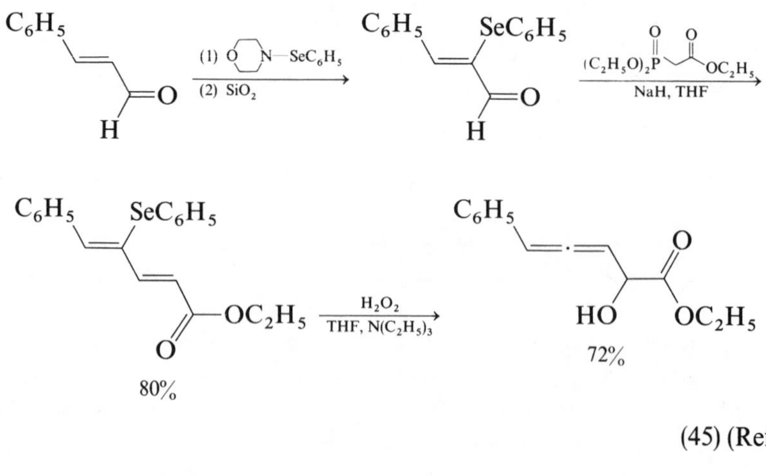

80%

72%

(45) (Ref. 65)

5. PROPARGYL SELENOXIDES

The rate data summarized in Table 1 illustrate that propargyl selenoxides rearrange at only slightly higher temperatures than do allyl selenoxides. Much less work has been done on them, although several useful transformations appear to be possible.

Aryl propargyl selenides (prepared by low temperature ozonization) are stable up to $-50-\!-30°C$, at which point they rearrange to the α-arylselenoacroleins [Eq. (46)]. The intermediacy of the expected allenol selenenate ester could be directly demonstrated only for the 2-nitrophenyl compound.[8] In the phenyl series, the allenol selenenate probably hydrolyzes to selenenic acid and allenol, which react rapidly with each other [Eq. (47)]. Thus, side reactions involving electrophilic selenium species, which occasionally plague all selenoxide to selenenate conversions,[15,20,30] are much more prevalent here because a reactive substrate is formed in the reaction mixture.

Phenyl propargyl selenide is converted to a dianion on treatment with two equivalents of lithium diisopropylamide. With suitable choices of electrophiles, two consecutive electrophilic groups can be introduced [Eq. (48)]. The new

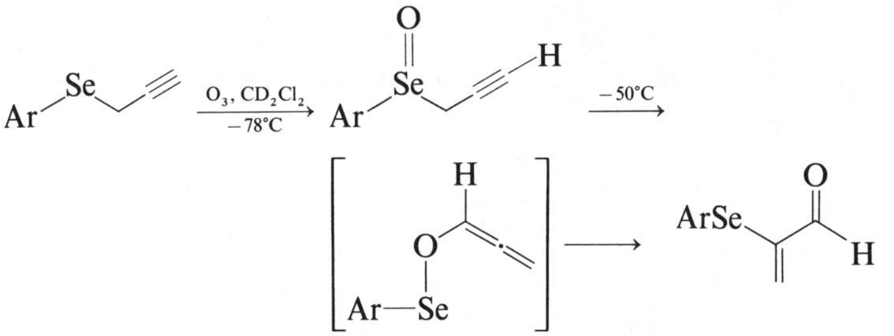

(46) (Ref. 8)

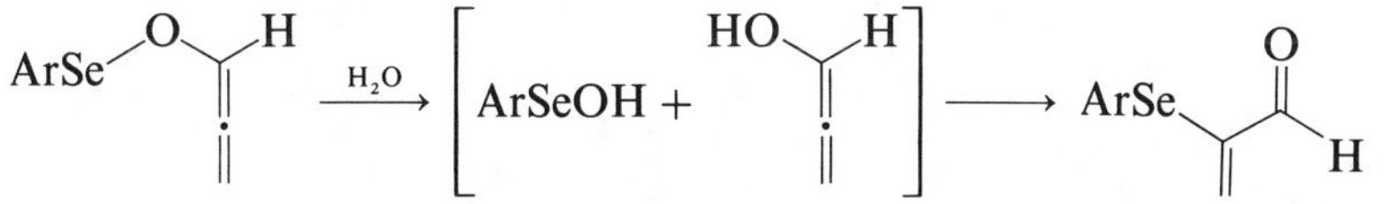

(47)

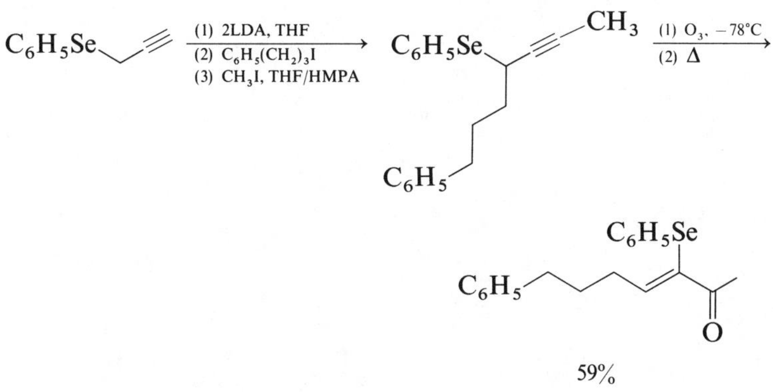

(48) (Ref. 8)

59%

propargyl selenide forms enone on oxidation. A more complex example is provided by Eq. (49), which summarizes the key steps in the synthesis of 7-hydroxymyoporone.[8,81,82]

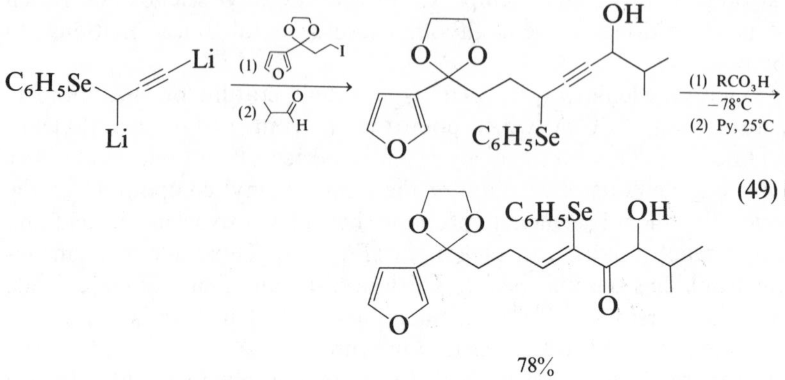

(49) (Ref. 80)

78%

6. ALLENYL SELENOXIDES

The [2,3]sigmatropic rearrangement of allenyl selenoxides provides propargyl alcohols. Examples reported in the literature are summarized in Eqs. (50)–(52).

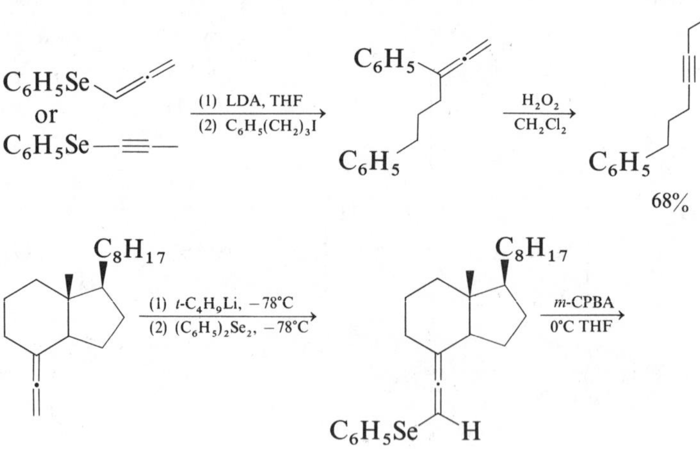

(50) (Ref. 18)

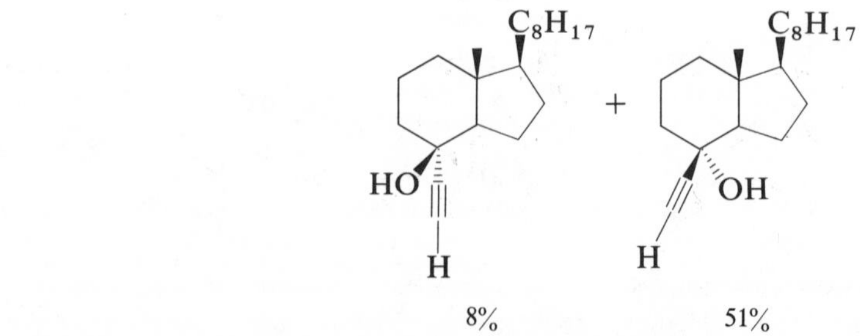

68%

(51) (Ref. 83)

8%   51%

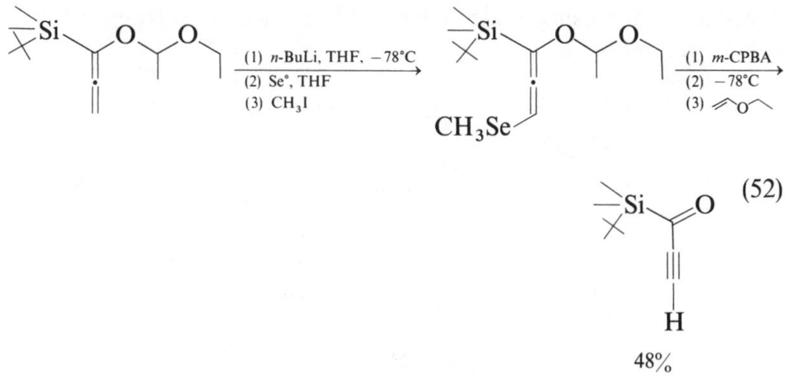

(52) (Ref. 81)

48%

Equation (50) illustrates how allenyl or propynyl phenyl selenide functions as a propargyl alcohol anion synthon, and Eqs. (51) and (52) show how selenenylation of an allenyl anion can be used to carry out an oxidation with allylic rearrangement. The ethyl vinyl ether in the last step of Eq. (52) was used to trap methaneselenenic acid.

7.   ALLYL SELENIMIDES

The nitrogen analogs of selenoxides, the selenimides, should in principle be capable of the same transformations as selenoxides. Their chemistry has, however, been little studied. Hori and Sharpless[84] reported that 10-(phenylseleno)-β-pinene afforded the expected rearranged allylic sulfonamido compound on treatment with anhydrous chloramine-T [Eq. (53)]. Improved experimental procedures have recently been developed[51,85–87] and applied to a series of allyl

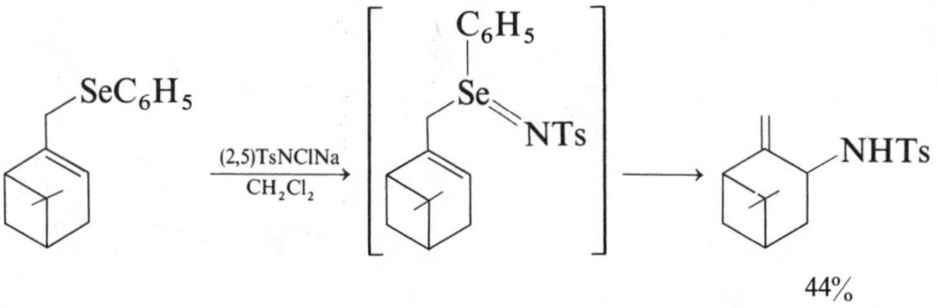

44%

(53) (Ref. 84)

selenides [Eqs. (54)–(56)]. The reaction shows all of the characteristics of a [2,3]sigmatropic rearrangement: suprafacial migration of the N—Se grouping, high selectivity in favor of products with a trans double bond, and complete allylic rearrangement when two isomeric selenides are subjected to the reaction conditions. A synthesis of α-amino acids (78–84% enantiomeric excess) starting with ethyl-(S)-lactate and using a selenimide [2,3]sigmatropic rearrangement as a key step has been reported.[87]

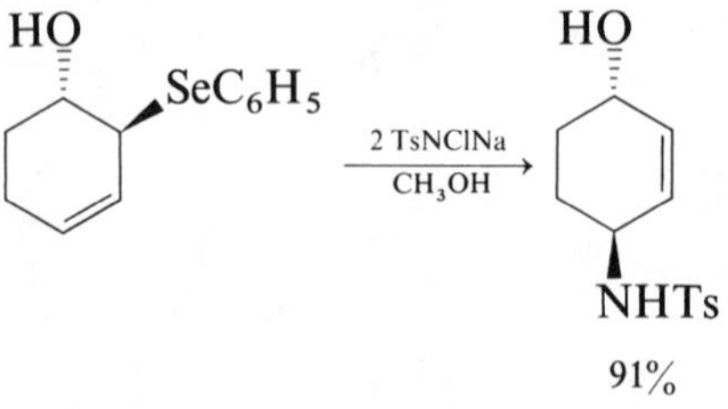

$$(54) \text{ (Ref. 85)}$$

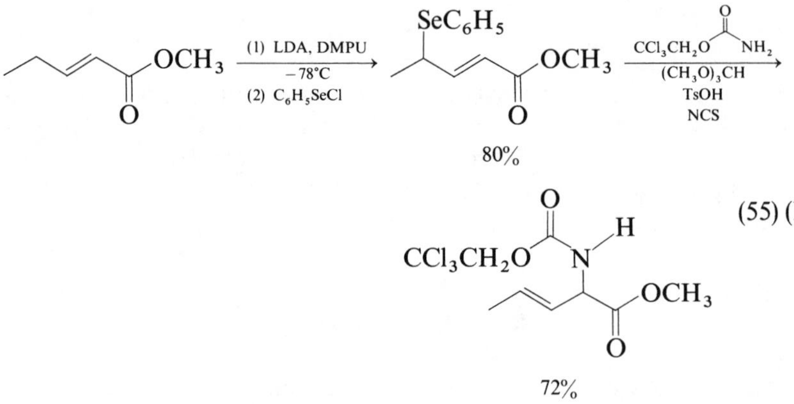

91%

80%

$$(55) \text{ (Ref. 51)}$$

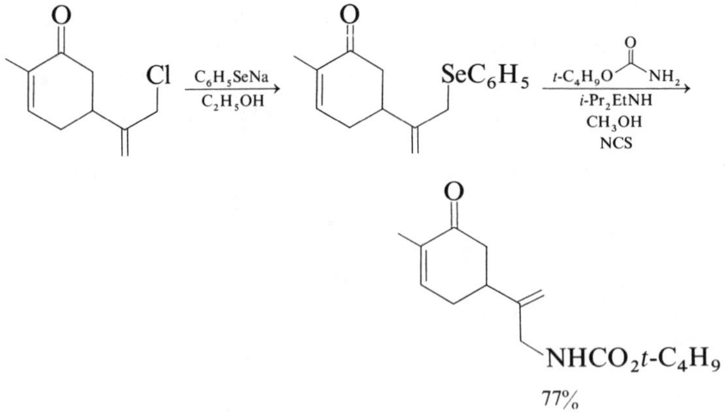

72%

$$(56) \text{ (Ref. 86)}$$

77%

8. ALLYL SELENONIUM YLIDES

Most of the common methods for the preparation of sulfonium and phosphonium ylides have also been tested with selenium compounds. Thus the reaction of diazo compounds[3,88] [Eq. (57)] and the Simmons–Smith reagent [Eq.

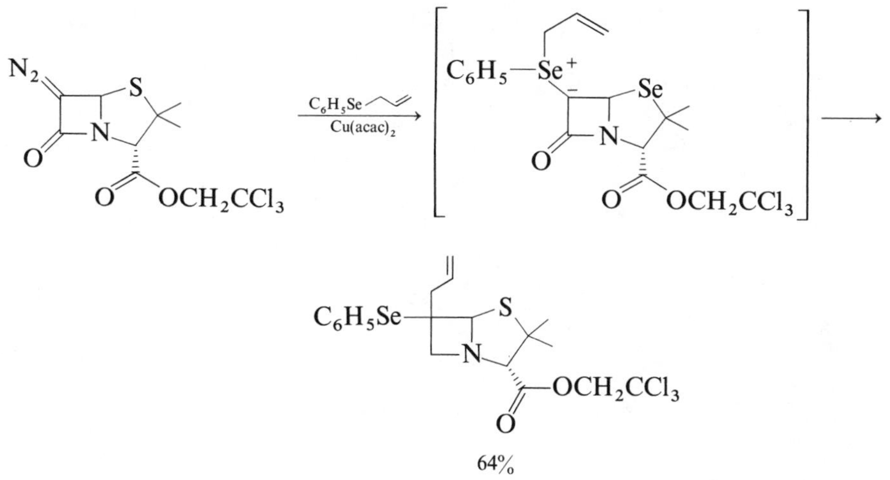

64%

(57) (Refs. 3, 88)

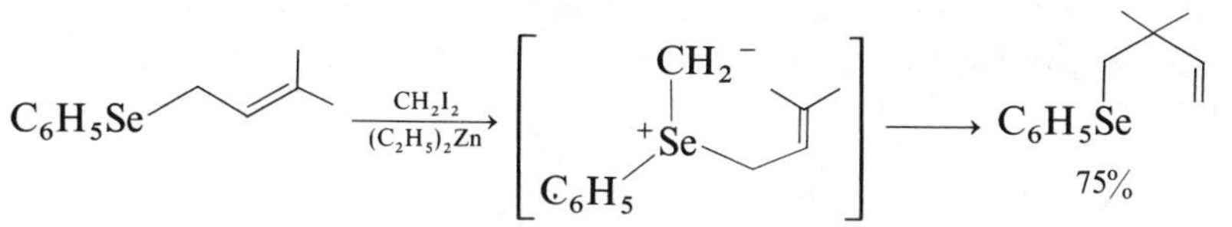

75%

(58) (Ref. 89)

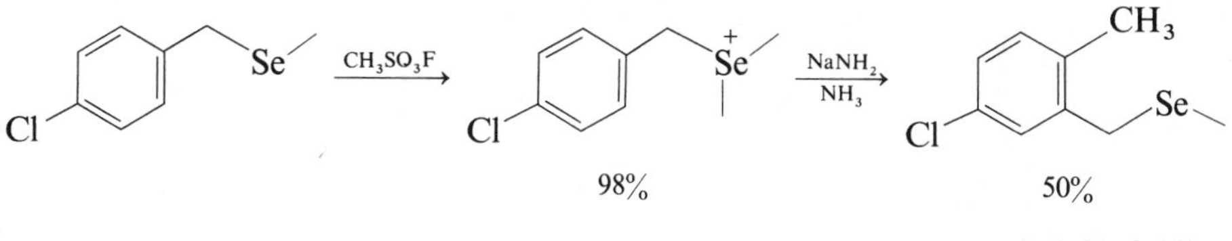

98%   50%

(59) (Ref. 90)

85%

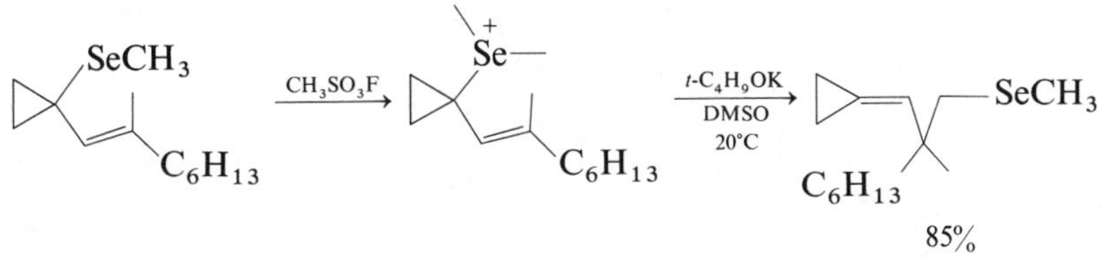

(60) (Ref. 22)

(58)] with selenides gives products expected from the formation of unstable selenium ylides. The deprotonation of allyl[22] or benzyl[90] selenonium salts also leads to characteristic rearrangement products. The allylation of some types of selenium substituted enolates forms selenium ylides [Eq. (61)].

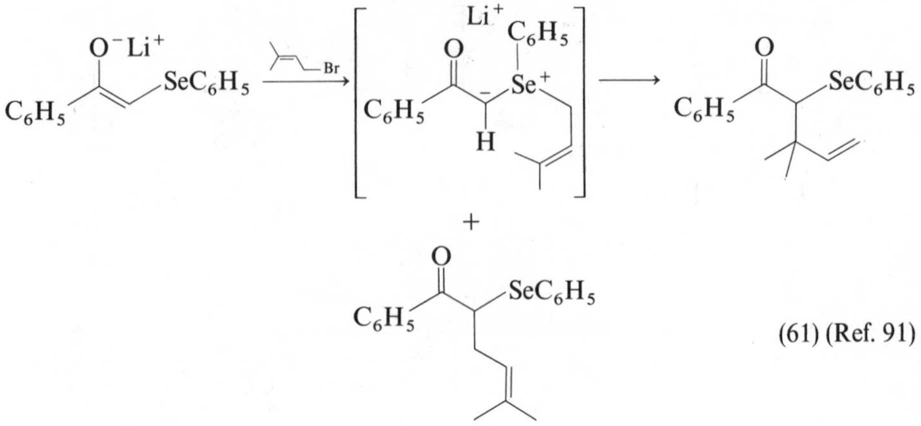

(61) (Ref. 91)

9. MISCELLANEOUS [2,3]SIGMATROPIC REARRANGEMENTS

A number of hydroxylation reactions involving benzeneseleninic anhydride or seleninic acid have been postulated to proceed by [2,3]sigmatropic rearrangement of intermediate seleninate esters [e.g., Eqs. (62) and (63)].[92–94]

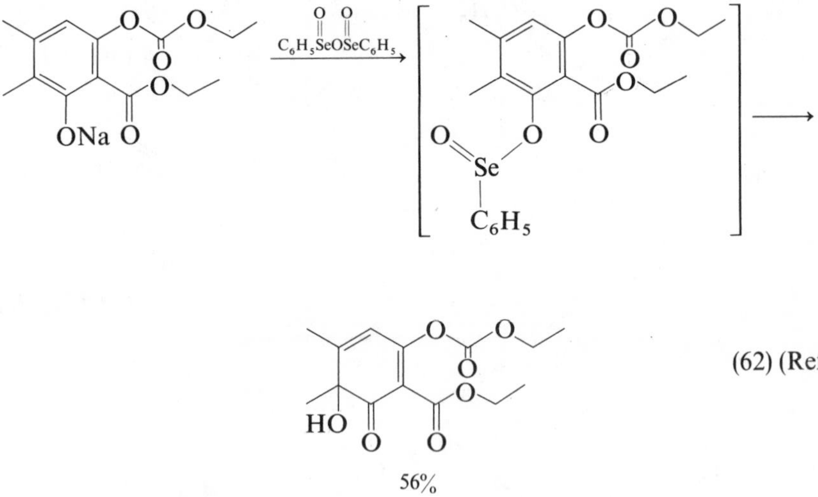

(62) (Ref. 92)

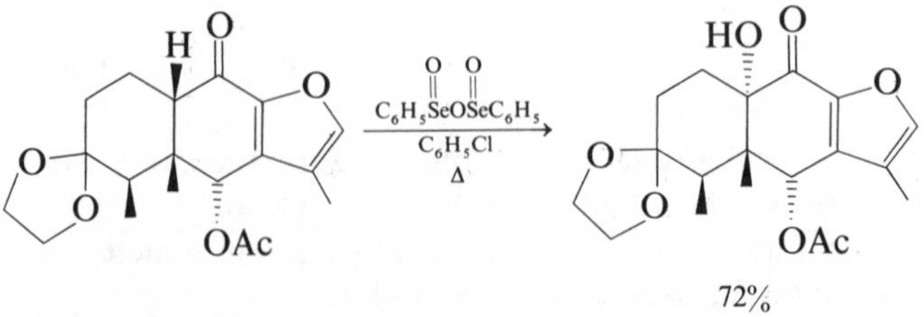

56%

(63) (Ref. 93)

72%

The examples cited in the various sections of this chapter serve to illustrate the favorable properties of selenoxide and selenimide [2,3]sigmatropic rearrangements for the introduction of diverse functionality into organic molecules. The advantage of extraordinarily mild reaction conditions for both oxidation and rearrangement steps, as well the general absence of serious side reactions assures further development in the applications of this reaction.

REFERENCES

1. H. J. Reich, J. M. Renga, and I. L. Reich, *J. Am. Chem. Soc.*, **97**, 5434 (1975).

2. H. J. Reich and S. K. Shah, *J. Org. Chem.*, **42**, 1773 (1977).

3. K. B. Sharpless, K. M. Gordon, R. F. Lauer, D. W. Patrick, S. P. Singer, and M. W. Young, *Chem. Scr.* **8A**, 9 (1975).

4. The Pummerer Reaction will not be discussed in detail here. Some leading references are N. Ikota and B. Ganem, *J. Org. Chem.*, **43**, 1607 (1978); J. A. Marshall and P. D. Royce, Jr., *J. Org. Chem.*, **47**, 693 (1982); I. D. Entwhistle, R.A.W. Johnstone, and J. H. Varley, *Chem. Commun.*, 61 (1976); F. C. Brown, *Tetrahedron*, **24**, 1845 (1978); Y. Tsuda, S. Hosoi, A. Nakai, T. Ohshima, Y. Sakai, and F. Kiuchi, *Chem. Commun.*, 1216 (1984); G. Galambos and V. Simonidesz, *Tetrahedron Lett.*, **23**, 4371 (1982); S. L. Schreiber and C. Santini, *J. Am. Chem. Soc.*, **106**, 4038 (1984); Y. Okamoto, R. Homsany, and T. Yano, *Tetrahedron Lett.*, 2529 (1972); and Y. Okamoto, K. L. Chellappa, and R. Homsany, *J. Org. Chem.*, **38**, 3172 (1973).

5. K. B. Sharpless and R. F. Lauer, *J. Am. Chem. Soc.*, **94**, 7154 (1972).

6. R. Tang and K. Mislow, *J. Am. Chem. Soc.*, **92**, 2100 (1970).

7. D. A. Evans and G. C. Andrews, *Acct. Chem. Res.*, **7**, 147 (1974) and references therein.

8. H. J. Reich, S. K. Shah, P. M. Gold, and R. E. Olson, *J. Am. Chem. Soc.*, **103**, 3112 (1981).

9. K. B. Sharpless and R. F. Lauer, *J. Org. Chem.*, **37**, 3973 (1972).

10. H. J. Reich, M. C. Schroeder, and I. L. Reich, *Isr. J. Chem.*, **24**, 157 (1984).

11. H. Nishiyama, K. Itagaki, K. Sakuta, and K. Itoh, *Tetrahedron Lett.*, **22**, 5285 (1981).

12. H. Nishiyama, S. Narimatsu, and K. Itoh, *Tetrahedron Lett.*, **22**, 5285 (1981).

13. H. Nishiyama, H. Yokoyama, S. Narimatsu, and K. Itoh, *Tetrahedron Lett.*, **23**, 1267 (1982).

14. T. Di Giamberardino, S. Halazy, W. Dumont, and A. Krief, *Tetrahedron Lett.*, **24**, 3413 (1983).

15. H. J. Reich, S. Wollowitz, J. E. Trend, F. Chow, and D. F. Wendelborn, *J. Org. Chem.*, **43**, 1697 (1978).

16. S. Wollowitz, "Reactivity and Rearrangement of Selenoxides and Sulfoxides," unpublished doctoral dissertation, University of Wisconsin, Madison, 1980.

17. H. J. Reich, K. E. Yelm, and S. Wollowitz, *J. Am. Chem. Soc.*, **105**, 2503 (1983).

18. H. J. Reich, S. K. Shah, P. M. Gold, and R. E. Olson, *J. Am. Chem. Soc.*, **103**, 3112 (1981).

19. H. J. Reich and S. Wollowitz, *J. Am. Chem. Soc.*, **104**, 7051 (1982).

20. H. J. Reich, I. L. Reich, and S. Wollowitz, *J. Am. Chem. Soc.*, **100**, 5981 (1978).

21. I. L. Reich and H. J. Reich, *J. Org. Chem.*, **46**, 3721 (1981).

22. S. Halazy and A. Krief, *Tetrahedron Lett.*, **22**, 2135 (1981).

23. H. J. Reich, *J. Org. Chem.*, **40**, 2570 (1975).

24. W. G. Salmond, M. A. Barta, A. M. Cain, and M. C. Sobala, *Tetrahedron Lett.*, 1683 (1977).

25. T. Wakamatsu, K. Akasaka, and Y. Ban, *J. Org. Chem.*, **44**, 2008 (1979); and G. Quinkert, G. Dürner, E. Kleiner, F. Adam, E Haput, and D. Leibfritz, *Chem. Ber.*, **113**, 2227 (1980).

26. A. B. Smith, III, P. A. Levenberg, P. J. Jerris, R. M. Scarborough, Jr., and P. M. Wovkulich, *J. Am. Chem. Soc.*, **103**, 1501 (1981).

27. H. E. Zimmerman and R. R. Diehl, *J. Am. Chem. Soc.*, **101**, 1841 (1979).

28. P. A. Grieco, Y. Yokoyama, S. Gilman and M. Nishizawa, *J. Org. Chem.*, **42**, 2034 (1977).

29. H. J. Reich, F. Chow, and S. L. Peake, *Synthesis*, 299 (1978).

30. F. A. Davis, O. D. Stringer, and J. M. Billmers, *Tetrahedron Lett.*, **24**, 1213 (1983).

31. F. A. Davis, O. D. Stringer, and J. P. McCauley, Jr., *Tetrahedron*, **41**, 4747 (1985).

32. B. B. Branchaud and C. T. Walsh, *J. Am. Chem. Soc.*, **107**, 2153 (1985).

33. K. B. Sharpless and R. F. Lauer, *J. Am. Chem. Soc.*, **95**, 2697 (1973).

34. W. C. Still and D. Mobilio, *J. Org. Chem.* **48**, 4785 (1983).

35. M. Clarembeau and A. Krief, *Tetrahedron Lett.*, **25**, 3025 (1984).

36. R. H. Wollenberg, *Tetrahedron Lett.*, **21**, 3139 (1980).

37. K. Tanaka, M. Terauchi, and A. Kaji, *Bull. Chem. Soc., Jpn.*, **55**, 3935 (1982); and S.-Y. Chen and M. M. Joullié, *Tetrahedron Lett.*, **24**, 5027 (1983).

38. S. I. Pennanen, *Synth. Commun.*, **10**, 373 (1980).

39. T. Mukaiyama, *Angew. Chem. Int. Ed.*, **15**, 94 (1976).

40. P. A. Grieco, S. Gilman, and M. Nishizawa, *J. Org. Chem.*, **41**, 1485 (1976); T. G. Back, and D. J. McPhee, *J. Org. Chem.*, **49**, 3842 (1984).

41. D. L. J. Clive, G. Chittattu, N. J. Curtis, and S. M. Menchen, *Chem. Commun.*, 770 (1978).

42. T. Kametani, H. Nemoto, and K. Fukumoto, *Heterocycles*, **6**, 1365 (1977); and *Bioorg. Chem.*, **7**, 215 (1978).

43. S.-I. Murahashi and T. Yano, *J. Am. Chem. Soc.*, **102**, 2456 (1980).

44. J. P. Marino and J. C. Jaén, *Tetrahedron Lett.*, **24**, 441 (1983).

45. R. S. Brown, S. C. Eyley, and P. J. Parsons, *Chem. Commun.*, 438 (1984).

46. A. J. Pearson, S. L. Kole, and T. Ray, *J. Am. Chem. Soc.*, **106**, 6060 (1984).

47. A. J. Pearson, T. Ray, I. C. Richards, J. Clardy, and L. Silveira, *Tetrahedron Lett.*, **24**, 5827 (1983).

48. W. Kulik, H. D. Verkruijsse, R. L. P. de Jon, H. Hommes, and L. Brandsma, *Tetrahedron Lett.*, **24**, 2203 (1983).

49. H. J. Reich and C. Montes, unpublished results.

50. M. J. Kelley, "Synthesis and Reactions of α,β-Unsaturated Silyl Ketones," unpublished doctoral dissertation, University of Wisconsin, Madison, 1983.

51. J. N. Fitzner, D. V. Pratt, and P. B. Hopkins, *Tetrahedron Lett.*, **26**, 1959 (1985).

52. J. A. Oakleaf, M. T. Thomas, A. Wu, and V. Snieckus, *Tetrahedron Lett.*, 1645 (1978).

53. H. Nishiyama, T. Kitajima, A. Yamamoto, and K. Itoh, *Chem. Commun.*, 1232 (1982).

54. H. Nishiyama, K. Itagaki, N. Osaka, and K. Itoh, *Tetrahedron Lett.*, **23**, 4103 (1982).

55. P. Magnus, F. Cooke, and T. Sarkar, *Organometallics*, **1**, 562 (1982).

56. Y. Yamamoto, Y. Saito, and K. Maruyama, *J. Org. Chem.*, **48**, 5408 (1983).

57. Y. Yamamoto, Y. Saito, and K. Maruyama, *Tetrahedron Lett.*, 4597 (1982).

58. H. J. Reich, M. C. Clark, and W. W. Willis, Jr., *J. Org. Chem.*, **47**, 1618 (1982).

59. K. B. Sharpless, R. F. Lauer, and A. Y. Teranishi, *J. Am. Chem. Soc.*, **95**, 6137 (1973).

60. M. Jefson and J. Meinwald, *Tetrahedron Lett.*, **22**, 3561 (1981).

61. D. R. Williams and K. Nishitani, *Tetrahedron Lett.*, **21**, 4417 (1980).

62. S. Halazy and A. Krief, *Tetrahedron Lett.*, **22**, 1833 (1981).

63. S. Halazy and A. Krief, *Tetrahedron Lett.*, **22**, 2135 (1981).

64. P. Lerouge and C. Paulmier, *Tetrahedron Lett.*, **25**, 1983 (1984).

65. P. Lerouge and C. Paulmier, *Tetrahedron*, **25**, 1987 (1984).

66. H. J. Reich and F. Chow, *Chem. Commun.*, 790 (1975).

67. S. Halazy and A. Krief, *Tetrahedron Lett.*, **22**, 1833 (1981).

68. T. G. Back, S. Collins, V. Gokhale, and K.-W. Law, *J. Org. Chem.*, **48**, 4776 (1983).

69. J. V. Comasetto, *J. Organomet. Chem.*, **253**, 131 (1983).

70. Y.-H. Kang and J. L. Kice, *Tetrahedron Lett.*, **23**, 5373 (1982).

71. S. Torii, K. Uneyama, M. Ono, and T. Bannov, *J. Am. Chem. Soc.*, **103**, 4606 (1981).

72. M. A. Warpehoski, B. Chabaud, and K. B. Sharpless, *J. Org. Chem.*, **47**, 2897 (1982).

73. D. Arigoni, A. Vasella, K. B. Sharpless, and H. P. Jensen, *J. Am. Chem. Soc.*, **95**, 7917 (1973).

74. N. Rabjohn, "Selenium Dioxide Oxidation" in W. G. Dauben (ed.), *Organic Reactions*, Wiley, New York (1976).

75. W.-D. Woggon, F. Ruther, and H. Egli, *Chem. Commun.*, 706 (1980).

76. L. M. Stephenson and D. R. Speth, *J. Org. Chem.*, **44**, 4683 (1979).

77. C. A. Hoeger, "I. Formation, Observation, and Reactions of a Selenolseleninate. II. An Investigation into the Chemistry of β-Ketoseleninic and Allylseleninic Acids," unpublished doctoral dissertation, University of Wisconsin, Madison, 1983.

78. M. A. Umbreit and K. B. Sharpless, *J. Am. Chem. Soc.*, **99**, 5526 (1977).

79. K. B. Sharpless, T. Hori, L. K. Truesdale, and C. O. Dietrich, *J. Am. Chem. Soc.*, **98**, 269 (1976).

80. K. B. Sharpless and S. P. Singer, *J. Org. Chem.*, **41**, 2504 (1976).

81. H. J. Reich, M. J. Kelley, R. E. Olson, and R. C. Holtan, *Tetrahedron*, **39**, 949 (1983).

82. H. J. Reich, P. M. Gold, and F. Chow, *Tetrahedron Lett.*, 4433 (1979).

83. A. Haces, E. M. G. A. van Kruchten, and W. H. Okamura, *Tetrahedron Lett.*, **23**, 2707 (1982).

84. T. Hori and K. B. Sharpless, *J. Org. Chem.*, **44**, 4208 (1979).

85. J. E. Fankhauser, R. M. Peevey, and P. B. Hopkins, *Tetrahedron Lett.*, **25**, 15 (1984).

86. R. G. Shea, J. N. Fitzner, J. E. Fankhauser, and P. B. Hopkins, *J. Org. Chem.*, **49**, 3647 (1984).

87. J. N. Fitzner, R. G. Shea, J. E. Fankhauser, and P. B. Hopkins, *J. Org. Chem.*, **50**, 417 (1985).

88. P. J. Giddings, D. I. John, and E. J. Thomas, *Tetrahedron Lett.*, **21**, 395 (1980).

89. Z. Kosarych and T. Cohen, *Tetrahedron Lett.*, **23**, 3019 (1982).

90. P. G. Gassman, T. Miura, and A. Mossman, *Chem. Commun.*, 558 (1980).

91. H. J. Reich and M. L. Cohen, *J. Am. Chem. Soc.*, **101**, 1307 (1979).

92. D. H. R. Barton, P. D. Magnus, and M. N. Rosenfeld, *Chem. Commun.*, 301 (1975); D. H. R. Barton, S. V. Ley, P. D. Magnus, and M. N. Rosenfeld, *J. Chem. Soc. Perkin Trans. 1*, 567 (1977).

93. K. Yamakawa, T. Satoh, N. Ohba, R. Sakaguchi, S. Takita, and N. Tamura, *Tetrahedron*, **37**, 473 (1981).

94. T. Frejd and K. B. Sharpless, *Tetrahedron Lett.*, 2239 (1978).

9

Organic Metals Based on Selenium Compounds

FRED WUDL

Institute for Polymers and Organic Solids
Department of Physics,
University of California, at Santa Barbara, Santa Barbara,
California

CONTENTS

1. INTRODUCTION

Since the field of organic metals is now a rather broad one and is being reviewed (in sections) almost yearly, this chapter will deal only with organic metals based on organoselenium compounds. In terms of organization, a brief background on organic metals will be presented first, followed by preparation of organoselenium donors, preparation of organoselenium-conducting polymers, preparation of conducting salts, properties of organoselenium-based conductors, and future directions.

1.1. Organic Metals

The emphasis in this chapter is organoselenium *chemistry* as applied to organic metals; perforce, this section will be very brief on the *physical* aspects of organic metals. The readers are directed to extensive treatises on the subject.[1]

The most recent, balanced (*chemistry* and *physics*), review in this area of research is that of Bryce and Murphy.[1] Organic metals belong to two broad families of materials; crystalline charge transfer (CT) complexes and partially oxidized (or reduced) polymers.

Information on CT complexes has been gathered through the 1950s and thoroughly reviewed by Gutmann and Lyons.[2] Up to 1973, organic conductors could only be categorized as semiconductors; room temperature resistivities on the order of $10^{-1} -> 10^6$ Ω cm and energy of activation (ΔE_g) of 0.01–several electron volts. In the 1960s it was discovered that there was a structure property relationship; in order to observe high conductivity the planar molecules need to stack uniformly forming segregated positively and negatively charged columns of donors and acceptors, respectively. In that decade the research centered around principally the radical anion salts derived from tetracyanoquinodimethane (TCNQ, $\underline{1}$).

$\underline{1}$ $\underline{2}$

The first organic metal resulted from the combination of $\underline{1}$ and tetrathiofulvalene (TTF, $\underline{2}$). This CT salt exhibited a room temperature resistivity of $\simeq 2 \times 10^{-2}$ Ω cm and the latter decreased upon cooling to a value of $\simeq 10^{-4}$ Ω cm at 56 K; below that temperature, the resistance increased dramatically to the point where the material became an insulator below 40 K. By comparison, copper has a room temperature resistivity of $\simeq 1.7 \times 10^{-6}$ Ω cm.

In the same decade, another important aspect of organic CT complexes, *mixed valence* (partial charge transfer or partial charge per stack), was confirmed.[3] The significance of this phenomenon to the chemist interested in designing new donors or acceptors is that donors with very low ionization potentials or acceptors with very high electron affinities will in fact lead to *insulating* CT salts. Also in the 1970s, the electrical properties of partially oxidized or reduced poly(acetylene) were discovered.

In the 1980s, the most important discovery was superconductivity,[4] first in an organoselenium material and later in an organosulfur, bis-(ethylenedithio)-tetrathiofulvalene (BEDTTTF) salt.[5,6] As mentioned previously in the case of TTF·TCNQ, organic metals appeared to be doomed to undergo a metal-to-insulator transition at some temperature below room temperature. This observation was a result of the fact that organic metals based on TTF (and its derivatives) and also on TCNQ (and its derivatives) were anisotropic (maximum conductivity along the stacks and minimum in the other two directions).[1] In order to achieve superconductivity, the metal-to-insulator transition had to be suppressed. One way in which this could be accomplished was by decreasing the anisotropy (increasing the *dimensionality*) of the solid; that is, increasing the interstack intermolecular interactions. This could be done by increasing the size of the heteroatom, which may have been one of the driving forces for Engler and Patel[7] to synthesize the first organoselenium-based metal, TSF·TCNQ (TSF is the selenium analog of TTF). While this modification had the effect of reducing the transition temperature to 42 K, it did not remove it. Another potential method for the removal of the insulating state could be by application of pressure as suggested by Weger and executed later by Jerome et al.[4]

Over the years, a number of organoselenium donors for the formation of highly conducting organic compounds have been prepared, these are collected in Fig. 1. Of these, only TMTSF, TSF, and TSeT have received most of the attention. Another interesting thing is that of all these donors, only HMTSF and TMTSF form metals with TCNQ, but all (except HMTSF and TSF) form metals with closed shell anions.

When TMTSF forms salts of stoichiometry $(TMTSF)_2X$ (where X is a univalent complex anion, e.g., PF_6), most of the salts become superconducting at low temperature probably because the solid state structure is a very special one where the TMTSF molecules form *nonuniform* zig–zag stacks. As a result of this kind of stacking (which is probably due to the methyl substituents),

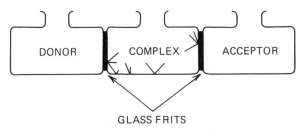

GLASS FRITS

FIGURE 1. Common donors in organic metals.

special selenium–selenium interactions are set up not only along the stacks but between stacks.[8] This same kind of special interchalcogen interaction is apparently responsible for the observation of superconductivity in the (BEDTTTF)$_2$X salts.[9] Further examination of the previous mode of packing reveals that the methylenes that are located in one of the mirror planes of HMTSF would be involved in serious nonbonded repulsive interactions with the anions; this would explain why HMTSF does not form the same type of salts as TMTSF when closed shell complex anions are involved.

In summary, the most salient features that donors should possess when organic metals are the goal, are (a) D$_{2h}$ symmetry and (b) as many chalcogens as possible in the periphery while avoiding potential steric repulsion between substituents on the donor and counterions.

1.2. Polymeric Conductors

The same principles that were developed for crystalline organic conductors apply (with minor modification) to polymers. That is, in order to observe high conductivity, the polymer must be partially oxidized or reduced. The most studied polymers are those with a conjugated backbone where removal or addition of an electron is facile. Thus polyenes, poly(phenylene), poly(phenylene-chalcogenides), and poly(heterocycles) are being thoroughly explored. So far, with only one exception, polymeric conductors containing selenium have had poorer characteristics (room temperature conductivity, etc.) than the other chalcogen polymers. Until very recently, conducting polymers were quite unattractive from a technical applications point of view because they were non-processable, but rapid strides are being made to remedy this problem.[10]

In principle, it should be possible to produce polymers that do not need to be partially ionized by external redox reagents to exhibit high conductivity, but to date these materials have not been made. The basic idea is to generate an organic material whose electronic structure is akin to that of graphite.

2. PREPARATION OF ORGANOSELENIUM DONORS

The preparation of these molecules will be discussed in order of increasing difficulty; the selenoacenes are the simplest to prepare and the fulvalenes the most difficult. In recent years other selenium-containing donors such as the bis-selenopyranilidenes[11] and different isomers of the tetraselenoanthracenes[12] were reported, but their syntheses will not be covered in detail here.

2.1. Tetraselenoacenes

Organoselenium donors based on acenes are usually prepared in one step from the acene halide and selenium or selenide.[13,14] Thus, tetraselenotetralene (TST) and tetraselenonaphthalene (TSN) are prepared from dichloro tetracene and tetrachloronaphthalene, respectively. The tetracene donor is prepared using selenium at high temperature and the naphthalene donor employing selenide in a polar aprotic solvent. The yields, particularly in the case of TSN, are poor. But since only one synthetic step is involved (provided one obtains 1,4,5,8-tetrachloronaphthalene as a gift, because its synthesis is rather tedious,[15] one can put up with small yields. Since, so far, no highly conducting salts of TSN have been reported, it may not be necessary to develop a better synthesis of this donor. There is a claim in the patent literature that TST can be prepared from tetracene and selenium in refluxing DMF (N,N-dimethylformamide).[16] That claim could not be substantiated by us in the laboratory.[17] One of the most efficient ways to isolate TST is by extraction of the crude reaction mixture with a formic acid solution of hydrogen peroxide; this selectively converts the TST into a soluble radical cation without oxidizing the rest of the reaction mixture. After filtration, the filtrate can be reduced with Zn to produce the neutral donor that can be further purified by crystallization from trichlorbenzene.

2.2. Tetraselenafulvalenes

The tetraselenafulvalenes are prepared according to procedures developed for the preparation of their sulfur analogs. Thus, the first preparation of the unsubstituted tetraselenafulvalene is that developed by Engler and Patel[7] (Scheme 1). The key step, the coupling of two dithiolium rings through the 2 positions, did not work in the selenium analog presumably because the diselenolium cation is not nearly as stable as the aromatic sulfur analog. As shown in Scheme I the coupling was accomplished by the use of a P^{III} reagent (usually a phosphite, although phosphines can also be used).

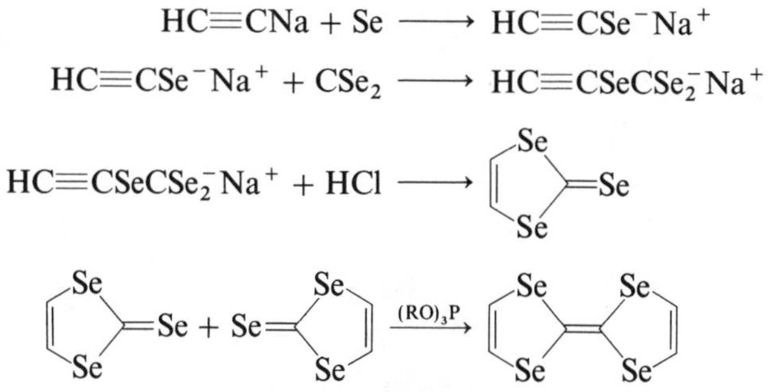

$$HC\equiv CNa + Se \longrightarrow HC\equiv CSe^-Na^+$$

$$HC\equiv CSe^-Na^+ + CSe_2 \longrightarrow HC\equiv CSeCSe_2^-Na^+$$

SCHEME 1

Another procedure for the preparation of TSF, also based on the sulfur analog, was developed by Lakshmikantham and Cava.[18] This procedure differs in the method of preparation of the intermediate selone (see Scheme 2).

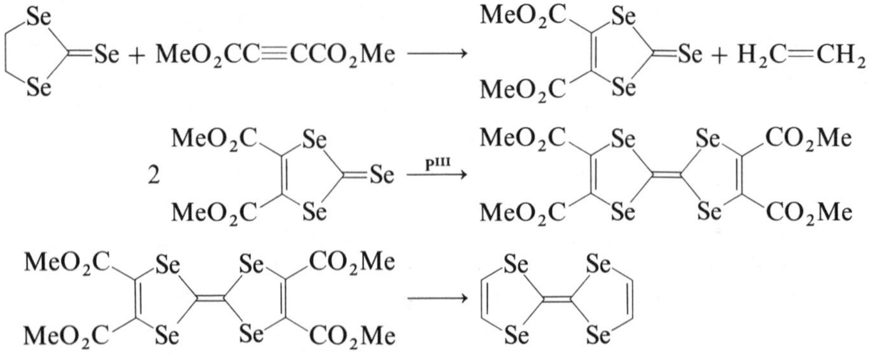

SCHEME 2

The last step involves the simultaneous demethylation and decarboxylation by treatment with lithium bromide in HMPA (hexamethylphosphoramide). The conditions for this last reaction depend on the history of the HMPA (and perhaps other unknown factors) since a small amount of water is necessary but too much water is detrimental. Several groups have had difficulties achieving resonable yields of TSF due to the near failure of the last step. The ethylene triselenocarbonate is prepared from carbon diselenide and ethylene bromide.[18] The simplest synthesis of TSF (in terms of number of steps) is that reported by Okamoto and Wojciechowski;[19] it involves high pressure treatment of a mixture of carbon diselenide and dimethyl acetylenedicarboxylate. Unfortunately, two problems associated with this preparation are (a) not everyone has access to high pressure equipment and (b) the capricious decarboxymethylation mentioned previously.

The selones required for the preparation of substituted tetraselenafulvalenes are prepared via diselenocarbamates and haloketones as shown in Scheme 3:

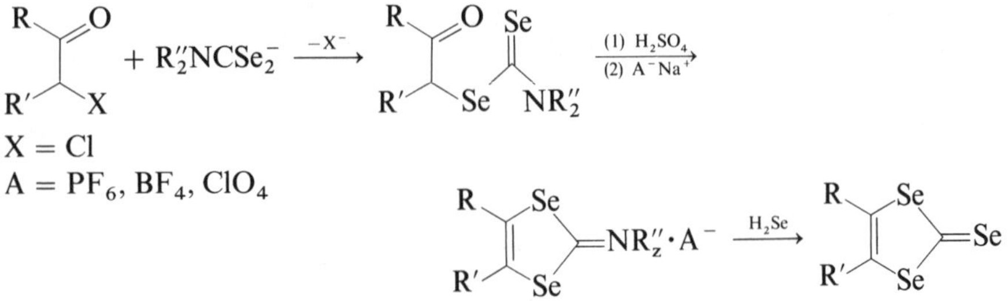

SCHEME 3

Another approach to the construction of substituted selones was reported by Berg et al.[20] and was based on a previous application by Spencer et al.[21] of an electrocyclic reaction to the preparation of thiaselones; it consists of the mild pyrolysis (refluxing toluene) of substituted 1,2,3-selenadiazoles in the presence of carbon diselenide.

In all the preparations shown, carbon diselenide was employed as the main selenium-containing synthon. The commercial availability of this reagent varies with time (and demand); also, it is a fetid as well as toxic material. In order to avoid this reagent, a procedure involving the use of a "masked" diselenocarbamate intermediate was developed;[22] the procedure involved the hydroselenation of a dimethylimminium chloride:

$$Me_2 \overset{+}{N}{=}CCl_2Cl^- + 2\,HSe^-M^+ \xrightarrow{\;xsB:\;} Me_2NCSe_2^-\,M^+ + 2\,\overset{+}{B}HCl^- + MCl$$

$$M^+ \text{ usually} = \overset{+}{B}H, \quad B\text{: usually} = Et_3N\text{:}$$

Later is was discovered that certain commercial phosgene imminium chloride samples were contaminated with some sulfur contaminant, probably dimethylthiocarbamoyl chloride or $[(CH_3)_2N{=}C(SCl)Cl]^+\cdot Cl^-$. This small incorporation of sulfur in the final tetraselenafulvalene (particularly TMTSF) had only a small effect on the physical properties of materials derived from it.[23]

3. ORGANOSELENIUM POLYMERS

There are a number of known organoselenium polymers[24] but the ones of interest for their transport properties are only poly(p-phenylene selenide), poly-(thiophene selenide), poly(selenophene) and polycarbon diselenide.

3.1. Poly(p-phenylene selenide)

The preparation of this polymer was inspired by the results with poly(p-phenylene sulfide) (PPS) as described by Rabolt et al.[25] and by Chance et al.[26] The latter two groups found that the sulfide polymer could be partially oxidized by arsenic pentafluoride to yield a material of modest room temperature conductivity. In the interim it was discovered that (a) during partial oxidation, PPS rearranges to large segments containing dibenzothiophene units[27] and (b) much higher conductivities could be achieved if solutions of PPS in AsF_3 were partially oxidized with AsF_5. This latter discovery reported by Frommer et al.[10] constitutes a substantial advance in the area of conducting polymers in general because it allows *fabrication* in the form of film casting.

The polymer poly(p-phenylene selenide) (PPSe) was prepared by two groups. The first synthesis which was reported by Sandman et al.[28] had its basis in PPS syntheses. Sandman prepared PPSe by nucleophilic aromatic substitution of bromide with sodium selenide in DMF at 120–140°C for 20 h. The material was reported to be a yellow powder of composition: $C_6H_{3.9-4.2}Se_{1.06-1.15}Br_{0.015}$. The excess Se was suggested by the authors to be in the form of diselenide linkages.

The second synthesis, developed by Cava and co-workers,[29] yielded PPSe of correct stoichiometry in excellent agreement with the empirical formula

C_6H_4Se.[30] Their approach was also based on previous work on PPS and involved the Cu mediated reaction of p-bromophenylselenolate in pyridine at 200°C over 24 h. This material was cream colored after treatment with hydrochloric acid, water, ammonia, water, and warm DMF.

3.2. Poly(2,5-thienylene selenide)

This novel polymer together with its sulfur analog [poly(2,5-thienylene sulfide)] was prepared recently by Cava via an interesting new polymerization reaction:

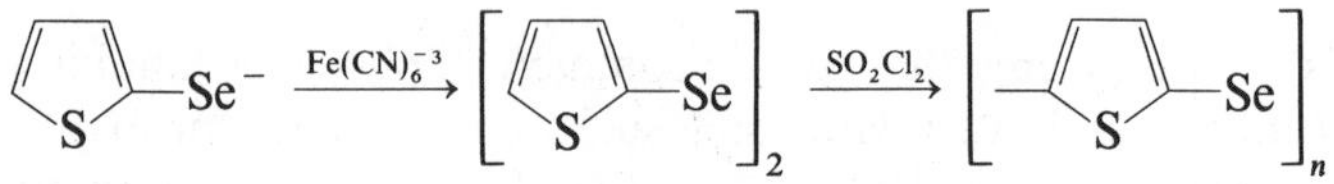

The starting selenolate was prepared from 2-lithio-thiophene and selenium and the polymerization reaction was carried out in methylene chloride at ambient temperature followed by reflux overnight. The resulting polymer was also cream colored upon extraction with carbon disulfide and hot DMF.

3.3. Poly(selenophene)

This polymer was prepared by the same procedures employed for the preparation of its sulfur analog; that is, chemical coupling via a Grignard reagent of 2,5-dibromo selenophene[31] and electrochemical anodic polymerization of selenophene.[32]

In the chemical polymerization the usual conditions required for the preparation of high purity material as applied, for example, to poly(thiophene)[33] were apparently not observed. This may explain why the molecular weight of the product was relatively low. Also, no selenium analyses were given; it is possible that the same kind of reductive dechalogenation reaction that takes place during polymerization of thiophene applies in this case.

The electrochemical polymerization occurred under essentially the same conditions as those employed for thiophene[34] yielding a uniform blue–black film that was less stable to the atmosphere than its sulfur analog.

4. PREPARATION OF CONDUCTING SALTS

All organic conductors that consist of neutral, insulating, starting materials, whether polymeric or not, are converted to conductors by removal or addition of electrons. In the special case of organoselenium compounds, addition of electrons, not surprisingly, leads to reductive decomposition. For that reason this section will deal only with the methodology that was developed over the years for the controlled partial oxidation of organosulfur and organoselenium compounds.

There are four methods that have become established for the partial oxidation of organic materials with potential for high conductivity: (a) reaction in homogeneous medium with an organic or inorganic acceptor, (b) conversion to a soluble, usually insulating, salt followed by metathesis to a conducting salt, (c) direct electrochemical oxidation with concomitant separation (crystallization or polymerization) at the electrode, and (d) gas solid reaction with a strong oxidizing gas.

4.1. Charge Transfer Complexes from Homogenous Medium

The simplest case consists of mixing hot solutions of donor and acceptor followed by slow cooling. The solvent of choice for sulfur-containing donors is acetonitrile. Unfortunately most organoselenium donors are insoluble in polar aprotic solvents such as acetonitrile so that less convenient solvents such as benzonitrile, chlorobenzene, o-dichlorobenzene, and polychlorinated alkanes must be used. A detailed description of variations on this theme was published by Andersen et al.[35] This procedure is particularly useful for the growth of organic alloys (e.g. TTF·TSF·TCNQ).

A slightly more complicated homogeneous medium technique is that of slow diffusion. The most straightforward apparatus for this procedure is an "H" cell where solutions of donor and acceptor are placed in each of the arms until they meet at the horizontal branch. The product usually crystallizes out on, or near the horizontal tube. In most cases, the starting materials are only very moderately soluble in the working medium [which may consist of a single solvent (99% of the time) or a mixture (usually no more than two solvents)] and may be carefully deposited in the bottom of the vertical branches of the cell before it is filled with solvent(s). A slightly more sophisticated approach involves three compartments separated by medium glass frits (Fig. 2).

This is the ideal apparatus for the growth of TTF·TCNQ crystals[36] that grow best at 30°C from acetonitrile.

Sometimes vacuum is the best medium if the donor is sufficiently volatile.[35] For this method, a sealed tube with the acceptor at one end and the donor at

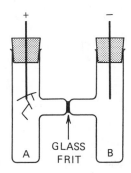

FIGURE 2.   A three-compartment slow diffusion apparatus.

the other is placed in a temperature gradient depending on the volatility of the starting components. The best procedure for growing tetrathiotetracene iodide is under vacuum. In this case, the tetracene region of an evacuated system is heated and the iodine region cooled.

4.2. Methathesis

Although many TTF salts can be prepared by this method,[37] it has not been exploited with the selenium analogs even though a soluble salt of TMTSF can be prepared.[38] In this approach, TTF is oxidized with a solution of hydrogen peroxide and fluoroboric acid in acetonitrile to produce moderately soluble, insulating $(TTF)_3(BF_4)_2$. In order to prepare, for example, $TTF_{12}SCN_7$, one mixes hot solutions of the fluoroborate and tetrabutylammonium thiocyanate in acetonitrile and harvests the product crystals upon slow (overnight) cooling to room temperature.

4.3. Electrochemical Oxidation

In this procedure, a modified "H" cell is employed (Fig. 3).

A solution of the donor (usually 6 mg per 25 mL) in preformed electrolyte (usually the tetrabutylammonium salt of the desired anion in a concentration

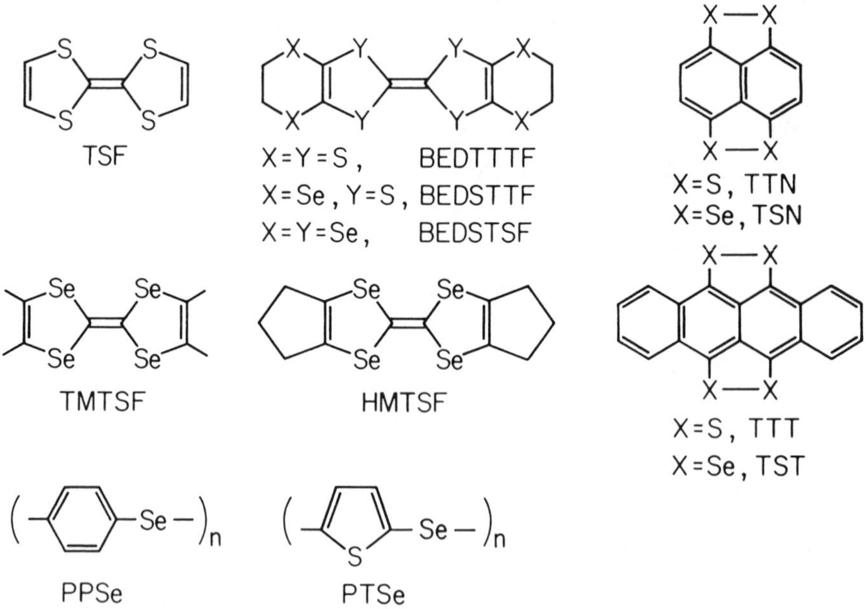

TSF

X=Y=S, BEDTTTF
X=Se, Y=S, BEDSTTF
X=Y=Se, BEDSTSF

X=S, TTN
X=Se, TSN

TMTSF

HMTSF

X=S, TTT
X=Se, TST

PPSe

PTSe

FIGURE 3. A modified "H" cell.

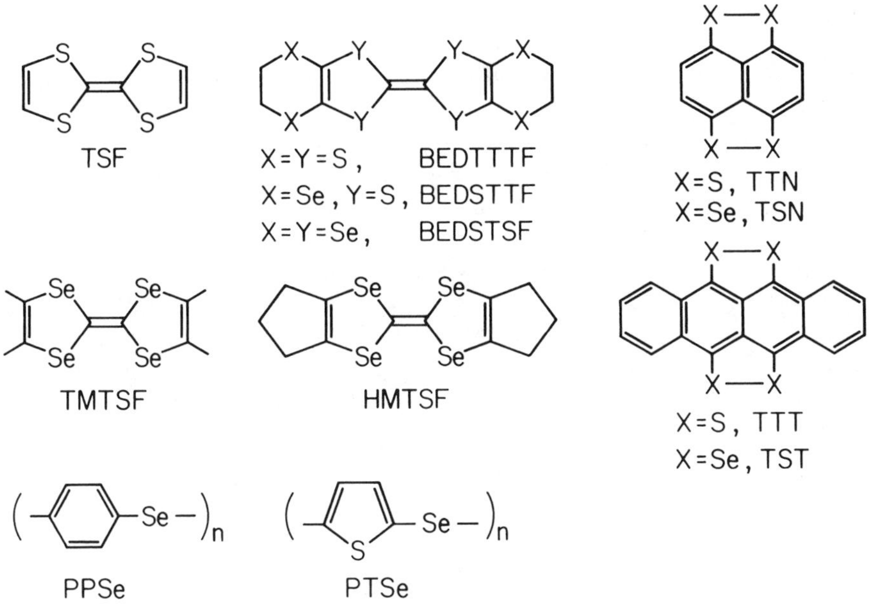

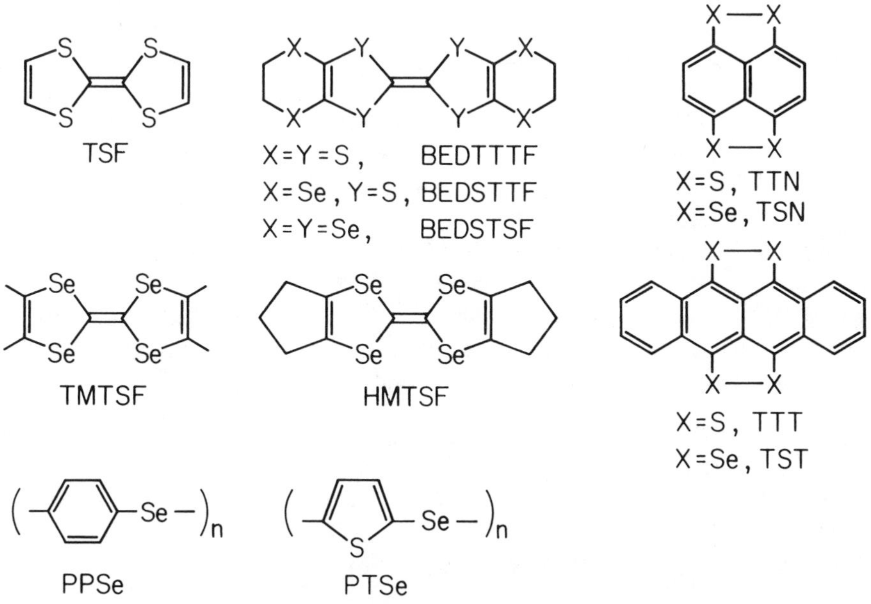

TSF

X=Y=S, BEDTTTF
X=Se, Y=S, BEDSTTF
X=Y=Se, BEDSTSF

X=S, TTN
X=Se, TSN

TMTSF

HMTSF

X=S, TTT
X=Se, TST

$\left(-\underset{}{\bigcirc}-Se-\right)_n$

PPSe

$\left(-\underset{S}{\bigcirc}-Se-\right)_n$

PTSe

TSF

X=Y=S, BEDTTTF
X=Se, Y=S, BEDSTTF
X=Y=Se, BEDSTSF

X=S, TTN
X=Se, TSN

TMTSF

HMTSF

X=S, TTT
X=Se, TST

$\left(-\underset{}{\bigcirc}-Se-\right)_n$

PPSe

$\left(-\underset{S}{\bigcirc}-Se-\right)_n$

PTSe

FIGURE 3. (*Continued*)

405

of 0.1–0.005 M) is placed in compartment A and the same volume of pure electrolyte solution is added to compartment B. The electrode entering compartment A (usually Pt) is connected to the positive pole of a direct current source (a 6-V battery is adequate but a source that has both constant current or constant potential capability is desirable) and the electrode from compartment B (in general Pt but flame oxidized graphite is preferable since the surface carbonyls can be reduced before the electrolyte is decomposed) is connected to the negative pole. Crystal deposition starts, in most cases, within $\frac{1}{2}$ h–several days. When this technique is applied to polymer deposition, the monomer (thiophene, selenophene, azulene, pyrrole, etc.) is added to the A compartment of the same type of cell filled with electrolyte (usually 0.1–1 M tetrabutylammonium fluoroborate or lithium perchlorate). The A compartment electrode is usually a platinum sheet, carbon sheet, or ITO glass. Polymer deposition occurs in the form of a deeply colored film within 0.5 to several hours. As grown, these films are already partially oxidized (p-doped) and in most cases can be reduced to a neutral state (compensated) by simply reversing the polarity of the cell. The neutral polymers are usually much lighter in color (yellow orange to red).

4.4. Gas Solid Reaction

When the organic material to be converted to a conductor is completely insoluble, it may be partially oxidized by either compaction against platinum mesh followed by electrolysis or by treatment with a strong gaseous oxidizing agent such as AsF_5, Cl_2, Br_2, I_2, and so on. Fortunately, for not too obvious reasons, the oxidation does take place homogeneously throughout the sample, particularly if it is submitted to several exposure–pump–exposure cycles. The amount of doping can be determined by weight uptake or elemental analysis or both. In the case of the poly(heterocycles) it can be determined by gravimetric electron voltage spectroscopy (GEVS),[39] a recently developed technique that is in essence gravimetric cyclic voltammetry. A polymer sample to be "doped" is grown on a substrate that is allowed to vibrate during the redox process; weight uptake changes the vibration frequency in a quantitative manner so that very small changes in weight can be detected.

5. PROPERTIES OF ORGANOSELENIUM CONDUCTORS

In the interval between 1974 and 1982, the most exciting results of research in organic conductors were those produced by organoselenium solids either in the form of charge transfer complexes or salts with closed shell anions. The reasons for the excitement can be summarized as follows:[1]

1. TSF·TCNQ had a higher room temperature conductivity than TTF·TCNQ.
2. The metal-to-insulator transition temperature (T_{MI}) of TSF·TCNQ was 14 K lower than that of TTF·TCNQ.

3. HMTSF·TCNQ had a very high room temperature conductivity (2000 Ω^{-1} cm^{-1}) and a remarkably shallow metal-to-insulator transition such that the material was still very highly conducting at 1 K.

4. A number of $(TMTSF)_2X$ salts were superconductors under pressure and $X = ClO_4$ was superconductor at 1 K and atmosphere pressure.

These series of spectacular discoveries with the selenium compounds induced research into selenium substituted polymers with the hope that a parallel set of discoveries would ensue but as can be seen in the following list that was not the case:

1. When partially oxidized PPSe has a room temperature conductivity that is anywhere from 10^{-5} times [28] to 10^{-10} times[29] that of PPS. In fact, if the latter result is correct, PPSe is among the best *unsaturated insulating polymers* made.

2. Poly(selenophene) is a poorer conductor (10^{-2} times) than poly(thiophehe) and in the doped state is *less* stable than its sulfur analog.[32]

But all is not bad with selenium substitution as shown by Cava and co-workers in the case of poly(thiophene sulfide) versus poly(selenophene sulfide) where the latter, in the form of its AsF_6 salt, was a 10^4 times better conductor than the former under identical conditions.[37]

6. RECENT RESULTS AND FUTURE DIRECTIONS

6.1. Recent Results

With The IBM group's discovery of superconductivity in $(BEDTTTF)_4(ReO_4)_2$, the new selenium derivatives of this organosulfur donor (originally prepared independently by Cava and co-workers[38a] and Cowan and co-workers[38b] were prepared by Schumaker et al.,[39] probably with the intent to increase the the superconducting transition (T_c). Preliminary results[40] indicate that none of the selenium containing derivatives have yielded a superconducting Phase. One cannot predict, however, if a more unconventional anion may someday yield a salt with a remarkably higher T_c. Carbon diselenide was reported to polymerize to a semiconductor or metallic polymer.[42] Superconductivity was reported under high pressure but it may be due to selenium generated in the decomposition of CSe_2 into graphite and selenium metal.[43]

6.2. Future Directions

It is clear from Section 6 that the synthetic methodology for the preparation of organoselenium conductors is already adequate. The sulfur contamination in the commercial phosgene imminium chloride should not really be a problem since it should be possible to remove it by exhaustive chlorination.

As far as the future of organoselenium metals is concerned, more organic conductors that are devoid of a metal-to-insulator transition will have to be developed. To date only $(TMTSF)_2ClO_4$,[1] $(BEDTTTF)_n(I_3)_m$,[6] and dithiapyrene·$TCNQ$[41] have that property.

As far as polymers are concerned, the search is concentrating on a polymer that will be highly conducting in its neutral state; i.e., will be a zero-band-gap semiconductor. Whether the candidate will be organoselenium polymer is open to speculation.

REFERENCES

1. M. R. Bryce and L. C. Murphy, *Nature*, **309**, 119 (1984); (a) *J. Pys. Paris Colloq.*, **C3** (1983); (b) K. Bechgaard and D. Jerome, *Sci. Am.*, **247**, 50 (1982); (c) *Mol. Cryst. Liq. Cryst.*, **79**, 1–4 (1982; (d) *Mol. Cryst. Liq.*, **85**, 1–4 (1982); (e) *Mol. Cryst. Liq.*, **86**, 1–4 (1982); (f) *Chem. Scr.*, **17**, 1–5 (1981); (g) L. Alcacer, Ed., *The Physics and Chemistry of Low Dimensional Solids*, D. Reidel., Dordrecht, Holland, 1980; (h) W. E. Hatfield, Ed., *Molecular Metals*, Plenum, New York, 1979; (i) J. T. Devresse, R. P. Edward, and V. E. vanDoren, Eds., *Highly Conducting One-Dimensional Solids*, Plenum New York, 1979; (j) H. J. Keller, Ed., *Chemistry and Physics of One-Dimensional Metals*, Plenum, New York, 1977.

2. F. Gutmann and L. E. Lyons, *Organic Semiconductors*, Wiley, New York-London, 1967.

3. S. Megtert, J. P. Pouget, and R. Comes, *Ann. N.Y. Acad. Sci.*, **313**, 235 (1978).

4. D. Jerome, A. Mazaud, M. Ribault, and K. Bechgaard, *J. Phys. Lett.*, **41**, L-95 (1980).

5. S. S. P. Parkin, E. M. Engler, R. R. Schumaker, R. Lagier, V. Y. Lee, J. Voiron, K. Carneiro, J. C. Scott, and R. L. Greene, Ref. 1(a), pp. 791–797.

6. E. B. Yagubskii, I. F. Shchegolev, V. N. Laukhin, P. S. Kononovich, A. K. Kartsovnik, A. V. Zvarykina, and L. I. Buravov, *JETP Lett.*, **39**, 12 (1984).

7. E. M. Engler and V. V. Patel, *J. Am. Chem. Soc.*, **96**, 7376 (1974).

8. F. Wudl, *J. Am. Chem. Soc.*, **103**, 7064 (1981).

9. P. M. Grant, Ref. 1(a), pp. 847–857. P. M. Grant, *Phys. Rev. B*, **26**, 6888 (1982)

10. J. E. Frommer, R. L. Elsenbaumer, H. Eckhardt, and R. R. Chance, *J. Polym. Sci., Polym. Lett. Ed.*, **21**, 39, 1983. M. Aldissi and R. Liepins, *J. Chem. Soc. Chem. Commun.*, 255 (1984).

11. S. Es-Seddiki, G. Le Coustumer, Y. Mollier, and M. Devaud, Ref. 1(e), pp. 71–77.

12. H. Endres, H. J. Keller, J. Queckbörner, J. Weigl, and D. Schweitzer, Ref. 1(e), pp. 111–122.

13. C. Marschalk, *Bull. Soc. Chim. Fr.*, 800 (1952).

14. A. Yamahira, T. Nogami, and H. Mikawa, *J. Chem. Soc. Chem. Commun.*, 904 (1983).

15. H. Reimlinger and G. King, *Chem. Ber.*, **95**, 1043 (1962).

16. E. A. Perez-Alberne, U.S. Patent No. 3,723,417, Mar. 27, 1973.

17. F. Wudl, unpublished results.

18. M. V. Lakshmikantham and M. P. Cava, *J. Org. Chem.*, **41**, 882 (1976).

19. Y. Okamoto and P. S. Wojciechowski, *J. Chem. Soc. Chem. Commun.*, 669 (1981).

20. C. Berg, K. Bechgaard, J. R. Andersen, and C. S. Jacobsen, *Tetrahedron Lett.*, 1719 (1976).

21. H. K. Spencer, M. V. Lakshmikantham, M. P. Cava, and A. F. Garito, *J. Chem. Soc. Chem. Commun.*, 867 (1975).

22. F. Wudl and D. Nalewajek, *J. Chem. Soc. Chem. Commun.*, 866 (1980). L. Chiang, T. O. Poehler, A. N. Bloch, and D. O. Cowan, *J. Chem. Soc. Chem. Commun.*, 866 (1980).

23. A. Moradpour, K. Bechgaard, M. Barrie, C. Lenoir, K. Murata, R. C. Lacoe, M. Ribault, and and D. Jerome, *Mol. Cryst. Liq. Cryst.*, **120**, 69–72 (1985).

24. D. L. Klayman and W. H. H. Günther, *Organic Selenium Compounds: Their Chemistry and Biology*, Wiley—Interscience, New York—London, 1973, pp. 815–834.

25. J. F. Rabolt, T. C. Clarke, K. K. Kanazawa, J. R. Reynolds, and G. B. Street, *J. Chem. Soc. Chem. Commun.*, 347 (1980).

26. R. R. Chance, L. W. Shacklette, G. G. Miller, D. M. Ivory, J. M. Sowa, R. L. Elsenbaumer, and R. H. Baughman, *J. Chem. Soc. Chem. Commun.*, 348 (1980).

27. L. W. Shacklette, R. L. Elsenbaumer, R. R. Chance, H. Eckhardt, J. E. Frommer, and R. H. Baughman, *J. Chem. Phys.*, **75**, 1919 (1981).

28. D. J. Sandman, M. Rubner, and L. Samuelson, *J. Chem. Chem. Soc. Chem. Commun.*, 1133 (1982).

29. K.-Y. Jen, M. V. Lakshmikantham, M. Albeck, M. P. Cava, W.-S. Huang, and A. G. Mac-Diarmid, *J. Polym. Sci., Polym. Lett. Ed.*, **21**, 441 (1983).

30. Actually there is something wrong with this result because from the synthetic method there should be some Br left as one end group. In fact from the Br content one could derive a degree of polymerization (DP).

31. M. D. Benzoari, P. Kovacic, S. Gronowitz, and A.-B. Hörnfeldt, *J. Polym. Sci., Polym. Lett. Ed.*, **19**, 347 (1981).

32. D. Ginley, Sandia Laboratories, Albuquerque, New Mexico, private communication.

33. M. Kobayashi, J. Chen. T.-C. Chung, F. Moraes, A. J. Heeger, and F. Wudl, *Synth. Metals*, **9**, 77(1984).

34. M. A. Druy and R. J. Seymour, Ref. 1(a), pp. 595–598.

35. J. R. Andersen, E. M. Engler, and K. Bechgaard, *Ann. N.Y. Acad. Sci.*, **313**, 293 (1978).

36. M. L. Kaplan, *J. Cryst. Growth.*, **33**, 161 (1976). J. Anzai, *J. Cryst. Growth.*, **33**, 185 (1976).

37. K. Y. Jen, N. Benfaremo, M. P. Cava, W.-S. Huang, and A. G. MacDiarmid, *J. Chem. Soc. Chem. Commun.*, 633 (1983).

38. M. Mizuno, A. F. Garito, and M. P. Cava, *J. Chem. Soc. Chem. Commun.*, 18 (1978). W. Krug, A. N. Bloch, T. O. Poehler, and D. O. Cowan, *Ann. N.Y. Acad. Sci.*, **313**, 366 (1978).

39. R. R. Schumaker, V. Y. Lee, and E. M. Engler, Ref. 1(a), pp. 1139–1145.

40. E. M. Engler, private communication.

41. K. Bechgaard, *Mol. Cryst. Liq. Cryst.*, **125**, 81–89 (1985).

42. Y. Okamoto and P. S. Wojclechowski, *J. Chem. Soc. Chem. Commun.*, 386, (1982); H. Kobayashi, A. Kobayashi, and Y. Sasaki, *Mol. Cryst. Liq. Cryst.*, **118**, 427 (1985).

43. H. Kobayashi, R. Kato, A. Kobayashi, and H. Kawamura, Proceedings of the International Conference on Synthetic Metals, Kyoto, Japan, 1986.

Index